DESIGNERS' GUIDES TO THE EUROCODES

DESIGNERS' GUIDE TO EN 1993-2 EUROCODE 3: DESIGN OF STEEL STRUCTURES.

PART 2 : STEEL BRIDGES

Eurocode Designers' Guide Series

Designers' Guide to EN 1990. Eurocode: Basis of Structural Design. H. Gulvanessian, J.-A. Calgaro and M. Holický. 0 7277 3011 8. Published 2002.

Designers' Guide to EN 1994-1-1. Eurocode 4: Design of Composite Steel and Concrete Structures. Part 1.1: General Rules and Rules for Buildings. R. P. Johnson and D. Anderson. 0 7277 3151 3. Published 2004.

Designers' Guide to EN 1997-1. Eurocode 7: Geotechnical Design – General Rules. R. Frank, C. Bauduin, R. Driscoll, M. Kavvadas, N. Krebs Ovesen, T. Orr and B. Schuppener. 0 7277 3154 8. Published 2004.

Designers' Guide to EN 1993-1-1. Eurocode 3: Design of Steel Structures. General Rules and Rules for Buildings. L. Gardner and D. Nethercot. 0 7277 3163 7. Published 2004.

Designers' Guide to EN 1992-1-1 and EN 1992-1-2. Eurocode 2: Design of Concrete Structures. General Rules and Rules for Buildings and Structural Fire Design. A. W. Beeby and R. S. Narayanan. 0 7277 3105 X. Published 2005.

Designers' Guide to EN 1998-1 and EN 1998-5. Eurocode 8: Design of Structures for Earthquake Resistance. General Rules, Seismic Actions, Design Rules for Buildings, Foundations and Retaining Structures. M. Fardis, E. Carvalho, A. Elnashai, E. Faccioli, P. Pinto and A. Plumier. 0 7277 3348 6. Published 2005.

Designers' Guide to EN 1994-2. Eurocode 4: Design of Composite Steel and Concrete Structures. Part 2: General Rules and Rules for Bridges. C. R. Hendy and R. P. Johnson. 0 7277 3161 0. Published 2006.

Designers' Guide to EN 1995-1-1. Eurocode 5: Design of Timber Structures. Common Rules and for Rules and Buildings. C. Mettem. 0 7277 3162 9. Forthcoming: 2007 (provisional).

Designers' Guide to EN 1991-4. Eurocode 1: Actions on Structures. Wind Actions. N. Cook. 0 7277 3152 1. Forthcoming: 2007 (provisional).

Designers' Guide to EN 1996. Eurocode 6: Part 1.1: Design of Masonry Structures. J. Morton. 0 7277 3155 6. Forthcoming: 2007 (provisional).

Designers' Guide to EN 1991-1-2, 1992-1-2, 1993-1-2 and EN 1994-1-2. Eurocode 1: Actions on Structures. Eurocode 3: Design of Steel Structures. Eurocode 4: Design of Composite Steel and Concrete Structures. Fire Engineering (Actions on Steel and Composite Structures). Y. Wang, C. Bailey, T. Lennon and D. Moore. 0 7277 3157 2. Forthcoming: 2007 (provisional).

Designers' Guide to EN 1992-2. Eurocode 2: Design of Concrete Structures. Part 2. Concrete Bridges. C. R. Hendy and D. A. Smith. 0 7277 3159 3. Published 2007.

Designers' Guide to EN 1991-2, 1991-1-1, 1991-1-3 and 1991-1-5 to 1-7. Eurocode 1: Actions on Structures. Traffic Loads and Other Actions on Bridges. J.-A. Calgaro, M. Tschumi, H. Gulvanessian and N. Shetty. 0 7277 3156 4. Forthcoming: 2007 (provisional).

Designers' Guide to EN 1991-1-1, EN 1991-1-3 and 1991-1-5 to 1-7. Eurocode 1: Actions on Structures. General Rules and Actions on Buildings (not Wind). H. Gulvanessian, J.-A. Calgaro, P. Formichi and G. Harding. 0 7277 3158 0. Forthcoming: 2007 (provisional).

www.eurocodes.co.uk

DESIGNERS' GUIDES TO THE EUROCODES

DESIGNERS' GUIDE TO EN 1993-2 EUROCODE 3: DESIGN OF STEEL STRUCTURES

PART 2: STEEL BRIDGES

C. R. HENDY and C. J. MURPHY

thomas telford

eurocodes expert

Published by Thomas Telford Publishing, Thomas Telford Ltd, 1 Heron Quay, London E14 4JD
URL: http://www.thomastelford.com

Distributors for Thomas Telford books are
USA: ASCE Press, 1801 Alexander Bell Drive, Reston, VA 20191-4400
Japan: Maruzen Co. Ltd, Book Department, 3–10 Nihonbashi 2-chome, Chuo-ku, Tokyo 103
Australia: DA Books and Journals, 648 Whitehorse Road, Mitcham 3132, Victoria

First published 2007

Eurocodes Expert

Structural Eurocodes offer the opportunity of harmonized design standards for the European construction market and the rest of the world. To achieve this, the construction industry needs to become acquainted with the Eurocodes so that the maximum advantage can be taken of these opportunities

Eurocodes Expert is a new ICE and Thomas Telford initiative set up to assist in creating a greater awareness of the impact and implementation of the Eurocodes within the UK construction industry

Eurocodes Expert provides a range of products and services to aid and support the transition to Eurocodes. For comprehensive and useful information on the adoption of the Eurocodes and their implementation process please visit our website or email eurocodes@thomastelford.com

A catalogue record for this book is available from the British Library

ISBN: 978-0-7277-3160-9

Typeset by Academic + Technical, Bristol
Printed and bound in Great Britain by MPG Books, Bodmin

Preface

Aims and objectives of this guide

The principal aim of this book is to provide the user with guidance on the interpretation and use of EN 1993-2 and to present worked examples. It covers topics that will be encountered in typical steel bridge designs, and explains the relationship between EN 1993-2 and the other Eurocodes.

EN 1993-2 is not a 'stand alone' document and refers extensively to other Eurocodes. Its format is based on EN 1993-1-1 and generally follows the same clause numbering. It identifies which parts of EN 1993-1-1 are relevant for bridge design and adds further clauses which are specific to bridges. It is therefore not useful to produce guidance on EN 1993-2 in isolation and this guide covers material in a variety of other parts of Eurocode 3 which will need to be used in bridge design.

This book also provides background information and references to enable users of Eurocode 3 to understand the origin and objectives of its provisions.

Layout of this guide

EN 1993-2 has a foreword, ten sections and five annexes. This guide has an introduction which corresponds to the foreword of EN 1993-2, Chapters 1 to 10 which correspond to Sections 1 to 10 of EN 1993-2 and Annexes A to E which again correspond to Annexes A to E of EN 1993-2.

The guide generally follows the section numbers and first sub-headings in EN 1993-2 so that guidance can be sought on the code on a section-by-section basis. The guide also follows the format of EN 1993-2 to lower levels of sub-heading in cases where this can conveniently be done and where there is sufficient material to merit this. The need to use many Eurocode parts can initially make it a daunting task to locate information in the order required for a real design. In some places, therefore, additional sub-sections are included in this guide to pull together relevant design rules for individual elements, such as transverse stiffeners. Additional sub-sections are identified as such in the sub-section heading.

The following parts of Eurocode 3 will typically be required in a steel bridge design:

EN 1993-1-1: General rules and rules for buildings
EN 1993-1-5: Plated structural elements
EN 1993-1-8: Design of joints
EN 1993-1-9: Fatigue strength of steel structures
EN 1993-1-10: Selection of steel for fracture toughness and through-thickness properties

The following may also be required:

EN 1993-1-7: Strength and stability of planar plated structures transversely loaded
EN 1993-1-11: Design of structures with tension components made of steel

In this guide, the above are sometimes referred to by using 'EC3' for EN 1993, so EN 1993-1-1 is referred to as EC3-1-1. Where clause numbers of the various parts of EN 1993 are referred to in the text, they are prefixed by the number of the relevant part of EN 1993. Hence:

- 3-1-1/clause 5.2.1(3) means clause 5.2.1, paragraph (3) of EN 1993-1-1
- 3-1-5/expression (3.1) means equation (3.1) in EN 1993-1-5
- 3-2/clause 3.2.3 means clause 3.2.3 of EN 1993-2.

Note that, unlike other guides in this series, even clauses in EN 1993-2 itself are prefixed with '3-2'. There are so many references to other parts of Eurocode 3 required that to do otherwise would be confusing.

Expressions repeated from the ENs retain their number and are referred to as expressions. Where additional equations are provided in the guide, they are numbered sequentially within each sub-section of a main section so that, for example, the third additional equation within sub-section 6.1 would be referenced equation (D6.1-3). Additional figures and tables follow the same system. For example, the second additional figure in section 6.4 would be referenced Fig. 6.4-2.

Acknowledgements

Chris Hendy would like to thank his wife, Wendy, and two boys, Peter Edwin Hendy and Matthew Philip Hendy, for their patience and tolerance of his pleas to finish 'just one more section'. He would also like to thank Jessica Sandberg and Rachel Jones for their efforts in checking many of the Worked Examples.

Chris Murphy would like to thank his wife, Nicky, for the patience and understanding that she constantly displayed during the preparation of this guide.

Both authors would also like to thank their employer, Atkins, for providing both facilities and time for the production of this guide.

Chris Hendy
Chris Murphy

Contents

Introduction

The provisions of EN 1993-2 are preceded by a foreword, most of which is common to all Eurocodes. This *Foreword* contains clauses on:

- the background to the Eurocode programme
- the status and field of application of the Eurocodes
- national standards implementing Eurocodes
- links between Eurocodes and harmonized technical specifications for products
- additional information specific to EN 1993-2
- National Annex for EN 1993-2.

Guidance on the common text is provided in the introduction to the *Designers' Guide to EN 1990, Eurocode: Basis of Structural Design*[1] and only background information relevant to users of EN 1993-2 is given here.

It is the responsibility of each national standards body to implement each Eurocode part as a national standard. This will comprise, without any alterations, the full text of the Eurocode and its annexes as published by the European Committee for Standardization, CEN (from its title in French). This will usually be preceded by a National Title Page and a National Foreword, and may be followed by a National Annex.

Each Eurocode recognizes the right of national regulatory authorities to determine values related to safety matters. Values, classes or methods to be chosen or determined at national level are referred to as nationally determined parameters (NDPs). Clauses of EN 1993-2 in which these occur are listed in the *Foreword.*

NDPs are also indicated by notes immediately after relevant clauses. These Notes give recommended values. It is expected that most of the member states of CEN will specify the recommended values, as their use was assumed in the many calibration studies done during drafting. Recommended values are used in this guide, as the National Annex for the UK was not available at the time of writing. Comments are made regarding the likely values to be adopted where different.

Each National Annex will give or cross-refer to the NDPs to be used in the relevant country. Otherwise the National Annex may contain only the following:[2]

- decisions on the use of informative annexes, and
- references to non-contradictory complementary information to assist the user to apply the Eurocode.

The set of Eurocodes will supersede the British bridge code, BS 5400, which is required (as a condition of BSI's membership of CEN) to be withdrawn by early 2010, as it is a 'conflicting national standard'.

Additional information specific to EN 1993-2

The information specific to EN 1993-2 emphasizes that this standard is to be used with other Eurocodes. The standard includes many cross-references to other parts of EN 1993 and does not itself reproduce material which appears in other parts of EN 1993. This guide however is intended to be self-contained for the design of steel bridges and therefore provides commentary on other parts of EN 1993 as necessary.

The *Foreword* lists the clauses of EN 1993-2 in which National choice is permitted. Elsewhere, there are cross-references to clauses with NDPs in other codes. Otherwise, the Normative rules in the code must be followed, if the design is to be 'in accordance with the Eurocodes'.

In EN 1993-2, Sections 1 to 10 are Normative. Its Annexes A, B, C, D and E are 'Informative' as alternative approaches may be used in these cases. Annexes A and B, concerning bearings and expansion joints respectively, are scheduled to be moved to EN 1990 in the near future as their provisions are not specific to steel bridges. A National Annex may make Informative provisions Normative in the country concerned, and is itself normative in that country, but not elsewhere. The 'non-contradictory complementary information' referred to above could include, for example, reference to a document based on provisions of BS 5400 covering matters not treated in the Eurocodes. Each country can do this, so some aspects of the design of a bridge will continue to depend on where it is to be built.

CHAPTER 1

General

This chapter is concerned with the general aspects of EN 1993-2, *Eurocode 3: Design of Steel Structures, Part 2: Steel Bridges*. The material described in this chapter is covered in section 1 of EN 1993-2 in the following clauses:

- Scope *Clause 1.1*
- Normative references *Clause 1.2*
- Assumptions *Clause 1.3*
- Distinction between principles and application rules *Clause 1.4*
- Terms and definitions *Clause 1.5*
- Symbols *Clause 1.6*
- Conventions for member axes *Clause 1.7*

1.1. Scope

1.1.1. Scope of Eurocode 3

The scope of EN 1993 is outlined in 3-2/clause 1.1.1 by reference to 3-1-1/clause 1.1.1. It is to be used with EN 1990, *Eurocode: Basis of Structural Design*, which is the head document of the Eurocode suite and has an Annex A2, 'Application for bridges'. ***3-1-1/clause 1.1.1(2)*** emphasizes that the Eurocodes are concerned with structural behaviour and that other requirements, e.g. thermal and acoustic insulation, are not considered. *3-1-1/clause 1.1.1(2)*

The basis for verification of safety and serviceability is the partial factor method. EN 1990 recommends values for load factors and gives various possibilities for combinations of actions. The values and choice of combinations are to be set by the National Annex for the country in which the structure is to be constructed.

Eurocode 3 is also to be used in conjunction with EN 1991, *Eurocode 1: Actions on Structures* and its National Annex, to determine characteristic or nominal loads. When a steel structure is to be built in a seismic region, account needs to be taken of EN 1998, *Eurocode 8: Design of Structures for Earthquake Resistance*.

3-1-1/clause 1.1.1(3), as a statement of intention, gives undated references. It supplements the Normative rules on dated reference standards, given in 3-2/clause 1.2, where the distinction between dated and undated standards is explained. The Eurocodes are concerned with design and not execution, but minimum standards of workmanship and material specification are required to ensure that the design assumptions are valid. For this reason, 3-1-1/clause 1.1.1(3) lists the European standards for steel products and for the execution of steel structures. The remaining paragraphs of 3-1-1/clause 1.1.1 list the various parts of EN 1993. *3-1-1/clause 1.1.1(3)*

1.1.2. Scope of Part 2 of Eurocode 3

EN 1993-2 covers structural design of steel bridges and steel parts of composite bridges. Its format is based on EN 1993-1-1 and generally follows the same clause numbering.

It identifies which parts of EN 1993-1-1 are relevant for bridge design and which parts need modification. It also adds provisions which are specific to bridges. The majority of *3-2/clause 1.1.2* re-emphasizes the requirements discussed in section 1.1.1 above.

3-2/clause 1.1.2

1.2. Normative references

References are given only to other European standards, all of which are intended to be used as a package. Formally, the Standards of the International Organisation for Standardisation (ISO) apply only if given an EN ISO designation. National standards for design and for products do not apply if they conflict with a relevant EN standard. As Eurocodes may not cross-refer to national standards, replacement of national standards for products by EN or ISO standards is in progress, with a time-scale similar to that for the Eurocodes.

During the period of change-over to Eurocodes and EN standards, it is possible that an EN referred to, or its National Annex, may not be complete. Designers who then seek guidance from national standards should take account of differences between the design philosophies and safety factors in the two sets of documents.

Cross-references from EN 1993-2 to EN 1993-1

The parts of EN 1993 most likely to be referred to in the design of a steel bridge are listed in Table 1.2-1. General provisions on serviceability limit states and their verification will be found in EN 1990.

Table 1.2-1. References to EN 1993, Eurocode 3: Design of steel structures

Title of Part	Subjects referred to from EN 1993-2
EN 1993-1-1, *General Rules and Rules for Buildings*	Stress–strain properties of steel; γ_M for steel General design of unstiffened steelwork Classification and resistances of cross-sections Non-linear global analysis Buckling of members and frames; column buckling curves
EN 1993-1-5, *Plated Structural Elements*	Design of cross-sections in slenderness Class 3 or 4 Effect on stiffness of shear lag in steel plate elements Design where transverse, longitudinal, or bearing stiffeners are present Transverse distribution of stresses in a wide flange Shear buckling; flange-induced web buckling In-plane transverse forces on webs
EN 1993-1-7, *Transversely Loaded Planar Plated Structures*	Design of deck plates with transverse loading (although this requires supplementary guidance – see section 6.5.2 of this guide)
EN 1993-1-8, *Design of Joints*	Modelling of flexible joints in analysis Design of joints in steel and composite members Design of splices between main bridge beams Design using structural hollow sections
EN 1993-1-9, *Fatigue Strength of Steel Structures*	Fatigue loading Classification of details into fatigue categories Limiting stress ranges for damage-equivalent stress verification Fatigue verification in welds and connectors
EN 1993-1-10, *Material Toughness and Through-thickness Properties*	Selection of steel grade (Charpy test, and Z quality)
EN 1993-1-11, *Design of Structures with Tension Components*	Design of bridges with prestressing or cable support, such as cable-stayed bridges

1.3. Assumptions

It is assumed in using EN 1993-2 that the provisions of EN 1990: *Basis of Structural Design* will be followed. It is also essential to note that various clauses in Eurocode 3 assume that EN 1090 will be followed in the fabrication and erection processes. This is particularly important for the design of slender elements where the imperfections for analysis and buckling resistance formulae depend on imperfections from fabrication and erection being limited to the levels in EN 1090. EN 1993-2 should not therefore be used for design of bridges that will be fabricated and erected to specifications other than EN 1090 without a very careful comparison of the respective tolerance and workmanship requirements.

1.4. Distinction between principles and application rules

Reference has to be made to EN 1990 for the distinction between 'Principles' and 'Application Rules'. Essentially, Principles comprise general statements and requirements which must be followed and Application Rules are rules which comply with these Principles. There may however be other ways to comply with the Principles and these methods may be substituted if it is shown that they are at least equivalent to the Application Rules with respect to safety, serviceability and durability. This however presents the problem that such a design could not then be deemed to comply wholly with the Eurocodes.

According to EN 1990, Principles are supposed to be marked with a 'P' adjacent to the paragraph number. Eurocode 3 does not consistently follow this requirement and the distinction between Principles and Application Rules according to EN 1990 is therefore lost. Principles can generally still be identified by the use of 'shall' within a clause, while 'should' and 'may' are generally used for Application Rules but this is not completely consistent.

1.5. Terms and definitions

Reference is made to the definitions given in clauses 1.5 of EN 1990 and EN 1993-1. Further bridge-specific definitions are provided.

Many types of analysis are defined in clause 1.5.6 of EN 1990. It should be noted that an analysis based on the deformed geometry of a structure or element under load is termed 'second-order', rather than 'non-linear'. The latter term refers to the treatment of material properties in structural analysis. Thus, according to EN 1990, 'non-linear analysis' includes 'rigid-plastic'. There is no provision for use of the latter in bridges other than by reference to EN 1993-1-1 by way of a National Annex for accidental situations only.

Concerning use of words generally, there are significant differences from British codes. These arose from the use of English as the base language for the drafting process, and the resulting need to improve precision of meaning, to facilitate translation into other European languages. In particular:

- 'action' means a load and/or an imposed deformation
- 'action effect' and 'effect of action' have the same meaning: any deformation or internal force or moment that results from an action
- 'resistance' is used for matters relating to strength, such as shear resistance
- 'capacity' is used for matters relating to deflection or deformation, such as slip capacity of a shear connector.

1.6. Symbols

The symbols in the Eurocodes are all based on ISO standard 3898: 1997.[3] Each code has its own list, applicable within that code. Some symbols have more than one meaning, the particular meaning being stated in the clause. There are a few important changes from previous practice in the UK. For example, a section modulus is W, with subscripts to denote elastic or plastic behaviour.

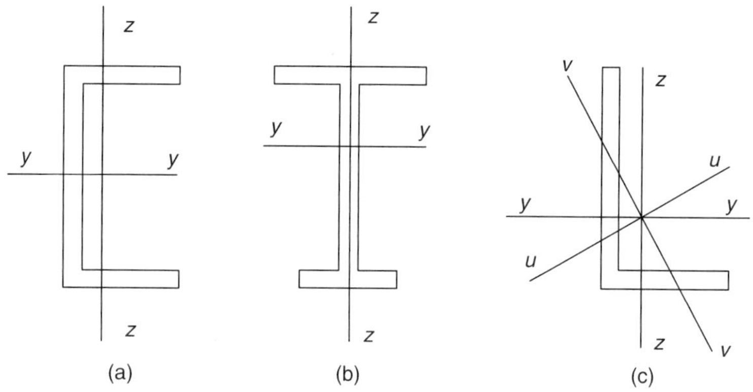

Fig. 1.7-1. Sign convention for axes of members

The use of upper-case subscripts for γ factors for materials implies that the values given allow for two types of uncertainty, i.e. in the properties of the material and in the resistance model used.

1.7. Conventions for member axes

3-1-1/clause 1.7(2)

There is an important change from previous practice in the UK. An x–x axis is along a member and a y–y axis is parallel to the flanges of a steel section – ***3-1-1/clause 1.7(2)***. The y–y axis generally represents the major principal axis, as shown in Fig. 1.7-1(a) and (b). This convention for member axes is more compatible with most commercially available analysis packages than that used in previous UK bridge codes. Where the y–y axis is not a principal axis, the major and minor principal axes are denoted u–u and v–v, as shown in Fig. 1.7-1(c).

CHAPTER 2

Basis of design

This chapter discusses the basis of design as covered in section 2 of EN 1993-2 in the following clauses:

- Requirements *Clause 2.1*
- Principles of limit state design *Clause 2.2*
- Basic variables *Clause 2.3*
- Verification by the partial factor method *Clause 2.4*
- Design assisted by testing *Clause 2.5*

2.1. Requirements

3-2/clause 2.1.1 makes reference to EN 1990 for the basic principles and requirements for the design process for steel bridges. This includes the limit states and combinations of actions to consider, together with the required performance of the bridge at each limit state. These basic performance requirements are deemed to be met if the bridge is designed using actions in accordance with EN 1991, combination of actions and load factors at the various limit states in accordance with EN 1990 and the resistances, durability and serviceability provisions of EN 1993. *3-2/clause 2.1.1*

3-2/clause 2.1.2, by reference to 3-1-1/clause 2.1.2(1), identifies that different levels of reliability are required for different types of structures. The required level of reliability depends on the consequences of structural collapse. For example, the collapse of a major bridge would be potentially much more severe in terms of loss of life than would collapse of an agricultural building. In recognition of this, EN 1990 identifies four 'execution classes', from 1 to 4, which reflect an increasing level of reliability required from the structure. Most bridges will require execution Class 3 or 4. The execution class is then invoked in EN 1090-2 and this dictates the level of testing and the acceptance criteria required in fabrication. *3-2/clause 2.1.2*

3-2/clause 2.1.3.2 gives requirements for design working life, durability and robustness. The design working life for bridges and components of bridges is also covered in EN 1990. This predominantly affects detailing of the corrosion protection system and requirements for maintenance and inspection (***3-1-1/clause 2.1.3.1(1)***) and calculations on fatigue (***3-2/clause 2.1.3.1(2)P***). Temporary structures (that will not be dismantled and reused) have an indicative design life of 10 years, while bearings have a life of 10–25 years and a permanent bridge has an indicative design life of 100 years. The design lives of temporary bridges and permanent bridges can be varied in project specifications and the National Annex respectively via ***3-2/clause 2.1.3.2(1)***. For political reasons, it is likely that the UK will adopt a design life of 120 years for permanent bridges for consistency with previous national design standards. *3-2/clause 2.1.3.2* *3-1-1/clause 2.1.3.1(1)* *3-2/clause 2.1.3.1(2)P* *3-2/clause 2.1.3.2(1)*

3-2/clause 2.1.3.3(1) to ***3-2/clause 2.1.3.3(3)*** cover general durability requirements which are elaborated on in 3-2/clause 4 and discussed in more detail in Chapter 4 of this guide. In *3-2/clause 2.1.3.3(1) to 3-2/clause 2.1.3.3(3)*

general, to achieve the design working life, bridges and bridge components should be designed against corrosion, fatigue and wear and should be regularly inspected and maintained. Where components cannot be designed for the full working life of the bridge, they need to be replaceable. To prevent slip and consequential possible wear and ingress of moisture between plates in connections, ***3-2/clause 2.1.3.3(4)*** requires permanent connections to be made using one of the following:

3-2/clause 2.1.3.3(4)

- Category B preloaded bolts (no slip at serviceability limit state – SLS)
- Category C preloaded bolts (no slip at ultimate limit state – ULS)
- fit bolts
- rivets
- welding.

3-2/clause 2.1.3.3(5)

3-2/clause 2.1.3.3(5) is intended to cover the situation of loads being transmitted in direct bearing, such as at the bottom of a bearing stiffener. The implication is that loads may be carried in this way at ULS as long as the connecting welds are designed to carry fatigue loading. This is usually done by ignoring any transmission of forces in bearing for the fatigue calculation.

Accidental actions should also be considered in accordance with EN 1991-1-7. As a general principle, parts of bridges which support containment devices, such as parapets, should be designed to be stronger than the containment device so that the bridge is not itself damaged in an impact. ***3-2/clause 2.1.3.4*** requires that where a structural component, such as a stay cable, is damaged by an accidental action, the remaining bridge should be capable of carrying the relevant actions in the accidental combination. This is discussed further for cable-supported structures in section 5.1.4 of this guide.

3-2/clause 2.1.3.4

2.2. Principles of limit state design

3-2/clause 2.2(1)

3-2/clause 2.2(1) is a reminder that the material resistance formulae given in EC3 assume that the specified requirements for materials, such as ductility, fracture toughness and through-thickness properties are met. These are covered in section 3 of EN 1993-2. It is also assumed that the requirements of EN 1090, such as tolerances in the fabrication and erection processes, will be followed as these assumptions are also included in some resistance formulae, such as those for buckling.

3-2/clause 2.2(3)

3-2/clause 2.2(4)

Elastic global analysis generally has to be used in bridge design (***3-2/clause 2.2(3)***) but plastic analysis can be used in accidental situations, such as impact on a parapet. This is discussed further in section 5.4.1 of this guide. ***3-2/clause 2.2(4)***, together with 3-2/clause 9.2.1(1), suggests that adequate fatigue life can be achieved by using 'appropriate detailing', without explicit calculation, and cites 3-2/Annex C on orthotropic decks as an example. 'Appropriate detailing' is intended to mean details which have shown themselves to be adequate in the past through in-service performance on similar structures or through testing. Although 3-2/clause 9.1.2(1) allows member states to specify situations which do not need a fatigue check, the UK National Annex requires a fatigue check for all components subject to cyclic loading and does not adopt the deemed-to-satisfy approach. In particular, the details in Annex C are not regarded in the UK as sufficiently proven to mitigate the need for explicit fatigue calculation.

2.3. Basic variables

Combinations of actions

3-2/clause 2.3.1(1)

3-2/clause 2.3.1(1) refers to Annex A2 of EN 1990 for combinations of actions. For each permanent action, such as self-weight, the unfavourable (adverse) or favourable (relieving) partial load factor as applicable can generally be used throughout the entire structure when calculating each particular action effect. There can however be some exceptions prompted by EN 1990 clause 6.4.3.1(4) which states that 'where the results of a verification are very sensitive to variations of the magnitude of a permanent action from place to place in the structure, the unfavourable and the favourable parts of this action shall be considered as individual actions. Note – This applies in particular to the verification of static equilibrium and analogous limit states.' One such exception is intended to be the verification of uplift at bearings on continuous beams, where each span would be treated separately when applying unfavourable and favourable values of load. The same applies to holding-down bolts. EC3 makes a specific recommendation to do this in 3-1-1/clause 2.4.4.

3-1-1/clause 2.3.1(4) requires the effects of uneven settlement, imposed deformations and prestressing (denoted by 'P') to be grouped with other permanent actions 'G' to form a single permanent action 'G + P'. Favourable or unfavourable load factors are then applied to this single action as appropriate without considering any differential effect of factoring the imposed deformation and the permanent load separately. Combination of 'G' + 'P' into a single permanent action 'G + P' would not always appear to be appropriate and contradicts the general format for combinations of actions in EN 1990 which requires *3-1-1/clause 2.3.1(4)*

$$\sum_{j} \gamma_{G,j} G_{k,j} + \gamma_{p} P + \text{etc.}$$

1. For uneven settlements, EN 1990 Annex A2 identifies uneven settlements as a permanent action, G_{set} and gives it a separate partial factor $\gamma_{G,set}$. The recommended value when linear elastic analysis is used is 1.2 which is less than the recommended value of 1.35 for other permanent loads. In this situation, the use of a single permanent load factor would be more conservative.
2. For imposed deformations (e.g. lowering a bearing in continuous construction), the effect of the imposed deformation is not related to the magnitude of the bridge self-weight and there therefore seems no reason to group them together and apply the same favourable or unfavourable factor to both. This would not allow the possibility of a differential effect between them to be considered.

Combinations of actions for installation of cables, replacement of cables or accidental removal of cables in cable-supported bridges are discussed in section 5.1.4 of this guide. Similar problems of combining 'G' + 'P' into a single permanent action 'G + P' are identified for cable structures.

Actions to consider

The actions to consider are given in EN 1991. Actions to consider in erection stages are given in EN 1991-1-6. Actions which are essentially imposed deformations (such as differential settlement) rather than imposed forces can sometimes be neglected where there is adequate ductility in cross-sections and the overall member is restrained against buckling. This is discussed in section 5.4.3 of this guide.

2.4. Verification by the partial factor method

Generally, the 'nominal' dimensions of the structure to be used for modelling and section analysis may be assumed to be equal to those which are put on the project drawings or which are quoted in product standards; ***3-1-1/clause 2.4.2(1)*** refers. Where EN 1993-2 requires allowance to be made for equivalent geometric imperfections, either in buckling resistance formulae or for use in global analysis, ***3-1-1/clause 2.4.2(2)*** clarifies that the imperfections provided in EN 1993 allow for geometric tolerances, structural imperfections, residual stresses and variations in yield stress. This is discussed further in section 5.3 of this guide. *3-1-1/clause 2.4.2(1)* *3-1-1/clause 2.4.2(2)*

3-1-1/clause 2.4.3(1) clarifies that cross-section resistances are based on the nominal dimensions above, together with nominal or characteristic values of the material properties as specified in the relevant sections of EN 1993. The design resistance to a *3-1-1/clause 2.4.3(1)* particular effect, R_d, is determined from the characteristic or nominal resistance, R_k, as follows:

$$R_d = R_k/\gamma_M \qquad \text{3-1-1/(2.1)}$$

where γ_M is the relevant material factor for that resistance given in EC3.

3-1-1/clause 2.4.4(1)

For permanent load calculation, the favourable or unfavourable partial load factor as applicable can generally be used throughout the entire structure, but as discussed in section 2.3 above, there are exceptions for design situations which are analogous to verifications of static equilibrium (EQU). This is referred to also in ***3-1-1/clause 2.4.4(1)***.

2.5. Design assisted by testing

3-1-1/clause 2.5(2)

The characteristic resistances in EN 1993 have, in theory, been derived using Annex D of EN 1990. EN 1990 allows two alternative methods of calculating design values of resistance. Either the characteristic resistance is first determined and the design resistance determined from this, using appropriate partial factors, or the design resistance is determined directly. EN 1993 uses the latter approach and hence ***3-1-1/clause 2.5(2)*** states that the characteristic resistances have been obtained from:

$$R_k = R_d\gamma_{Mi} \qquad \text{3-1-1/(2.2)}$$

where γ_{Mi} is the relevant material factor such that R_k represents the lower 5% fractile for infinite tests. Where it is necessary to determine the characteristic resistance for prefabricated products, this same method of determination of R_k has to be used. Discussion on the use of EN 1990 is outside the scope of this guide and is not considered further here.

CHAPTER 3

Materials

This chapter discusses material selection as covered in section 3 of EN 1993-2 in the following clauses:

- General *Clause 3.1*
- Structural steel *Clause 3.2*
- Connecting devices *Clause 3.3*
- Cables and other tension elements *Clause 3.4*
- Bearings *Clause 3.5*
- Other bridge components *Clause 3.6*

3.1. General

3-1-1/clause 3.1(1)

3-1-1/clause 3.1(2)

3-1-1/clause 3.1(1) requires the nominal values of material properties provided in section 3 of EN 1993-1-1 to be adopted as characteristic values in all design calculations. The resistances and calculation methods in EN 1993-2 and 1993-1-1 are limited to use with the steel grades listed in 3-1-1/Table 3.1, which covers steels with yield strength up to 460 MPa – see ***3-1-1/clause 3.1(2)***. A country's National Annex may give guidance on using steel to designations other than those in 3-1-1/Table 3.1. The use of steel grades with yield strength greater than 460 MPa for structural design, including bridge design, is covered by EN 1993-1-12; it does so by providing further requirements and modifications to the rules in the other parts of EN 1993.

3.2. Structural steel

3.2.1. Material properties

3-1-1/clause 3.2.1(1)

As the rules in EN 1993 use both the yield strength (f_y) and ultimate tensile strength (f_u) of the steel, the designer must establish a suitable strength for both. For commercially available steel, strengths vary with plate thickness and this variation must be included in resistance calculations. Two options for selecting material strength are provided in ***3-1-1/clause 3.2.1(1)***:

1. Obtain the f_y and f_u values from the product standard of the material grade being used. f_y is obtained as the R_{eH} value and f_u is obtained as the R_m value. The values appropriate to the actual plate thickness should be selected.
2. Use the simplified values of f_y and f_u provided in 3-1-1/Table 3.1. These allow the designer to use the maximum f_y and f_u up to 40 mm thick plate which will generally give a less conservative resistance than that using the product standards. The product standards tend to reduce the allowable values of f_y and f_u for plates above 16 mm thick.

The National Annex may specify which option should be used. (The UK National Annex specifies option 1.)

3.2.2. Ductility requirements

3-1-1/clause 3.2.2(1)

Many design clauses in EC3 assume the material used in steel components will be sufficiently ductile to enable redistribution and ductile behaviour after yield. ***3-1-1/clause 3.2.2(1)*** requires a minimum acceptable ductility to be specified and recommends the following:

(i) The ratio f_u/f_y of the specified minimum ultimate tensile strength f_u to the specified minimum yield strength f_y should be greater than or equal to a limiting value, recommended to be 1.10.
(ii) The elongation at failure on a test piece with gauge length $= 5.65\sqrt{A_0}$ (where A_0 is the cross-sectional area of the test piece) should not be less than a limiting value, recommended to be 15%.
(iii) The ultimate strain ε_u, (where ε_u corresponds to the strain when the ultimate strength f_u is reached) should be greater or equal to $15\varepsilon_y$ (where ε_y is the strain at yield).

Steel grades in 3-1-1/Table 3.1 will automatically provide the levels of ductility required above.

The above ductility recommendations may be modified by the National Annex. In the past in the UK, the minimum value of the ratio f_u/f_y was set at 1.2 with a view to protecting against brittle fracture and providing adequate ductility. There is however little evidence that this ratio is important to these characteristics or that a ratio more than the recommended one of 1.1 is required, particularly as separate checks on brittle fracture (2-2/clause 3.2.3) and ductility (item (ii) above) must also be made. It should be noted however that the plastic shear resistance (discussed in section 6.2.6.1 of this guide) makes allowance for some strain hardening, so the actual provided ratio f_u/f_y cannot be allowed to get too low. This latter point clearly does not relate to ductility provision. A specified minimum value of the ratio f_u/f_y of 1.2 would effectively prohibit the use of S500 to S700 steel grades, although the limiting ratio for the use of such steel may again be set in the National Annex to EN 1993-1-12. The use of S500 to S700 steel grades is not covered in this guide.

3.2.3. Fracture toughness

3-2/clause 3.2.3(1)
3-2/clause 3.2.3(2)

3-2/clause 3.2.3(1) requires all steel material to have sufficient toughness to prevent brittle fracture from occurring during the design life of the bridge. ***3-2/clause 3.2.3(2)*** allows EN 1993-1-10 to be used to select the required steel grade to give adequate toughness and deems its use to be sufficient to guard against brittle fracture. Note 2 of 3-2/clause 3.2.3(2) was included as a result of German comment with a view to ensuring that, at welded details, the parent metal has adequate toughness in the upper shelf region of the toughness–temperature transition curve. This suggested that higher Charpy requirements than derived from EN 1993-1-10 should be specified at welded joints to guarantee adequate ductility. 3-2/Table 3.1 gives some suggested additional requirements for welded structures but they are not mandatory and can be varied in the National Annex. These additional recommendations have not been adopted in the UK National Annex. The provisions of EN 1993-1-10 are discussed below.

The main factors in assessing brittle fracture resistance to EN 1993-1-10 are the minimum temperature that the steel component could experience in service and the maximum tensile stress that may occur in the component under this temperature. EN 1993-1-10 deals with these main factors by listing in 3-1-10/Table 2.1 the maximum allowable thicknesses of steel components of different grades in relation to their minimum temperature and associated stress level. These are by no means the only factors influencing brittle fracture as discussed below.

For each steel bridge component the general design approach is to calculate the reference minimum temperature T_{Ed}, and the associated stress σ_{Ed} in the component at T_{Ed}. The designer can then establish suitable steel grades for the component from 3-1-10/Table 2.1.

Other parameters which affect a component's brittle fracture resistance, such as crack type, component shape, strain rate, residual stress and degree of cold forming, are dealt with in EN 1993-1-10 by converting each parameter into a correction of the reference minimum temperature.

Providing all fatigue details on the steel component are covered by a detail category in EN 1993-1-9, the particular detail itself does not have to be considered in the simple brittle fracture assessment to EN 1993-1-10. This can be unconservative for details in a low detail category, as such details are more likely to trigger a brittle fracture. This was recognized in BS 5400: Part 3: 2000[4] and the UK National Annex makes allowance for this effect in the ΔT_R parameter below. Gross stress concentrations (such as an abrupt change of section next to the particular detail) are also not covered by EN 1993-1-10. The UK National Annex again makes specific allowance for gross stress concentrations in the ΔT_R parameter.

The approach in EN 1993-1-10 is only intended to be used for the selection of steel material for new construction. It is not intended to cover the brittle fracture assessment of steel materials in service. EN 1993-1-10 also gives guidelines for assessing brittle fracture resistance with fracture mechanics methods. These may be of benefit where there is no welding, tension or fatigue loading as the maximum allowable thicknesses from 3-1-10/Table 2.1 may be conservative in such cases.

Procedure to EN 1993-1-10

Calculation of T_{Ed}:

T_{Ed} is derived from the following expression given in ***3-1-10/clause 2.2(5)***: *3-1-10/clause 2.2(5)*

$$T_{Ed} = T_{md} + \Delta T_r + \Delta T_\sigma + \Delta T_R + \Delta T_{\dot{\varepsilon}} + \Delta T_{\varepsilon_{cf}} \qquad \text{3-1-10/(2.2)}$$

where:

T_{md} is the lowest air temperature with a 'specified' return period as defined in EN 1991-1-5. EN 1991-1-5 uses an annual probability of exceedance of 0.02 as the default. Isotherms for different locations are not given directly in EN 1991-1-5 and reference has to be made to the National Annex or other data.

ΔT_r is an adjustment temperature to take account of radiation loss. Although reference is made to EN 1991-1-5 for its determination, it is not defined there. The radiation loss allows both for the difference between shade air temperature and bridge effective temperature and also for any temperature difference across the cross-section. The latter is represented in EN 1991-1-5 by a non-linear temperature variation across the cross-section; 1-1-5/clause 6.1.4.2 refers. This temperature variation however also includes a small part of the uniform temperature component (1-1-5/clause 6.1.4.2(1) Note 2) so full addition of this variation to the minimum bridge uniform temperature is too conservative. Conversely, neglect of the non-linear temperature variation altogether is slightly on the unsafe side. However, given that the actual contribution of the temperature difference profile, when its uniform temperature component is removed, is small, it is reasonable to ignore its contribution. Therefore it is reasonable for ΔT_r to be determined simply as the difference between the minimum air temperature, T_{min}, and the minimum bridge uniform temperature, $T_{e,min}$ as defined in EN 1991-1-5. This effectively means that $T_{md} + \Delta T_r = T_{e,min}$. For steel decks, ΔT_r will generally be negative, thus reducing the temperature below that of the air temperature. For concrete decks, ΔT_r will generally be positive thus increasing the temperature above that of the air temperature. It is suggested here that ΔT_r is not taken greater than zero.

ΔT_σ is an adjustment temperature to take account of the stress, yield strength, type of crack imperfection, shape and dimensions of the steel component. If the maximum permissible element thicknesses are derived from 3-1-10/Table 2.1, EN 1993-1-10 recommends a value of 0 K for ΔT_σ.

ΔT_R is an adjustment temperature which enables the designer to allow for different reliability levels. Again, if the minimum permissible element thicknesses are derived from 3-1-10/Table 2.1, EN 1993-1-10 recommends a value of 0 K for ΔT_R. This is however an NDP and the UK National Annex uses it to include for the effects of fatigue detail type and gross stress concentration, which are not otherwise addressed by EN 1993-1-10. The UK National Annex also uses ΔT_R to make corrections for steel grades greater than S355. It would be more appropriate to do this via ΔT_σ, but it is not itself an NDP.

3-1-10/clause 2.3.1(2)

$\Delta T_{\dot{\varepsilon}}$ is an adjustment temperature to allow for unusual rates of loading. ***3-1-10/clause 2.3.1(2)*** states that most transient and persistent design situations are covered by a reference strain rate ($\dot{\varepsilon}_0$) of 4×10^{-4}/s. For other strain rates $\dot{\varepsilon}$ (e.g. for impact loads), $\Delta T_{\dot{\varepsilon}}$ can be calculated from the following formula:

$$\Delta T_{\dot{\varepsilon}} = \frac{1440 - f_y(t)}{550} \times \left(\ln \frac{\dot{\varepsilon}}{\dot{\varepsilon}_0}\right)^{1.5} [^\circ\mathrm{C}] \qquad \text{3-1-10/(2.3)}$$

where $\dot{\varepsilon}$ is the anticipated strain rate due to impact loads and $f_y(t)$ is the yield stress of the steel component in question. $f_y(t)$ is either taken from the R_{eH} values of the relevant product standard or taken from $f_y(t) = f_{y,nom} - 0.25(t/t_0)$ where:

$f_{y,nom}$ is the yield strength of the minimum thickness specified in the relevant product standard
t is the thickness of the plate in mm
t_0 = 1 mm.

Care should be taken with the sign of $\Delta T_{\dot{\varepsilon}}$. Expression 3-1-10/(2.3) will return a positive value of $\Delta T_{\dot{\varepsilon}}$ if $\dot{\varepsilon}$ is greater than $\dot{\varepsilon}_0$. Contrary to the sign convention used in expression 3-1-10/(2.2), the positive value of $\Delta T_{\dot{\varepsilon}}$ needs to be *deducted* from T_{Ed} in expression 3-1-10/(2.2) as the increased rate of loading will be detrimental to the component's ability to withstand brittle fracture. It would have been preferable to add a minus sign in front of expression 3-1-10/(2.3) for compatibility with expression 3-1-10/(2.2). Strain rates for impact will typically be two orders of magnitude greater than the value of $\dot{\varepsilon}_0$ for normal loading, although clearly the calculation is complex and involves consideration of the deformation characteristics of both the impacting vehicle and the part of the structure being hit. In the absence of a strain rate to use for impact loading, the approach of BS 5400: Part 3: 2000[4] could be followed. This would mean first calculating the allowable steel thickness ignoring impact and then halving this thickness to allow for impact.

$\Delta T_{\varepsilon_{cf}}$ is an adjustment temperature to take account of any cold forming applied to the steel component. $\Delta T_{\varepsilon_{cf}}$ is to be calculated from the following formula:

$$\Delta T_{\varepsilon_{cf}} = 3\varepsilon_{cf}[^\circ\mathrm{C}] \qquad \text{3-1-10/(2.4)}$$

where ε_{cf} is the permanent strain from cold forming measured as a percentage.

Calculation of σ_{Ed}:

The stress in the component, σ_{Ed}, at the reference temperature, should strictly be based on principal stress (although this is not stated) and should be calculated from the following combination of actions:

$$E_d = E\left(A[T_{Ed}] \text{`+'} \sum G_K \text{`+'} \psi_1 Q_{K1} \text{`+'} \sum \psi_{2,i} Q_{Ki}\right) \qquad \text{3-1-10/(2.1)}$$

where '$A[T_{Ed}]$' is the leading action which is basically the temperature T_{Ed}. Expression 3-1-10/(2.1) is essentially an accidental combination with temperature taken as the leading action. The effects of the temperature action $E(A[T_{Ed}])$ should include restraint to temperature movement. Combination and load factors should be taken appropriate to the serviceability limit. $\sum G_K$ is the permanent load, $\psi_1 Q_{K1}$ is the frequent value of the most onerous variable action (e.g. traffic) and $\sum \psi_{2,i} Q_{Ki}$ are the quasi-permanent values of any other applicable variable actions.

During drafting, concern was expressed in the UK over the potential excessive benefit allowed in 3-1-10/Table 2.1 at low applied stress. This concern arises because residual stresses from fabrication dominate at low applied stress, but 3-1-10/Table 2.1 continues to give a large benefit with reducing applied stress. As a consequence, the UK National Annex requires σ_{Ed} to always be taken as $0.75 f_y(t)$, but where the actual applied tensile stress is less than $0.5 f_y(t)$ the value of ΔT_R can be increased to compensate. This is more consistent with the approach previously used in BS 5400: Part 3.

3-1-10/clause 2.1(2)

The Note to ***3-1-10/clause 2.1(2)*** permits elements in compression to not be checked for fracture toughness. This is misleading as residual stresses and locked-in stresses, due to lack of fit in erection and fabrication, will often produce net tensile stresses. Additionally, slender members subject to compressive force may develop tension at one fibre due to growth of an initial bow imperfection. It is because of these secondary sources of tensile stress that ***3-2/clause 3.2.3(3)*** recommends that compression members in bridges are checked for fracture toughness using $\sigma_{Ed} = 0.25 f_y(t)$ for bridges. This value of stress can be varied in the National Annex.

3-2/clause 3.2.3(3)

A further UK concern was that 3-1-10/Table 2.1 in some cases permits up to 70°C temperature difference between T_{Ed} and the test temperature at which the Charpy energy was determined. A National Annex provision was therefore added in Note 3 of 3-1-10/clause 2.2(5) to allow countries to limit this temperature difference. The UK National Annex to EN 1993-1-10 sets a limit of 20°C between the test temperature and the application temperature, $T_{md} + \Delta T_r$, for bridges.

Worked Example 3.2-1: Selection of suitable steel grade for bridge bottom flanges

Select suitable steel grades for the bottom flanges of a series of motorway overbridges at a location in the UK where $T_{md} + \Delta T_r = -20°C$ (see discussions on radiation loss in the main text). Impact loading does not have to be considered and there are no gross stress concentrations. The proposed flange thicknesses are as follows:

Bridge 1 = 20 mm $\sigma_{Ed} = 259$ MPa $f_y(t) - 345$ MPa for 20 mm
Bridge 2 = 30 mm $\sigma_{Ed} = 259$ MPa $f_y(t) = 345$ MPa for 30 mm
Bridge 3 = 40 mm $\sigma_{Ed} = 259$ MPa $f_y(t) = 345$ MPa for 40 mm
Bridge 4 = 50 mm $\sigma_{Ed} = 251$ MPa $f_y(t) = 335$ MPa for 50 mm
Bridge 5 = 60 mm $\sigma_{Ed} = 251$ MPa $f_y(t) = 335$ MPa for 60 mm
Bridge 6 = 63 mm $\sigma_{Ed} = 251$ MPa $f_y(t) = 335$ MPa for 63 mm

The stresses in the bottom flanges σ_{Ed} all equate to $0.75 f_y(t)$ as recommended in the main text. From expression 3-1-10/(2.2):

$\Delta T_\sigma = 0°C$ (3-1-10/clause 2.2(5) Note 2 – Using tabulated values according to 3-1-10/clause 2.3)
$\Delta T_R = 0°C$ (3-1-10/clause 2.2(5) Note 1)

$\Delta T\dot{\varepsilon} = 0°C$ (Impact loading does not apply)
$\Delta T_{\varepsilon_{cf}} = 0°C$ (No cold formed steel components to be used)

$T_{Ed} = (T_{md} + \Delta T_r) + \Delta T_\sigma + \Delta T_R + \Delta T_{\dot{\varepsilon}} + \Delta T_{\varepsilon_{cf}}$
$T_{Ed} = -20°C + 0°C + 0°C + 0°C + 0°C = -20°C$

From 3-1-10/(Table 2.1), maximum permissible thicknesses for various grades are as follows ($T_{Ed} = -20°C$, $\sigma_{Ed} = 0.75 f_y(t)$):

S355JR = 20 mm, S355J0 = 35 mm, S355J2 = 50 mm, S355K2 = 60 mm, S355NL = 90 mm.

Therefore the following steel grades would be allowed to EN 1993-1-10:

Bridge 1 = 20 mm Use S355JR

The UK National Annex prevents the use of the JR grade for bridges through Note 3 of 3-1-10/clause 2.2(5) by setting a limit of 20°C between the test temperature (20°C in this case) and the application temperature, $T_{md} + \Delta T_r$ (−20°C in this case). This would then require S355J0 to be used for Bridge 1.

Bridge 2 = 30 mm Use S355J0
Bridge 3 = 40 mm Use S355J2
Bridge 4 = 50 mm Use S355J2
Bridge 5 = 60 mm Use S355K2
Bridge 6 = 63 mm Use S355NL

Further reference should be made to the National Annex to ensure that the steel will also meet any additional requirements at welded details.

Worked Example 3.2-2: Selection of a suitable steel grade for a bridge bottom flange subject to impact load

Select a suitable steel grade for the bottom flange of an overbridge which will be susceptible to impact load from high-sided vehicles. The bottom flange thickness = 40 mm, there are no gross stress concentrations and $T_{md} + \Delta T_r = -12°C$.

Project-specified strain rate under impact loading = 1.7×10^{-2}/s (see, however, the discussions on impact load above).

The stress in the bottom flange σ_{Ed} is taken as $0.75f_y(t)$ as discussed in the main text.

From 3-1-10/clause 2.2:

$\Delta T_\sigma = 0°C$ (3-1-10/clause 2.2(5) Note 2 – Using tabulated values according to 3-1-10/clause 2.3)

$\Delta T_R = 0°C$ (3-1-10/clause 2.2(5) Note 1)

$$\Delta T_{\dot\varepsilon} = \frac{1440 - f_y(t)}{550} \times \left(\ln\frac{\dot\varepsilon}{\dot\varepsilon_0}\right)^{1.5} \text{ [°C] where } f_y(t) = 345\,\text{MPa for 40 mm plate.}$$

where:

ε = impact strain rate = 1.7×10^{-2}/s
ε_0 = reference strain rate = 4.0×10^{-4}/s (3-1-10/clause 2.3.1)

$$\Delta T_{\dot\varepsilon} = \frac{1440 - 345}{550} \times \left(\ln\frac{1.7 \times 10^{-2}}{4 \times 10^{-4}}\right)^{1.5} = 14.5°C$$

$\Delta T_{\varepsilon_{cf}} = 0°C$ (No cold formed steel components to be used)

$$T_{Ed} = (T_{md} + \Delta T_r) + \Delta T_\sigma + \Delta T_R + \Delta T_{\dot\varepsilon} + \Delta T_{\varepsilon_{cf}}$$
$$T_{Ed} = -12°C + 0°C + 0°C - 14.5°C + 0°C = -26.5°C$$

From 3-1-10/(Table 2.1), maximum permissible thicknesses (t) may be interpolated from the table. Take S355J2 for example:

$\sigma_{Ed} = 0.75f_y(t)$, $T_{Ed} = -20.0°C$, $t = 50$ mm
$\sigma_{Ed} = 0.75f_y(t)$, $T_{Ed} = -30.0°C$, $t = 40$ mm

By interpolation, $\sigma_{Ed} = 0.75f_y(t)$, $T_{Ed} = -26.5°C$, $t = 43.5\,\text{mm} > 40\,\text{mm}$, so S355J2 is adequate.

Further reference should be made to the National Annex to ensure that the grades will also meet any additional guidelines at welded details.

3.2.4. Through-thickness properties

During fabrication, rapidly cooling and shrinking weld metal can lead to the development of large tensile strains through the thickness of plates. The magnitude of the strain in the through-thickness direction is a function of the weld size, weld orientation, plate thickness, degree of shrinkage restraint and the amount of preheating used in the weld procedure. Steel contains micro defects in the form of inclusions, particularly sulphur, and these defects can initiate cracks under the action of through-thickness tension, leading to tearing as shown in Fig. 3.2-1. This phenomenon is known as 'lamellar tearing'. The micro imperfections, prior to any lamellar tearing occurring, are too small to be detected by ultrasonic testing so no useful information can be derived from such testing prior to welding. Ultrasonic testing can however be used after welding to check that lamellar tearing has not occurred.

In order to successfully resist these weld shrinkage strains without lamellar tearing occurring, steel plates must have sufficient ductility in the through-thickness direction. The measure of ductility perpendicular to the plane of a steel plate is referred to as its 'through-thickness ductility'.

In order to assess whether the through-thickness properties of a plate are acceptable for a given configuration, ***3-2/clause 3.2.4(1)*** refers to EN 1993-1-10. The measure of through-thickness ductility is the 'Z' value. The 'Z' value is essentially the percentage reduction in area obtained at failure in a through-thickness tensile test specimen. *3-2/clause 3.2.4(1)*

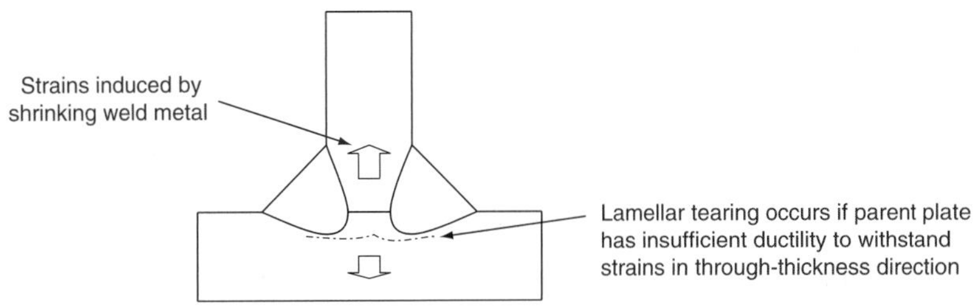

Fig. 3.2-1. Lamellar tearing

Assessing through-thickness ductility to EN 1993-1-10

From 3-1-10/clause 3.2 lamellar tearing can be neglected if $Z_{Ed} \leq Z_{Rd}$ where:

Z_{Ed} is the required through-thickness ductility ('Z value') resulting from the effect of weld size, weld orientation, plate thickness, restraint and degree of preheating.

Z_{Rd} is the available through-thickness ductility ('Z' value to EN 10164) of the parent plate.

Z_{Ed} is calculated from $Z_{Ed} = Z_a + Z_b + Z_c + Z_d + Z_e$

where:

Z_a is the Z value taken from 3-1-10/Table 3.2(a) to represent the effect of the fillet weld depth.

Z_b is the Z value taken from 3-1-10/Table 3.2(b) to represent the effect of the shape and arrangement of the welds. Table 3.2 does not explicitly cover cruciform joints in the Z_b value section. Cruciform joints should be assessed on the basis of the geometry of a Tee joint (based on the worst side of the cruciform if not symmetric). The greater restraint to shrinkage that may result in a cruciform joint should be considered in the Z_d value.

Z_c is the Z value taken from 3-1-10/Table 3.2(c) to represent the effect of parent plate thickness on the probability of lamellar tearing occurring. For cruciform and Tee joints there appears to be an incentive to make the thinner plate continuous to minimize the value. This should not generally be done and the thinner plate should generally be made discontinuous at the thicker plate to minimize the size of welds required.

Z_d is the Z value taken from 3-1-10/Table 3.2(d) to take account of the amount that free shrinkage of the weld metal will be restrained.

Z_e is the Z value from 3-1-10/Table 3.2(e) to take account of the effect that preheating before welding has on the probability of lamellar tearing occurring. In EN 1993-1-10, the effects of preheating are found to be beneficial. However, concerns have been expressed by some in the UK steel industry that preheating can actually increase susceptibility to lamellar tearing, so it is recommended here that benefit is not taken from preheating.

Having calculated Z_{Ed}, the required through-thickness ductility to EN 10164 is obtained from EN 1993-2 Table 3.2. The limits of Table 3.2 may be modified by the National Annex.

There is concern within the steel industry that the provisions in EN 1993-1-10 may lead to an unnecessary increase in quantities of steel being specified with '*Z*' requirements. It should be borne in mind that the most important consideration is to provide good detailing that is least prone to through-thickness problems, such as passing a thicker plate continuously through a thinner one to minimize the size of welds required. 3-1-10/Table 3.1 introduces two quality classes: Class 1 and 2. Class 1 requires a specification of through-thickness properties to control lamellar tearing in all cases. Class 2 requires specification of through-thickness properties only for the most high-risk details, with post-fabrication inspection to check that lamellar tearing has not occurred. Since, in most cases, the fabricator is best placed to choose the method of controlling lamellar tearing, the UK National Annex opts for Class 2 with specification of 'Z' requirements only for certain details prone to lamellar tearing such as, for example, cruciform joints with large welds.

Worked Example 3.2-3: Assessment of whether steel with enhanced through-thickness properties (to EN 10164) needs to be specified at a halving joint detail

The middle flange plate is slotted around the girder web in Fig. 3.2-2. This has been done despite normal good practice to slot the thicker plate through the thinner one because, in this case, the stress in the web is very high and would lead to a larger weld if the web were slotted.

(i) $a_{eff} = 10$ mm (3-1-10/Fig. 3.2), therefore $Z_a = 3$ (3-1-10/Table 3.2)
(ii) $Z_b = 0$ (multi-run fillet welds)
(iii) $Z_c = 4$ (half joint web = 16 mm)
(iv) $Z_d = 0$ (free-shrinkage possible)
(v) $Z_e = 0$ (no pre-heating specified)

From 3-1-10/section 3.2: $Z_{Ed} = Z_a + Z_b + Z_c + Z_d + Z_e$ therefore $Z_{Ed} = 3 + 0 + 4 + 0 + 0 = 7$

From 3-2/Table 3.2, for $Z_{Ed} \leq 10$ there is no need to specify steel with through-thickness properties to EN 10164.

3.2.5. Tolerances

3-2/clause 3.2.5(1)
3-2/clause 3.2.5(1) requires that the dimensional tolerances on rolled steel sections, hollow sections and plates comply with those stated in the relevant product standards. This is to ensure that the variations from nominal dimensions are adequately catered for by the EC3 material partial factors. For sections fabricated by welding, additional tolerances are given in EN 1090-2 – *3-2/clause 3.2.5(2)* refers. Tolerances on plate thickness and cross-section dimensions do not need to be considered in structural analysis – *3-1-1/clause 3.2.5(3)* refers. Other fabrication tolerances, such as straightness of struts and verticality of supports, are also specified in EN 1090. These fabrication imperfections, as distinct

3-2/clause 3.2.5(2)
3-1-1/clause 3.2.5(3)

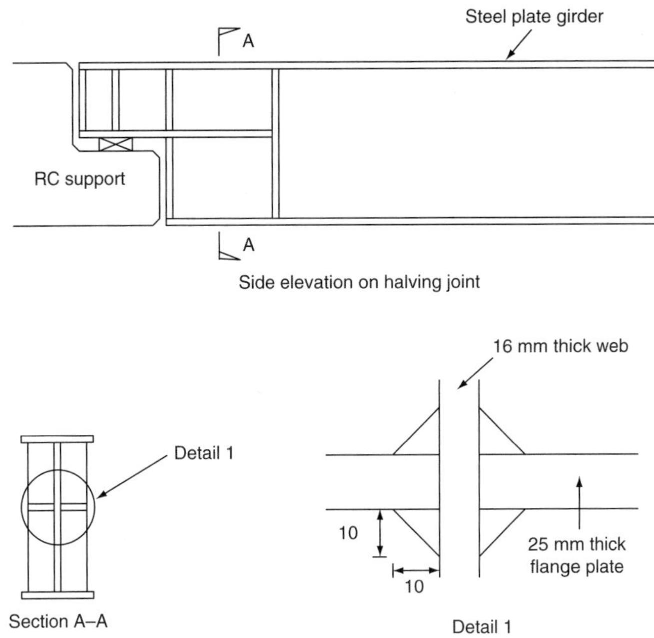

Fig. 3.2-2. Figure for Worked Example 3.2-3

from tolerances on cross-section dimensions, must be included in structural analysis where second-order effects are significant as discussed in sections 5.2 and 5.3 of this guide. The equivalent geometric imperfections for use in structural analysis given in 3-2/clause 5.3 are greater than the allowable geometric imperfections specified in EN 1090 because they also include the effects of welding residual stresses.

Additional guidance regarding the allowable tolerances and inspection requirements for steel orthotropic decks are provided in 3-2/Annex C.

3.2.6. Design values of material coefficients

The following material coefficients should be used in calculations for steels listed in 3-1-1/ Table 3.1:

Modulus of elasticity	$E = 2.10 \times 10^6$ MPa
Shear modulus	$G = 8.10 \times 10^5$ MPa
Poisson's ratio	$\nu = 0.3$
Coefficient of linear thermal expansion	$\alpha = 12 \times 10^{-6}$ per °C

For simplicity, EN 1994 generally allows the coefficient of linear thermal expansion for steel in composite bridges to be taken as $\alpha = 10 \times 10^{-6}$ per °C, which is the same as for concrete. This avoids the need to calculate internal restraint stresses from uniform temperature change, which otherwise result from different coefficients of thermal expansion for steel and concrete. The overall movement from uniform temperature change (or force due to restraint of movement) should however be calculated using $\alpha = 12 \times 10^{-6}$ per °C throughout.

'E' values for tension rods and cables of different types are not covered by this clause and are given in section 3.4.2 of this guide.

Table 3.3-1. Strengths of bolt grades covered by EC3-2

Bolt grade	4.6	5.6	6.8	8.8	10.9
f_{yb} (MPa)	240	300	480	640	900
f_{ub} (MPa)	400	500	600	800	1000

3.3. Connecting devices

3.3.1. Fasteners

The design of bolted and riveted connections is covered in section 8.1 of this guide.

3.3.1.1. Bolts, nuts and washers

3-2/clause 3.3.1.1(1) The rules in EC3-2 for designing bolts assume that the bolts, nuts and washers comply with the product standards (Group 4) in 3-1-8/clause 2.8 – ***3-2/clause 3.3.1.1(1)*** refers. This is a long list, which is not reproduced here, but it covers the most commonly used components previously used in the UK.

3-2/clause 3.3.1.1(2) ***3-2/clause 3.3.1.1(2)*** states that the bolt grades covered by the EC3-2 rules are limited to those in 3-2/Table 3.3, reproduced above as Table 3.3-1.

3-2/clause 3.3.1.1(3) Table 3.3-1 contains nominal values of the yield strength f_{yb} and ultimate tensile strength f_{ub}. ***3-2/clause 3.3.1.1(3)*** requires these values to be used as characteristic values in the design calculations.

3.3.1.2. Preloaded bolts

Grade 8.8 and 10.9 high-strength bolts for preloaded connections can also be used in accordance with EN 1993-1-8 provided that they comply with the reference standards of Group 4 in 3-1-8/clause 2.8. Tightening must be carried out in accordance with EN 1090.

3.3.1.3. Rivets

Should the designer wish to specify rivets as an alternative to bolts, they may be designed in accordance with EN 1993-1-8 provided the rivets comply with reference standards in Group 6 of 3-1-8/clause 2.8.

3.3.1.4. Anchor bolts

3-2/clause 3.3.1.4(1) Anchor bolts which are being designed in accordance with EN 1993-1-8 must comply with either EN 10025 or the reference standards in Group 4 of 3-1-8/clause 2.8. Reinforcing bars may also be used as anchor bolts provided that they comply with EN 10080. ***3-2/clause 3.3.1.4(1)*** requires that the nominal yield strength for anchor bolts does not exceed 640 MPa. (This presumably takes priority over 3-1-8/clause 3.3 which restricts yield strength to 640 MPa for shear but allows 900 MPa otherwise.)

3.3.2. Welding consumables

3-2/clause 3.3.2(1) *3-2/clause 3.3.2(2)* The design of welded connections is covered in section 8.2 of this guide. Welded connections designed in accordance with EN 1993-1-8 assume that all the welding consumables comply with reference standards Group 5 of 3-1-8/clause 2.8. This is required by ***3-2/clause 3.3.2(1)***. Additionally, ***3-2/clause 3.3.2(2)*** requires all mechanical properties of the weld to be not less than those of the parent plate. This ensures that no special consideration in design is needed for butt welded connections between plates and rolled sections. For high-strength steels, with yield strength greater than 460 MPa, this rule is modified by EN 1993-1-12 which gives methods of designing welds with lower strength than the parent plate.

3.4. Cables and other tension elements

3-2/clause 3.4(1) ***3-2/clause 3.4(1)*** refers to EN 1993-1-11 for the design of tension components. Relevant provisions are discussed under the following additional sub-sections.

3.4.1. Types of cables covered (additional sub-section)

EN 1993-1-11 covers bridges with **adjustable and replaceable** steel tension components. The types of tension components covered fall into three groups as follows:

1. **Tension rod systems (Group A).** These generally comprise prestressing bars of solid round cross-section connected to end anchorages by threading of the bar. They are

typically proprietary systems. A typical use would be for holding down girders subject to uplift forces.

2. **Ropes (Group B).** These include spiral strand ropes, fully locked coil ropes and strand ropes which are composed of wires which are anchored in sockets or other end terminations.

   - Spiral strand ropes comprise a series of round wires laid helically in two or more layers around a centre, usually a wire. They are fabricated mainly in the diameter range 5 mm to ~160 mm and are typically used as stay cables and hangers for bridges.
   - Fully locked coil ropes comprise a series of wires laid helically in two or more layers around a centre, usually a wire and with an outer layer of Z-shaped wires which lock together. They are fabricated in the diameter range 20 to ~180 mm and are mainly used as stay cables, suspension cables and hangers for bridges.
   - Strand ropes comprise a series of multi-wire strands laid helically around a centre. They are mainly used as hangers for suspension bridges.

3. **Bundles of parallel wires or strands (Group C).** These include bundles of parallel wires and bundles of parallel strands which need individual or collective anchoring and individual or collective protection. They are mainly used as stay cables and external tendons. Bundles of parallel wires are also used for main cables for suspension bridges.

Typical cross-sections for these cable types are given in 3-1-11/Annex C but are not reproduced here.

3.4.2. Cable stiffness (additional sub-section)

For cable-supported bridges, the stiffness of the cables has to be derived in accordance with EN 1993-1-11. 3-1-11/clause 3.2 gives guidance on values of modulus of elasticity 'E', for use in analysis. Three situations are identified for the different cable groups above:

1. **Tension rod systems (Group A):** E can be taken as 210 000 MPa.

2. **Ropes (Group B):** E varies with stress level and repeated loading. A secant value should be determined by testing over the range of stress expected in the cable within the bridge. For preliminary design, E can however be taken from 3-1-11/Table 3.1. It should be noted that 'E' values are considerably lower than for tension rods.

3. **Bundles of parallel wires or strands (Group C):** E can be obtained from EN 10138 or 3-1-11/Table 3.1. The latter leads to values of E of 205 000 ± 5000 MPa for bundles of parallel wires and 195 000 ± 5000 MPa for bundles of parallel strands.

EN 1993-1-11 also covers the analysis of cable-supported bridges, including treatment of load combinations and non-linear effects. This is discussed in section 5.1.4 of this guide. The non-linear effects of cable sag can be accounted for without formal non-linear analysis by using a reduced modulus of elasticity, E_t, according to the Ernst equation given in expression 3-1-11/(5.1):

$$E_t = \frac{E}{1 + \dfrac{w^2 l^2 E}{12\sigma^3}} \qquad \text{3-1-11/(5.1)}$$

where:

E is the actual modulus of elasticity
w is the unit weight of the cable (from 3-1-11/Table 2.2)
l is the horizontal span of the cable
σ is the stress in the cable.

For short cables, the apparent modulus will normally be very close to the full modulus unless the cables are particularly heavy or particularly lightly stressed.

3.4.3. Other material properties and corrosion protection (additional sub-section)

Detailed guidance is given in 3-1-11/clause 3 and 3-1-11/clause 4 on other material properties and corrosion protection respectively. These are not discussed further here.

3.5. Bearings

3-2/clause 3.5(1)

3-2/clause 3.5(1) requires that all steel bridge bearings comply with EN 1337. EN 1337 comprises 11 parts. Part 1 is entitled 'General design rules' and gives requirements common to all bearings. The remaining parts cover the design of different types of bearings and requirements for their protection, installation, inspection and maintenance.

3.6. Other bridge components

In order to ensure consistent good quality, all ancillary items (such as waterproofing, expansion joints, parapets, crash barriers) should comply with the relevant technical specifications and product standards. The National Annex may limit the types of components that may be used. It is more likely that individual Clients will specify such limitations for their individual projects.

CHAPTER 4

Durability

This chapter discusses durability as covered in section 4 of EN 1993-2. It introduces two additional sub-sections as follows:

- Durable details *Section 4.1*
- Replaceability *Section 4.2*

Bridges must be sufficiently durable so that they remain serviceable throughout their design life. ***3-2/clause 4(1)*** refers the designer, by way of EN 1993-1-1, to EN 1990 clause 2.4(1)P where the following requirement is given: *3-2/clause 4(1)*

The structure shall be designed such that deterioration over its design working life does not impair the performance of the structure below that intended, having due regard to its environment and the anticipated level of maintenance.

Steel components should either be designed to function adequately for the full design life of the bridge, with appropriate levels of inspection and maintenance being carried out as provided for in the design, or should be designed to be replaceable as required by 3-2/clause 4(6) – see section 4.1, (item 6) below. To achieve the former, parts susceptible to corrosion, mechanical wear or fatigue should have access for inspection and maintenance commensurate with the assumptions made in the design – ***3-1-1/clause 4(3)*** refers. Ideally all parts should be accessible but if a part cannot be inspected for signs of corrosion, a corrosion allowance on the thickness of the part should be made in accordance with ***3-2/clause 4(4)*** and a suitable fatigue check performed, reflecting the lack of accessibility – see section 4.1 (item 4) below. ***3-2/clause 4(5)*** however requires that all components should be checked for fatigue, whether accessible for inspection or not. *3-1-1/clause 4(3)* *3-2/clause 4(4)* *3-2/clause 4(5)*

4.1. Durable details (additional sub-section)

In order to meet durability requirements, some suggested guidelines are given below:

1. **Specifying a steel grade that does not require painting.** As the majority of steel bridge durability problems involve corrosion of the steel after failure of the protective paint system, 'weathering steel' can often be an effective alternative to ordinary painted steels. 'Weathering steel' is a low-alloy steel that corrodes at a much slower rate than standard steel grades. The corrosion induces a stable patina of fine-grained rust which remains adhered to the base metal and slows the rate of corrosion to a level which enables the steel to be left in standard atmospheric conditions unpainted. A small corrosion allowance on thickness, whose magnitude depends on environment, still has to be made.

   Weathering steel has advantages for health and safety (by eliminating the risks of maintenance painting at height or inside box girders), for the environment (by eliminating emissions of solvents into the atmosphere when the paint cures) and for reducing costs (by eliminating whole-life maintenance costs associated with repainting the structure). It should not however be used in coastal or aggressive chemical environments or other areas where a high concentration of chloride ions is present, as the functioning of the patina is inhibited.

Guidance on the use of weathering steel is available directly from producers and also in Reference 5.

An even more effective, but very expensive, alternative to weathering steel is stainless steel.

2. **Avoidance of corrosion traps in detailing.** Durability problems tend to start at corrosion traps on the steel structure. Durability can therefore be much improved if the detailing avoids corrosion traps as far as possible. This issue is particularly important for non-painted weathering steels. In addition, it is recommended that water contaminated with de-icing salts is kept well away from steel components by effective fail-safe drainage systems.

3. **Avoidance of details that cannot be easily painted.** For structures that contain painted steelwork, many durability problems can be avoided by ensuring that there are no areas where access is difficult for applying paint.

4. **Sacrificial thickness and fatigue checks for inaccessible components.** If areas are totally inaccessible during the design life then they can be increased in thickness so that they are not overstressed if part of the section is lost due to corrosion. In the absence of guidance in EC3 (a National Annex may give guidance), it is recommended that designers use the provisions in BS 5400: Part 3.[4] For a design life of 120 years, this gave recommended values of sacrificial thickness to apply to each inaccessible surface as follows:

   (i) 6 mm at industrial or marine sites
   (ii) 4 mm at other inland sites
   (iii) 1 mm in addition to the excess under (i) and (ii) where free drainage cannot be specified.

   In addition, EN 1993-1-9 requires that inaccessible components are checked for fatigue using the 'safe life' concept. Potentially, this would require more onerous partial factors in the fatigue check of the inaccessible component, although it is likely that the 'safe life' approach will be used in the UK for all details, whether accessible for inspection or not, as discussed in Chapter 9 of this guide.

5. **Careful specification of the painting system.** The designer is recommended to ensure that the protective paint system is carefully and accurately specified. Of particular importance is the specification of the initial surface preparation works as these works form the foundation for the rest of the paint system.

6. **Careful specification of the fabrication and erection works.** Some durability problems can be caused by poor fabrication and erection procedures. Steel bridge structures designed to EN 1993-2 should be fabricated to EN 1090-2 in which the fabrication procedures are designed to ensure durable steel components.

7. **Elimination of slip in joints.** To prevent slip and consequential possible wear and ingress of moisture between plates in connections, 3-2/clause 2.1.3.3 requires permanent connections to be made using one of the following:

   - Category B preloaded bolts (no slip at serviceability limit state (SLS)
   - Category C preloaded bolts (no slip at ultimate limit state (ULS)
   - fit bolts
   - rivets
   - welding.

4.2. Replaceability (additional sub-section)

3-2/clause 4(6) requires that components which cannot be designed with sufficient reliability to achieve the design working life should be replaceable. Typical components which should be replaceable, along with suggestions for complying with 3-2/clause 4(6), are as follows: *3-2/clause 4(6)*

1. **The corrosion protection system.** Ensure that the corrosion protection system can be replaced safely at the end of its design life.
2. **Stays, cables, hangers.** Carry out design checks to ensure that the structure is still adequate if a cable is removed. Ensure that the cable connection detail allows the cables to be replaced in the future. This is discussed in more detail in section 5.1.4 of this guide.
3. **Bearings.** Ensure that bearings are detailed so that they can be simply removed from the structure without excessive effort. Provide jacking stiffeners so that the structure can be safely jacked up to enable replacement of the bearing.
4. **Expansion joints.** Ensure that the expansion joints can be replaced without damage to the bridge deck.
5. **Asphalt layer and waterproofing.** Ensure that the structure can withstand replacement of the surfacing and waterproofing.
6. **Guardrails, parapets, wind shields and noise barriers.** Ensure that these components can be easily removed from the structure without damage occurring to the main bridge. Components, such as parapets, which may be susceptible to errant vehicle impact should be designed so that their foundation (e.g. deck cantilevers) and anchorage is stronger than the parapet post. This will ensure that repairs, if required, are only required for the parapet and not the bridge deck – 3-2/clause 2.1.3.3(2) refers.
7. **Drainage devices.** Ensure that drainage systems are able to be cleared at regular intervals by providing sufficient rodding eyes at accessible locations. Ensure that the drainage system can be easily replaced if needed.

CHAPTER 5

Structural analysis

This chapter discusses structural analysis as covered in section 5 of EN 1993-2 in the following clauses:

- Structural modelling for analysis — *Clause 5.1*
- Global analysis — *Clause 5.2*
- Imperfections — *Clause 5.3*
- Methods of analysis considering material non-linearities — *Clause 5.4*
- Classification of cross-sections — *Clause 5.5*

This section of EN 1993-2 covers the structural idealization of bridges and the methods of analysis required in different situations, including the section properties to be used. It also covers the section classification of members for cross-section checks contained in 3-2/clause 6. Much reference has to be made to other parts of EN 1993 to pull together all the relevant information required for analysis. In particular, reference has to be made to EN 1993-1-5 for the effects of shear lag and plate buckling.

5.1. Structural modelling for analysis

5.1.1. Structural modelling and basic assumptions

3-2/clause 5.1.1(1)
3-2/clause 5.1.1(4)

The basic requirement of ***3-2/clause 5.1.1(1)*** for analysis is that it should realistically model the true behaviour. The Note to ***3-2/clause 5.1.1(4)*** acknowledges that reference may be necessary to other parts of EN 1993 to achieve this. Where stiffness in analysis might be affected by shear lag or plate buckling effects, reference needs to be made to 3-1-5/clause 2.2. This gives rules for when and how to take these effects into account. For steel-only bridges, it will generally only be necessary to consider these effects for box girders with an orthotropic deck or other steel beams with a common steel top flange. For steel and concrete composite members, slightly different rules for shear lag apply for concrete flanges. These are given in EN 1994-2.

The Note to 3-2/clause 5.1.1(4) also refers to EN 1993-1-11 for the design of cable-supported structures. Specific guidance on modelling joints, ground–structure interaction and cable-supported structures is given in sections 5.1.2 to 5.1.4 respectively below.

Shear lag

In wide flanges, in-plane shear flexibility leads to a non-uniform distribution of bending stress across the flange width. This effect is known as shear lag and is illustrated in Fig. 5.1-1 for a simply supported box girder with knife edge load applied at midspan. The elastic distribution of shear stress across the box top flange leads to a transverse strip of flange deforming as shown. The free ends of the box top flange therefore adopt a similar deflected shape arising from this shear deformation together with axial shortening from the compressive bending stresses. The distorted box top flange is shorter along the webs

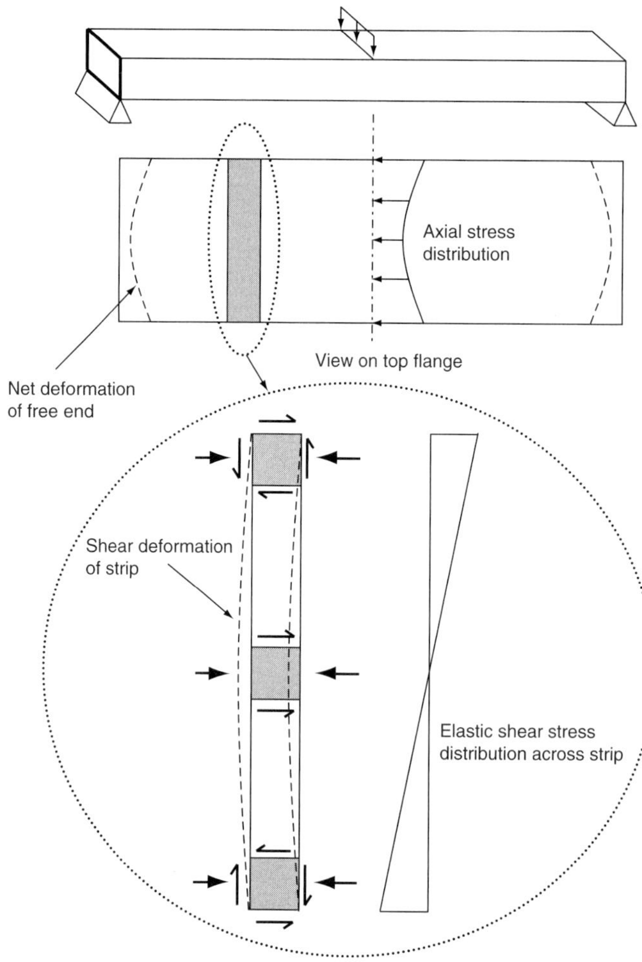

Fig. 5.1-1. Illustration of shear lag in simply supported box girder

than along its centre so the axial compressive stress must therefore be greater at the webs than in the middle of the flange. The stress in the flange adjacent to the web is consequently found to be greater than expected from analysis with gross cross-sections, while the stress in the flange remote from the web is lower than expected. Similar results are produced with continuous beams with the maximum in-plane shear lag displacements occurring at points of contraflexure. This shear lag also leads to a loss of stiffness of a section in bending, which can be important in determining realistic distributions of moments in analysis.

The determination of the actual distribution of stress is a complex problem which depends on the loading configuration, the stiffening to the flanges and any plasticity occurring. The stress distribution at the serviceability limit state can be modelled using elastic finite-element analysis with shell elements. At the ultimate limit state, plasticity usually occurs and non-linear finite-element analysis is required to produce an accurate representation of the stress distribution.

The Eurocodes account for both the loss of stiffness and localized increase in flange stresses by the use of an effective width of flange which is less than the actual available flange width. The effective flange width concept is artificial but, when used with engineering bending theory, leads to uniform stresses across the whole reduced flange width that are equivalent to the peak values adjacent to the webs in the true situation. It follows from the above that if finite-element modelling of flanges is performed with sufficient detail for the flange elements, shear lag will be taken into account and the additional use of an effective flange in accordance with this clause would be unnecessary.

3-1-5/clause 2.2(3)

For global analysis, ***3-1-5/clause 2.2(3)*** allows the effective width of flange acting on each side of a web to be taken as the lower of the full available width and $L/8$ where L is the span

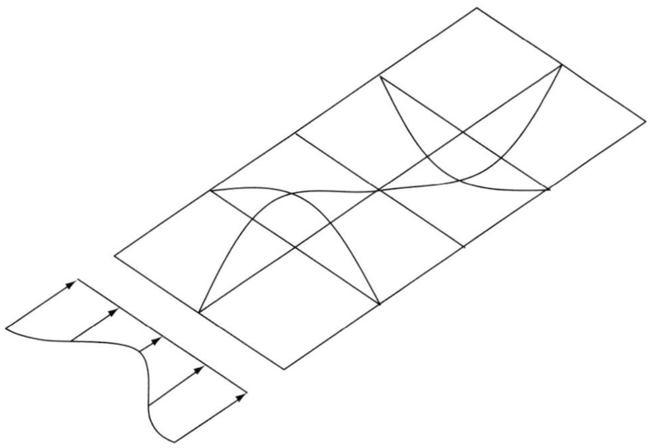

Fig. 5.1-2. Stress distribution across width of slender plate

or twice the length of a cantilever. This width may be taken as constant throughout the entire span. Alternatively, the values for serviceability limit state (SLS) cross-section design from 3-1-5/clause 3 could be used. These are discussed later in section 6.2.2.3, together with worked examples.

Plate buckling

Slender plates (Class 4 according to 3-1-1/clause 5.5) also exibit a loss of stiffness under load. The stiffness of perfectly flat plates suddenly reduces when the elastic critical buckling load is reached. In 'real' plates that have imperfections, there is an immediate reduction in stiffness from that expected from the gross plate area because of the growth of geometric imperfections under load. This stiffness continues to reduce with increasing load. This arises because non-uniform stress develops across the width of the plate as shown in Fig. 5.1-2. The non-uniform stress arises because the development of the buckle along the centre of the plate leads to a greater developed length of the plate along its centreline than along its edges. Thus the shortening due to membrane stress, and hence the membrane stress itself, is less along the centreline of the plate.

This loss of stiffness must be considered in the global analysis, where significant, and can also be represented by an effective width of plate. The reduction in ultimate strength (caused both by the non-uniform axial membrane stress and the out-of-plane bending stresses due to the deflections in Fig. 5.1-2) is also accounted for by using effective widths for the plate panels, but these widths are smaller than those appropriate for stiffness in global analysis; the reduction in strength due to plate buckling is greater than the reduction in stiffness. The same effective widths as used for strength calculation can however be used for global analysis (***3-1-5/clause 2.2(4)***) or more accurate effective widths for global analysis can be determined from 3-1-5/Annex E. Alternatively, ***3-1-5/clause 2.2(5)*** allows the effects of plate buckling to be ignored in global analysis where the effective areas of compression elements at the ultimate limit state are greater than ρ_{lim} times the gross area. ρ_{lim} is a limiting value of the ultimate limit state (ULS) reduction factor for plate buckling discussed in section 6.2.2.5 of this guide and is a nationally determined parameter whose recommended value is 0.5. This value will ensure that plate buckling effects rarely need to be considered in global analysis.

3-1-5/clause 2.2(4)
3-1-5/clause 2.2(5)

A similar loss of stiffness occurs from bowing of any longitudinal stiffeners present and further modifications to the effective areas are used to model this effect also. The rules in 3-1-5/clause 4.3 are used to do this and these are discussed later in sections 6.2.2.5 and 6.2.2.6 of this guide where strength is also discussed.

Shear lag combined with plate buckling effects

Since the concept of effective widths for both shear lag and plate buckling can be confusing, EN 1993-1-5 distinguishes between effective widths for shear lag and for plate buckling and for the combined effective widths using the following notation:

effective^p – effective width for plate buckling
effective^s – effective width for shear lag
effective – effective width for plate buckling and shear lag.

The combination of the two effects is achieved by first calculating the effective^p width for plate buckling and then considering only that part of the area which is in the effective^s width for shear lag.

5.1.2. Joint modelling

3-2/clause 5.1.2(1) *3-1-1/clause 5.1.2(1)* *3-1-1/clause 5.1.2(2)*

3-2/clause 5.1.2(1) refers to both EN 1993-1-1 and EN 1993-1-8. ***3-1-1/clause 5.1.2(1)*** and ***3-1-1/clause 5.1.2(2)*** state that it is generally permissible to ignore detailed considerations of joint stiffness in analysis of bridges, with joints treated as either pinned or rigid as appropriate. One exception to this is where 'semi-continuous' joints, as defined in EN 1993-1-8, are used. These are joints which are neither rigid nor pinned but have a certain amount of flexibility when resisting load. An example of such a joint might include a connection made via bolted end plates, where flexure of the end plates gives joint flexibility but the joint still is capable of carrying moment. It is recommended that semi-continuous joints are not used for bridges so that fatigue can be assessed using the detail categories in EN 1993-1-9. This is the reason for the Note to ***3-2/clause 5.1.2(5)***. Semi-continuous joints may still, in some cases, be unavoidable, such as end plate connections in some U-frame bridges. In this latter specific case, the flexibility would have to be considered in deriving the restraint provided to the compression flange by the U-frame. EN 1993-1-8 provides methods of determining the joint stiffness.

3-2/clause 5.1.2(5)

Another apparent exception to the above rule, where joint behaviour must be considered, is in the consideration of bolt slip discussed in section 5.2.1 of this guide.

5.1.3. Ground–structure interaction

3-1-1/clause 5.1.3(1)

3-1-1/clause 5.1.3(1) refers to 'deformation characteristics of supports', so the stiffness of the bearings, piers, abutments and ground have to be taken into account in analysis. This also includes consideration of stiffness in determining effective lengths for buckling or in calculating buckling resistances directly from the analysis. For further guidance on the latter, see section 5.2 of this guide.

5.1.4. Cable-supported bridges (additional sub-section)

A detailed treatment of the design of cable-supported bridges is outside the scope of this guide but a few salient points are noted here. The general guidance in sections 5.2 to 5.4 of this guide are also relevant.

5.1.4.1. Analysis

EN 1993-1-11 covers the design of cable-supported bridges. The analysis of cable-supported bridges needs to consider non-linearities arising from second-order effects under axial load, from large deflections altering the overall bridge geometry and from the sag of cables. The latter may be covered by a simple correction to the 'E' value of the cables as discussed in section 3.4 of this guide. Where there are significant non-linearities, the design at the ultimate limit state needs to be performed by applying factored loads to the analysis model in the same way as discussed in section 5.2 for second-order effects.

In general, the analysis should consider the build-up of load effects throughout the construction sequence. An analysis should be performed using characteristic values of actions to determine an intended design profile. This allows the deformed shape to be monitored on site and cables adjusted to achieve this profile if necessary. An important distinction must therefore be made between bridges where the cables are to be adjusted on site to achieve the assumed design profile of the bridge and those where no adjustment is to be made as discussed below.

Design for in-service condition

The intention in ***3-1-11/clause 5.3*** is that if cables are to be adjusted to achieve the assumed design deflection profile, then the self-weight and cable preloads are combined into a single permanent action, '$G + P$', whose application to the structure corresponds to the intended permanent profile of the bridge. For ultimate limit states, this single entity is then multiplied by either the favourable or unfavourable load factor γ_G as appropriate to determine action effects. It is however essential that cables are adjusted on site if necessary to achieve the intended design profile if the actions are to be combined in this way. This is because the combined effects of dead load and cable preload (e.g. bending moments) are formed from the difference between two large opposing actions whose net effect will typically be designed to be as near to zero throughout the bridge as possible. A load factor applied to a near zero effect will obviously still give a near zero effect at ULS. If there is no control on deflections, by adjusting cables, the real (as opposed to calculated) difference between these two large numbers can become very large if, for example, the bridge's self-weight is greater than the characteristic value assumed in the design. *3-1-11/clause 5.3*

There are however some problems with this approach in certain structures and it additionally contradicts the general format for effects of actions in EN 1990 as discussed in section 2.3 of this guide. In some situations, the required deflection control will be automatically achieved through normal site controls on profile. For example, it would not be possible to achieve an unintended differential between 'G' and 'P' in a large cable-stayed bridge with a flexible deck because the deflections during construction would become excessive and the cables would have to be adjusted. Application of separate favourable and unfavourable partial factors to self-weight and prestress in this situation would be unrealistic as the deflections and stresses found from such an analysis could not actually occur in practice due to site profile controls.

If the deck was however very stiff compared to the cables, such as might occur in a short-span cable-stayed bridge with a stiff concrete deck, an unintended differential between 'G' and 'P' might not be noticed as the difference from predicted deflections would be less measurable. The same would apply to a bridge with external prestressing, where it is unlikely that cables would be adjusted in any case. In both the latter cases, combination of '$G + P$' into one entity with a common load factor is potentially unsafe.

The authors would prefer that the presumption should always initially be for separate combination unless there is a demonstrable reason to do otherwise. A cautionary note is therefore given as follows. For some structural types, combination of P and G into a single action ($G + P$) is not appropriate because normal site monitoring of deflections and adjustment of cables will be insufficient to guarantee that there is no significant unintended imbalance between G and P. This will be the case for structures where the deflections from an unintended out of balance of P and G would be small, where the bridge deck is stiff in flexure compared to the support offered by the cables or where the profile of the structure is unaffected by the prestressing force. Such structures could include cable-stayed bridges with stiff decks, externally post-tensioned bridges and guyed towers and masts. In such cases, the actions P and G should have partial factors applied to them separately. In all cases, the method of applying partial factors should be agreed with the appropriate Overseeing Authority.

3-1-11/clause 2.3.5(3) does acknowledge that if cable adjustment is not intended, the effects of possible variations in prestress force should be considered. No numerical guidance is however provided so the above approach is recommended.

Design during construction

Further to the discussion above, it would also be necessary to treat self-weight and cable preloads separately with separate favourable and unfavourable load factors to determine the possible differential effects for ultimate limit states for stages of construction before cables have been adjusted or where it is not possible to detect the differential effect. This is the basis of ***3-1-11/clause 5.2(3)***, which requires the partial factor γ_P for prestressing to be defined for this situation in the National Annex. *3-1-11/clause 5.2(3)*

Cable replacement

Cables should normally be replaceable and the design should consider both a controlled replacement and an accidental removal. The load combination for controlled replacement can be defined in the National Annex to EN 1993-1-11 via clause 2.3.6. Often, these conditions will be project-specific. The load combination for accidental removal should be considered in an accidental combination but the National Annex may again define the relevant loading.

3-1-11/clause 2.3.6(2)

The dynamic effect of a sudden accidental cable removal should be considered. ***3-1-11/ clause 2.3.6(2)*** suggests this can be done by calculating the design effects for the structure with the cable in place, E_{d1}, and with the cable removed, E_{d2}, and calculating a dynamic design effect to add to E_{d1} given by:

$$E_d = kE_{d2} - E_{d1} \qquad \text{3-1-11/(2.4)}$$

EN 1993-1-11 sets the value of k at 1.5.

This formula produces incorrect results, particularly for cables remote from the removed cable where there are no effects from the cable removal so that $E_{d1} = E_{d2}$. In this case, the formula still predicts that the additional dynamic force to consider is $0.5E_{d2}$. It is suggested here that a more appropriate formula is:

$$E_d = k(E_{d2} - E_{d1}) \qquad \text{(D5.1-1)}$$

This ensures the system is designed for additional effects equal to the change in static internal effects caused by cable removal, multiplied by a dynamic factor. $k = 2.0$ corresponds to zero damping and $k = 1.8$ would be a reasonable value for most structures to make allowance for some damping. $k = 1.5$ would probably be too optimistic with this formulation.

5.2. Global analysis

5.2.1. Effects of deformed geometry of the structure

Second-order effects with axial force

Second-order effects in the context of 3-2/clause 5.2 are additional action effects caused by the interaction of axial forces and deflections under load. First-order deflections lead to additional moments caused by the eccentricity of the axial forces and these in turn lead to further increases in deflection. Such effects are also sometimes called P–Δ effects because additional moments are generated from the product of the axial load and element or system deflections. The simplest case is a cantilevering pier with axial and horizontal loads applied at the top as in Fig. 5.2-1. Second-order effects can be calculated by second-order analysis, as noted in ***3-1-1/clause 5.2.1(1)***, which takes into account this additional deformation.

3-1-1/clause 5.2.1(1)

Second-order effects apply to in-plane and out-of-plane modes of buckling, including lateral torsional buckling. The latter behaviour is more complex and requires a finite-element analysis using shell elements to properly model second-order effects and instability. In this case, lateral displacements in the compression flange from initial imperfections and/or

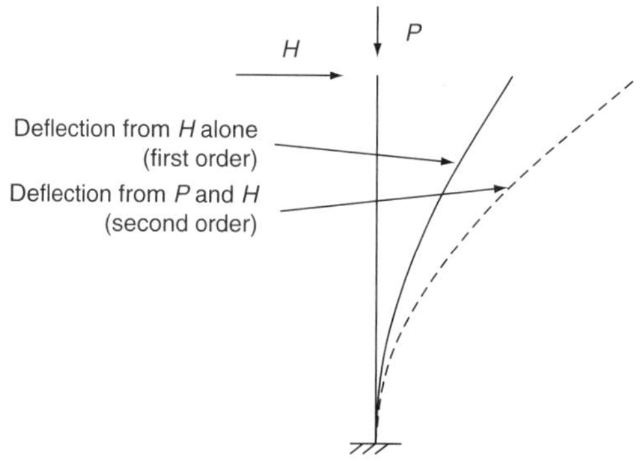

Fig. 5.2-1. Deflections for an initially straight pier with transverse load

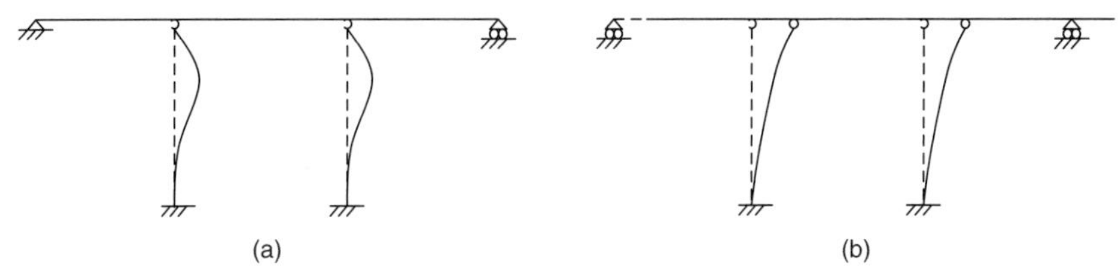

Fig. 5.2-2. Examples of local and global instability: (a) local second-order effects; (b) global second-order effects

transverse load are increased by the flange compression arising from overall bending of the beam. A method of checking beams for out-of-plane instability while modelling only in-plane second-order effects is given in clause 6.3.4 of EN 1993-1-1.

Second-order effects apply to both 'isolated' members (e.g. as in Fig. 5.2-1 or Fig. 5.2-2(a) and to overall bridges which can sway involving several members in a mutually dependent mode (Fig. 5.2-2(b)). ***3-1-1/clause 5.2.1(2)*** requires second-order effects to be considered if they significantly increase the action effects in the structure. 3-2/clause 5.2.1(4) gives guidance on what is 'significant' as discussed below. *3-1-1/clause 5.2.1(2)*

Second-order analysis is essentially the default analysis in the Eurocodes. First-order analysis may only be used if the relaxation in 3-2/clause 5.2.1(4) applies. A disadvantage of having to perform second-order analysis is that the principle of superposition is no longer valid and all loads must be applied to the bridge in combination with all their respective load and combination factors. Consequently it will still usually be necessary to use first-order theory initially to determine influence lines (or surfaces) and critical load cases for application in a second-order analysis. Fortunately, there will mostly be no need to do such analysis as alternative methods are discussed in this section and frequently second-order effects will, in any case, be small and may therefore be neglected.

A criterion is given in ***3-2/clause 5.2.1(4)*** (by reference to EN 1993-1-1) for when global second-order effects can be neglected: *3-2/clause 5.2.1(4)*

$$\alpha_{cr} = \frac{F_{cr}}{F_{Ed}} \geq 10 \qquad \text{3-2/(5.1)}$$

where F_{cr} is the elastic critical buckling load for the structure and F_{Ed} is the design load on the structure. The ratio is the factor by which all loads must be increased to cause elastic instability. 3-2/clause 5.2.1(4) also allows this criterion to be applied to individual elements of the bridge whereupon F_{Ed} and F_{cr} then relate to forces in these elements. For most bridges, it should however be possible to avoid both verifying this criterion and having to do second-order analysis by using first-order analysis and subsequent member stability checks with effective lengths that cover both local member and overall bridge behaviour. This is discussed in section 5.2.2 of this guide.

Notwithstanding the point made above that expression 3-2/(5.1) should rarely need to be used, it may not be convenient to perform elastic critical buckling analysis for its verification should it be required. An earlier draft of EN 1993-2 recognized this and had an alternative statement thus:

$$\alpha_{cr} = M_I/\Delta M_I \geq 10 \qquad \text{(D5.2-1)}$$

where M_I is the moment from first-order analysis, including the effects of initial imperfections. ΔM_I is the increase in bending moments calculated from the deflections obtained from first-order analysis (the P–Δ moments). This criterion avoids the need for elastic critical buckling analysis and, for the case of a pin-jointed strut with sinusoidal bow, is the same as expression 3-2/(5.1) which can be shown as follows.

The extra deflection from a first-order analysis can easily be shown to be given by:

$$\Delta v = a_0 F_{Ed}/F_{cr} \qquad \text{(D5.2-2)}$$

It follows that the extra moment from the first-order deflection is therefore:

$$\Delta M_I = F_{Ed}(a_0 F_{Ed}/F_{cr}) \qquad \text{(D5.2-3)}$$

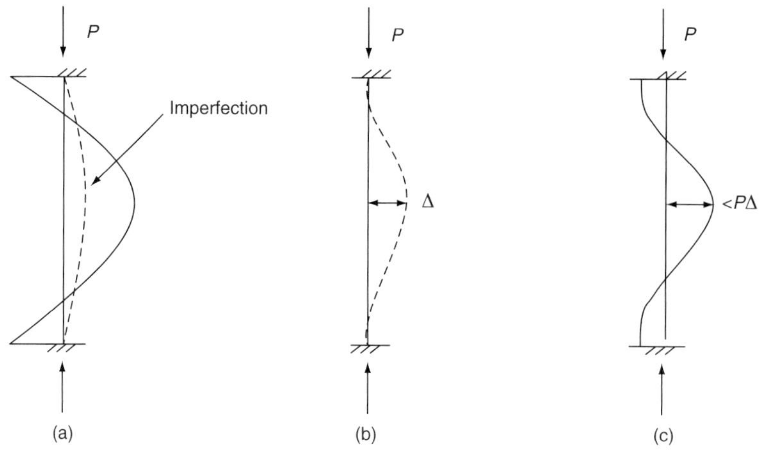

Fig. 5.2-3. Extra moments from deflection in built-in bowed strut: (a) first-order moment due to imperfections; (b) first-order deflections; (c) additional moment from deflection

Putting equation (D5.2-3) into equation (D5.2-1) gives expression 3-2/(5.1):

$$\alpha_{cr} = M_I/\Delta M_I = \frac{F_{Ed}a_0}{F_{Ed}(a_0 F_{Ed}/F_{cr})} = \frac{F_{cr}}{F_{Ed}} \geq 10$$

The equivalence is only valid for a pin-ended strut with a sinusoidal bow and hence sinusoidal curvature, but it generally remains sufficiently accurate. Note that it is found that for constant curvature (equal end moments)

$$M_I/\Delta M_I = \frac{8}{\pi^2}\frac{F_{cr}}{F_{Ed}}$$

For anything other than a pin-ended strut or statically determinant structure, it will not be easy to determine ΔM_I from the deflections found by first-order analysis. This is because in indeterminate structures, the extra moment cannot be calculated at all sections directly from the local '$P \times \Delta$' because of the need to maintain compatibility as illustrated in Fig. 5.2-3. (This is similar to secondary effects of prestressing in prestressed structures.) In Fig. 5.2-3, it would be conservative to take ΔM_I as $P\Delta$ at mid-height when calculating α_{cr} for the mid-height position if the actual distribution of additional moment is not obtained from the deflection by a further analysis which models the first-order deflected shape. Another problem is that equation (D5.2-1) is unlikely to be satisfied if applied near a point of contra-flexure in indeterminate structures. To avoid this problem, equation (D5.2-2) should be applied only at the peak moment positions between each adjacent point of contraflexure. ΔM_I can again be conservatively based on the maximum P–Δ in the member. These problems led to equation (D5.2-1) being removed from EN 1993-2 but the equivalent expression is provided still in clause 5.2.1(3) of EN 1994-2.

Slip of bolts

Bolt slip needs to be included in analysis, whether first order or second order, where it is significant as stated in ***3-1-1/clause 5.2.1(6)***. No specific guidance is however given in EN 1993-2.

3-1-1/clause 5.2.1(6)

It is recommended here that bolt slip should be taken into account for bracing members in the analysis of braced systems. This is because a sudden loss of stiffness arising from bolt slip leads to an increase in deflection of the main member and an increased force on the bracing member, which could lead to overall failure. Ideally therefore, bracing members should be designed as non-slip at ULS (Category C to EN 1993-1-8) to avoid this consideration.

Slip can also occur in main beam splices. It has been UK practice to design bolts to slip at ULS (Category B to EN 1993-1-8) without consideration of slip in global analysis. This is justifiable as, although slip could alter the moment distribution in the beam, splices are usually positioned near to points of contraflexure and therefore slip will not shed significant moment to either adjacent hog or sag zones. Also, the loading that gives maximum moment at the splice will not be fully coexistent with that for either the maximum hogging moment or maximum sagging moment in adjacent regions.

5.2.2. Structural stability of frames and second-order analysis

This section has been split into three sub-sections in this guide for convenience.

5.2.2.1. General

Where it is necessary to take second-order effects and imperfections into account, this may be achieved in one of three ways according to ***3-1-1/clause 5.2.2(3)*** and ***3-1-1/clause 5.2.2(7)***: *3-1-1/clause 5.2.2(3)* *3-1-1/clause 5.2.2(7)*

1. Use of second-order analysis including both 'global' system imperfections and 'local' member imperfections as discussed in section 5.3. Where a beam is susceptible to lateral torsional buckling, imperfections must also be modelled to cater for second-order effects from this mode of buckling as discussed in section 5.3.4. If this method is followed, no individual checks of member stability are required using 3-2/clause 6.3 and members are checked for cross-section resistance only. Rather than superimposing local and global imperfections, it is possible to apply a unique overall imperfection to the structure based on the shape of the lowest mode of buckling of the structure. This method is given in 3-1-1/clause 5.3.2(11) and is discussed in section 5.3.2 of the guide.
2. Use of second-order analysis including 'global' system imperfections only with stability checks according to 3-2/clause 6.3 subsequently carried out for individual members using the end moments and axial loads from the analysis. Since the member end forces and moments contain second-order effects from global behaviour, the effective length of individual members is then based on the member length, rather than a greater effective length that includes the effects of global sway deformations. It should be noted that when 3-1-1/clause 6.3.3 is used for member checks, the member moments will be further multiplied by the 'k_{ij}' parameters. Since the second-order analysis will already have amplified these moments (providing sufficient nodes have been included along the member in the analysis model), this is conservative and it would be permissible to limit the values of the calculated 'k_{ij}' parameters to unity where they exceed unity. However, the imperfections within the members have not been considered or amplified by the second-order analysis. These are included by way of the first term in the equations in this clause

   $$\frac{N_{Ed}}{\chi N_{Rk}/\gamma_{M1}}$$

3. Use of first-order analysis without modelled imperfections. Members are then checked to 3-2/clause 6.3 using appropriate effective lengths covering the lowest buckling mode of the bridge involving the element under consideration. All second-order effects are then included in the relevant resistance formulae in 3-1-1/clause 6.3. This latter method will be most familiar to UK bridge engineers, as tables of effective lengths for members with varying end conditions of rotational and positional fixity have commonly been used. The use of effective lengths for this method is discussed later.

Second-order analysis itself can be done either by direct analysis that accounts for the deformed geometry (computer programs are readily available to do this) or by amplification of the moments from a first-order analysis (including the effects of imperfections) as discussed below – ***3-1-1/clause 5.2.2(4)*** refers. Where either approach is used, it should only be performed by, or under the guidance of, experienced engineers because the guidance on the use of imperfections in terms of shapes, combinations and directions of application are not comprehensive in EC3; judgement is required. *3-1-1/clause 5.2.2(4)*

5.2.2.2. Second-order analysis by the use of moment magnifiers

Although the elastic critical buckling load or moment itself has little direct relevance to real member strength, it gives a good indication of susceptibility to second-order effects and can

3-2/clause 5.2.2(5)

also be used as a parameter in determining second-order effects from the results of a first-order analysis. The method of ***3-2/clause 5.2.2(5)*** is based on the elastic theory that total moments in a pin-ended strut, including second-order effects, can be derived by multiplying first-order moments (including moments arising from initial imperfections) by a magnifier that depends on the axial load and the Euler buckling load of the member. The simplest example of this is a pin-ended column, length L, under axial load only with an initial sinusoidal bow imperfection of maximum displacement a_0. The Euler buckling load is given by:

$$F_{cr} = \pi^2 EI/L^2$$

If the axial load is F_{Ed} then the final deflection is given by:

$$a = a_0 \left[\frac{1}{1 - (F_{Ed}/F_{cr})}\right]$$

(This is obtained from simple elastic theory by solving

$$EI \frac{d^2(v - v_0)}{dx^2} + F_{Ed} v = 0$$

where v is the lateral displacement as a function of height x up the column and $v_0 = a_0 \sin \pi x/L$.)

The corresponding final maximum moment including second-order effects, $M_{Ed}^{II} = F_{Ed} a$, is then given by:

$$M_{Ed}^{II} = F_{Ed}\left[\frac{a_0}{1 - (F_{Ed}/F_{cr})}\right] = M_{Ed}^{I}\left[\frac{1}{1 - (F_{Ed}/F_{cr})}\right] \qquad \text{(D5.2-4)}$$

where $M_{Ed}^{I} = F_{Ed} a_0$ is the first-order moment. The magnifier here is $1/(1 - F_{Ed}/F_{cr})$, which assumes that the initial imperfection is sinusoidal. Similar results are produced for the magnification of moments in pin-ended struts with applied end moments or transverse load, but the magnifier varies depending on the distribution of the first-order moment. For uniform moment, the amplifier above is slightly unconservative, but it will generally suffice with sufficient accuracy.

The pin-ended strut case is not itself an application of great practical significance as second-order effects and imperfections for pin-ended struts are covered in the resistance formulae for flexural buckling in 3-1-1/clause 6.3. It does however illustrate the basis of expression 3-2/(5.2), which allows total moments in bridges and bridge components, including second-order effects, to be found by increasing the first-order moments (including the effects of all imperfections) as follows:

$$M_{II} = M_I \left[\frac{1}{1 - (1/\alpha_{cr})}\right] \qquad \text{3-2/(5.2)}$$

with $\alpha_{cr} = F_{cr}/F_{Ed}$ defined in section 5.2.1.1 above. For uniform isolated members, $\alpha_{cr} = F_{cr}/F_{Ed}$ is safe to use for sinusoidal or triangular distribution of curvature but is slightly unconservative for uniform curvature, although not unduly so. A similar expression is given in EN 1992-1-1 thus:

$$M_{II} = M_I \left[1 + \frac{\beta}{(F_{cr}/F_{Ed}) - 1}\right] \qquad \text{(D5.2-5)}$$

with $\beta = \pi^2/c_0$ and $F_{cr} = \pi^2 EI/L_{cr}^2$. c_0 depends on the distribution of moment and hence curvature in the column. For uniform curvature, $c_0 = 8$. For sinusoidal curvature (and approximately for triangular curvature or parabolic curvature), $c_0 = \pi^2$ and the expression for moment simplifies to the simple form of expression 3-1-1/(5.4). L_{cr} is the effective length for buckling which can be determined as discussed in section 5.2.2.3 below. Alternatively, F_{cr}/F_{Ed} can be determined directly by computer elastic critical buckling analysis.

The above expressions all assume that the peak first-order moment occurs at the same section as the peak moment from the P–Δ effect. Considering an integral pier, with end

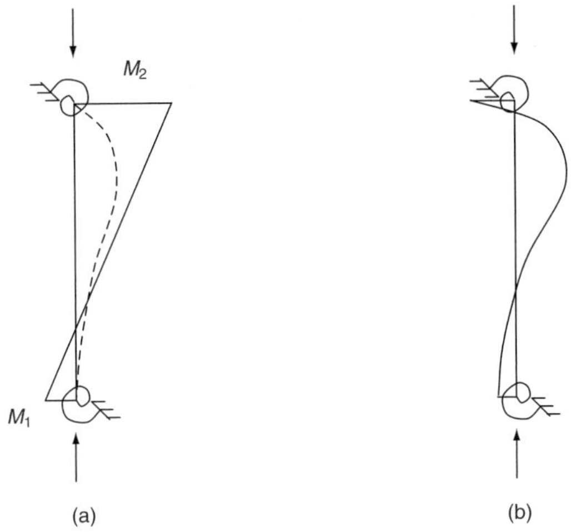

Fig. 5.2-4. Amplification of applied first order moments (imperfections excluded for clarity): (a) first-order applied moment and resulting deflection; (b) additional moments from second-order effects

rotational restraint arising from connection to the foundations at one end and the deck at the other, Fig. 5.2-4 shows that the P–Δ moment actually reduces the peak first-order end moment at the top. EN 1992 overcomes this conservatism for concrete elements by allowing an equivalent first-order moment to be used, but only where there is **no transverse load** applied in the height of the column **and the members cannot sway**. A more detailed discussion on this is provided in the *Designers' Guide to EN 1992-2*.[6]

The limitations on use and accuracy of this method mean that it will usually be better to perform an elastic second-order computer analysis where it is necessary to consider second-order effects, or to include them by means of appropriate effective lengths and resistance formulae.

5.2.2.3. *Effective lengths*

Where second-order effects need to be accounted for but it is not desired to carry out a second-order analysis, the concept of effective length can be used together with the resistance formulae and interactions in 3-1-1/clause 6.3. In this case, imperfections need not be modelled if local and global effects are included in the effective length as stated in 3-1-1/clause 6.3.3(3) and ***3-1-1/clause 5.2.2(8)***. This method will be most familiar to UK bridge engineers. Effective lengths can also be used in the moment magnification method described above. *3-1-1/clause 5.2.2(8)*

3-2/Annex D gives methods of calculating effective lengths for isolated bridge members in trusses and for buckling of arch bridges. (It also gives imperfections for arches for use in second-order analysis.)

Further relevant guidance on effective lengths for axially loaded members can be found in EN 1992-1-1. Typical examples of isolated members include:

- piers with free sliding bearings at their tops (Fig. 5.2-5(b)), assuming the load moves with the pier
- piers with fixed bearings at their tops, but where the deck itself provides no positional restraint and moves with the pier (Fig. 5.2-5(b) again)
- piers with fixed (pinned) bearings at their tops which are restrained in position by connection via the deck to a rigid abutment or other stocky pier (Fig. 5.2-5(c)).

The effective lengths given in the cases (a) to (e) of Fig. 5.2-5 assume that the foundations (or other restraints) providing rotational restraint are infinitely stiff. In practice, this will never be the case and the effective length will always be somewhat greater than the theoretical value for rigid restraints and 3-1-1/clause 5.2.2(8) requires any flexibility to be considered. 2-1-1/clause 5.8.3 gives a method of accounting for this rotational flexibility in the effective

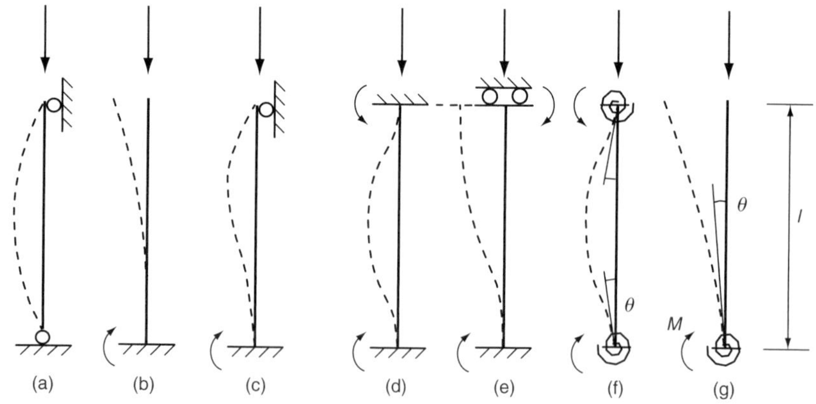

Fig. 5.2-5. Examples of different buckling modes and corresponding effective lengths for isolated members: (a) $L_{cr} = l$; (b) $L_{cr} = 2l$; (c) $L_{cr} = 0.7l$; (d) $L_{cr} = l/2$; (e) $L_{cr} = l$; (f) $l/2 < L_{cr} < l$; (g) $L > 2l$

length using equation (D5.2-6) for braced members (Fig. 5.2-5(f)) and equation (D5.2-7) for unbraced members (Fig. 5.2-5(g)):

$$L_{cr} = 0.5l\sqrt{\left(1 + \frac{k_1}{0.45 + k_1}\right)\left(1 + \frac{k_2}{0.45 + k_2}\right)} \qquad \text{(D5.2-6)}$$

$$L_{cr} = l\max\left\{\sqrt{1 + 10 \times \frac{k_1 k_2}{k_1 + k_2}}; \left(1 + \frac{k_1}{1 + k_1}\right)\left(1 + \frac{k_2}{1 + k_2}\right)\right\} \qquad \text{(D5.2-7)}$$

where k_1 and k_2 are the flexibilities of the rotational restraints at ends 1 and 2 respectively relative to the flexural stiffness of the member itself such that:

$k = (\theta/M)(EI/l)$ where:

k $= (\theta/M)(EI/l)$
θ is the rotation of the restraint for a bending moment M;
EI is the bending stiffness of the compression member;
l is the clear height of compression member between end restraints.

As can be seen from the formulae, equation (D5.2-7) can also be used for members with different rotational restraints at both ends but no lateral restraint at the top. This is useful for piers which are integral with a deck where deck and pier can sway. Quick inspection of equation (D5.2-7) shows that the theoretical case of a member with ends built in rigidly for moment ($k_1 = k_2 = 0$), but free to sway in the absence of positional restraint at one end, gives an effective length $L_{cr} = l$ as expected. The value of end stiffness to use for piers in integral construction can be determined from a plane frame model by deflecting the pier to give the deflection relevant to the mode of buckling and determining the moment and rotation produced in the deck at the connection to the pier. Alternatively, the analytical method described below could be used. Cracking of concrete should be considered in deriving the stiffness of the foundation or other members if relevant. The Note to 2-1-1/clause 5.8.3.2(3) recommends that no value of k is taken less than 0.1.

It should be noted that the cases in Fig. 5.2-5 do not allow for any rigidity of positional restraint in the sway cases. If significant lateral restraint is available, as might be the case in an integral bridge where one pier is very much stiffer than the others, ignoring this restraint will be very conservative as the more flexible piers may actually be 'braced' by the stiffer one. In this situation, a computer elastic critical buckling analysis will give a reduced value of effective length. (In many cases, however, it will be possible to see by inspection that a pier is braced.)

For more complex situations (such as for a member with varying section along its length), it is preferable to work directly from F_{cr}. F_{cr} can be calculated from a computer elastic critical

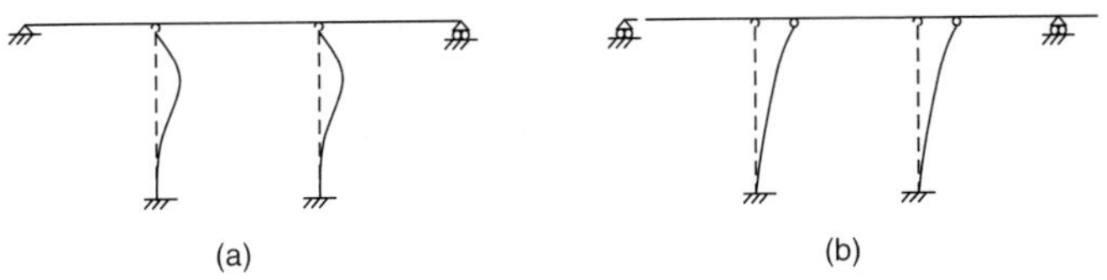

Fig. 5.2-6. 'Local' and 'global' buckling modes: (a) buckling of individual piers (braced); (b) overall buckling in sway mode (unbraced)

buckling analysis and then used either to perform a moment magnification calculation using expression 3-2/(5.2) or to determine the slenderness from expression 3-1-1/(6.50) for use with the member resistance curves in 3-2/clause 6.3.1.

Effective lengths can also be derived for piers in integral bridges and other bridges where groups of piers of varying stiffness are connected to a common deck. In this instance, the buckling load, and hence effective length, of any one pier depends on the load and geometry of the other piers also. All piers may sway in sympathy and act as unbraced (Fig. 5.2-6(b)) or a single stiffer pier or abutment might prevent sway and give braced behaviour for the other piers (Fig. 5.2-6(a)). The analytical method above could also be used in this situation to produce an accurate effective length by applying coexisting loads to all piers and increasing all loads proportionately until a buckling mode involving the pier of interest is found. P_{cr} is then taken as the axial load in the member of interest at buckling.

5.3. Imperfections

5.3.1. Basis

3-1-1/clause 5.3.1(1)

Imperfections comprise geometric imperfections and residual stresses – see ***3-1-1/clause 5.3.1(1)***. The term 'geometric imperfection' is used to describe departures from the exact centreline setting out dimensions found on drawings which occur during fabrication and erection. This is inevitable as all construction work can only be executed to certain tolerances. Geometric imperfections include lack of verticality, lack of straightness, lack of fit and minor joint eccentricities. The behaviour of members under load is also affected by residual stresses within the members. Residual stresses can lead to yielding at lower applied external load than predicted from stress analysis ignoring such effects. The effects of these residual stresses can be modelled by additional equivalent geometric imperfections. The equivalent geometric imperfections referred to in ***3-1-1/clause 5.3.1(2)*** therefore cover both geometric imperfections and residual stresses.

3-1-1/clause 5.3.1(2)

3-1-1/clause 5.3.1(3)

3-1-1/clause 5.3.1(3) identifies that imperfections can apply to overall structure geometries (global imperfection) or locally to members (local imperfection). Imperfections must be included in global analysis unless they are included by use of the appropriate resistance formulae in clause 6.3 when checking the members; discussion is given in section 5.2. For example, the flexural buckling curves provided in 3-1-1/Fig. 6.4 include all imperfections for a given member effective length of buckling.

5.3.2. Imperfections for global analysis of frames

3-1-1/clause 5.3.2(1)

3-1-1/clause 5.3.2(2)

As a general method, ***3-1-1/clause 5.3.2(1)*** allows the shape of imperfections to be derived from the shape of the elastic buckling mode being considered. In-plane and out-of-plane buckling modes, including symmetric and asymmetric modes, should be considered as required by ***3-1-1/clause 5.3.2(2)***. Several modes should be considered rather than just the one with lowest load factor. The rules in EN 1993-1-1 cover the overall analysis of beam elements only and do not consider local plate buckling. EN 1993-1-5 gives other rules for modelling imperfections in plate elements. This is discussed in section 5.3.5. The remainder of section 5.3.2 of this guide is split into two additional sub-sections dealing with the use of a unique global plus local imperfection and the use of a combination of local and global imperfections respectively.

5.3.2.1. Imperfections based on overall buckling mode shape

3-1-1/clause 5.3.2(11)

3-1-1/clause 5.3.2(11) allows a unique distribution of global and local imperfection to be applied, based on the mode shape of buckling being considered for the bridge and having the same shape, using expressions 3-1-1/(5.9) and (5.10). They are reproduced here as a single formula:

$$\eta_{\text{init}} = \frac{\alpha(\bar{\lambda} - 0.2)}{\bar{\lambda}^2} \frac{1 - \dfrac{\chi \bar{\lambda}^2}{\gamma_{M1}}}{1 - \chi \bar{\lambda}^2} \frac{M_{\text{Rk}}}{EI\eta''_{\text{cr,max}}} \eta_{\text{cr}} \qquad \text{(D5.3-1)}$$

where:

η_{cr} represents the local ordinates of the mode shape and η'' is the curvature produced by the mode shape such that $EI\eta''_{\text{cr,max}}$ is the greatest bending moment due to η_{cr} at the critical cross-section. Other terms are as follows:

α is the imperfection factor taken from 3-1-1/Tables 6.1 and 6.2 for the relevant mode of buckling. For varying cross-section, the greatest value can conservatively be taken. $\bar{\lambda} = \sqrt{\alpha_{\text{ult,k}}/\alpha_{\text{cr}}}$ where $\alpha_{\text{ult,k}}$ is the load amplifier to reach the characteristic squash load N_{Rk} of the most axially stressed section and α_{cr} is the load amplifier for elastic critical buckling;

χ is the reduction factor for the above slenderness determined using the relevant buckling curve appropriate to α.

The imperfections of equation (D5.3-1) are based on the same imperfections implicit in the strut design formula in 3-1-1/clause 6.3.1.2. The use and derivation of this expression is illustrated most simply by considering a pin-ended strut, as in Fig. 5.3-1, for which elastic analysis using equation (D5.3-1) produces the same results.

As discussed in section 6.3.1.2 of this guide, the imperfection parameter found from the Perry–Robertson analysis is $y\bar{\eta}_{\text{init}}/i^2$ where $\bar{\eta}_{\text{init}}$ is the magnitude of the initial imperfection bow assumed and y is the distance from the relevant centroidal axis to the extreme fibre. EC3 makes this imperfection parameter equal to $\alpha(\bar{\lambda} - 0.2)$. Consequently, equating $y\bar{\eta}_{\text{init}}/i^2$ to $\alpha(\bar{\lambda} - 0.2)$, the amplitude of imperfection to use in analysis is given by:

$$\bar{\eta}_{\text{init}} = \alpha(\bar{\lambda} - 0.2)\frac{i^2}{y} \qquad \text{(D5.3-2)}$$

For a strut of length L_{cr}, the radius of gyration can be found from:

$$i^2 = \frac{\sigma_{\text{crit}} L_{\text{cr}}^2}{\pi^2 E} \qquad \text{(D5.3-3)}$$

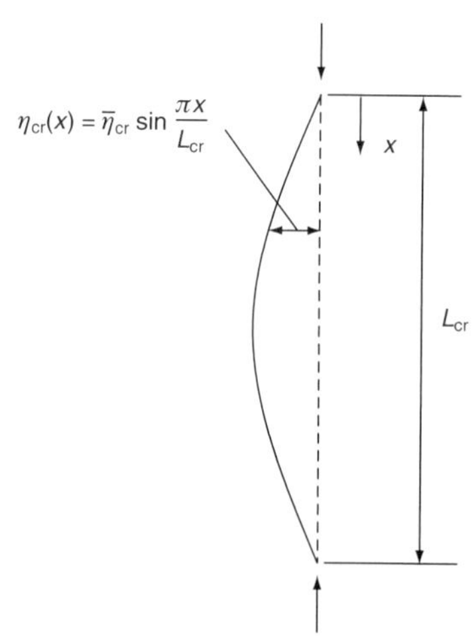

Fig. 5.3-1. Buckling mode shape for pin-ended strut in compression

For an elastic moment resistance M_{Rk}, y is given by:

$$y = \frac{I f_{yd}}{M_{Rk}} \qquad \text{(D5.3-4)}$$

The slenderness ratio for axial load can also be found from:

$$\bar{\lambda}^2 = \frac{\alpha_{ult,k}}{\alpha_{cr}} = \frac{f_{yd}}{\sigma_{crit}} \qquad \text{(D5.3-5)}$$

Substitution of equations (D5.3-3) to (D5.3-5) into equation (D5.3-2) to eliminate i and y gives:

$$\bar{\eta}_{init} = \frac{\alpha(\bar{\lambda} - 0.2)}{\bar{\lambda}^2} \frac{M_{Rk}}{\pi^2 EI/L_{cr}^2} \qquad \text{(D5.3-6)}$$

For a pin-ended strut, the mode shape is as follows:

$$\eta_{cr}(x) = \bar{\eta}_{cr} \sin\frac{\pi x}{L_{cr}} \qquad \text{(D5.3-7)}$$

where $\bar{\eta}_{cr}$ is the peak amplitude of the mode shape, usually scaled to unity.

The curvature of the mode shape is obtained by differentiation:

$$|\eta''_{cr}| = \frac{\pi^2}{L_{cr}^2} \bar{\eta}_{cr} \sin\frac{\pi x}{L_{cr}} \qquad \text{(D5.3-8)}$$

therefore

$$\eta''_{cr,max} = \frac{\pi^2}{L_{cr}^2} \bar{\eta}_{cr} \qquad \text{(D5.3-9)}$$

Introducing equation (D5.3-9) into equation (D5.3-6) gives the following expression for the amplitude:

$$\bar{\eta}_{init} = \frac{\alpha(\bar{\lambda} - 0.2)}{\bar{\lambda}^2} \frac{M_{Rk}}{EI\eta''_{cr,max}} \bar{\eta}_{cr} \qquad \text{(D5.3-10)}$$

The imperfection is therefore distributed as:

$$\eta_{init} = \frac{\alpha(\bar{\lambda} - 0.2)}{\bar{\lambda}^2} \frac{M_{Rk}}{EI\eta''_{cr,max}} \eta_{cr} \qquad \text{(D5.3-11)}$$

This can be seen to be essentially the same as equation (D5.3-1) but without the term

$$\frac{1 - \dfrac{\chi\bar{\lambda}^2}{\gamma_{M1}}}{1 - \chi\bar{\lambda}^2}$$

which is a correction to allow for the material factor γ_{M1} which in EN 1993-2 is equal to 1.1. It is required because γ_{M1} is used with the resistance curves in 3-1-1/clause 6.3 whereas γ_{M0} is used in cross-section resistance checks.

The general procedure is thus to first determine the mode shape assuming some maximum ordinate (usually 1.0 as the mode shapes are usually normalized), and then to determine the greatest moment from this mode shape assuming the same maximum ordinate. The imperfection is then calculated from equation (D5.3-1) assuming the same distribution as the buckled shape.

For arch bridges, the imperfections given in 3-2/clause D.3.5 can be used directly.

5.3.2.2. Separate local and global imperfections

In general, imperfections can be applied as a combination of a global sway imperfection and local member imperfections – ***3-1-1/clause 5.3.2(3)***. *3-1-1/clause 5.3.2(3)*

The sway imperfection is applied as an angular lean, ϕ, given by expression 3-1-1/(5.5) as follows:

$$\phi = \phi_0 \alpha_h \alpha_m \qquad \text{(D5.3-12)}$$

where:

ϕ_0 is the basic value of lean of 1/200;
α_h is a reduction factor for height, h, given by $\alpha_h = 2/\sqrt{h}$ but not less than $\frac{2}{3}$ or greater than 1.0;
α_m is a reduction factor to allow for the reduced probability of all piers leaning in the same direction by the same amount given by $\alpha_m = \sqrt{0.5(1 + 1/m)}$ where m is the number of piers that are capable of actually resisting the sway and which carry an axial load not less than 50% of the average pier load.

Local member imperfections are applied as a bow over the member length, L, with magnitude e_0/L. e_0 is determined from 3-1-1/Table 5.1 according to the type of cross-section defined in 3-1-1/Table 6.2. In some cases, it is also advisable to try a case where the local imperfection is distributed in the same manner as the shape of the member buckling mode obtained if sway were prevented, although the need for this is somewhat mitigated by the conservatism of the imperfections in 3-1-1/Table 5.1. If this is done, the amplitude $e_{0,\text{mod}}$ over the half wavelength of buckling L_{cr} (measured from a line joining points of contraflexure) can be determined from 3-1-1/Table 5.1 using $e_{0,\text{mod}}/L_{\text{cr}}$. This is illustrated in Fig. 5.3-2 for the extreme case of infinitely stiff end rotational restraint. In this case, the imperfection shown can lead to greater moments than occur if the single half wave bow imperfection is used. In all cases, care should be taken with the direction of the local bow to ensure the maximum combined effect from local and global imperfections is obtained.

3-1-1/clause 5.3.2(7)

The above imperfections can be taken into account either by modelling them directly in the structural system or by replacing them by equivalent forces as noted in ***3-1-1/clause 5.3.2(7)***. The latter is a useful alternative, as the same model can be used to apply different imperfections, but the disadvantage is that the axial forces in members must first be known before the equivalent forces can be calculated. The equivalent forces are shown in 3-1-1/Fig. 5.4; they are not reproduced here.

3-1-1/clause 5.3.2(8)

3-1-1/clause 5.3.2(8) requires sway imperfections to be considered in all relevant directions but they need not be considered to act in more than one direction at a time. This illustrates that judgement will always be needed in determining the critical distribution of imperfections.

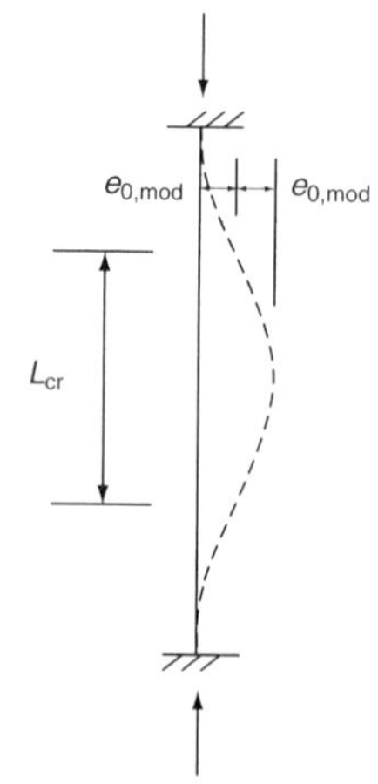

Fig. 5.3-2. Example of possible additional local imperfection to consider where there are rotationally fixed-ended conditions

5.3.3. Imperfections for analysis of bracing systems

This section relates to plan bracing systems for beams, although 3-2/clause 5.3.3 relates to both beams and compression members. The analysis of torsional (vertical) bracing is discussed in section 6.3.4.2 of this guide. When plan bracing systems are present, the relevant imperfections for analysis of the bracing system are not necessarily the same as those for the bridge beams themselves. Bracing is usually required to the compression flanges of bridge beams. This may be in the form of plan bracing alone, as shown in 3-1-1/Fig. 5.6, or may be a combination of plan bracing and torsional bracing. The latter is found typically in steel and concrete composite bridges in hogging zones where the deck slab forms plan bracing to the tension flange and the bottom flange is connected to the deck plan bracing via torsional bracing. Design of plan bracing in combination with torsional bracing is discussed in section 6.3.4.2.

Plan bracing systems may be analysed by applying a bow of magnitude $e_0 = \alpha_m L/500$ to the braced members (if members are in compression) or to braced flanges (if members are in bending) – ***3-1-1/clause 5.3.3(1)*** refers. L is the span of the bracing system and $\alpha_m = \sqrt{0.5(1 + 1/m)}$ is a reduction factor to allow for the reduced probability of all flanges bowing in the same direction by the same maximum amount; m is the number of flanges being braced. As an alternative, ***3-1-1/clause 5.3.3(2)*** permits equivalent uniformly distributed forces to be applied to the bracing system through each flange with magnitude $8N_{Ed}(e_0 + \delta_q)/L^2$ per unit length along the beam, where N_{Ed} is the maximum flange force defined in 3-1-1/clause 5.3.3(3). The total lateral force applied to the bracing per unit length along the beam is then given by: *3-1-1/clause 5.3.3(1)* *3-1-1/clause 5.3.3(2)*

$$q = \sum N_{Ed} 8 \frac{e_0 + \delta_q}{L^2} \qquad \text{3-1-1/(5.13)}$$

where δ_q is the in-plane deflection in the bracing system under the load q and any other imposed loads, calculated from first-order analysis. Since the applied force depends on the first-order deflection, this is an iterative calculation unless second-order analysis is used, whereupon the Note to 3-1-1/clause 5.3.3(2) allows δ_q to be taken as zero.

For bracing systems to compression flanges, ***3-1-1/clause 5.3.3(3)*** allows N_{Ed} to be taken as M_{Ed}/h where M_{Ed} is the maximum beam moment and h is the overall beam depth. The Note to the clause however clarifies that if the beam carries compressive load, the part of the compression carried by the flange should be included in the calculation of N_{Ed}. *3-1-1/clause 5.3.3(3)*

The design of bracing systems is discussed further in section 6.3.4.2.6 of this guide.

5.3.4. Member imperfections

3-1-1/clause 5.3.4(1) reminds the designer that the effects of member imperfections are included within the buckling resistance formulae of 3-1-1/clause 6.3. Conversely, ***3-1-1/clause 5.3.4(2)*** is a reminder that if member imperfections are included in the second-order analysis, there is no need to do additional member stability checks to 3-1-1/clause 6.3. It will usually be possible to use effective lengths for members together with 3-1-1/clause 6.3 and avoid the use of second-order analysis as discussed above. *3-1-1/clause 5.3.4(1)* *3-1-1/clause 5.3.4(2)*

If lateral torsional buckling is to be taken into account by second-order analysis, a bow imperfection about the beam minor axis of $0.5e_0$ is recommended in ***3-1-1/clause 5.3.4(3)*** where e_0 is again taken from 3-1-1/Table 5.1. If lateral-torsional buckling is to be covered totally by second-order analysis, appropriate finite-element analysis capable of modelling the behaviour will be required. This will usually require modelling of the beam with shell elements unless a simplified model can be developed, e.g. by considering buckling of the compression chord alone between rigid restraints in a manner similar to that proposed in 3-2/clause 6.3.4.2(2). *3-1-1/clause 5.3.4(3)*

5.3.5. Imperfections for use in finite-element modelling of plate elements (additional sub-section)

EN 1993-2 does not give guidance on imperfections for use in buckling checks of plate elements. 3-1-5/clause C.5 however gives some guidance on imperfections and the Annex C as a whole gives advice on finite-element modelling of plate elements.

In general, the distribution (or shape) of the imperfections to be used can be determined by one of four methods:

1. **Using the same distribution as the mode shapes found from elastic critical buckling analysis.** Elastic critical buckling analysis can be used to determine a unique imperfection distribution, with the same form as the buckling mode shape, in the same manner as discussed in section 5.3.2.1 above. It is often assumed that this method of applying imperfections will maximize the reduction in resistance but this is not always true and there are difficulties in implementation. The imperfection distribution will vary with each load case and it is difficult to specify the imperfection magnitude for coupled modes involving both overall stiffened panel buckling and local sub-panel buckling. The elastic buckling mode with the lowest load factor may not also be the critical mode shape for reducing ultimate strength. Often, a slightly lower resistance is produced using method (4).
2. **Using assumed imperfection shapes based on buckling under direct stress.** The imperfection distribution can be based on the local and global plate buckling mode shapes for compression acting alone in the longitudinal direction. This method will not necessarily maximize the loss of resistance, but the resulting resistance will usually not be far from the true resistance and its use can be justified by the use of partial safety factors.
3. **Applying transverse loading.** A variation on (2) above is to apply transverse loading so that the first order effects of such loading replicate the first order effects of imperfections.
4. **Application of the deformed shape at failure.** In this method, the deformed shape of the structure obtained at failure from a previous analysis is used as the initial imperfection shape. This frequently gives the lowest resistance (but rarely significantly lower than the other methods). It has the disadvantage that the method is iterative, as an initial analysis to failure is required to produce the imperfection shape.

3-1-5/clause C.5(1)

3-1-5/clause C.5(2)

EN 1993-1-5 gives recommendations for imperfections broadly based on method (2) but the general statement of the required approach to modelling imperfections in Note 1 of 3-1-5/clause C.5(2) is based on method (1). As in EN 1993-1-1, ***3-1-5/clause C.5(1)*** requires both geometric imperfections and structural imperfections (residual stresses) to be considered, but equivalent geometric imperfections, containing both types, may be used in accordance with ***3-1-5/clause C.5(2)***. These are given in 3-1-5/Table C.2 and 3-1-5/Fig. C.1. These include bow imperfections for out-of-plane buckling of stiffeners between transverse stiffeners, imperfections for plate sub-panels based on the elastic critical mode shape of buckling, and twist imperfections for torsional buckling of stiffener outstands. Bow imperfections for the overall member are covered by 3-1-1/Table 5.1.

The imperfections for sub-panel buckling and stiffener out-of-plane buckling are shown in Fig. 5.3-3. For sub-panel buckling, the recommended maximum imperfection e_0 is the minimum of $a/200$ or $b/200$ and the distribution is sinusoidal in both directions as shown in Fig. 5.3-3(a). For longitudinal stiffeners, the recommended maximum global bow imperfection e_0 is the minimum of $a/400$ or $b/400$. The limitation to $b/400$ is not easy to justify as the actual geometrical tolerance on longitudinal stiffeners in EN 1090 is $a/500$ and not dependent on b. For stiffened panels where the length is only moderately greater than the width, say $a < 2b$, it is unlikely that the plate panel will have any significant restraining effect transversely on the stiffener. It is therefore recommended that the stiffener imperfection is generally based on $a/400$ as shown in Fig. 5.3-3(b). Where the panel is very long, it should be noted that several half wavelengths of buckling might be possible for the stiffeners between transverse stiffeners but this is not covered by the imperfection suggested. Elastic critical buckling analysis would be required to check if this mode occurred at a lower load factor.

3-1-5/clause C.5(3)

The direction of application of the imperfection shape must be selected to minimize the resistance – ***3-1-5/clause C.5(3)*** refers. This is typically important for compression in longitudinal stiffener effective sections which are generally asymmetric and thus the moment from the axial load and imperfection generates different stresses at the two extreme fibres.

Overall imperfections for the whole structure and for the whole member should be considered in addition to the plate imperfections above so as to correctly model the

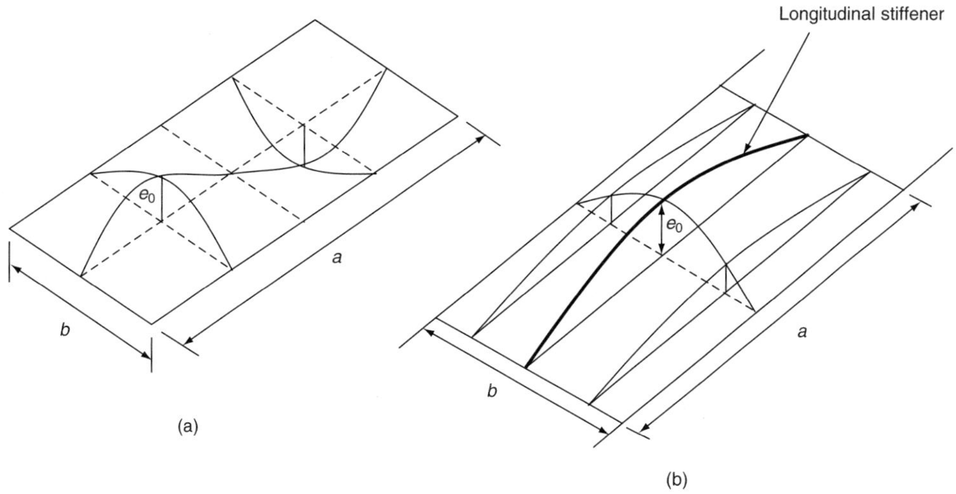

Fig. 5.3-3. Equivalent geometric imperfections in plate panels: (a) sub-panel imperfections; (b) overall stiffened panel imperfections

overall behaviour of the system. When the various different types of plate imperfection discussed above and the global structure and member imperfections are combined, one imperfection is identified as being the 'leading imperfection' and the others may be reduced to 70% of their tabulated values in accordance with ***3-1-5/clause C.5(5)***. *3-1-5/clause C.5(5)*

5.4. Methods of analysis considering material non-linearities

5.4.1. General

3-2/clause 5.4.1(1) requires that internal forces and moments for all non-accidental situations are determined by elastic analysis. Elastic global analysis is therefore generally required for bridges. This is in contrast to the rules for buildings where rigid plastic global analysis may be used where members are Class 1 at hinge locations and other requirements are met according to 3-1-1/clause 5.6. *3-2/clause 5.4.1(1)*

For accidental situations, such as vehicular impact on a bridge pier or impact on a parapet, the National Annex may give guidance on when 'plastic' global analysis can be used. A source of confusion is that the term 'plastic analysis' is used in EN 1993-1-1 to cover both non-linear analysis and rigid plastic analysis in its clause 5.4.3(1); no distinction is made between these two very different types of analysis.

EN 1993-1-5 Annex C gives rules for non-linear finite-element modelling of plates. To determine the resistance of plates, the analysis must be second order (geometrically non-linear) and consider imperfections. From 3-1-5/Table C.1, the material behaviour can either be elastic, in which case failure occurs at first yield somewhere in the plate, or it can be non-linear, in which case some redistribution can occur and a greater load obtained. 3-2/clause 5.4.1(1) appears to prohibit the use of the latter for non-accidental situations for bridges on the basis that it is a 'plastic' analysis, although this was not the intention and results from the all-encompassing definition of plastic analysis above. The resistances for plates in 3-1-5/clause 4.4 reflect this non-linear behaviour. It would however be unusual to use such an analysis in design. Further considerations of 3-1-5/Annex C are beyond the scope of this guide.

Rules for rigid plastic analysis are given in 3-1-1/clause 5.4.3 and 3-1-1/clause 5.6. Two essential general criteria are that members must have Class 1 cross-sections (unless an explicit check of rotation capacity is made) and that members must not be susceptible to overall instability, such as flexural or lateral torsional buckling.

5.4.2. Elastic global analysis

3-1-1/clause 5.4.2(1) requires that linear elastic global analysis, based on the material properties for steel given in clause 3, is used regardless of the stress level in the members. *3-1-1/clause 5.4.2(1)*

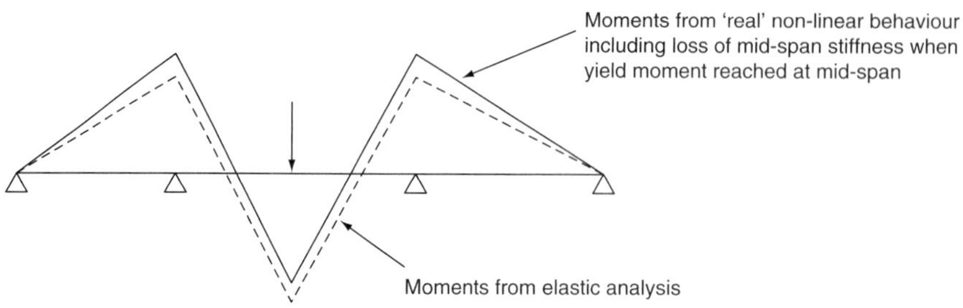

Fig. 5.4-1. Effect of mixed class section design

3-1-1/clause 5.4.2(2)

This applies even where the cross-section resistance of local sections is based on their plastic resistances – ***3-1-1/clause 5.4.2(2)*** refers. This is essentially consistent with UK practice but some care should be taken with mixing section classes within a bridge when elastic analysis is used. For example, if a mid-span section of a continuous bridge is designed in bending as Class 2 and the section at an internal support is Class 3, then the Class 3 section may become overstressed due to the elastic moments shed from mid-span while the plastic section resistance develops there and stiffness is lost. This is illustrated in Fig. 5.4-1.

Mixed class design has rarely been found to be a problem as the load cases producing maximum moment at mid-span and at a support rarely coexist except where adjacent spans are very short compared to the span considered. To safeguard against this problem, EN 1994-2 clause 6.2.1.3(2) provides a rule whereby the moment at a Class 1 or 2 section should not exceed 90% of its plastic bending resistance when there are adjacent sections in Class 3 or 4 with a bending moment of the opposite sign, unless account is taken of the redistribution of moments to the adjacent sections due to inelastic behaviour. It is suggested that a similar limitation should be used when designing bridges to EN 1993-2.

If redistribution is to be explicitly checked, a conservative method is illustrated in Fig. 5.4-2. In this example, a Class 2 section is at mid-span of the middle span and the support sections are Class 3. A simplified load case is shown to produce maximum sagging moment. Elastic analysis is used up to a fraction α of the entire applied load such that first yield of the Class 2 section is reached. The remaining fraction $(1 - \alpha)$ of the load is then applied to a model with a hinge placed at the yielded location and the resulting moments added to those from the first part of the analysis. The resistances of the Class 3 sections at the adjacent supports would then be checked for this total moment. It will often not actually be necessary to carry out such an analysis as it will usually be possible simply to redistribute the moments by 'lifting' the elastic moment diagram so that the first yield moment is not exceeded at the Class 2 section and then to check that the elastic resistance moment is not exceeded at the support.

3-1-1/clause 5.4.2(3)

Elastic global analysis may also be used where local cross-sections are susceptible to local buckling – ***3-1-1/clause 5.4.2(3)*** refers. However, the loss of elastic stiffness due to local plate buckling may need to be accounted for as discussed in section 5.1.1 of this guide. Similar considerations apply to shear lag effects which are also discussed in section 5.1.1.

3-2/clause 5.4.2(4)

It is permissible to neglect some effects of actions at the ultimate limit state in accordance with ***3-2/clause 5.4.2(4)*** and these are discussed in section 5.4.3 below.

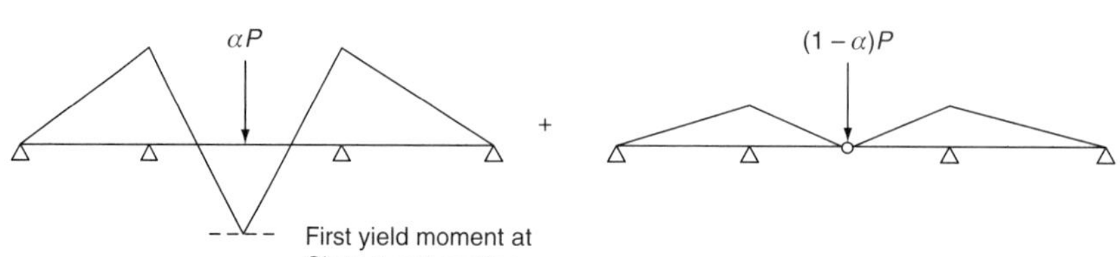

Fig. 5.4-2. Illustration of determination of total moment at supports due to shedding from mid-span

5.4.3. Effects which may be neglected at the ultimate limit state (additional sub-section)

Effects from global analysis

Large plastic strains are possible for beams where cross-sections are Class 1. This permits the formation of plastic hinges and the use of a rigid plastic global analysis. The elastic effects of indirect actions (which impose displacements and/or rotations) can be relieved through plastic deformation for Class 1 sections. 3-2/clause 5.4.2(4) therefore allows such effects to be neglected at the ultimate limit state where all sections are Class 1. These include the effects of:

- differential temperature
- differential shrinkage
- differential settlement.

The same capacity for plastic strain should also mean that the effects of staged construction could safely be neglected at the ultimate limit state, although this is not explicitly stated in EN 1993. It would not be common to do this however, as a separate analysis considering the staged construction would then be required for the serviceability limit state.

3-2/clause 5.4.2(4) does not permit the effects of imposed deformations to be ignored where all sections are Class 2. Class 2 sections exhibit sufficient plastic strain to attain the plastic section resistance but have limited rotation capacity beyond this point. This is however normally considered adequate to relieve the effects of imposed deformations. EN 1994-2 does permit these effects to be ignored where all sections are in either Class 1 or 2, so there is an inconsistency at present.

If the effects of indirect actions are to be ignored, it is not sufficient for all sections of a beam to be Class 1 if the beam is susceptible to overall instability, such as lateral torsional buckling. In this instance, the forces caused by the imposed deformations could lead to premature failure by buckling. Consequently, the above effects should additionally only be ignored where the beam is not prone to lateral torsional buckling. A statement to this effect is not given in EN 1993-2, which is an omission. It is suggested here that this condition be achieved by ensuring that the reduction factor, χ_{LT}, for lateral torsional buckling in accordance with 3-1-1/clause 6.3.2.2 is less than 0.2 throughout.

Effects from local analysis

In section design, restraint of torsional warping may be neglected for box sections at the ultimate limit state according to 3-1-1/clause 6.2.7(7). This is because torsional warping in boxes does not contribute to carrying the torsion, so the effects may be relieved by local yielding; section 6.2.7 of this guide refers. The effects must however be considered at the serviceability limit state.

For open sections, 3-1-1/clause 6.2.7(7) allows St Venant torsion to be neglected at ultimate limit state. This is because it will often be more efficient to carry an imposed torsional load through warping torsion. It would however seem illogical not to alternatively permit the neglect of warping torsion, which is often done in design. If the effects of St Venant torsion are neglected in open sections, the imposed torsional load must be carried by warping torsion. In general, torsion must be carried by one or a combination of the resisting mechanisms.

5.5. Classification of cross-sections

5.5.1. Basis

The local buckling resistance of webs and flanges in compression will have a significant effect on the loads and rotations that a member can withstand. The ability of a steel component to resist local buckling in compression is categorized by its 'section classification'. The classification of cross-sections is the established method of taking account in design of local buckling of plane steel elements in compression. It determines the available methods of global analysis and the basis for resistance to bending. The section classification is a function of the cross-sectional geometry (plate edge support conditions and b/t ratio), the stress distribution across the plate and the plate yield strength.

5.5.2. Classification

3-1-1/clause 5.5.2(1)

Steel components are grouped into the following four classifications according to ***3-1-1/clause 5.5.2(1)***:

- **Class 1** cross-sections are those that can form a plastic hinge and then carry on rotating without loss of resistance. It is a requirement of EN 1993-1-1 for the use of rigid plastic global analysis that the cross-sections at all plastic hinges are in Class 1. For steel bridges, EN 1993-2 does not permit rigid-plastic analysis other than for accidental combinations.
- **Class 2** cross-sections are those that can develop their plastic moment resistance, but have limited rotation capacity after reaching it because of local buckling. The ultimate limit state is assumed to occur in a fully restrained Class 2 cross-section when a plastic hinge develops and therefore rigid plastic analysis is inappropriate.
- **Class 3** cross-sections are those in which the stress in the extreme compression fibre of the steel member, assuming an elastic distribution of stresses, can reach the yield strength but will become susceptible to local buckling before development of the plastic resistance moment. The ultimate limit state occurs in a fully restrained Class 3 cross-section when yielding occurs in the extreme compression fibre.
- **Class 4** cross-sections are those in which local buckling will occur before the attainment of yield stress in one or more parts of the cross-section. The ultimate limit state occurs in a Class 4 cross-section when local buckling occurs. EN 1993-1-5 is used to determine effective widths for the panels of Class 4 members as discussed in section 6.2.2.5 of this guide – ***3-1-1/clause 5.5.2(2)*** refers.

3-1-1/clause 5.5.2(2)

The four types of idealized behaviour are illustrated for bending only in Fig. 5.5-1. In reality, the moment continues to rise to a peak beyond the plastic moment, M_{pl}, in both the Class 1 and 2 cases due to strain hardening and there is a loss of stiffness as soon as the elastic moment, M_{el}, is reached. The Class of cross-section is determined from the width-to-thickness limits given 3-1-1/Table 5.2 for webs and flanges in compression – ***3-1-1/clause 5.5.2(3)*** refers. ***3-1-1/clause 5.5.2(4)*** clarifies that a compression part is any part that is totally or partially in compression. If a steel component has different section classifications for the web and the flange, then the cross-section should be classified according to its least favourable class of compression parts – see ***3-1-1/clause 5.5.2(6)***.

3-1-1/clause 5.5.2(3)
3-1-1/clause 5.5.2(4)
3-1-1/clause 5.5.2(6)

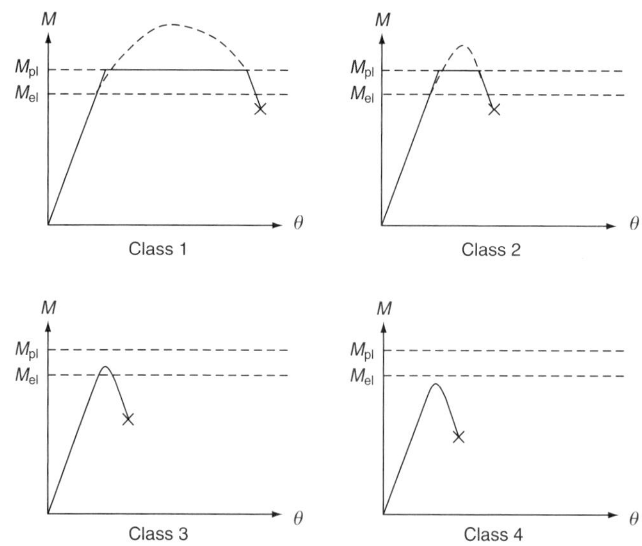

Fig. 5.5-1. Idealised moment–rotation relationships for Class 1 to 4 sections

The use of 3-1-1/Table 5.2 is fairly self-explanatory. A plastic stress block is used to check for compliance with Class 1 or 2 requirements and if this cannot be demonstrated, elastic stress blocks are used to check that the section is Class 3 rather than Class 4 – ***3-1-1/ clause 5.5.2(8)*** refers. Where both axial load and moment are present, these need to be combined when deriving the plastic stress block or, alternatively, the web Class can conservatively be determined on the basis of axial load alone. Examples of determining section classification where axial load is present are given in sections 6.2.10 and 6.2.11 of this guide. *3-1-1/clause 5.5.2(8)*

The numbers in 3-1-1/Table 5.2 appear different from those in BS 5400: Part 3: 2000[4] because the coefficient that takes account of yield strength, ε, is defined as $\sqrt{(235/f_y)}$ in the Eurocodes, and as $\sqrt{(355/f_y)}$ in BS 5400. After allowing for this, the limits for webs at the Class 2–Class 3 boundary agree closely with those in BS 5400, but there are differences for flanges. For outstand flanges, EN 1993 is more liberal at the Class 2–Class 3 boundary, and slightly more severe at the Class 3–Class 4 boundary. For internal flanges of boxes, EN 1993 is considerably more liberal for all Classes.

EN 1993-1-5 is used to determine effective widths for the panels of Class 4 members. Where a member is longitudinally stiffened, it should be classified as Class 4 unless it can be classified in a higher class by ignoring the longitudinal stiffeners. It is noted in section 6.2.2.5.2.1 of this guide that there is a small discontinuity in the Class 3–Class 4 boundary for internal plates in compression as assessed by 3-1-1/Table 5.2 and EN 1993-1-5. The former leads to slightly more slender parts being classed as Class 3 than the latter. Alternatively, a Class 4 member can be treated as Class 3 and the limiting stress method discussed in section 6.2.2.6 can be used.

3-1-1/clause 5.5.2(9) provides a method of treating a Class 4 section as an equivalent Class 3 section if the maximum design stress calculated on the gross cross-section, $\sigma_{com,Ed}$, is less than yield and if the section width-to-thickness ratios satisfy the increased limits allowed in the clause, using the calculated stress $\sigma_{com,Ed}$. Where second-order effects are significant, these should either be included in the global analysis when determining $\sigma_{com,Ed}$ or the section should be checked using the member rules of EN 1993-2 clause 6.3 and the member treated as Class 4 without applying 3-1-1/clause 5.5.2(9), as required by ***3-1-1/ clause 5.5.2(10)***. The effective Class 3 approach of 3-1-1/clause 5.5.2(9) should not be used in conjunction with 3-2/clause 6.3 because second-order effects considered via the resistance formulae may lead to a stress greater than $\sigma_{com,Ed}$. *3-1-1/clause 5.5.2(9)* *3-1-1/clause 5.5.2(10)*

Another way of treating a Class 4 section as an equivalent Class 3 section is to replace the yield stress by a reduced stress, σ_{limit}, in all calculations. This method is discussed in sections 6.2.4, 6.2.5 and 6.2.10 of this guide, covering resistance to compression, bending moment and combined compression and bending respectively.

5.5.3. Flange-induced buckling of webs (additional sub-section)

It should be noted that further limits on the slenderness of webs may also arise from considerations of flanged-induced buckling. This is discussed in section 6.10 of this guide.

CHAPTER 6

Ultimate limit states

This chapter discusses ultimate limit states as covered in section 6 of EN 1993-2 in the following clauses:

- General — *Clause 6.1*
- Resistance of cross-sections — *Clause 6.2*
- Buckling resistance of members — *Clause 6.3*
- Built-up compression members — *Clause 6.4*
- Buckling of plates — *Clause 6.5*

The following sections have also been added in this guide to deal with certain elements and situations where the relevant rules are scattered around the various parts of Eurocode 3.

- Intermediate transverse stiffeners — *Section 6.6*
- Bearing stiffeners and beam torsional restraint — *Section 6.7*
- Loading on cross-girders of U-frames — *Section 6.8*
- Torsional buckling of stiffeners – outstand limitations — *Section 6.9*
- Flange-induced buckling and effects due to curvature — *Section 6.10*

6.1. General

The partial factors for materials referred to in ***3-2/clause 6.1(1)P*** take account of both variations in the material strength and also the scatter of test results from the particular design resistance model used; the shear buckling model, for example. Consequently, different factors apply to different resistance mechanisms. To take account of this, EN 1993-2 recommends values of seven different partial material factors which cover different failure modes. The recommended values are provided in 3-2/Table 6.1, reproduced here as Table 6.1-1. They may be amended in the National Annex. Recommended values of material factors have been derived as discussed in section 2.5 of this guide. *3-2/clause 6.1(1)P*

One salient point to note is the use of the material factor $\gamma_{M0} = 1.00$ for the cross-section resistance of members. This has arisen because a studies of steels produced to European standards demonstrated that their actual characteristic strengths were well in excess of the required values. This might not however always be the case. In some cases, strain hardening of steel also means that resistances can exceed values based on the yield strength. This gives some further justification for a unity material factor, but only where the effects of strain hardening have not already been included in the resistance model.

Table 6.1 of EN 1993-2 states that the factor γ_{M1} relates to the resistance of members to instability. It also however applies to shear buckling (3-1-5/clause 5), resistance to patch loads (3-1-5/clause 6) and cross-section resistance where the limiting stress method is used (3-1-5/clause 10).

Table 6.1-1. Partial factors for materials

Resistance type	Factor	Recommended value
(a) Resistance of members and cross-section		
– resistance of cross-sections to excessive yielding including local buckling	γ_{M0}	1.00
– resistance of members to instability assessed by member checks	γ_{M1}	1.10
– resistance to fracture of cross-sections in tension	γ_{M2}	1.25
(b) Resistance of joints		
– resistance of bolts	γ_{M2}	1.25
– resistance of rivets		
– resistance of pins		
– resistance of welds		
– resistance of plates in bending		
– slip resistance:		
– at ultimate limit state	γ_{M3}	1.25
– at serviceability limit state	$\gamma_{M3,ser}$	1.10
– bearing resistance of an injection bolt	γ_{M4}	1.10
– resistance of joints in hollow section lattice girders	γ_{M5}	1.10
– resistance of pins at serviceability limit state	$\gamma_{M6,ser}$	1.00
– preload of high-strength bolts	γ_{M7}	1.10

6.2. Resistance of cross-sections

6.2.1. General

Checks on members are typically carried out in two parts when using the rules in EN 1993. First, critical sections are checked within the member for cross-section resistance. Although these are referred to as 'cross-section checks', the rules for cross-section resistance also make provision for local buckling effects which affect a certain finite length of the member rather than just a single cross-section, e.g. shear buckling. Second, the overall member stability is checked using the buckling rules in section 6.3. The exception to this is where second-order analysis, with member and global imperfections fully accounted for, has been used to determine the effects within the member. In this case only cross-section checks as described in this section are required.

3-1-1/clause 6.2.1(2)

Rules are given within section 6.2 for the combination of different stress resultants such as bending, shear and axial load. EN 1993-2 generally refers to the corresponding sections of EN 1993-1-1 for these interactions. However ***3-1-1/clause 6.2.1(2)*** requires reference to be made to EN 1993-1-5 where sections are in Class 4 or when there is shear buckling or transverse loading. Most of these interaction formulae involve some degree of plastic redistribution that has been validated by testing. Where the stress resultants are not known, as might be the case where stresses have been taken directly from a finite-element model, an alternative verification given by the Von Mises equivalent stress criterion in ***3-1-1/clause 6.2.1(5)*** can be used:

3-1-1/clause 6.2.1(5)

$$\left(\frac{\sigma_{x,Ed}}{f_y/\gamma_{M0}}\right)^2 + \left(\frac{\sigma_{z,Ed}}{f_y/\gamma_{M0}}\right)^2 - \left(\frac{\sigma_{x,Ed}}{f_y/\gamma_{M0}}\right)\left(\frac{\sigma_{z,Ed}}{f_y/\gamma_{M0}}\right) + 3\left(\frac{\tau_{Ed}}{f_y/\gamma_{M0}}\right)^2 \leq 1.0 \qquad \text{3-1-1/(6.1)}$$

where $\sigma_{x,Ed}$ is the longitudinal direct stress, $\sigma_{z,Ed}$ is the direct transverse stress and τ_{Ed} is the shear stress in the plane of the plate.

This criterion may always be used where there is no local buckling (including shear buckling) and may sometimes be necessary where a suitable interaction formula is not provided. This equivalent stress criterion does not however allow for any plastic redistribution, when used with elastically derived stresses, and corresponds to first yielding. It is therefore conservative compared to other interaction formulae provided in EN 1993. BS 5400: Part 3[4] made some allowance for flexural plasticity in its Von Mises equation by splitting the longitudinal stress into elastic axial and bending components and making a reduction to the bending component.

If it is desired to apply expression 3-1-1/(6.1) to members which are Class 4 (rather than using the interactions for Class 4 sections), then two approaches are possible. One possibility is to use effective section properties when calculating stresses (as discussed in detail in section 6.2.2.5 of this guide) but the section must not be prone to shear buckling as this is not included within expression 3-1-1/(6.1). Alternatively, the method of 3-1-5/ clause 10 can be used to check stresses on the gross cross-section, but the allowable stresses in expression 3-1-1/(6.1) are modified to allow for local buckling. In this latter case, shear buckling effects can be included by way of the reduction to allowable stress. This is discussed in section 6.2.2.6 of this guide.

A more general version of expression 3-1-1/(6.1) may be required in some situations where, for example, there is through-thickness stress or there are shear stresses in more than one plane as occurs with distortion of box girders:

$$\frac{\sqrt{2}}{2f_y/\gamma_{M0}}[(\sigma_{x,Ed}-\sigma_{y,Ed})^2+(\sigma_{y,Ed}-\sigma_{z,Ed})^2+(\sigma_{z,Ed}-\sigma_{x,Ed})^2+6(\tau_{xy,Ed}^2+\tau_{yz,Ed}^2+\tau_{xz,Ed}^2)] \leq 1.0 \qquad \text{(D6.1-1)}$$

where $\sigma_{x,Ed}$ is the longitudinal direct stress, $\sigma_{z,Ed}$ is the direct transverse stress and $\sigma_{y,Ed}$ is the through-thickness stress, if any. $\tau_{xy,Ed}$ is the shear stress in the plane of the plate and $\tau_{yz,Ed}$ and $\tau_{xz,Ed}$ are shear stresses acting on two perpendicular planes transverse to the plane of the plate.

One further convenient alternative to the interactions presented in 3-2/clause 6.2 is provided in ***3-1-1/clause 6.2.1(7)***: *3-1-1/clause 6.2.1(7)*

$$\frac{N_{Ed}}{N_{Rd}}+\frac{M_{y,Ed}}{M_{y,Rd}}+\frac{M_{z,Ed}}{M_{z,Rd}} \leq 1.0 \qquad \text{3-1-1/(6.2)}$$

where N_{Rd}, $M_{y,Rd}$ and $M_{z,Rd}$ are the design resistances for each effect acting individually but with reductions for shear where the shear force is sufficiently large. This can be used for Class 1, 2 and 3 cross-sections but is particularly useful for the case of axial load, shear and bending (uniaxial or biaxial) in Class 1 and 2 cross-sections. In this case, use of expression 3-1-1/(6.2) makes it unnecessary to compute the resultant plastic stress block for axial load and bending. The use of this interaction is discussed in sections 6.2.10 and 6.2.11 of this guide.

For cross-section checks, the relevant recommended value of material partial factor is generally $\gamma_{M0} = 1.0$, including for Class 4 sections in bending and compression (except where the reduced stress method of 3-1-5/clause 10 is used). However, for shear and transverse loads, where the resistance of the section is reduced by local buckling, the recommended material factor is $\gamma_{M1} = 1.1$. The recommended material factor is always $\gamma_{M1} = 1.1$ for member buckling checks in accordance with 3-2/clause 6.3.

A further point to note is that 'extreme fibres' for Class 3 cross-section checks may be taken as the centre of the flanges according to ***3-1-1/clause 6.2.1(9)***, rather than the actual outer fibres. The difference can be significant for shallow members. Class 3 cross-sections can just develop compressive yield at their extreme fibres but will fail by local buckling if this compressive yielding starts to spread further into the cross-section. The maximum resistance is therefore reached when the extreme compression fibre reaches yield. In design, the moment resistance of a Class 3 section is usually taken to be the moment which produces yield at either fibre. However, if the tension fibre reaches yield first, a plastic stress block can start to develop in the tension zone before yield is reached at the compression fibre and the assumption of fully elastic behaviour is conservative. ***3-1-1/clause 6.2.1(10)*** calls this effect 'partial plastification' of the tension zone and permits it to be considered in determining the resistance of a Class 3 section. This is discussed further in section 6.2.5 of this guide.

3-1-1/clause 6.2.1(9)

3-1-1/clause 6.2.1(10)

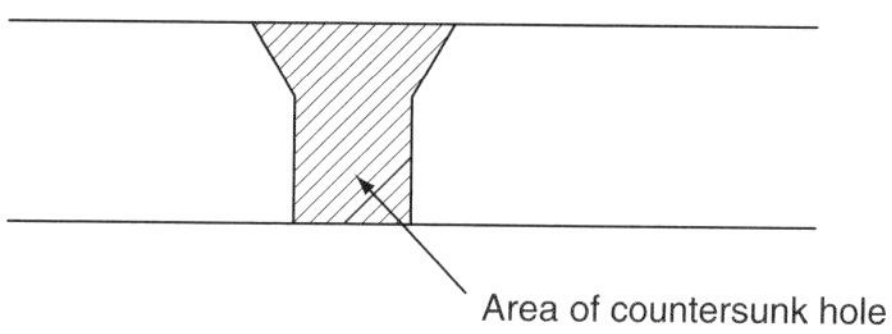

Fig. 6.2-1. Area of countersunk hole

6.2.2. Section properties

6.2.2.1. *Gross cross-section*

3-2/clause 6.2.1.1(1)

3-2/clause 6.2.1.1(1) defines the gross cross-section as the whole cross-section ignoring bolt holes but including larger holes, such as a cut-out for a drainage pipe.

6.2.2.2. *Net area*

3-1-1/clause 6.2.2.2(1)
3-1-1/clause 6.2.2.2(2)

Some resistances require consideration of net sections. The 'net' area of a steel component is defined in ***3-1-1/clause 6.2.2.2(1)*** as its gross area less appropriate deductions for all holes and other openings. The area of a hole is the maximum area removed from the steel component in cross-section. ***3-1-1/clause 6.2.2.2(2)*** reminds the designer that the countersunk portion of a hole should also be deducted if countersunk bolts are to be used as fasteners, as shown in Fig. 6.2-1.

3-1-1/clause 6.2.2(3)
3-1-1/clause 6.2.2(4)

If fastener holes are not staggered then the net area of the steel component will be the gross area minus the area of all the holes at that section – ***3-1-1/clause 6.2.2(3)*** refers. If the fasteners are staggered then, in accordance with ***3-1-1/clause 6.2.2(4)***, the net area of the steel component will be the greater of the following:

1. The gross area of the steel component minus the area of holes at any cross-section perpendicular to the member axis (e.g. Section 1–1 in Fig. 6.2-2).
2. The gross area of the steel component minus an effective area allowing for staggered holes as follows:

$$t\left(nd - \sum \frac{s^2}{4p}\right) \qquad \text{3-1-1/(6.3)}$$

where: s is the staggered pitch parallel to the member axis;
p is the spacing of centres of the same two holes measured perpendicular to the member axis;
t is the thickness of the steel component;
n is the number of holes in any diagonal or zig-zag line extending progressively across the component;
d is the diameter of the hole.

Surface 2–2 in Fig. 6.2-2 indicates a typical application of expression 3-1-1/(6.3) where $n = 2$. The net area from expression 3-1-1/(6.3) should not be taken greater than the gross area, although other resistance checks effectively stop this from being done.

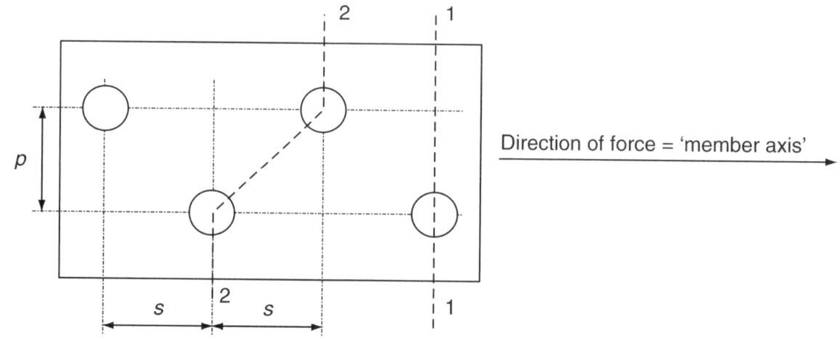

Fig. 6.2-2. Parameters for use in expression 3-1-1/(6.3)

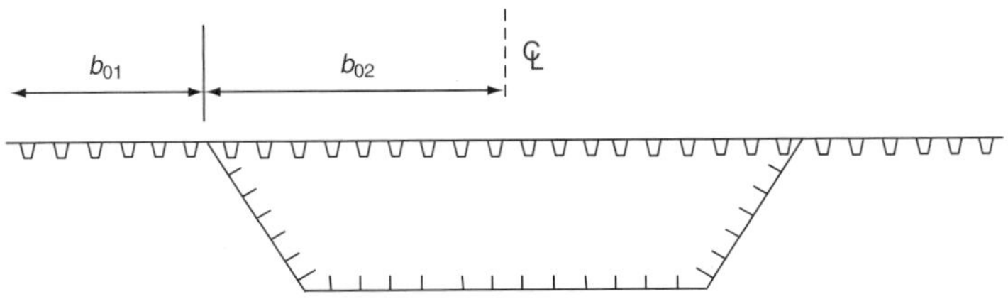

Fig. 6.2-3. Definition of b_0 for internal and outstand flanges

3-1-1/clause 6.2.2.2(5)

If expression 3-1-1/(6.3) is applied to an angle or other member with holes on several faces, ***3-1-1/clause 6.2.2.2(5)*** requires p to be measured along the centre of the thickness of the plates when the dimension extends around a corner. If a member is connected eccentrically, this eccentricity needs to be considered. EN 1993-1-8 gives a method for tension connections which is discussed in section 6.2.3 of this guide. Where an unequal angle is connected by way of holes on its smaller leg only, 3-1-8/clause 3.10.3 requires the net area for tension calculations to be based on a fictitious equal angle with leg size based on the smaller of those for the real unequal angle.

6.2.2.3. *Effective widths for shear lag*

6.2.2.3.1. Shear lag for members in bending at SLS and ULS (additional sub-section)

3-2/clause 6.2.2.3(1)

A description of the causes and idealization of shear lag effects is given in section 5.1.1 of this guide. This section describes the calculation procedure for determining effective widths for shear lag at both serviceability limit states (SLS) and ultimate limit states (ULS). ***3-2/clause 6.2.2.3(1)*** makes reference to EN 1993-1-5 for this calculation, both directly and through EN 1993-1-1.

The effect of shear lag is greatest in locations of high shear where the force in the flanges is changing rapidly. Consequently, effective widths for shear lag at intermediate supports will be smaller than those for the span regions. Shear lag must be considered in section design at both SLS and ULS in EN 1993. This is unlike design to BS 5400: Part 3[4] where it was permissible to neglect shear lag at ULS on the basis that stresses could redistribute across the cross-section with a little plasticity. Different effective widths are however obtained for SLS and ULS in EN 1993-1-5 and the reduction at ULS will typically be quite small because allowance is made for plastic redistribution within the rules of EN 1993-1-5.

3-1-5/clause 3.2.1(1)

Effective widths are calculated as a function of the available width, the distance between points of main beam zero bending moment adjacent to the location considered and the amount of stiffening. The effective width at SLS is given by ***3-1-5/clause 3.2.1(1)***:

$$b_{\text{eff}} = \beta b_0 \qquad \text{3-1-5/(3.1)}$$

Table 6.2-1. Effective width factors, β from 3-1-5/Table 3.1

K	Location for bending	β-value
0.02		$\beta = 1.0$
$0.02 < k < 0.70$	Sagging bending	$\beta = \beta_1 = \dfrac{1}{1 + 6.4k^2}$
	Hogging bending	$\beta = \beta_2 = \dfrac{1}{1 + 6.0\left(k - \dfrac{1}{2500k}\right) + 1.6k^2}$
>0.07	Sagging bending	$\beta = \beta_1 = \dfrac{1}{5.9k}$
	Hogging bending	$\beta = \beta_2 = \dfrac{1}{8.6k}$
All k	End support	$\beta_0 = (0.55 + 0.025/k)\beta_1$, but $\beta_0 < \beta_1$
All k	Cantilever	$\beta = \beta_2$ at support and at end

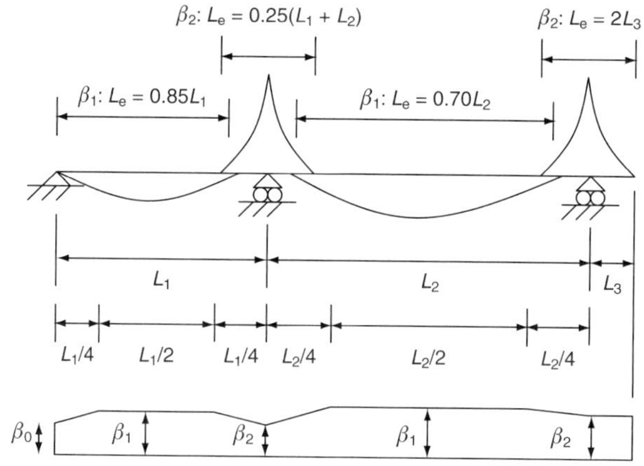

Fig. 6.2-4. Length L_e for continuous beam and distribution of effective[s] width

where b_0 is the physical width available equal to the full width of outstands and half the width of internal plates between webs as shown in Fig. 6.2-3. β is a factor accounting for width-to-span ratio and stiffening and is found from 3-1-5/Table 3.1, reproduced here as Table 6.2-1, and depends on:

$$k = \alpha_0 b_0 / L_e \text{ and } \alpha_0 = \sqrt{1 + \frac{A_{sl}}{b_0 t}}$$

where L_e represents the distance between points of zero bending moment and can be determined from 3-1-5/Fig. 3.1 (reproduced as Fig. 6.2-4) provided that adjacent internal spans do not differ by more than 50% and a cantilever span is not longer than half the adjacent span – ***3-1-5/clause 3.2.1(2)*** refers. A_{sl} is the total area of longitudinal stiffeners in the width b_0. Figure 6.2-4 also shows the distribution of effective widths.

3-1-5/clause 3.2.1(2)

The limitations on span length ratios for use of Fig. 6.2-4 are made so that the bending moment distributions within spans are of similar shape to those in Fig. 6.2-4. The simple rules do not cater for other cases such as spans that are permanently hogging. If spans or moment distributions do not comply with the above requirements, then the distance between points of zero bending moment, L_e, should be calculated for the actual moment distribution. This is less desirable for design because analysis will have to be done first with gross cross-section properties to determine the likely distribution of moment.

At ULS, the effective width is much greater than at SLS, due to a certain amount of plastic redistribution, and will often approach the full available width for typical width-to-span ratios. (The difference to previous UK practice is therefore less than first appears.) The effective width at ULS can conservatively be taken as the SLS value or may optimally be calculated according to Note 3 of ***3-1-5/clause 3.3(1)***:

3-1-5/clause 3.3(1)

$$A_{eff} = \beta^{\kappa} A_{c,eff} \geq \beta A_{c,eff} \qquad \text{3-1-5/(3.5)}$$

A_{eff} is used here rather than b_{eff} to include the effects of reduction in area from plate buckling effects as well (see sections 6.2.2.5 and 6.2.2.6 of this guide) but the equation has the effect of reducing the available width in the same way as expression 3-1-5/(3.1) so that $b_{eff} = \beta^{\kappa} b_0$. The effective area accounting for both plate buckling and shear lag is the effective plate area within the width b_{eff}.

Figures 6.2-5 and 6.2-6 show the fraction of the full available width obtained for support and mid-span zones of a multi-span continuous bridge with equal internal spans of L. Results are produced for cases with no longitudinal stiffeners (Fig. 6.2-5) and for an amount of longitudinal stiffeners equal to the deck plate area (Fig. 6.2-6). It can be seen that there is considerably more width available at ULS than at SLS. Also, support zones, where the shear is high, suffer a much greater reduction in effectiveness. Typical values of b_0/L are unlikely to exceed 0.1 so it can be seen that shear lag will not usually have a great effect at ULS. The acting flange width is unlikely to be reduced for most bridges,

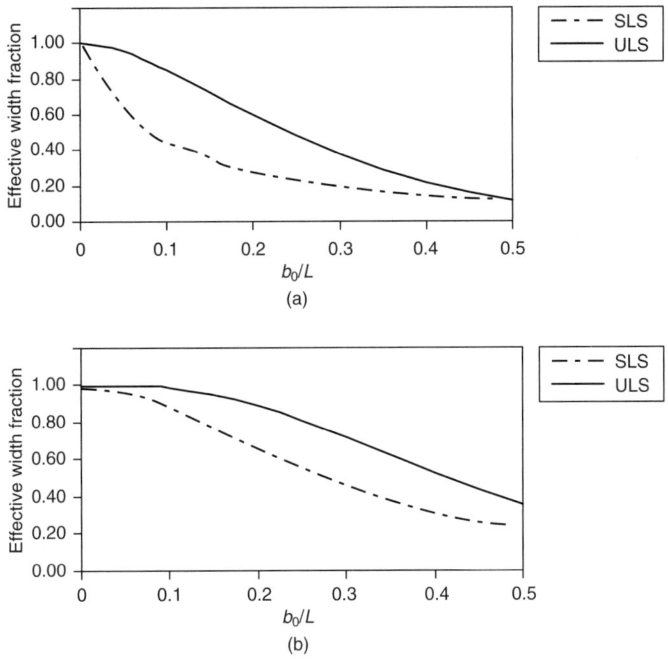

Fig. 6.2-5. No longitudinal stiffeners ($\alpha_0 = 1$): (a) support; (b) mid-span

other than stiffened box girders or steel beam bridges with a common orthotropic deck, as flanges will not generally be sufficiently wide. The values obtained at SLS are, in fact, very similar to those that were obtained from BS 5400: Part 3.[4]

Where it is necessary to determine a more realistic distribution of longitudinal stress across the width of the flange, as may be required in a check of combined local and global effects in a deck plate, the formulae in 3-1-5/clause 3.2.2 Fig. 3.3 (not reproduced here) may be used to estimate stresses. A typical location where this might be necessary would be in checking a deck plate at a transverse diaphragm between main beams where the deck plate has overall longitudinal direct stress from global bending and is also subjected to a local

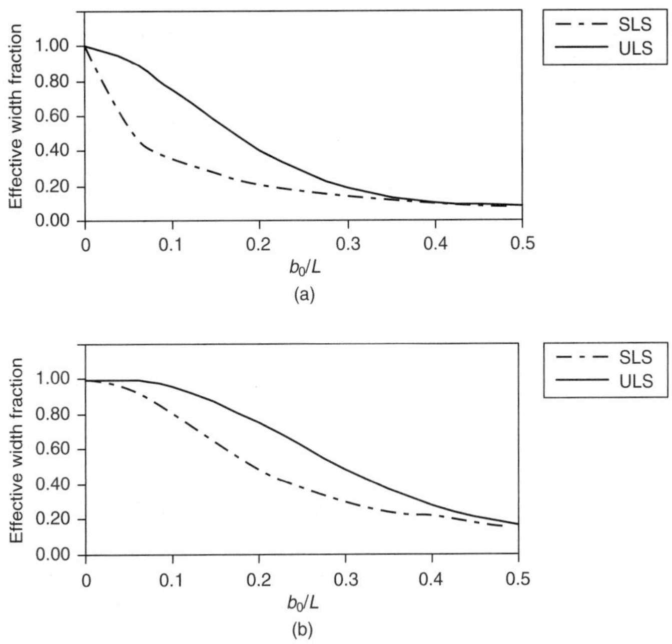

Fig. 6.2-6. Equal longitudinal stiffeners and plate areas ($\alpha_0 = 1.41$): (a) support; (b) mid-span

hogging moment from wheel loads. The use of the formula in EN 1993-1-5 can be beneficial here as the global and local effects in the deck plate do not occur at the same location; the greatest local effects occur in the middle of the plate remote from the webs, while the global longitudinal stresses are greatest adjacent to the webs.

Worked Example 6.2-1: Effective widths of a box girder

A box girder bridge has the span layout and cross-section shown in Fig. 6.2-7. The top flange has trough stiffeners such that $A_{sl}/b_0t = 0.5$. Determine the effective width of top flange acting with each web at mid-span and over the supports for the main span at both SLS and ULS.

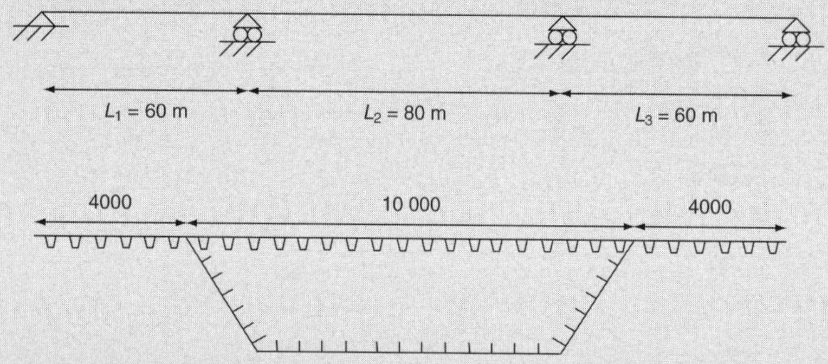

Fig. 6.2-7. Bridge deck for Worked Example 6.2-1

Considering mid-span first:

SLS

From 3-1-5/Fig. 3.1, $L_e = 0.7L_2 = 0.7 \times 80\,000 = 56\,000$ mm

From 3-1-5/Table 3.1, the cantilever portion has effective width as follows:

$$\alpha_0 = \sqrt{1 + \frac{A_{sl}}{b_0 t}} = \sqrt{1 + 0.5} = 1.225$$

$$k = \frac{\alpha_0 b_0}{L_e} = \frac{1.225 \times 4000}{56\,000} = 0.0875$$

$$\beta = \beta_1 = \frac{1}{1 + 6.4k^2} = \frac{1}{1 + 6.4 \times 0.0875^2} = 0.953$$

From expression 3-1-5/(3.1): $b_{eff} = \beta \times b_0 = 0.953 \times 4000 = 3813$ mm

From 3-1-5/Table 3.1, the internal portion has effective width as follows:

$$\alpha_0 = \sqrt{1 + \frac{A_{sl}}{b_0 t}} = \sqrt{1 + 0.5} = 1.225$$

$$k = \frac{\alpha_0 b_0}{L_e} = \frac{1.225 \times 5000}{56\,000} = 0.1094$$

$$\beta = \beta_1 = \frac{1}{1 + 6.4k^2} = \frac{1}{1 + 6.4 \times 0.1094^2} = 0.929$$

From expression 3-1-5/(3.1): $b_{eff} = \beta \times b_0 = 0.929 \times 5000 = 4645$ mm

Hence the total width attached to each web at SLS = 3813 + 4645 = **8458 mm**

ULS

For the cantilever, from expression 3-1-5/(3.5): $b_{eff} = \beta^k \times b_0 = 0.953^{0.0875} \times 4000 =$ 3983 mm

For the inner part, from expression 3-1-5/(3.5): $b_{eff} = \beta^k \times b_0 = 0.929^{0.1094} \times 5000 =$ 4959 mm

Hence the total width attached to each web at ULS = 3983 + 4959 = **8942 mm**

Considering an internal support:

SLS

From 3-1-5/Fig. 3.1, $L_e = 0.25(L_1 + L_2) = 0.25(60\,000 + 80\,000) = 35\,000$ mm

From 3-1-5/Table 3.1, the cantilever portion has effective width as follows:

$$\alpha_0 = \sqrt{1 + \frac{A_{sl}}{b_0 t}} = \sqrt{1 + 0.5} = 1.225$$

$$k = \frac{\alpha_0 b_0}{L_e} = \frac{1.225 \times 4000}{35\,000} = 0.140$$

$$\beta = \beta_2 = \frac{1}{1 + 6.0\left(k - \frac{1}{2500k}\right) + 1.6k^2}$$

$$= \frac{1}{1 + 6.0\left(0.140 - \frac{1}{2500 \times 0.140}\right) + 1.6 \times 0.140^2} = 0.539$$

From expression 3-1-5/(3.1): $b_{eff} = \beta \times b_0 = 0.539 \times 4000 = 2157$ mm

From 3-1-5/Table 3.1, the internal portion has effective width as follows:

$$\alpha_0 = \sqrt{1 + \frac{A_{sl}}{b_0 t}} = \sqrt{1 + 0.5} = 1.225$$

$$k = \frac{\alpha_0 b_0}{L_e} = \frac{1.225 \times 5000}{35\,000} = 0.1750$$

$$\beta = \beta_2 = \frac{1}{1 + 6.0\left(k - \frac{1}{2500k}\right) + 1.6k^2}$$

$$= \frac{1}{1 + 6.0\left(0.175 - \frac{1}{2500 \times 0.175}\right) + 1.6 \times 0.175^2} = 0.480$$

From expression 3-1-5/(3.1): $b_{eff} = \beta \times b_0 = 0.480 \times 5000 = 2398$ mm

Hence the total width attached to each web at SLS = 2157 + 2398 = **4555 mm**

ULS

For the cantilever, from expression 3-1-5/(3.5): $b_{eff} = \beta^k \times b_0 = 0.539^{0.140} \times 4000 =$ 3668 mm

For the inner part, from expression 3-1-5/(3.5): $b_{eff} = \beta^k \times b_0 = 0.480^{0.1750} \times 5000 =$ 4397 mm

Hence the total width attached to each web at ULS = 3668 + 4397 = **8065 mm**

6.2.2.3.2. Dispersion of concentrated loads (additional sub-section)

The effective flange width according to expression 3-1-5/(3.1) does not apply to the calculation of stress dispersal from concentrated axial forces. Shear lag still affects the rate of dispersal of local concentrated loads, but this rate is not connected to the bending moment profile. Consequently, where concentrated axial loads are applied to a section, such as in a cable-stayed bridge, separate calculation must be made of the effective area over which this force acts at each cross-section throughout the span.

3-1-5/clause 3.2.3(1)

3-1-5/clause 3.2.3(1) covers the dispersal of stress from concentrated loads in its expression (3.2). It is mainly intended for determining the distribution of stress in webs subjected to concentrated patch loads applied locally through a flange (e.g. local wheel loads or reactions during a bridge launch), but could be used to determine the dispersal of stress from longitudinal axial forces, such as from prestressing. For patch loading, the

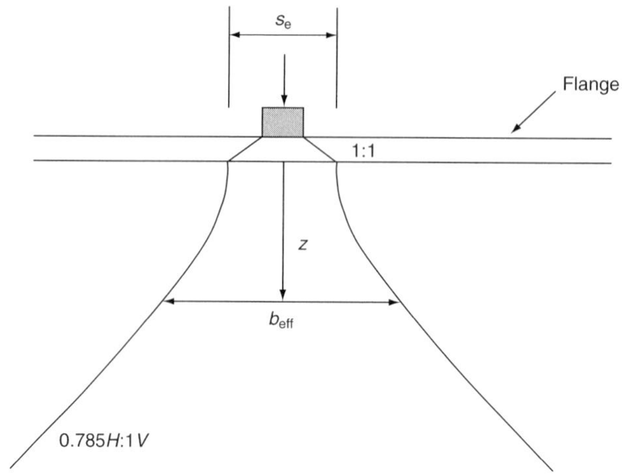

Fig. 6.2-8. Idealised spread for unstiffened web

spread of load through a flange from expression (3.2) is at $1H:1V$, which is less rapid than assumed in previous UK practice. The calculated spread width, b_{eff}, is not the full extent of spread, but is an equivalent width such that the mean stress calculated with this width equates to the peak elastic stress in the 'real' distribution. Since expression 3-1-5/(3.2) represents an elastic distribution of stress, it may be used for fatigue calculations as well as for ULS ones.

For an unstiffened flange, with a load applied through the flange, the spread width simplifies to:

$$b_{\mathrm{eff}} = \sqrt{s_{\mathrm{e}}^2 + (z/0.636)^2} \tag{D6.2-1}$$

and the design transverse stress at depth z below the loaded flange is:

$$\sigma_{z,\mathrm{Ed}} = \frac{F_{\mathrm{Ed}}}{b_{\mathrm{eff}} t} \tag{D6.2-2}$$

where s_{e} is the loaded width at the top of the web under the loaded flange and t is the web thickness. The angle of spread through an unstiffened web tends to a constant value of $0.785H:1V$ when remote from the loaded area (which is approximately at a distance equal to twice the loaded width at flange level) as shown in Fig. 6.2-8. However, the initial stress trajectory beneath the flange is vertical, so there is no simple idealized spread angle that can be used throughout as was previous UK practice.

Care is needed when using expression 3-1-5/(3.2) for stiffened plates where the stiffener spacing is large compared to the loaded width, as the formula is derived assuming the stiffeners to be closely spaced and smeared. The Note to 3-1-5/clause 3.2.3(1) consequently limits its use to situations where $s_{\mathrm{st}}/s_{\mathrm{e}} \le 0.5$, where s_{st} is the stiffener spacing. Outside this limit, equation (D6.2-2) above for unstiffened plates should be used.

6.2.2.4. Effective properties of cross-sections with Class 3 webs and Class 1 or 2 flanges

3-1-1/clause 6.2.2.4(1)

The method given in ***3-1-1/clause 6.2.2.4(1)*** is often referred to as 'the hole in the web' method. In beams subjected to hogging bending, it often happens that the bottom flange is in Class 1 or 2, and the web is in Class 3. The initial effect of local buckling of the web would be a small reduction in the bending resistance of the section. The assumption that a defined depth of web, the 'hole', is not effective in bending enables the reduced section to be upgraded from Class 3 to Class 2, and removes the sudden change in the bending resistance that would otherwise occur. The method is analogous to the use of effective areas for Class 4 sections, to allow for local buckling. The *Designers' Guide to EN 1994-2*[7] gives more detail on this method and an example of its use.

It should be noted that if a Class 3 cross-section is treated as an equivalent Class 2 cross-section for section design, it should still be treated as Class 3 when considering the actions to consider in its design. Indirect actions, such as differential settlement, which may be neglected for true Class 2 sections, should not be ignored for effective Class 2 sections. When indirect actions contain both primary and secondary components, such as differential shrinkage acting on statically indeterminate structures, the primary self-equilibrating stresses could reasonably be neglected, but not the secondary effects.

6.2.2.5. Class 4 members – general and effective section method

6.2.2.5.1. Methods of approach

Class 4 members are those that are unable to attain the full yield stress under the loading considered because of the onset of local buckling. Plate buckling is discussed generally in section 5.1.1. The method for dealing with Class 4 cross-sections is given in EN 1993-1-5. Two methods are presented and ***3-2/clause 6.2.2.5(1)*** requires that one of these methods is followed: *3-2/clause 6.2.2.5(1)*

(i) Use of section properties based on effective[P] widths to allow for both plate and stiffener buckling – 3-1-5/clause 4 covers this method.
(ii) Use of section properties based on the gross cross-section but with a reduced allowable stress limit (less than yield) – 3-1-5/clause 10 covers this method.

EN 1993-2 allows the National Annex to choose which method to use, but there are restrictions on the applicability of method (i) given in EN 1993-1-5 in some cases. It is therefore logical to permit both methods to be used. In this guide, the effective section method is discussed in detail under section 6.2.2.5 and the reduced stress method is discussed in section 6.2.2.6, although a brief comparison of the methods is first given below.

(i) Effective sections to EN 1993-1-5 clause 4

The first method differs significantly from UK practice to date. This is because the use of effective[P] widths for web and flange elements allows load shedding between all the various elements such that their combined strength is optimally used.

The load shedding implicit in the effective width model of EN 1993-1-5 implies that there is sufficient post-buckling strength and ductility to permit this redistribution. Figure 6.2-9 gives definitions of panel components. Unstiffened plates and sub-panels can maintain their peak resistance for a reasonable strain increase after their maximum resistance is reached, so such load shedding is possible. The post-buckling strength stems from an unstiffened plate panel's ability for load to concentrate along its longitudinal supported edges after elastic buckling. Stiffened panels undergoing overall buckling generally have less post-buckling strength however, and for short wide panels, buckling is largely column-like where the elastic critical buckling load is an upper bound to the resistance. The effective width method still implicitly assumes there is adequate deformation capacity to shed load to other plate elements. Details of the test results that were used by the EN 1993-1-5 Project Team in the calibration of this method are not known to the authors of this guide. It represents a significant change from previous UK practice.

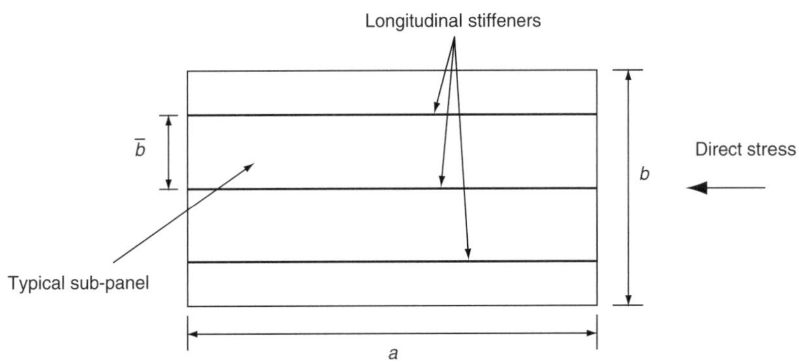

Fig. 6.2-9. Stiffened panel with sub-panels

The above assumptions of post-buckling strength and ductility certainly do not apply where local torsional buckling (sometimes known as tripping) of open stiffeners occurs, as there is insufficient post-buckling strength in such an element with a free edge to maintain its load over any strain increase. The load drops off rapidly when buckling occurs, which can lead to progressive failure. It is therefore essential to prevent torsional buckling when the method of effective sections is used. A method for ensuring its prevention is given in 3-1-5/clause 9.2.1. Torsional buckling is discussed in section 6.9 of this guide.

3-1-5/clause 4.1(1)

There are further restrictions on method (i) given in ***3-1-5/clause 4.1(1)***:

(a) The panels should nominally be rectangular and the flanges should be parallel (to within 10°). However, it is possible to square off panels based on their largest dimensions to calculate a lower bound on the effective width fraction, ρ, to overcome this limitation.
(b) Stiffeners must be provided longitudinally and/or transversely, i.e. not skewed.
(c) An unstiffened open hole in a panel should not have diameter exceeding 5% of the panel width, b. This is because large holes can limit post-buckling strength and ductility of panels. Secondary bending stresses are also set up, particularly around web openings, which should be accounted for. No rules are given as to how heavily a hole would have to be stiffened (both transversely and longitudinally) to permit a relaxation of this limit or how to consider the secondary bending stresses. This is therefore a matter for judgement by individual designers.
(d) Members must be of uniform cross-section. Haunched members with haunch angle less than 10° can be treated as uniform for consistency with (a) above. If flanges are continuously curved in elevation, the resulting pressure imposed on the web can be dealt with using 3-1-5/clause 8, but EN 1993 provides no means of considering the interaction with other effects. It is difficult therefore to use the effective section method for beams with continuously curved flanges without some judgement – see the discussion in section 6.10.1.1 of this guide.
(e) The web should be adequate to prevent buckling of the compression flange into the plane of the web. Rules are given in 3-1-5/clause 8 which are discussed and extended in section 6.10 of this guide.

Another restriction not specifically mentioned in EN 1993-1-5 is that the effective section method cannot be used (without modification) where there is a uniform transverse direct stress accompanying the longitudinal stress. The rules and interactions for transverse loading in 3-1-5/clause 6 and 3-1-5/clause 7 may be applied for concentrated loads, but the effect of more uniform transverse stress would need to be evaluated using the method of reduced stresses in 3-1-5/clause 10.

3-1-5/clause 4.3(6)

The effective section method may be used where the flange has a greater yield strength than the web, provided that the flange yield stress is not more than a recommended limit of twice that of the web – ***3-1-5/clause 4.3(6)*** refers. The web stresses must then not exceed the yield strength of the web and the effective widths of the web should be determined using the higher flange yield strength.

(ii) Reduced stress limits to EN 1993-1-5 clause 10

Where the conditions above for the use of effective widths are not met, a method based on stress analysis with gross cross-section properties and subsequent plate buckling checks may be used according to 3-1-5/clause 10. This method may always be used as an alternative to the effective width approach, but it takes no account of the beneficial shedding of load from overstressed panels. The method is discussed further in section 6.2.2.6 of this guide.

For greatest structural economy, it is generally better to use 3-1-5/clause 4, although there are some exceptions as discussed in section 6.2.2.6 below.

3-1-5/clause 4.3(3)
3-1-5/clause 4.3(4)

6.2.2.5.2. Method using effective sections

Effective widths are determined on the basis of the distribution of stresses acting on the individual parts of the cross-section. ***3-1-5/clause 4.3(3)*** and ***3-1-5/clause 4.3(4)*** allow section properties to be developed separately for axial loads and for bending, or alternatively they may be based on the overall stress distribution caused by combined axial load and bending. The latter option is less convenient because the section properties will vary with each load case.

The basic procedure, outlined in ***3-1-5/clause 4.4(3)***, is to determine the effective section for the flanges first, based on stresses computed with gross-section properties but allowing for shear lag if relevant. The effective section for a web should then be calculated using section properties comprising the gross web and the effective flanges (including shear lag effects). If the cross-section has longitudinal stiffeners, then the derivation of the effective section has to consider both local buckling of the plate sub-panels and overall buckling of the stiffened plates. If the stress in a cross-section builds up in stages with the cross-section changing throughout (as in steel–concrete composite construction), 3-1-5/clause 4.4(3) allows the stresses to first be built up with effective flanges and gross web. The total stress distribution so derived in the web may then be used to determine an effective section for the web and the resulting effective cross-section can be used for all stages of construction to build up the final stresses. This is a convenient approximation which overcomes the problem that otherwise the effective section of the web would keep changing throughout construction. *3-1-5/clause 4.4(3)*

Where there is biaxial bending, what constitutes a flange or a web is not defined. However, the precise classification matters less with the uniform approach to webs and flanges in EN 1993-1-5 than it would have done to BS 5400: Part 3.[4]

6.2.2.5.2.1. Effective widths for unstiffened plates and plate sub-panels

Effective widths for unstiffened plate panels, including sub-panels between stiffeners, are calculated using 3-1-5/clause 4.4. According to ***3-1-5/clause 4.4(1)***, the effective area of the plate is given by: *3-1-5/clause 4.4(1)*

$$A_{c,eff} = \rho A_c \qquad \text{3-1-5/(4.1)}$$

where ρ is a reduction factor which depends on whether the plate panel considered is internal (and therefore has both longitudinal edges stiffened) or is an outstand (and therefore has only one longitudinal edge stiffened). The distribution of the effective area within the plate panel is determined from either 3-1-5/Table 4.1 or 4.2 for internal or outstand elements respectively. ***3-1-5/clause 4.4(2)*** gives formulae for the reduction factors which are reproduced below. *3-1-5/clause 4.4(2)*

For internal elements:

$$\rho = \frac{\bar{\lambda}_p - 0.055(3+\psi)}{\bar{\lambda}_p^2} \leq 1.0 \text{ but } \rho = 1.0 \text{ for } \bar{\lambda}_p \leq 0.673 \qquad \text{3-1-5/(4.2)}$$

and for outstands:

$$\rho = \frac{\bar{\lambda}_p - 0.188}{\bar{\lambda}_p^2} \leq 1.0 \qquad \text{but } \rho = 1.0 \text{ for } \bar{\lambda}_p \leq 0.748 \qquad \text{3-1-5/(4.3)}$$

where ψ is the stress ratio across the plate shown in 3-1-5/Tables 4.1 and 4.2. The format of expression 3-1-5/(4.2) for pure compression ($\psi = 1$) was originally proposed by Winter.[8] The definition of slenderness, $\bar{\lambda}_p$, follows the usual Eurocode notation of being the square root of a ratio of a yield resistance to an elastic critical buckling resistance and is therefore:

$$\bar{\lambda}_p = \sqrt{\frac{f_y}{\sigma_{cr}}} = \sqrt{\frac{f_y}{\left(\dfrac{k_\sigma \pi^2 E t^2}{12(1-\upsilon^2)b^2}\right)}} = \frac{\bar{b}/t}{28.4\varepsilon\sqrt{k_\sigma}}$$

where $\varepsilon = \sqrt{235/f_y}$, k_σ is a buckling coefficient, determined from 3-1-5/Tables 4.1 and 4.2, which depends on stress distribution and panel edge support conditions, and $\bar{b}/t$ is the plate width-to-thickness ratio. The values of k_σ assume simply supported edges (except at free edges), but benefit could be taken in deriving higher values where significant edge rotational support stiffness could be guaranteed. They also assume infinitely long plates, which is discussed further below.

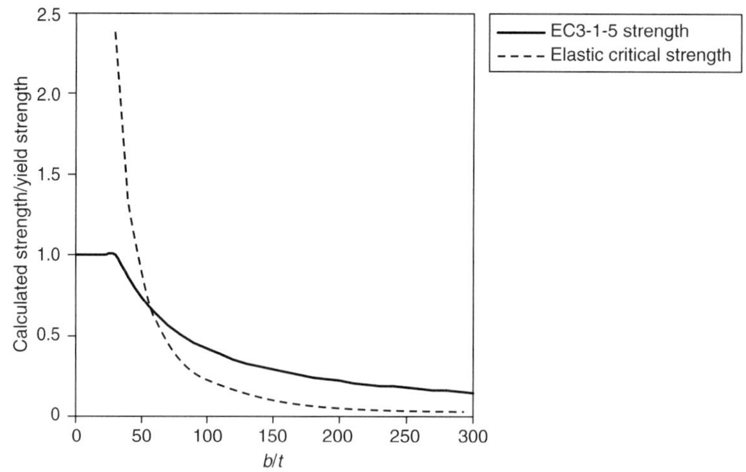

Fig. 6.2-10. Comparison of elastic critical and real strength of internal plates in S355 steel

It can be seen from Fig. 6.2-10 that the real plate strength is less than the elastic critical buckling load at low slenderness due to imperfections and occurrence of full plasticity. However, at high slenderness, the real plate strength exceeds the elastic critical value because of the post-buckling strength of the parts of the plate near to the supported edges.

The elastic critical buckling values presented in 3-1-5/clause 4 assume that the plate panels are much longer than they are wide. For internal plates, the lowest mode of buckling will have one transverse half wave of buckling and an integral number of half waves in the longitudinal direction. Minimum buckling load occurs where the length of the panel is an integer multiple of the width as shown in Fig. 6.2-11. For uniform compression, this results in $k_\sigma = 4$ for $a/b = 1, 2, 3$, etc. For length-to-width ratio, $1 < a/b < 3$, the buckling load is affected slightly by non-integer values of a/b and k_σ rises to approximately 4.5 at $a/b = 1.42$. For non-integer values of a/b greater than 3, any fluctuation in buckling load is minimal.

For panels that are shorter than they are wide, the buckling load begins to rise (although EN 1993-1-5 does not provide a formula for k_σ in this case) and the buckling mode becomes more and more like strut buckling of an isolated strip of plate without transverse edge restraint. For very low values of $a/b < 1$, the restraint from transverse bending of the plate is small and the idealized strut buckling mode is accurate and gives a critical stress $\sigma_{cr,c}$ approximately the same as would be obtained for plate buckling, $\sigma_{cr,p}$. As a/b increases towards 1.0, this approximation becomes more conservative as the restraint from transverse bending of the plate increases and $\sigma_{cr,p}$ is greater than the column critical stress $\sigma_{cr,c}$.

The reduction factor needed for column-type buckling is greater than for plate buckling at a given slenderness (because plates have some reserve of strength beyond the elastic critical buckling load whereas for struts the elastic critical load is an upper bound on strength), so

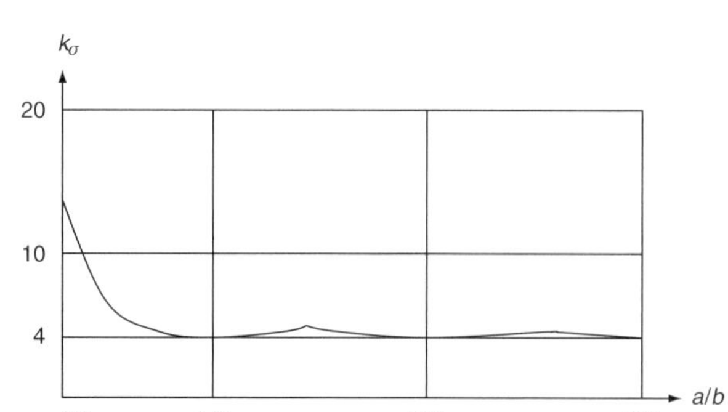

Fig. 6.2-11. Illustrative variation of k_σ with a/b for pure compression

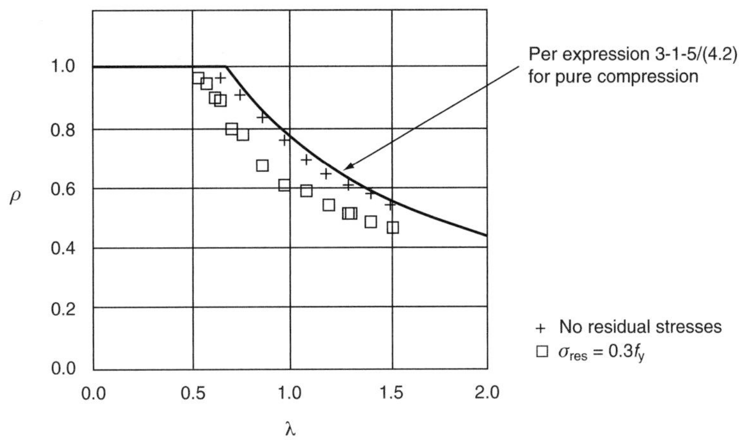

Fig. 6.2-12. Comparison of finite-element simulation with EN 1993-1-5 formula for internal plates in pure compression

the two situations have to be considered where $a/b < 1$. It is however always safe to ignore this column-type buckling behaviour for low a/b if ρ is derived using the slenderness for long panels. ***3-1-5/clause 4.4(6)*** allows column type buckling to be considered for plates by using ***3-1-5/clause 4.5.3(2)***, where the column buckling load for an unstiffened plate is given as: *3-1-5/clause 4.4(6)* *3-1-5/clause 4.5.3(2)*

$$\sigma_{\text{cr,c}} = \frac{\pi^2 E t^2}{12(1 - \nu^2)a^2} \qquad \text{3-1-5/(4.8)}$$

If benefit is taken from the increased buckling resistance associated with short panel length, it is important that the transverse stiffeners providing the reduced length must be checked for their ability to provide such support in accordance with 3-1-5/clause 9.2.1. This is discussed in section 6.6 of this guide.

The slenderness for column-type buckling is then given by ***3-1-5/clause 4.5.3(4)***: *3-1-5/clause 4.5.3(4)*

$$\bar{\lambda}_c = \sqrt{\frac{f_y}{\sigma_{\text{cr,c}}}} \qquad \text{3-1-5/(4.10)}$$

The column-type reduction factor, χ_c, is then determined from the flexural buckling curves of 3-1-1/clause 6.3.1.2 using the imperfection parameter $\alpha = 0.21$ in accordance with ***3-1-5/clause 4.5.3(5)***. Finally ***3-1-5/clause 4.5.4(1)*** requires interpolation to be performed between the reduction for plate behaviour, ρ, and the reduction for column behaviour, ρ_c, according to: *3-1-5/clause 4.5.3(5)* *3-1-5/clause 4.5.4(1)*

$$\rho_c = (\rho - \chi_c)\xi(2 - \xi) + \chi_c \qquad \text{3-1-5/(4.13)}$$

with $\xi = \sigma_{\text{cr,p}}/\sigma_{\text{cr,c}} - 1 (0 \le \xi \le 1)$ where $\sigma_{\text{cr,p}}$ is the elastic critical buckling stress for plate behaviour and $\sigma_{\text{cr,c}}$ is the elastic critical buckling stress for column buckling. ρ_c can conservatively be taken as χ_c by assuming $\sigma_{\text{cr,p}} = \sigma_{\text{cr,c}}$. Within the application rules presented in EN 1993-1-5, this conservative approximation of taking $\rho_c = \chi_c$ will be necessitated by the absence of a formula for short plates that considers plate behaviour as in Fig. 6.2-11. Solutions can however be found, such as those by Bulson[9] or from IDWR[10] (Fig. 6.2-17), which give values of k_σ for short plates. For pure compression only and $a/b < 1$, the following formula can be used to determine k_σ for plate-type buckling of short internal plates:

$$k_\sigma = \left(\frac{b}{a} + \frac{a}{b}\right)^2 \qquad \text{(D6.2-3)}$$

The overall reduction factor from expression 3-1-5/(4.13) should not be taken as less than that corresponding to a long plate. It is not therefore necessary to use expression 3-1-5/(4.13) for plate sub-panels unless benefit is to be taken from short panel length. This is generally not worth the effort (as illustrated in Worked Example 6.2-2) other than in verifying very highly stressed areas where the intention is to place transverse stiffeners very closely to prevent buckling.

For internal plates under pure compression (which will be typical for flanges), the limiting value of $\bar{b}/t$ for a fully effective plate in S355 steel is 31. This is higher than the ratio of 24 which was obtained using BS 5400: Part 3.[4] A non-linear finite-element study by B. Johansson and M. Veljkovic[11] showed that the EN 1993-1-5 plate reduction factor for plates in pure compression gave satisfactory predictions for plates without significant residual stresses, but could overestimate strength where there were large welds without stress relief. The results are indicated in Fig. 6.2-12. The results were deemed to support the use of the EN 1993-1-5 reduction factors for two reasons:

1. The slight over-prediction of strength in the case of low residual stress can be justified by the fact that the results from the non-linear analyses were themselves conservative compared to the results of physical tests on equivalent specimens.
2. Welds to plates in stiffened structures are usually small fillet welds which do not induce large residual stresses and butt welds between plates usually occur at wide intervals. The lower set of data with higher residual stresses was therefore ignored. Some caution would therefore be advised when using the EN 1993-1-5 plate rules with unusually large welds in close proximity; reduced effective widths might then be appropriate.

The ultimate resistance of plates under axial stress (but not the elastic critical stress) is influenced by whether or not the longitudinal plate edges can ripple in-plane. Panels bounded by longitudinal stiffeners with other plate panels surrounding them are automatically 'restrained' from such in-plane displacement due to the constraint of the surrounding panels. Web panels adjacent to flanges are only 'restrained' if the flange possesses adequate flexural stiffness and strength (about its weak axis) to prevent the in-plane displacement. EN 1993 does not distinguish between 'restrained' and 'unrestrained' conditions. The EN 1993-1-5 effective widths for internal panels were based on tests on square boxes where the panels were essentially 'unrestrained'.

It is also interesting to note that the limiting value of $\bar{b}/t$ for S355 steel at the Class 3–Class 4 boundary according to 3-1-1/Table 5.2 is $42\varepsilon = 34 > 31$ here, so there is a discontinuity in the design rules. No such discontinuity occurs for outstand elements and both methods give 14ε for pure compression.

Where the maximum stress in the plate, derived from analysis of the effective cross-section, is less than yield, a reduced value of slenderness (and hence greater effective width) may be derived by iteration using ***3-1-5/clause 4.4(4)***:

3-1-5/clause 4.4(4)

$$\bar{\lambda}_{p,red} = \bar{\lambda}_p \sqrt{\frac{\sigma_{com,Ed}}{f_y/\gamma_{M0}}} \qquad \text{3-1-5/(4.4)}$$

The stress is first calculated on an effective width based on $\bar{\lambda}_p$. $\bar{\lambda}_{p,red}$ is then calculated and a revised effective width is obtained. This iterative procedure continues until convergence occurs in determining $\sigma_{com,Ed}$. This method is of benefit in reducing the usage under interactions of direct stress with shear and transverse load. It may not however be used when checking overall member buckling to 3-2/clause 6.3 since the limiting loads for buckling by definition induce yield in the outer fibres of the cross-section. It would still be permissible to use $\bar{\lambda}_{p,red}$ however, if second-order analysis with imperfections were performed to allow for member buckling effects, but the iteration required would be even more prohibitive without purpose-developed software.

Biaxial stress in plates is not covered (other than by the rules for transverse loading in 3-1-5/clause 6 and the interactions in 3-1-5/clause 7). Where uniform transverse stress occurs (such as in the region of a transverse diaphragm at a support), it would either have to be included in the calculation of the reduction factor for longitudinal direct stress (but no method is given for this) or the method of individual panel checks given in 3-1-5/clause 10 would have to be used as discussed in section 6.2.2.6 of this guide.

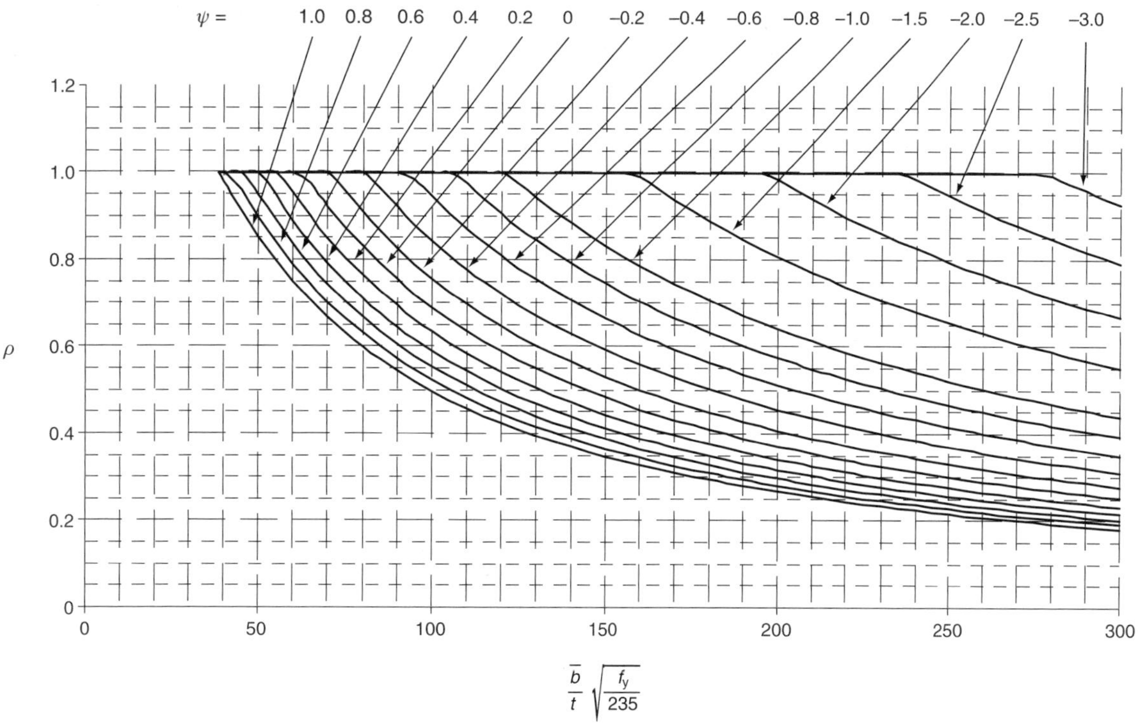

Fig. 6.2-13. Reduction factor for internal compression elements

For the simple case of long internal plates, graphs of reduction factor against $\overline{b}/t\sqrt{f_y/235}$ for different stress ratios, ψ, are given in Fig. 6.2-13.

Worked Example 6.2-2: Buckling of plate sub-panel

A plate of 10 mm thickness in S355 steel has a sub-panel which is 600 mm wide and which is under uniform compression. A transverse stiffener is added to reduce the panel length to 300 mm. The effective[p] width of the panel is calculated both with and without the transverse stiffener.

For a long panel, the reduction factor is as follows.

From expression 3-1-5/(4.2):

$$\overline{\lambda}_p = \frac{\overline{b}/t}{28.4\varepsilon\sqrt{k_\sigma}} = \frac{600/10}{28.4 \times 0.81 \times \sqrt{4}} = 1.304$$

$$\rho = \frac{\overline{\lambda}_p - 0.055(3+\psi)}{\overline{\lambda}_p^2} = \frac{1.304 - 0.055(3+1)}{1.304^2} = \mathbf{0.64}$$

When a transverse stiffener is added to restrict the panel length to 300 mm, column buckling should also be checked.

From expression 3-1-5/(4.8):

$$\sigma_{cr,c} = \frac{\pi^2 Et^2}{12(1-\nu^2)a^2} = \frac{\pi^2 \times 210 \times 10^3 \times 10^2}{12(1-0.3^2) \times 300^2} = 210.9\,\text{MPa}$$

$$\overline{\lambda}_c = \sqrt{\frac{f_y}{\sigma_{cr,c}}} = \sqrt{\frac{355}{210.9}} = 1.297$$

From curve a ($\alpha = 0.21$) of 3-1-1/Fig. 6.4, the reduction factor χ_c for column-type buckling = **0.47**.

As no rules are given on calculating the critical plate buckling stress for panels with $a/b < 1$, the calculation would normally stop here and the reduction would be limited to 0.64 for long panels. However, by using a formula for plate buckling of panels with $a/b < 1$, some improvement could be demonstrated as follows.

For $a/b = 300/600 = 0.5$, plate buckling behaviour gives $k_\sigma = (b/a + a/b)^2 = 6.25$ from equation (D6.2-3).

$$\sigma_{cr,p} = \left(\frac{k_\sigma \pi^2 E t^2}{12(1-\upsilon^2)\bar{b}^2}\right) = \left(\frac{6.25\pi^2 \times 210 \times 10^3 \times 10^2}{12(1-0.3^2)600^2}\right) = 330\,\text{MPa}$$

$$\bar{\lambda}_p = \frac{\bar{b}/t}{28.4\varepsilon\sqrt{k_\sigma}} = \frac{600/10}{28.4 \times 0.81 \times \sqrt{6.25}} = 1.043$$

$$\rho = \frac{\bar{\lambda}_p - 0.055(3+\psi)}{\bar{\lambda}_p^2} = \frac{1.043 - 0.055(3+1)}{1.043^2} = 0.757$$

From expression 3-1-5/(4.13), the final reduction factor interpolated between plate and column-type behaviour is:

$$\rho_c = (\rho - \chi_c)\xi(2-\xi) + \chi_c = (0.757 - 0.47) \times 0.564 \times (2 - 0.564) + 0.47 = \mathbf{0.70}$$

with $\xi = \sigma_{cr,p}/\sigma_{cr,c} - 1 = 330/211 - 1 = 0.564$.

Consequently, the use of equation (D6.2-3) for the calculation of critical buckling load for plate behaviour for short panels demonstrates a small improvement in effective width. It also illustrates that transverse stiffening would have to be very closely spaced to gain any significant benefit.

6.2.2.5.2.2. Stiffened plates

3-1-5/clause 4.5.1(1)
3-1-5/clause 4.5.1(2)

For stiffened plates, an overall reduction to the compression zone area is made to cater for both sub-panel buckling and overall buckling of the stiffened plate in accordance with ***3-1-5/clause 4.5.1(1)***. This effective area is obtained by reducing the gross area in two steps which is alluded to in ***3-1-5/clause 4.5.1(2)*** – further comment on this clause is made under the discussion on 3-1-5/clause 4.5.2(1). First, an effective area is derived for the sub-panels and any slender closed stiffeners to account for local buckling according to the rules for unstiffened plates as discussed above. Open stiffeners must also satisfy the limitations to prevent torsional buckling as discussed in section 6.9 of this guide. For a flat stiffener, the allowable b/t ratio is similar when calculated as a plate outstand to expression 3-1-5/(4.3) ($b/t = 11.3$ for S355 steel) or according to the torsional buckling rules ($b/t = 10.5$ for S355 steel) for the reasons discussed in section 6.9. It is strongly recommended that outstand parts of all open stiffeners be detailed to meet the outstand limits for full effectiveness as a plate from 3-1-5/Table 4.2 since there is little post-buckling strength in an outstand and buckling of an open stiffener can lead to sudden collapse. Second, a reduction factor for global buckling of the whole stiffened panel is determined and the effective cross-sectional area of the compression zone of the stiffened panel is then determined from ***3-1-5/clause 4.5.1(3)*** and ***3-1-5/clause 4.5.1(4)***:

3-1-5/clause 4.5.1(3)
3-1-5/clause 4.5.1(4)

$$A_{c,eff} = \rho_c A_{c,eff,loc} + \sum b_{edge,eff} t \qquad \text{3-1-5/(4.5)}$$

with

$$A_{c,eff,loc} = A_{sl,eff} + \sum_c \rho_{loc} b_{c,loc} t \qquad \text{3-1-5/(4.6)}$$

where:

$A_{sl,eff}$ is the sum of the effective cross-sectional area of all the longitudinal stiffeners (excluding attached web or flange plate) in the compression zone, reduced for plate buckling if relevant (as may occur for closed stiffeners);

$\sum_{c} \rho_{loc} b_{c,loc} t$ is the effective cross-sectional area of all the sub-panels in the compression zone, reduced for local plate buckling as discussed above, except for the effective parts of sub-panels which are supported by a web or a flange plate ($\sum b_{edge,eff} t$) as illustrated in Fig. 6.2-14 and Fig. 6.2-15. These are more general versions of Fig. 4.4 given in EN 1993-1-5. The edge pieces are excluded from expression 3-1-5/(4.6) as they are not influenced significantly by overall plate buckling. In Fig. 6.2-15, where there is stress reversal, the gross area could arguably be taken to stop at $0.4b_c$ from the last stiffener in the compression zone and the effective area similarly stopped $0.4b_{eff3}$ from this stiffener. However, EN 1993-1-5 clearly specifies the area shown;

ρ_c is the reduction factor for global buckling of the stiffened panel, ignoring local buckling of sub-panels.

It is important when doing this calculation for plates where the stress reverses and becomes tensile to not forget to include the tensile area in the section properties. *3-1-5/clause 4.5.1(8)*

3-1-5/clause 4.5.1(8) and ***3-1-5/clause 4.5.1(9)*** are a reminder that a further reduction to the effective area may be needed to allow for shear lag in accordance with 3-1-5/clause 3.3. *3-1-5/clause 4.5.1(9)* This further reduction, where needed, is best done after the effective area for plate buckling, $A_{c,eff}$, has been obtained.

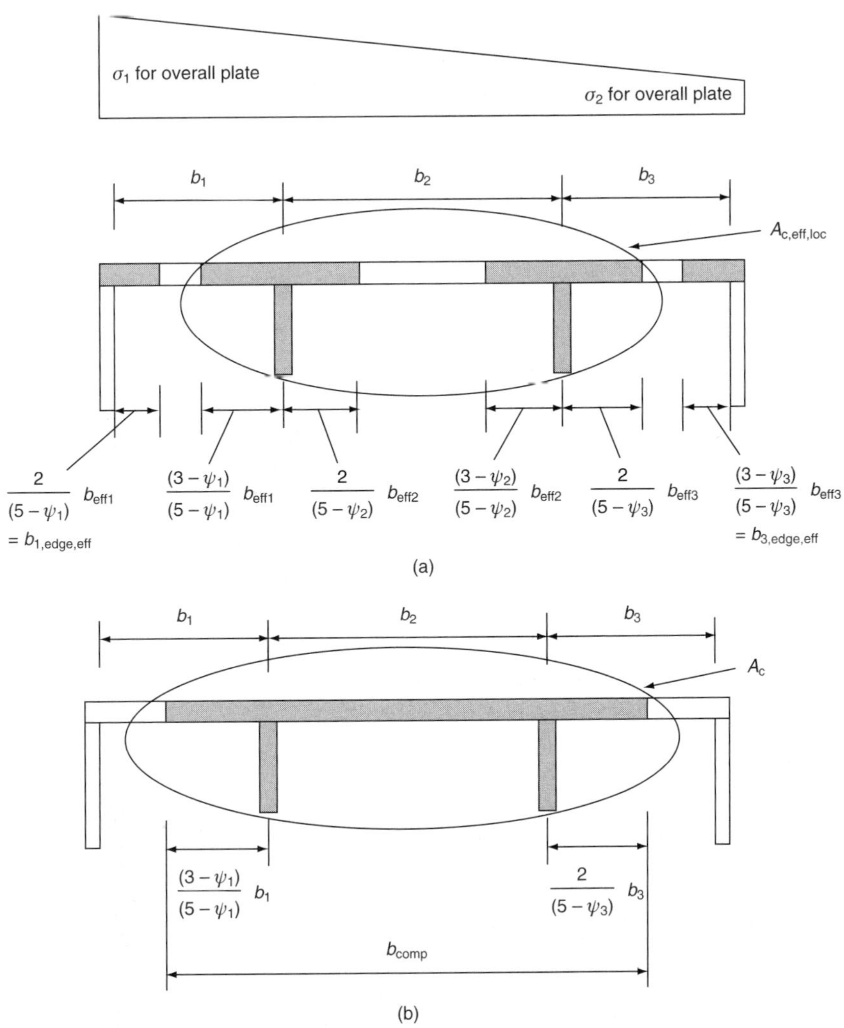

Fig. 6.2-14. Definition of (a) effective area, $A_{c,eff,loc}$ and (b) gross area, A_c, for stiffened plate under variable compression (no tension)

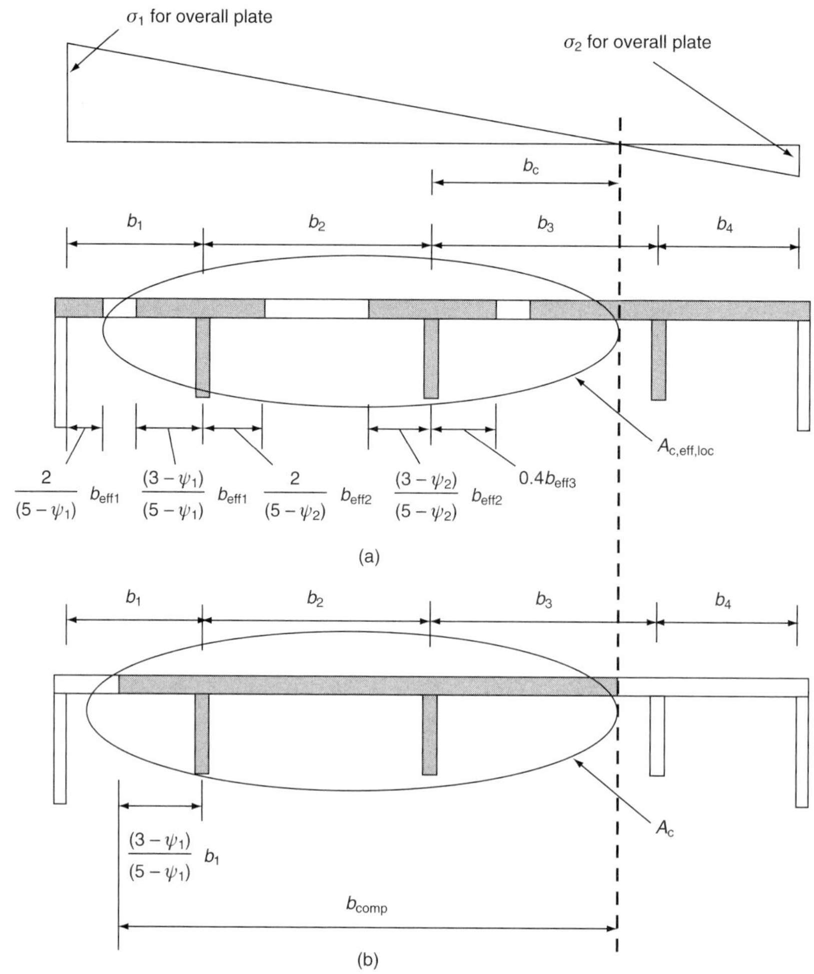

Fig. 6.2-15. Definition of (a) effective area, $A_{c,eff,loc}$ and (b) gross area, A_c, for stiffened plate under variable compression with stress reversal

The reduction factor for global buckling is determined from an empirical interpolation between the reduction factors for column-like buckling and for overall stiffened plate buckling in the same way as for unstiffened plates. This is because the reduction factor needed for a given slenderness is greater for column-like buckling. The formula in ***3-1-5/ clause 4.5.4(1)*** is used:

3-1-5/clause 4.5.4(1)

$$\rho_c = (\rho - \chi_c)\xi(2 - \xi) + \chi_c \qquad \text{3-1-5/(4.13)}$$

where:

- ρ is the reduction factor for overall stiffened plate buckling determined from expression 3-1-5/(4.2) or expression 3-1-5/(4.3) for slenderness $\bar{\lambda}_p$ according to 3-1-5/clause 4.5.2(1), as discussed below. The method of calculation of $\sigma_{cr,p}$ required to determine $\bar{\lambda}_p$ depends on the number of longitudinal stiffeners as discussed below.
- χ_c is the reduction factor for column buckling (by considering the stiffened plate as a strut with the support along its longitudinal edges removed) according to 3-1-5/clause 4.5.3 as discussed later.

$\xi = \sigma_{cr,p}/\sigma_{cr,c} - 1$, where $\sigma_{cr,p}$ is the elastic critical buckling stress for stiffened plate behaviour and $\sigma_{cr,c}$ is the elastic critical buckling stress for column buckling. Since $\sigma_{cr,p}$ should not be smaller than $\sigma_{cr,c}$, a lower limit of zero is placed on ξ. A further upper limit of 1.0 is also given to ensure that the reduction factor becomes that for stiffened plate behaviour when $\sigma_{cr,p}/\sigma_{cr,c} > 2$.

It will often not be worth the effort of calculating $\sigma_{cr,p}$ because it will typically be only slightly greater than $\sigma_{cr,c}$, unless the panel is significantly longer than it is wide. To be conservative therefore, $\sigma_{cr,p}$ may generally be taken equal to $\sigma_{cr,c}$, unless transverse restraints are very widely spaced, in which case the result will be excessively conservative. This is effectively what was assumed in BS 5400: Part 3.[4] It is advisable to determine the reduction for column buckling first since if the reduction factor $\chi_c = 1.0$ there will be no point considering plate action in any case. Additionally, for outstand stiffened plates, stiffened plate action will be very small as there is only support to the plate along one longitudinal edge and it will only therefore be necessary to calculate the column buckling load when deriving the reduction for overall buckling.

(a) Stiffened plate critical buckling stress

3-1-5/Annex A covers the calculation of $\sigma_{cr,p}$. A different method has to be used to determine $\sigma_{cr,p}$ depending on whether the stiffened plate has: *3-1-5/Annex A*

(i) many equal stiffeners in the compression zone, or
(ii) one or two stiffeners in the compression zone.

In the first method, stiffeners are smeared into an equivalent orthotropic (which is an abbreviation for orthogonally anisotropic) plate. This is adequate where there are three or more stiffeners, but smearing of the stiffeners becomes inaccurate for fewer stiffeners and account has to be taken of their actual location. The second method caters for unevenly spaced stiffeners of different sizes. Both methods assume that transverse stiffeners are 'rigid', which is automatically achieved if they are designed to 3-1-5/clause 9. No rules are given for design with flexible transverse stiffeners in EN 1993-1-5; their use is not prohibited but neither is it encouraged. Reference would have to be made to standard texts for critical buckling stresses where flexible stiffeners are to be used.

(i) Many stiffeners – equivalent orthotropic plate

In this method, the stiffened plate is treated as an orthotropic plate with stiffeners smeared. ***3-1-5/Annex A.1*** gives the formula for such a plate as: *3-1-5/Annex A.1*

$$\sigma_{cr,p} = k_{\sigma,p}\frac{\pi^2 E t^2}{12(1-\upsilon^2)b^2} \qquad \text{3-1-5/(A.1)}$$

where $k_{\sigma,p}$ is a coefficient from orthotropic plate theory that has to be determined ignoring sub-panel buckling such that $\sigma_{cr,p}$ is the critical stress at the edge of the panel with the greatest compressive stress. The calculation of $\sigma_{cr,p}$ should be based on the gross inertia of the stiffened plate as represented by the area A_c in Figs 6.2-14 and 6.2-15. (See discussion under 3-1-5/clause 4.5.2(1) below.) It is however always conservative to determine the effective section by assuming the panel to be in uniform compression. The critical stress can be determined from standard texts or by computer modelling, but either method must be able to ignore sub-panel buckling. The latter is a problem with using finite-element models.

3-1-5/Annex A.1 contains formulae for $k_{\sigma,p}$:

$$k_{\sigma,p} = \frac{2((1+\alpha^2)^2 + \gamma - 1)}{\alpha^2(\psi+1)(1+\delta)} \quad \text{if } \alpha \le \sqrt[4]{\gamma} \qquad \text{3-1-5/(A.2)}$$

$$k_{\sigma,p} = \frac{4(1+\sqrt{\gamma})}{(\psi+1)(1+\delta)} \quad \text{if } \alpha > \sqrt[4]{\gamma}$$

The values of $k_{\sigma,p}$ take into account smearing of stiffeners so can be used directly with expression 3-1-5/(A.1), taking t as the parent plate thickness. They are applicable only for evenly spaced (or approximately evenly spaced) identical stiffeners. The formulae are limited to an aspect ratio of $\alpha = a/b \geq 0.5$ but this is not too restrictive as benefit from orthotropic action will usually be negligible for $a/b < 0.5$ and $\sigma_{cr,p}$ will tend to the column critical buckling stress. The formulae are also limited to a stress ratio $\psi = \sigma_2/\sigma_1 \geq 0.5$.

The torsional inertia of the stiffeners is neglected in expression 3-1-5/(A.2), which has negligible effect for panels with open stiffeners but can have a more significant effect for panels with closed stiffeners, such as trough stiffeners on deck plates. The use of this method is illustrated in Worked Example 6.2-4. The definitions of A_p (area of *whole* parent plate $= bt$), $\sum I_{sl}$ (second moment of area of *whole* stiffened plate) and I_p (second moment of area of *whole* parent plate $= bt^3/10.92$) would, for consistency with the slenderness calculation in expression 3-1-5/(4.7), have been better to refer only to the part of the stiffened plate shown in Fig. 6.2-14 (i.e. the gross area but excluding the parts supported by the webs). This amendment is implemented in Worked Example 6.2-4 but the effect is small. Other definitions are: $\sum A_{sl}$ is the gross area of all the stiffener outstands, $\gamma = \sum I_{sl}/I_p$ and $\delta = \sum A_{sl}/A_p$.

A method based on the one in IDWR[10] could also be used for cases without intermediate flexible transverse stiffeners. This also deals with non-uniform stiffener spacings and sizings, panels with stress reversal, and considers the torsional inertia of stiffeners. The critical buckling stress at the edge of the panel with the greatest compressive stress is calculated, using the same notation for panels as EN 1993-1-5 (see Fig. 6.2-9), from:

$$\sigma_{cr,p} = \frac{\pi^2\sqrt{D_x D_y}}{b^2 t_{eff}}\left[k_0 + \frac{(k_i - k_0)H}{\sqrt{D_x D_y}}\right] \qquad \text{(D6.2-4)}$$

$$H = \frac{Gt^3}{6} + \frac{GI_T}{2\bar{b}}$$

where I_T is the St Venant torsional inertia of the stiffener outstand for open stiffeners or is the St Venant torsional inertia of the closed box formed by a stiffener and parent plate for a closed stiffener. $\bar{b}$ is the stiffener spacing.

$$D_x = \frac{E\sum I_{sc}}{b_{comp}}$$

$$D_y = \frac{Et^3}{12(1 - \nu\nu_y)}$$

where:

$\sum I_{sc}$ is the sum of the second moments of area of the stiffener effective sections in the compression zone, comprising stiffener and gross plating attached to that stiffener, where stiffeners are uniformly spaced and of equal size. This is the same as the inertia of the entire compression zone of the stiffened plate excluding the parts of sub-panels supported by webs or flanges as in Figs 6.2-14 and 6.2-15. Unequal stiffener spacings are discussed below.

b_{comp} is the width of the compression zone of the stiffened plate excluding the parts of sub-panels supported by webs or flanges as shown in Figs 6.2-14 and 6.2-15.

$$\nu_y = 0.3\left(\frac{\bar{b}t}{A_s + \bar{b}t}\right)$$

$$t_{eff} = t\left(1 + \frac{\sum A_s}{b_{comp}t}\right) \text{ i.e. the effective thickness in the width } b_{comp}$$

A_s is the gross area of an individual stiffener, excluding attached parent plate.

$\sum A_s$ is the sum of the gross areas of the stiffener attachments themselves, excluding the parent plate, within the compression zone of width b_{comp}.

k_i is the buckling coefficient for an unstiffened plate with aspect ratio $\varphi' = (a/b)(D_y/D_x)^{0.25}$, determined for stress ratio $\psi = \sigma_2/\sigma_1$, from Fig. 6.2-16.

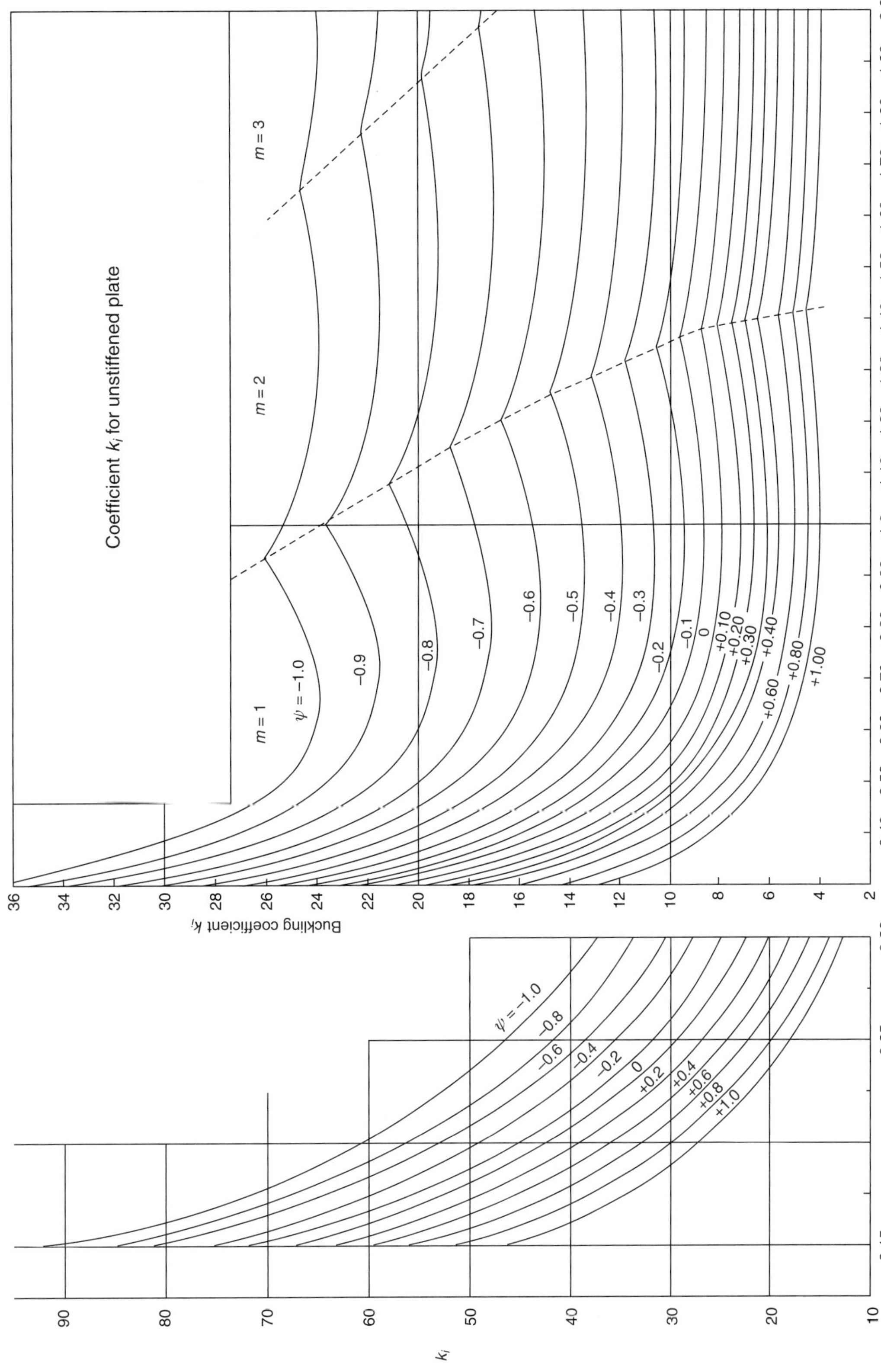

Fig. 6.2-16. Values of coefficient k_i

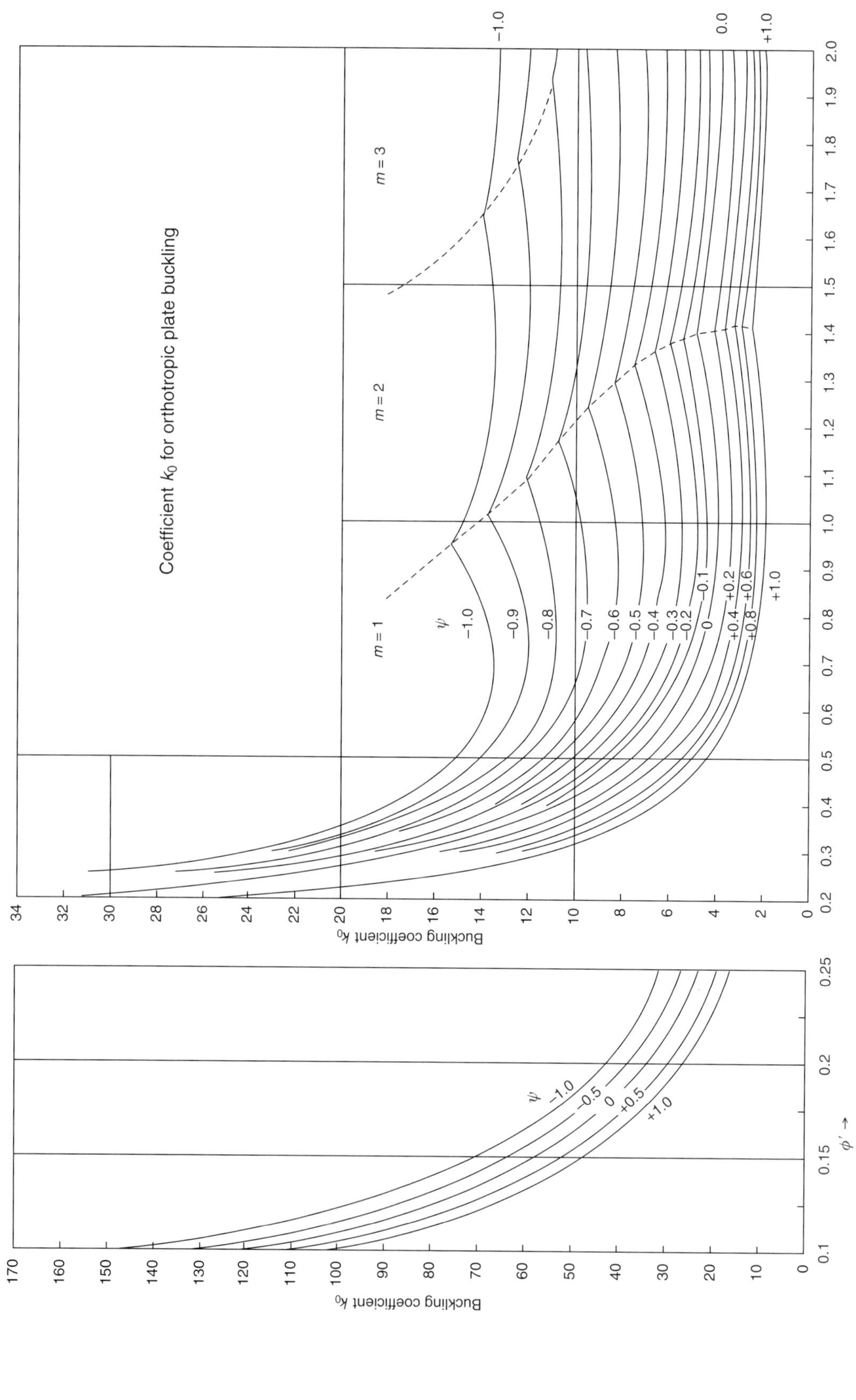

Fig. 6.2-17. Values of coefficient k_0

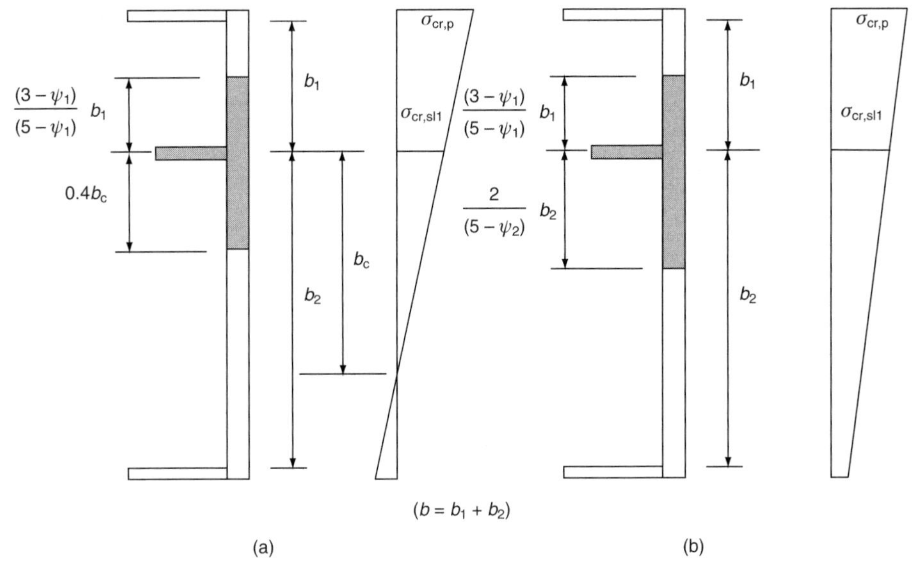

Fig. 6.2-18. Stiffener area, $A_{sl,1}$ for fictitious column method: (a) reversal of stress in panel; (b) no reversal of stress in panel

k_0 is the buckling coefficient for an orthotropic plate under in-plane bending and compression and having zero torsional stiffness with aspect ratio $\varphi' = (a/b)(D_y/D_x)^{0.25}$, determined for stress ratio $\psi = \sigma_2/\sigma_1$, from Fig. 6.2-17.

Where the stiffeners vary in spacing or size, an equivalent stiffness, $I_{s,eff}$ may be derived for each stiffener and attached parent plate in the compression zone. D_x may then be determined as $E\sum I_{s,eff}/b_{comp}$.

For uniform compression:

$$I_{s,eff} = 1.5I_{sc}\left(1 - \frac{4y_s^2}{b^2}\right)\left(\frac{N+1}{N+2}\right)$$

where N is the number of longitudinal stiffeners and y_s is the distance to the stiffener from the centre of the stiffened panel. I_{sc} is the second moment of area of the stiffener effective section considered, comprising stiffener and gross plating attached to that stiffener as shown in Fig. 6.2-18.

For bending or bending and compression where the stress on each edge of the plate is of opposite sign, for each stiffener, $I_{s,eff} = \beta I_{sc}$. β is an influence coefficient for stiffener location which varies from zero at the neutral axis and at the extreme panel compression fibre to 2.0 at a distance 80% of the way from the neutral axis to the extreme panel compression fibre. This is illustrated in Fig. 6.2-19. For bending and compression where the stresses on each edge of the plate are of the same sign, a similar weighting could be derived or the distribution for compression only could conservatively be used, providing the stiffeners

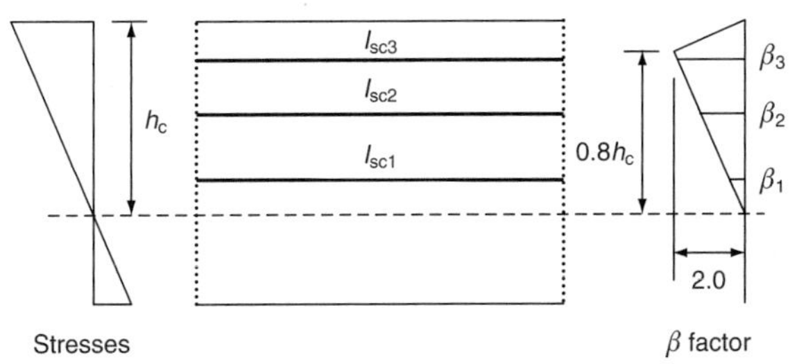

Fig. 6.2-19. Influence coefficient for stiffener inertia for bending or combined bending and axial with stress reversal

are not more widely spaced and/or smaller in the most heavily compressed part of the panel. Alternatively, and most simply, in all cases of variable stiffener size and spacing, the stiffness D_x may be based on the most flexible part of the plate.

Regardless of method used to determine the critical stress, the slenderness is then determined from ***3-1-5/clause 4.5.2(1)*** as follows:

3-1-5/clause 4.5.2(1)

$$\bar{\lambda}_p = \sqrt{\frac{\beta_{A,c} f_y}{\sigma_{cr,p}}} \qquad \text{3-1-5/(4.7)}$$

where $\beta_{A,c} = A_{c,eff,loc}/A_c$ and A_c is the gross area of the compression zone of the stiffened plate excluding the parts of sub-panels supported by a web or flange as shown in Figs 6.2-14 and 6.2-15. By way of the factor $\beta_{A,c} = A_{c,eff,loc}/A_c$, the slenderness is effectively the square root of the squash load of the stiffened plate, with allowance for sub-panel buckling, divided by the elastic critical buckling load of the overall gross stiffened plate. The latter use of gross area is intended to account for the fact that the stiffness of a locally buckled cross-section is larger than that of the effective area used for the resistance, i.e. the loss in stiffness is less than the loss of resistance. This gives a slightly lower slenderness than if effective areas (allowing for sub-panel buckling) are used for calculation of the overall buckling load, but there is not much difference. This latter fact was used to further justify the use of gross areas in the face of criticism from some quarters during drafting. The most important thing is that the area used in $\beta_{A,c}$ should be consistent with that used in the derivation of the critical buckling stress or the critical force for the plate is liable to be incorrect. This also applies to the modification of A_c and $A_{c,eff,loc}$ for shear lag required by 3-1-5/clause 4.5.2(1); the reduction for shear lag should essentially be the same for both areas so generally need not be considered in the calculation of these areas for slenderness calculation.

While 3-1-5/clause 4.5.2(1) and 3-1-5/Annex A.1 both specify gross areas to be used for calculation of critical stresses, the wording of 3-1-5/clause 4.5.1(2) adds some confusion. It states that 'The stiffened plate with effective[p] section areas for the stiffeners should be checked for global buckling…'. This was intended only to mean that a further reduction in area should be made to both the effective plate sub-panels and stiffeners for global buckling, not that effective areas should be used in determining the critical stresses.

A further comment on use of gross areas in critical stress calculation is that if the stiffeners were unusually widely spaced with short span, local shear lag could limit the effectiveness of the plating acting with the stiffener and thus use of the full gross plate width could overestimate the flexural stiffness and calculation of critical force. A reduced second moment of area for use in buckling critical stress calculation could be obtained by applying a transverse load to the stiffened panel and back-calculating a value from the deflection obtained. However, the reduction of stiffness from this effect is small for practical geometries and the use of gross properties can usually be justified. No requirement to consider this effect is given in the code.

The reduction factor for stiffened plate behaviour is found from the formulae for unstiffened plates in expressions 3-1-5/(4.2) and (4.3) using the slenderness in expression 3-1-5/(4.7).

(ii) One or two stiffeners in the compression zone

3-1-5/Annex A.2

The method given in ***3-1-5/Annex A.2*** is based on a model where the stiffener is treated as a fictitious column which is assumed to be restrained by elastic springs consisting of strips of the plate acting as beams at right angles to the stiffener. This method conservatively neglects torsion in the plate.

Two distinct situations of one and two stiffeners in the compression zone are covered in EN 1993-1-5. In either case, additional stiffeners in the tension zone are ignored. For webs with more than two stiffeners in the compression zone, either the web can be idealized as an orthotropic plate if the geometry is appropriate or the plate buckling load can conservatively be taken as that for the column buckling load.

One stiffener

For a single stiffener in the compression zone, the critical plate buckling stress can be calculated according to ***3-1-5/clause A.2.2***:

3-1-5/clause A.2.2

$$\sigma_{cr,sl} = \frac{1.05E}{A_{sl,1}} \frac{\sqrt{I_{sl,1}t^3 b}}{b_1 b_2} \quad \text{if } a \geq a_c \qquad \text{3-1-5/(A.4)}$$

$$\sigma_{cr,sl} = \frac{\pi^2 E I_{sl,1}}{A_{sl,1}a^2} + \frac{Et^3 ba^2}{4\pi^2(1-\nu^2)A_{sl,1}b_1^2 b_2^2} \quad \text{if } a \leq a_c$$

where:

$$a_c = 4.33\sqrt[4]{\frac{I_{sl,1}b_1^2 b_2^2}{t^3 b}}$$

which is the wavelength of buckling, assuming the rigid transverse stiffeners to be removed (and no flexible transverse stiffeners are present between rigid transverse stiffeners). b_1, b_2 are the distances from the stiffener to each plate edge such that their sum, b, equals the width (or height) of the whole stiffened plate. Longitudinal stiffeners in the tension zone are completely ignored in these calculations for global buckling.

$A_{sl,1}$ is the gross area of the stiffener and attached plating ignoring local plate buckling according to Fig. 6.2-18. (A similar figure is given in 3-1-5/Annex A1.) $I_{sl,1}$ is the second moment of area of this same area. The intention is that the area of attached plating is attributed to the stiffener in the same ratio as the effective width allowing for plate buckling from 3-1-5/Table 4.4. Consequently, the attached width of $[(3-\psi)/(5-\psi)]b_1$ for the higher stressed side of the stiffener is derived similarly to the value b_{e2} in 3-1-5/Table 4.4 i.e. $[1-2/(5-\psi)]b_1$. The attached width of $0.4b_c$ for the lower stressed side for panels where the stress reverses is also proportional to the value b_{e1} in 3-1-5/Table 4.4.

If there is a stress gradient, as in a web, the peak compressive stress at the plate boundary, $\sigma_{cr,p}$, exceeds that calculated at the stiffener effective section ($\sigma_{cr,sl}$) as shown in Fig. 6.2-18 and $\sigma_{cr,p}$ can be derived from Fig. 6.2-18 in the same way as discussed for column buckling below to avoid conservatism.

Two stiffeners

For two stiffeners in the compression zone, the procedure for a single stiffener is repeated three times, again completely ignoring any longitudinal stiffeners in the tension zone. First it is assumed that each stiffener buckles on its own with the other treated as rigid providing a rigid plate boundary. In this case, the value of b is taken equal to the sum of the resulting panel widths each side of the stiffener being considered. Then both stiffeners are treated as one combined stiffener with section properties equal to the sum of the two properties calculated for the individual stiffeners and with location based on the centre of force of the two separate stiffeners. The procedure is illustrated in 3-1-5/Fig. A.3 but is not reproduced here.

If there is a stress gradient, $\sigma_{cr,p}$ can be derived from $\sigma_{cr,sl}$ as discussed for the single stiffener case above.

In the case of either one or two stiffeners, the plate-type slenderness is again calculated from expression 3-1-5/(4.7).

(b) Stiffened plate column buckling load

The column buckling load can always be used on its own to determine a conservative value of the reduction factor, ρ_c. This avoids the need to determine the critical plate buckling load for a stiffened plate, which in many cases will produce very limited benefit anyway.

The elastic critical column buckling stress of the stiffener effective section with the highest compressive stress, and considering the supports along the longitudinal edges of the plate to

3-1-5/clause 4.5.3

be removed, is determined first from ***3-1-5/clause 4.5.3***:

$$\sigma_{cr,sl} = \frac{\pi^2 EI_{sl,1}}{A_{sl,1}a^2} \qquad \text{3-1-5/(4.9)}$$

where $A_{sl,1}$ is the gross area of the stiffener and attached plating ignoring local plate buckling according to Fig. 6.2-18, as for the stiffened plate buckling case of one stiffener above. $I_{sl,1}$ is the second moment of area of this same area.

If there is a stress gradient across the plate (as in a web), the peak compressive stress, $\sigma_{cr,c}$, does not occur at the location of the stiffener effective section. To overcome the conservatism here, the critical stress above is extrapolated to the peak value at the plate edge (in the same way as in Fig. 6.2-18) as follows:

$$\sigma_{cr,c} = \sigma_{cr,sl}\frac{b_c}{b_{sl,1}}$$

where b_c is the distance from the position of zero direct stress to the most compressive panel fibre and $b_{sl,1}$ is the distance from the position of zero direct stress to the stiffener. Note that this is not the same definition of b_c as in Fig. 6.2-18 as it is used in EN 1993-1-5 to refer to the stress distribution in both sub-panels and overall panels – the designer needs to think carefully which definition is relevant in each case until such a time as the document is improved editorially. For panels where the stress varies but is compressive throughout, b_c will be greater than the panel depth b. It is also unfortunate that $\sigma_{cr,sl}$ is used in expression 3-1-5/(4.9) to represent the column buckling load ignoring restraint from the parent plate transversely, while in expression 3-1-5/(A.4) the same symbol is used for the buckling load including restraint from the parent plate transversely.

3-1-5/clause 4.5.3(4)

The relative slenderness is calculated according to ***3-1-5/clause 4.5.3(4)*** as follows:

$$\bar{\lambda}_c = \sqrt{\frac{\beta_{A,c}f_y}{\sigma_{cr,c}}} \qquad \text{3-1-5/(4.11)}$$

where $\beta_{A,c} = A_{sl,1,eff}/A_{sl,1}$. It is again essential that the area $A_{sl,1}$ used here matches that assumed in the calculation of the column buckling stress as discussed for stiffened plate buckling above. Note that the definition of $A_{sl,1,eff}$ here is the effective area for one stiffener and effective plate rather than the entire panel. As a general point, it should always be checked that the two areas in $\beta_{A,c}$ correspond, i.e. they both refer to the entire compression zone or to just one stiffener.

3-1-5/clause 4.5.3(5)

The reduction factor χ_c is calculated for the above slenderness with the column buckling formulae in EN 1993-1-1, but the imperfection factor is increased in accordance with ***3-1-5/clause 4.5.3(5)*** in order to account for an assumed initial out-of-straightness of length/500 in contrast to length/1000 in EN 1993-1-1. This allows for the greater tolerance allowed for stiffeners in EN 1090. The imperfection factor is calculated as follows:

$$\alpha_e = \alpha + \frac{0.09}{i/e} \qquad \text{3-1-5/(4.12)}$$

where i is the radius of gyration of the stiffener and attached plating and e is the greatest of the distances from the centroid of the stiffened panel to the centre of the plate (e_2) or to the centroid of the longitudinal stiffener (e_1). EN 1993-1-5 Fig. A.1 illustrates this. $\alpha = 0.34$ for closed stiffeners and $\alpha = 0.49$ for open stiffeners.

Worked Example 6.2-3: Calculation of effective section for longitudinally stiffened footbridge

A steel footbridge fabricated from S355 steel has the cross-section shown in Fig. 6.2-20. The effective section properties for sagging bending calculation are calculated. Flange cross-girders and web transverse stiffeners are provided at 2000 mm centres. (Note that the longitudinal web stiffeners would not normally be economic for a web with this geometry. They have been added here to illustrate the design process.)

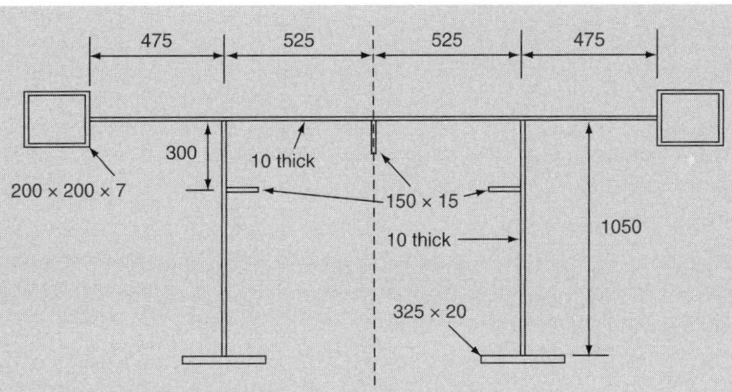

Fig. 6.2-20. Steel footbridge for Worked Example 6.2-3

Top flange between webs

For panels between main girders in uniform compression:

$$\bar{\lambda}_p = \frac{\bar{b}/t}{28.4\varepsilon\sqrt{k_\sigma}} = \frac{525/10}{28.4 \times 0.81 \times \sqrt{4}} = 1.141$$

(conservatively using panel centreline dimensions rather than width from face of web plate)

$$\rho = \frac{\bar{\lambda}_p - 0.055(3 + \psi)}{\bar{\lambda}_p^2} = \frac{1.141 - 0.055(3 + 1)}{1.141^2} = 0.71$$

The 150 × 15 stiffener has $h/t = 10 < 10.5$ which is the limit to prevent torsional buckling as discussed in section 6.9 of this guide.

The column buckling load is first calculated:

Since the stress is uniform, the stiffener effective section is simply stiffener plus half the plate width each side. The stiffener effective section on the deck plate has attached deck plate width = 525 mm. Therefore $A_{sl,1} = 525 \times 10 + 150 \times 15 = 7500\,\text{mm}^2$, $I_{sl,1} = 1.434 \times 10^7\,\text{mm}^4$ and the centroid of the effective section is 29 mm from the top of the flange. From expression 3-1-5/(4.9):

$$\sigma_{cr,c} = \sigma_{cr,sl} = \frac{\pi^2 EI_{sl,1}}{A_{sl,1}a^2} = \frac{\pi^2 \times 210 \times 10^3 \times 1.434 \times 10^7}{7500 \times 2000^2} = 991\,\text{MPa}$$

The effective area of the same stiffener effective section but allowing for plate buckling is $A_{sl,1,eff} = 0.71 \times 525 \times 10 + 150 \times 15 = 5978\,\text{mm}^2$.

$$\beta_{A,c} = \frac{A_{sl,1,eff}}{A_{sl,1}}$$

$$\bar{\lambda}_c = \sqrt{\frac{\beta_{A,c} f_y}{\sigma_{cr,c}}} = \sqrt{\frac{5978 \times 355}{7500 \times 991}} = 0.534$$

The reduction factor is then calculated from the column buckling curves using an imperfection:

$$\alpha_e = \alpha + \frac{0.09}{i/e} = 0.49 + \frac{0.09}{43.7/56} = 0.61$$

where:

$$i = \sqrt{\frac{I_{sl,1}}{A_{sl,1}}} = \sqrt{\frac{1.434 \times 10^7}{7500}} = 43.7\,\text{mm}$$

$e = 150/2 + 10 - 29.0 = 56$ mm (based on distance to stiffener outstand centroid)

$\alpha = 0.49$ for open stiffeners. From expression 3-1-1/(6.49):

$$\Phi = 0.5[1 + \alpha_e(\bar{\lambda} - 0.2) + \bar{\lambda}^2] = 0.5[1 + 0.61(0.534 - 0.2) + 0.534^2] = 0.744$$

$$\chi_c = \frac{1}{\Phi + \sqrt{\Phi^2 - \bar{\lambda}^2}} = \frac{1}{0.744 + \sqrt{0.744^2 - 0.534^2}} = \mathbf{0.80}$$

Next the global buckling reduction factor must be computed for stiffened plate action using the method for one stiffener in 3-1-5/clause A.2.2. Since the stress is uniform, the stiffener effective section is simply stiffener plus half the plate width each side as above for the column buckling check.

Therefore $A_{sl,1} = 525 \times 10 + 150 \times 15 = 7500\,\text{mm}^2$and $I_{sl,1} = 1.434 \times 10^7\,\text{mm}^4$ again.

The wavelength for buckling without transverse stiffeners is:

$$a_c = 4.33\sqrt[4]{\frac{I_{sl,1}b_1^2b_2^2}{t^3b}} = 4.33\sqrt[4]{\frac{1.434 \times 10^7 \times 525^2 \times 525^2}{10^3 \times 1050}} = 4370\,\text{mm} > a = 2000\,\text{mm}$$

(the actual panel length) which was expected here as the transverse plating alone is unlikely to restrict the buckling wavelength to such a short length. The critical stress is therefore:

$$\sigma_{cr,p} = \sigma_{cr,sl} = \frac{\pi^2 EI_{sl,1}}{A_{sl,1}a^2} + \frac{Et^3ba^2}{4\pi^2(1 - \nu^2)A_{sl,1}b_1^2b_2^2}$$

$$= \frac{\pi^2 \times 210 \times 10^3 \times 1.434 \times 10^7}{7500 \times 2000^2} + \frac{210 \times 10^3 \times 10^3 \times 1050 \times 2000^2}{4\pi^2(1 - 0.3^2)7500 \times 525^2 \times 525^2}$$

$$= 991 + 43 = 1034\,\text{MPa}$$

This is not significantly higher than that for column-like buckling. The effective areas of the gross compression zone and effective compression zone allowing for plate buckling needed for calculating $\beta_{A,c} = A_{c,eff,loc}/A_c$ are the same as those corresponding to the single stiffener in this case.

The slenderness is therefore:

$$\bar{\lambda}_p = \sqrt{\frac{\beta_{A,c}f_y}{\sigma_{cr,p}}} = \sqrt{\frac{5978 \times 355}{7500 \times 1034}} = 0.523$$

The slenderness is less than the critical value of 0.673 so there is no reduction for plate-type buckling, i.e. $\rho = 1.0$. The final reduction factor for global behaviour is given by expression 3-1-5/(4.13):

$$\rho_c = (\rho - \chi_c)\xi(2 - \xi) + \chi_c = (1.0 - 0.80) \times 0.04 \times (2 - 0.04) + 0.80 = \mathbf{0.82}$$

where $\xi = \dfrac{\sigma_{cr,p}}{\sigma_{cr,c}} - 1 = \dfrac{1034}{991} - 1 = 0.04$

This reduction factor is basically that for column-type buckling which illustrates that it is often not worth the extra effort of considering plate-type behaviour.

The effective plate areas and stiffener area therefore now need to be reduced by the factor 0.82.

$$A_{c,eff,loc} = A_{sl,eff} + \sum_c \rho_{loc}b_{c,loc}t = 150 \times 15 + 0.71 \times 525 \times 10 = 5978\,\text{mm}^2$$

$$A_{c,eff} = \rho_c A_{c,eff,loc} + \sum b_{edge,eff}t = 0.82 \times 5978 + 0.71 \times 525/2 \times 2 \times 10 = \mathbf{8629\,mm^2}$$

An effective width of $525 \times 0.71 \times 0.82/2 = 153$ mm is attached each side of the stiffener. The attached width adjacent to each web $= 525/2 \times 0.71 = 186$ mm. The stiffener has a reduced area $= 150 \times 15 \times 0.82 = 1845\,\text{mm}^2$. These are shown in Fig. 6.2-21.

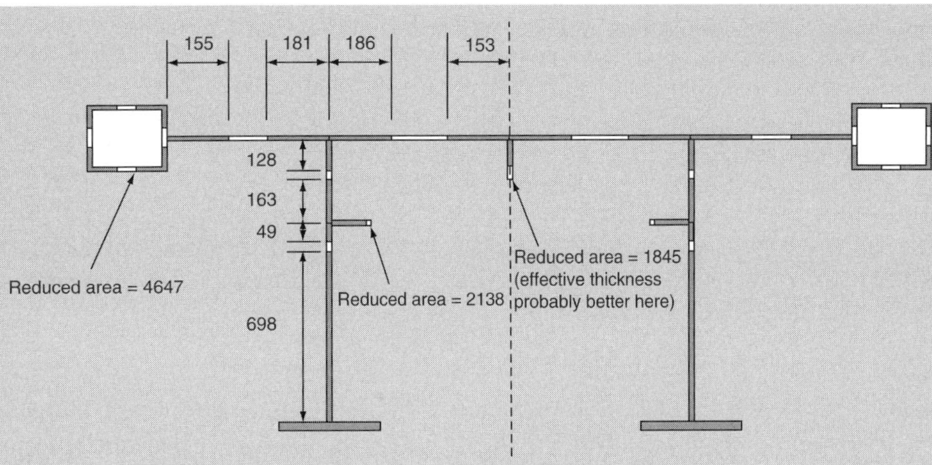

Fig. 6.2-21. Final effective section for section in Worked Example 6.2-3

Top flange cantilevers

The cantilever plate panels are stiffened by hollow sections at their edges so they will be treated as internal plate elements.

$$\bar{\lambda}_p = \frac{\bar{b}/t}{28.4\varepsilon\sqrt{k_\sigma}} = \frac{475/10}{28.4 \times 0.81 \times \sqrt{4}} = 1.032$$

(conservatively using panel centreline dimensions)

$$\rho = \frac{\bar{\lambda}_p - 0.055(3+\psi)}{\bar{\lambda}_p^2} = \frac{1.032 - 0.055(3+1)}{1.032^2} = 0.76$$

The 200 × 200 × 7 edge hollow section stiffeners have $\bar{b}/t = 27 < 31$ (the limiting ratio for full effectiveness for an internal plate in S355 steel) so will not be susceptible to local plate buckling.

The top flange cantilevers cannot generate any significant restraint to column-type buckling from plate action as the stiffened plates are only supported along one longitudinal edge. Consequently, the buckling load for global buckling will simply be taken as that due to column-type buckling.

For uniform compression, half the gross plating width is attached to the stiffener so $A_{sl,1} = 475/2 \times 10 + 4 \times 193 \times 7 = 7779\,\text{mm}^2$ and $I_{sl,1} = 3.36 \times 10^7\,\text{mm}^4$

$$\sigma_{cr,c} = \sigma_{cr,sl} = \frac{\pi^2 E I_{sl,1}}{A_{sl,1}a^2} = \frac{\pi^2 \times 210 \times 10^3 \times 3.36 \times 10^7}{7779 \times 2000^2} = 2238\,\text{MPa}$$

The effective area of the same stiffener effective section but allowing for plate buckling is $A_{sl,1,eff} = 0.76 \times 475/2 \times 10 + 4 \times 193 \times 7 = 7209\,\text{mm}^2$.

$$\beta_{A,c} = \frac{A_{sl,1,eff}}{A_{sl,1}}$$

$$\bar{\lambda}_c = \sqrt{\frac{\beta_{A,c} f_y}{\sigma_{cr,c}}} = \sqrt{\frac{7209 \times 355}{7779 \times 2238}} = 0.383$$

The reduction factor is then calculated from the column buckling curves using an imperfection, $\alpha_e = \alpha + 0.09/(i/e)$. By inspection this will give an imperfection somewhere between the value for curves 'c' and 'd' in 3-1-1/Fig. 6.4 so curve 'd' is conservatively used, whereupon:

$\chi_c = \mathbf{0.86}$

The effective plate areas and stiffener area therefore now need to be reduced by the factor 0.86.

$$A_{c,eff,loc} = A_{sl,eff} + \sum_{c} \rho_{loc} b_{c,loc} t = 4 \times 193 \times 7 + 0.76 \times 475/2 \times 10 = 7209\,\text{mm}^2$$

$$A_{c,eff} = \rho_c A_{c,eff,loc} + \sum b_{edge,eff} t = 0.86 \times 7209 + 0.76 \times 475/2 \times 10 = \mathbf{8005\,mm^2}$$

An effective width of $475 \times 0.76 \times 0.86/2 = 155$ mm is attached to the hollow section and $475 \times 0.76/2 = 181$ mm is attached to the web. The effective area of the hollow section $= 4 \times 193 \times 7 \times 0.86 = 4647\,\text{mm}^2$. These are shown in Fig. 6.2-21.

Webs

To determine the effective web, the neutral axis of the bridge with effective top flange and gross web is first determined. The neutral axis depth is found to be 639 mm from the bottom of the bottom flange as shown in Fig. 6.2-22.

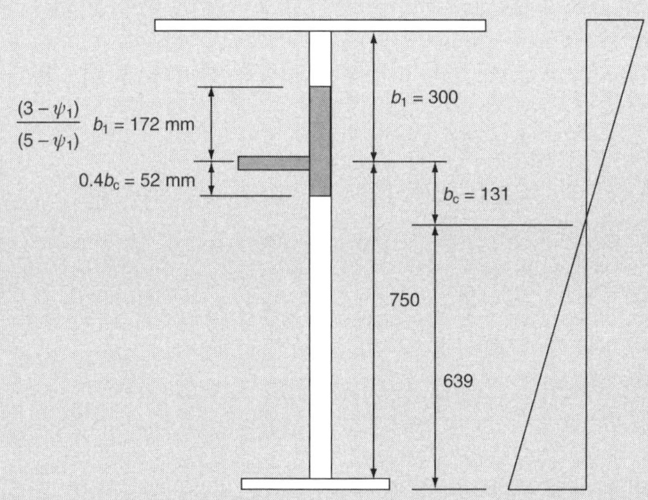

Fig. 6.2-22. Web stiffener effective section

For the top panel, $\psi_1 = 131/431 = 0.30$

Since $b_1/t = 300/10 = 30 < 31$ even for uniform compression, there is no reduction for plate buckling in the top panel.

For the bottom panel, $\psi_2 = -619/131 = -4.73$

Limiting ψ to -3.0 as in 3-1-5/Table 4.1, $k_\sigma = 5.98(1 - \psi)^2 = 5.98(1 + 3)^2 = 95.68$

$$\bar{\lambda}_p = \frac{\bar{b}/t}{28.4\varepsilon\sqrt{k_\sigma}} = \frac{750/10}{28.4 \times 0.81 \times \sqrt{95.68}} = 0.333 < 0.673$$

so there is no reduction for plate buckling in the lower panel.

The column buckling load only will be determined, as the analysis for the top flange gave little benefit from considering stiffened plate behaviour. The effective section for the column is shown in Fig. 6.2-22.

The upper attached width $= [(3 - \psi_1)/(5 - \psi_1)]b_1 = [(3 - 0.30)/(5 - 0.30)] \times 300 = 172$ mm

The lower attached width $= 0.4b_c = 0.4 \times 131 = 52$ mm

Therefore $A_{sl,1} = (172 + 52) \times 10 + 150 \times 15 = 4490\,\text{mm}^2$, $I_{sl,1} = 1.142 \times 10^7\,\text{mm}^4$ and the centroid is 45.1 mm from the back of the web plate.

$$\sigma_{cr,sl} = \frac{\pi^2 E I_{sl,1}}{A_{sl,1} a^2} = \frac{\pi^2 \times 210 \times 10^3 \times 1.142 \times 10^7}{4490 \times 2000^2} = 1318\,\text{MPa}$$

The critical stress based on the extreme compression fibre is therefore

$$\sigma_{cr,c} = \sigma_{cr,sl}\frac{b_c}{b_{sl,1}} = 1318 \times 431/131 = 4336\,\text{MPa}$$

noting the different definition of b_c here to that in Fig. 6.2-22.

Since there is no plate buckling, $A_{sl,1,eff} = A_{sl,1} = 4490\,\text{mm}^2$ so

$$\beta_{A,c} = \frac{A_{sl,1,eff}}{A_{sl,1}} = 1.0$$

$$\bar{\lambda}_c = \sqrt{\frac{\beta_{A,c}f_y}{\sigma_{cr,c}}} = \sqrt{\frac{355}{4336}} = 0.286$$

The reduction factor is then calculated from the column buckling curves using an imperfection

$$\alpha_e = \alpha + \frac{0.09}{i/e} = 0.49 + \frac{0.09}{50.5/40} = 0.56$$

where:

$$i = \sqrt{\frac{I_{sl,1}}{A_{sl,1}}} = \sqrt{\frac{1.142 \times 10^7}{4490}} = 50.5\,\text{mm}$$

$$e = 150/2 + 10 - 45.1 = 40\,\text{mm}$$

$\alpha = 0.49$ for open stiffeners

$$\Phi = 0.5[1 + \alpha_e(\bar{\lambda} - 0.2) + \bar{\lambda}^2] = 0.5[1 + 0.56(0.286 - 0.2) + 0.286^2] = 0.565$$

$$\chi = \frac{1}{\Phi + \sqrt{\Phi^2 - \bar{\lambda}^2}} = \frac{1}{0.565 + \sqrt{0.565^2 - 0.286^2}} = \mathbf{0.95}$$

A reduction must be made to the web area based on the locations of the attached widths for plate buckling and also to the stiffener area.

Above the stiffener, the effective width = 172 × 0.95 = 163 mm. No reduction is made to the web plate attached to the deck plate.

Below the stiffener, the effective width = 52 × 0.95 = 49 mm. No reduction is made to the web plate attached to the bottom flange.

The stiffener itself has reduced area = 150 × 15 × 0.95 = 2138 mm^2.

The final effective section for bending stress calculation is shown in Fig. 6.2-21.

Worked Example 6.2-4: Section properties for wide stiffened flange

A steel box girder has a bottom flange that is 4000 mm wide, 12 mm thick and has 9 no. 150 mm × 15 mm flat stiffeners at 400 mm centres, so all sub-panels are 400 mm wide. Diaphragms are provided in the box at 4000 mm centres. The effective area of the bottom flange when subjected to uniform compression is calculated. (The effects of shear lag are negligible in this example, but were they significant, the resulting effective^p area would need further reduction for shear lag.)

The reduction for local plate sub-panel buckling is calculated first.

$$\bar{\lambda}_p = \frac{\bar{b}/t}{28.4\varepsilon\sqrt{k_\sigma}} = \frac{400/12}{28.4 \times 0.81 \times \sqrt{4}} = 0.725$$

$$\rho = \frac{\bar{\lambda}_p - 0.055(3+\psi)}{\bar{\lambda}_p^2} = \frac{0.725 - 0.055(3+1)}{0.725^2} = 0.96, \text{ i.e. minimal reduction.}$$

The 150 × 15 stiffener has $h/t = 10 < 10.5$ which is the limit to prevent torsional buckling as discussed in section 6.9 of this guide, so torsional buckling is prevented. For overall buckling, the column buckling load and orthotropic plate buckling load need to be calculated. The column buckling load is first calculated:

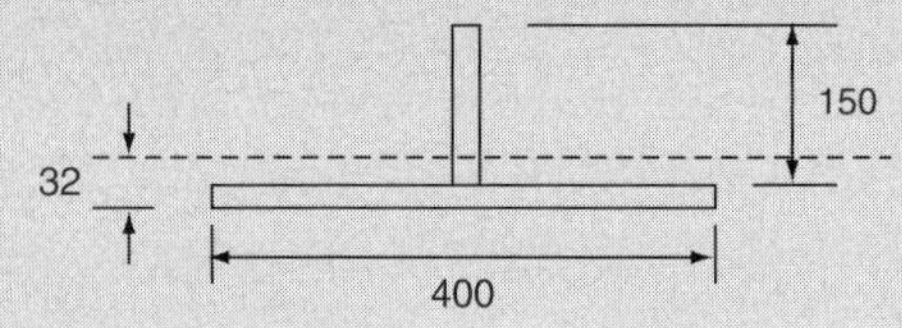

Fig. 6.2-23. Effective section for gross stiffener for Worked Example 6.2-4

$I_{sl,1}$ is simply equal to the inertia of one stiffener together with gross attached width of plate equal to the stiffener spacing $\bar{b} = 400\,\text{mm}$ (as shown in Fig. 6.2-23). $A_{sl,1}$ is the area for the above section.

$I_{sl,1} = 1.433 \times 10^7\,\text{mm}^4$ and $A_{sl,1} = 150 \times 15 + 400 \times 12 = 7050\,\text{mm}^2$

$$\sigma_{cr,sl} = \frac{\pi^2 E I_{sl,1}}{A_{sl,1} a^2} = \frac{\pi^2 \times 210 \times 10^3 \times 1.433 \times 10^7}{7050 \times 4000^2} = \mathbf{263.3\,MPa}$$

$$\beta_{A,c} = \frac{A_{sl,1,eff}}{A_{sl,1}}$$

where $A_{sl,1,eff}$ is the effective area of one stiffener and attached plate allowing for plate buckling. Effective width of plate per stiffener $= 0.96 \times 400 = 384\,\text{mm}$ so $A_{sl,1,eff} = 384 \times 12 + 150 \times 15 = 6858\,\text{mm}^2$.

$$\bar{\lambda}_c = \sqrt{\frac{\beta_{A,c} f_y}{\sigma_{cr,c}}} = \sqrt{\frac{6858 \times 355}{7050 \times 263.3}} = 1.145$$

The reduction factor is then calculated from the column buckling curves using an imperfection:

$$\alpha_e = \alpha + \frac{0.09}{i/e} = 0.49 + \frac{0.09}{45.1/55} = 0.60$$

where:

$$i = \sqrt{\frac{I_{sl,1}}{A_{sl,1}}} = \sqrt{\frac{1.433 \times 10^7}{7050}} = 45.1\,\text{mm}$$

$e = 150/2 + 12 - 32 = 55\,\text{mm}$ from Fig. 6.2-23

$\alpha = 0.49$ for open stiffeners

$\Phi = 0.5[1 + \alpha_e(\bar{\lambda} - 0.2) + \bar{\lambda}^2] = 0.5[1 + 0.60(1.145 - 0.2) + 1.145^2] = 1.439$

$$\chi = \frac{1}{\Phi + \sqrt{\Phi^2 - \bar{\lambda}^2}} = \frac{1}{1.439 + \sqrt{1.439^2 - 1.145^2}} = \mathbf{0.433}$$

The reduction factor for stiffened plate behaviour is next calculated:

For uniform compression, either the method of 3-1-5/Annex A.1 may be used to determine the overall stiffened plate buckling load or the alternative equation (D6.2-4) can be used. The calculation will be performed using both for illustration here.

Equation (D6.2-4):

$$I_T = \frac{150 \times 15^3}{3} = 1.6875 \times 10^5\,\text{mm}^4$$

$$H = \frac{Gt^3}{6} + \frac{GI_T}{2\bar{b}} = \frac{81 \times 10^3 \times 12^3}{6} + \frac{81 \times 10^3 \times 1.6875 \times 10^5}{2 \times 400} = 4.041 \times 10^7 \text{ Nmm}$$

Since the stiffeners have uniform spacing and size, the stiffness of the gross stiffened plate per metre $\Sigma I_{sc}/b_{comp}$ is simply equal to the stiffness of one effective section I_{sc} with attached width of plate equal to the stiffener spacing $\bar{b} = 400$ mm (as shown in Fig. 6.2-23), divided by the stiffener spacing $\bar{b} = 400$ mm.

$I_{sc} = 1.433 \times 10^7 \text{ mm}^4$

$$\frac{\sum I_{sc}}{b_{comp}} = \frac{I_{sc}}{\bar{b}} = \frac{1.433 \times 10^7}{400} = 35\,825 \text{ mm}^3$$

$$D_x = \frac{E\sum I_{sc}}{b_{comp}} = 210 \times 10^3 \times 35\,825 = 7.522 \times 10^9 \text{ Nmm}$$

$$D_y = \frac{Et^3}{12(1-\nu\nu_y)} = \frac{210 \times 10^3 \times 12^3}{12(1-0.3 \times 0.204)} = 3.221 \times 10^7 \text{ Nmm}$$

$$\text{where } \nu_y = 0.3\left(\frac{\bar{b}t}{A_s + \bar{b}t}\right) = 0.3\left(\frac{400 \times 12}{2250 + 400 \times 12}\right) = 0.204$$

with $A_s = 150 \times 15 = 2250 \text{ mm}^2$ as the area of an individual stiffener outstand.

$$t_{eff} = t\left(1 + \frac{\sum A_s}{b_{comp} t}\right) = 12.0\left(1 + \frac{9 \times 2250}{9 \times 400 \times 12}\right) = 17.63 \text{ mm}$$

$$\text{The aspect ratio } \varphi' = \frac{a}{b}\left(\frac{D_y}{D_x}\right)^{0.25} = \frac{4000}{4000}\left(\frac{3.221 \times 10^7}{7.522 \times 10^9}\right)^{0.25} = 0.26$$

For uniform compression, the stress ratio, $\psi = 1.0$. Therefore:

$k_i = 17.3$ from Fig. 6.2-16

$k_0 = 15.3$ from Fig. 6.2-17

$$\sigma_{cr,p} = \frac{\pi^2\sqrt{D_x D_y}}{b^2 t_{eff}}\left[k_0 + \frac{(k_i - k_0)H}{\sqrt{D_x D_y}}\right]$$

$$= \frac{\pi^2\sqrt{7.522 \times 10^9 \times 3.221 \times 10^7}}{4000^2 \times 17.63}\left[15.3 + \frac{(17.3 - 15.3) \times 4.041 \times 10^7}{\sqrt{7.522 \times 10^9 \times 3.221 \times 10^7}}\right]$$

$= \mathbf{266.4\ MPa}$

Annex A.1 method:

Panel aspect ratio $\alpha = a/b = 4000/4000 = 1.0$.

The ratio of second moment of area of the whole stiffened plate, ΣI_{sl}, to that of the parent plate alone, I_p is approximately the same as the ratio based on one stiffener effective section in this case, so base γ on one effective section. (See discussions in the main text as this approach is more consistent with the rules elsewhere in any case, rather than using the whole plate including the parts of the sub-panels attached to the webs.)

$$\gamma = \frac{\sum I_{sl}}{I_p} = \frac{1.433 \times 10^7}{400 \times 12^3/10.92} = 226.4$$

Similarly δ is based on one effective section:

$$\delta = \frac{\sum A_{sl}}{A_p} = \frac{2250}{400 \times 12} = 0.469$$

$\alpha = 1.0$, $\sqrt[4]{\gamma} = \sqrt[4]{226.4} = 3.88 > 1.0$ so the first formula is appropriate:

$$k_{\sigma,p} = \frac{2[(1+\alpha^2)^2 + \gamma - 1]}{\alpha^2(\psi+1)(1+\delta)} = \frac{2[(1+1.0^2)^2 + 226.4 - 1]}{1.0^2(1.0+1)(1+0.469)} = 156.2$$

From expression 3-1-5/(A.1):

$$\sigma_{cr,p} = k_{\sigma,p}\frac{\pi^2 Et^2}{12(1-v^2)b^2} = 156.2 \times \frac{\pi^2 \times 210 \times 10^3 \times 12^2}{12(1-0.3^2) \times 4000^2} = \mathbf{266.8\,MPa}$$

The answer is essentially the same as previously.

The critical stress for plate behaviour is only marginally higher than that for column buckling. Consequently, the extra effort involved in calculating it was not warranted here and this will often be the case. Benefit would only have been significant if the overall panel width was considerably reduced. This is illustrated below.

The slenderness for plate buckling is calculated from expression 3-1-5/(4.7):

$$\bar{\lambda}_p = \sqrt{\frac{\beta_{A,c} f_y}{\sigma_{cr,p}}} = \sqrt{\frac{0.973 \times 355}{266.8}} = 1.138$$

where the ratio $\beta_{A,c} = A_{c,eff,loc}/A_c$ refers to the entire stiffened plate compression zone width but, since the stiffeners are evenly spaced, it may be taken equal to the ratio for a single stiffener as used in the column check. Therefore $\beta_{A,c} = 6858/7050 = 0.973$.

$$\rho = \frac{\bar{\lambda}_p - 0.055(3+\psi)}{\bar{\lambda}_p^2} = \frac{1.138 - 0.055(3+1)}{1.138^2} = 0.71$$

The final reduction factor for global behaviour is given by expression 3-1-5/(4.13):

$$\rho_c = (\rho - \chi_c)\xi(2-\xi) + \chi_c = (0.71 - 0.433) \times 0.01 \times (2 - 0.01) + 0.433 = \mathbf{0.44}$$

where $\xi = \frac{\sigma_{cr,p}}{\sigma_{cr,c}} - 1 = \frac{266.8}{263.3} - 1 = 0.01$

(As predicted, this reduction factor is basically that for column-type buckling.)

Finally, the effective area of the whole compression zone is calculated by applying this reduction factor to the reduced area for local buckling according to expressions 3-1-5/(4.5) and 3-1-5/(4.6):

$$A_{c,eff,loc} = 9 \times 6858 = 61\,722\,\text{mm}^2$$

which excludes the part of the plate sub-panels attached to the web. From expression 3-1-5/(4.5):

$$A_{c,eff} = \rho_c A_{c,eff,loc} = 0.44 \times 61\,722 + 0.96 \times 12 \times 400 = 31\,766\,\text{mm}^2$$

There was no benefit from orthotropic action with the above flange aspect ratio. If the width of the panel is reduced to 2000 mm, there will be greater benefit as shown below:

Equation (D6.2-4):

The basic orthotropic properties of the plate remain the same. Only the aspect ratio changes.

The new aspect ratio $\varphi' = \frac{a}{b}\left(\frac{D_y}{D_x}\right)^{0.25} = \frac{4000}{2000}\left(\frac{3.221 \times 10^7}{7.522 \times 10^9}\right)^{0.25} = 0.51$

For uniform compression, the stress ratio, $\psi = 1.0$. Therefore:

$k_i = 6.1$ from Fig. 6.2-16

$k_0 = 4.1$ from Fig. 6.2-17

$$\sigma_{cr,p} = \frac{\pi^2\sqrt{D_x D_y}}{b^2 t_{eff}}\left[k_0 + \frac{(k_i - k_0)H}{\sqrt{D_x D_y}}\right]$$

$$= \frac{\pi^2\sqrt{7.522\times10^9\times3.221\times10^7}}{2000^2\times17.63}\left[4.1 + \frac{(6.1-4.1)\times4.041\times10^7}{\sqrt{7.522\times10^9\times3.221\times10^7}}\right]$$

$$= \mathbf{293.7\ MPa}$$

Annex A.1 method:
Similarly only the aspect ratio changes:

$$\alpha = a/b = 4000/2000 = 2.0$$

$$\alpha = 2.0 < \sqrt[4]{\gamma} = \sqrt[4]{226.4} = 3.88, \text{ so}$$

$$k_{\sigma,p} = \frac{2[(1+\alpha^2)^2+\gamma-1]}{\alpha^2(\psi+1)(1+\delta)} = \frac{2[(1+2.0^2)^2+226.4-1]}{2.0^2(1.0+1)(1+0.469)} = 42.6$$

From expression 3-1-5/(A.1):

$$\sigma_{cr,p} = k_{\sigma,p}\frac{\pi^2 E t^2}{12(1-\upsilon^2)b^2} = 42.6\times\frac{\pi^2\times210\times10^3\times12^2}{12(1-0.3^2)\times2000^2} = \mathbf{291.1\ MPa}$$

Both stresses are around 10% greater than for the 4000 mm wide flange, so orthotropic action would give some benefit in this case, but still not much.

6.2.2.6. Stress limits for Class 4 members according to EN 1993-1-5 clause 10

6.2.2.6.1. Introduction

In section 6.2.2.5, Class 4 members were treated using an effective section to allow for local buckling of sub-panels and overall buckling in stiffened panels. The allowable stress on such an effective section may then be taken as yield. The assumption in that method is that there is sufficient post-buckling strength to achieve the necessary redistribution of stress to allow all components to be stressed to their individual resistances. This approach is therefore not permitted (and is not appropriate) in a number of situations where there may not be sufficient post-buckling strength or where the geometry of the member is outside the limits for which the method has been tested. These exceptions are discussed in section 6.2.2.5.1 of this guide.

Where the conditions above for the use of effective widths are not met, a method based on gross properties and reduced stress limits may be used according to 3-1-5/clause 10. The inclusion of this method was proposed at a relatively late stage by a German delegation and, as such, has probably not been set out as clearly as is desirable. This leads to some ambiguity and therefore this section of the guide introduces some new terms in an attempt to improve clarity.

3-1-5/clause 10 may always be used as an alternative to the effective width approach, but no account is taken of the beneficial shedding of load from overstressed panels. It can therefore be conservative by comparison, although it is not always conservative where hand calculations are used – this is discussed in section 6.2.2.6.3 below. Additionally, since shear stresses and transverse direct stresses are considered directly in this method, no further interaction between these different effects needs to be considered. This is another potential area of conservatism as shear stresses and transverse stresses, whatever their magnitude, have an immediate effect on the resistance to direct stresses, whereas this is not the case when the interaction-based approach with effective sections is used. The distribution of transverse stress caused by local load application at a flange can be estimated using the method in 3-1-5/clause 3.2.3 which is discussed in section 6.2.2.3.2 of this guide.

Other parts of EN 1993-2 refer to this section for derivation of a reduced limiting stress, σ_{limit}, to be used in checks under bending and axial load. Generally, it will be better to use the full check of the section as outlined here, rather than derive σ_{limit} as an additional step, because the interaction with shear can be included at the same time. However, if σ_{limit} is to be evaluated for checks on bending and axial force alone, it is only necessary to perform the check described below for the compressive zone, i.e. $\alpha_{\text{ult,k}}$ is based on the compression zone only, even if the tensile stress at the tensile fibre is greater in magnitude. This is illustrated in the discussions on Note 2 of 3-1-5/clause 10(5)b) below and associated Fig. 6.2-27.

If the whole member is prone to overall buckling instability, such as flexural or lateral torsional buckling, these effects must either be calculated by second-order analysis and the additional stresses included when checking panels to 3-1-5/clause 10 (as discussed below) or by using a limiting stress σ_{limit} when performing the buckling checks to 3-2/clause 6.3. For flexural buckling, σ_{limit} can be calculated based on the lowest compressive value of axial stress $\sigma_{\text{x,Ed}}$, acting on its own, required to cause buckling failure in the weakest sub-panel or an entire panel, according to the verification formula in 3-1-5/clause 10 discussed below. This value of σ_{limit} is then used to replace f_y in all parts of the buckling check calculation. It is conservative, particularly when the critical panel used to determine σ_{limit} is not at the extreme compression fibre of the section where the greatest direct stress increase during buckling occurs. For lateral torsional buckling, σ_{limit} can be determined as the bending stress at the extreme compression fibre needed to cause buckling in the weakest panel. This would be very conservative if σ_{limit} were determined from buckling of a web panel which was not at the extreme fibre for the reason above; the web panel stress would not increase much during buckling.

The effects of shear could logically be excluded in deriving σ_{limit} for use in member buckling checks for consistency with the approach to checking member buckling elsewhere in EN 1993. A cross-section resistance check considering shear would then be necessary as discussed in the remainder of this section. Alternatively, a combined member buckling and cross-section check could conservatively be performed by including shear in the derivation of σ_{limit}.

A further method for considering overall buckling combined with local buckling is presented in 3-1-5/clause B.2. It is not discussed further here, but is essentially an extension of the rules in 3-1-5/clause 10 to include allowance for global buckling in the overall strength reduction factor.

6.2.2.6.2. Basic approach

The method is very similar in approach to that for verifying the out-of-plane buckling resistance of frames with bending and axial force discussed in section 6.3.4 of this guide.

The basic verification is performed by determining an overall slenderness for buckling of each plate element under all of the applied stresses acting together, i.e. direct stresses and shear stresses. This overall slenderness will generally need to be calculated for both plate-like and column-like buckling. Torsional buckling should be prevented through compliance with the requirements discussed in section 6.9 of this guide as torsional buckling is not otherwise easily catered for within this method. The slenderness definition in ***3-1-5/clause 10(3)*** takes the usual Eurocode form as follows:

3-1-5/clause 10(3)

$$\bar{\lambda} = \sqrt{\frac{\alpha_{\text{ult,k}}}{\alpha_{\text{cr}}}} \qquad \text{(D6.2-5)}$$

where $\alpha_{\text{ult,k}}$ is the minimum load factor applied to the design loads required to reach the characteristic resistance 'of the most critical point of the plate' ignoring any buckling effects. Where part of a plate is in tension, this definition is not satisfactory and separate checks of the tensile and compression zones will be required as discussed towards the end of this section.

α_{cr} is the minimum load factor applied to the design loads required to give elastic critical buckling of the panel considered under all stresses acting together. For stiffened plates, the lowest critical mode may be global plate buckling, local sub-panel buckling or a coupled mode. α_{cr} will need to consider both plate-like and column-like buckling as discussed below, which leads to slendernesses of $\bar{\lambda}_p$ and $\bar{\lambda}_c$ respectively.

Equation (D6.2-5) has been presented slightly differently from the expression provided in 3-1-5/clause 10(3), which is $\bar{\lambda}_p = \sqrt{\alpha_{ult,k}/\alpha_{cr}}$, as the latter is written for plate-like buckling only.

Two methods can be used to determine slenderness and calculate buckling reduction factors: (i) elastic critical buckling analysis with a finite-element model, or (ii) hand calculations. Both are discussed below, but method (ii) is generally more practical and can be carried out by spreadsheet. Worked examples 6.2-5 and 6.2-6 relate to the hand calculation method.

(i) Elastic critical buckling analysis with a finite-element model

This approach can be used for non-uniform panels which may also contain holes or have irregular stiffening. The stresses $\sigma_{x,Ed}$, $\sigma_{z,Ed}$ and τ_{Ed} in the individual plates (webs, flanges, etc.) are first determined using gross cross-section properties. (If calculated by hand, shear stresses in webs can be based on average values and flange shear stresses determined from classical elastic theory.) Finite-element models of the individual plates can then be generated with supports along edges supported by transverse stiffeners and web–flange junctions and the general stress field calculated above is applied to the edges of the individual plate models. A simplified case is shown in Fig. 6.2-24 where the direct and shear stresses have been made constant throughout.

The first stage of calculation requires the determination of $\alpha_{ult,k}$. ***3-1-5/clause 10(4)*** recommends that the criterion for reaching the characteristic resistance be taken as the Von Mises yield criterion such that: *3-1-5/clause 10(4)*

$$\frac{1}{\alpha_{ult,k}^2} = \left(\frac{\sigma_{x,Ed}}{f_y}\right)^2 + \left(\frac{\sigma_{z,Ed}}{f_y}\right)^2 - \left(\frac{\sigma_{x,Ed}}{f_y}\right)\left(\frac{\sigma_{z,Ed}}{f_y}\right) + 3\left(\frac{\tau_{Ed}}{f_y}\right)^2 \qquad \text{3-1-5/(10.3)}$$

where $\sigma_{x,Ed}$, $\sigma_{z,Ed}$ and τ_{Ed} are the direct stress in the longitudinal direction, the direct stress in the transverse direction and the shear stress respectively at a point in the plate which minimizes $\alpha_{ult,k}$. (This can be conservative where a panel is partly in tension as discussed in the discussion of hand calculations below.) Expression 3-1-5/(10.3) may be evaluated by hand or within the finite-element software such that

$$\alpha_{ult,k} = \frac{f_y}{\sigma_{eff}}$$

where σ_{eff} is the Von Mises equivalent stress:

$$(\sigma_{x,Ed}^2 + \sigma_{z,Ed}^2 - \sigma_{x,Ed}\sigma_{z,Ed} + 3\tau_{Ed}^2)^{0.5}$$

The second stage is to determine the lowest load factor α_{cr} to cause elastic critical buckling under the same applied stress field. This load factor needs to be determined for both plate-like and column-like behaviour.

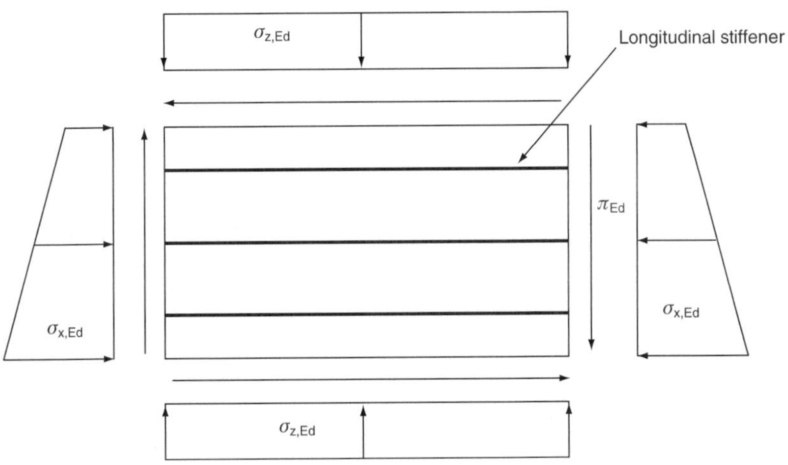

Fig. 6.2-24. Stresses in typical stiffened plate panel

The critical amplifier for plate-like buckling can be determined from a model with pinned supports along all supported edges under the complete stress field ($\sigma_{x,Ed}$, $\sigma_{z,Ed}$ and τ_{Ed}) above. This leads to a load factor $\alpha_{p,cr}$. Using the slenderness definition of equation (D6.2-5), the following reduction factors must be determined for plate-like behaviour.

$\rho_{p,x}$ is the plate-type reduction factor for longitudinal direct stress determined from 3-1-5/clause 4.4(2) determined with:

$$\bar{\lambda}_p = \sqrt{\frac{\alpha_{ult,k}}{\alpha_{p,cr}}}$$

$\rho_{p,z}$ is the plate-type reduction factor for transverse direct stress determined from 3-1-5/clause 4.4(2) determined with:

$$\bar{\lambda}_p = \sqrt{\frac{\alpha_{ult,k}}{\alpha_{p,cr}}}$$

χ_w is determined with:

$$\bar{\lambda}_p = \sqrt{\frac{\alpha_{ult,k}}{\alpha_{p,cr}}}$$

for shear stress from 3-1-5/clause 5.2(1).

The critical amplifier for column-like buckling must also be determined. For column-like buckling in the x direction, the supports in the model along the x direction edges have to be removed and the buckling analysis under the complete stress field repeated. This leads to a load factor $\alpha_{cx,cr}$. For column-like buckling in the z direction, the supports along the z direction edges have to be removed and the buckling analysis repeated. This leads to a load factor $\alpha_{cz,cr}$. From the slenderness $\bar{\lambda}_{cx} = \sqrt{\alpha_{ult,k}/\alpha_{cx,cr}}$, the column-type reduction $\chi_{c,x}$ is determined from 3-1-5/clause 4.5.3(3) for longitudinal direct stress and $\chi_{c,z}$ is similarly determined from $\bar{\lambda}_{cz} = \sqrt{\alpha_{ult,k}/\alpha_{cz,cr}}$ for transverse direct stress.

The final reduction factors for direct stress, ρ_x and ρ_z, can be obtained as an interpolation between plate- and column-like behaviour according to 3-1-5/clause 4.5.4(1) as follows:

$$\rho_x = (\rho_{p,x} - \chi_{c,x})\xi_x(2 - \xi_x) + \chi_{c,x}$$

$$\rho_z = (\rho_{p,z} - \chi_{c,z})\xi_z(2 - \xi_z) + \chi_{c,z}$$

where:

$$\xi_x = \frac{\alpha_{p,cr}}{\alpha_{cx,cr}} - 1 \text{ where } 0 \le \xi_x \le 1 \text{ and } \xi_z = \frac{\alpha_{p,cr}}{\alpha_{cz,cr}} - 1 \text{ where } 0 \le \xi_z \le 1$$

The final verification can be written independently of the method of cross-section verification as:

$$\frac{\rho\alpha_{ult,k}}{\gamma_{M1}} \ge 1.0 \qquad \text{3-1-5/(10.1)}$$

3-1-5/clause 10(5)

3-1-5/clause 10(5) provides two options for performing this verification. Most simply, ρ is conservatively taken as the lowest reduction factor for shear or direct stresses (i.e. the lower of ρ_x, ρ_z and χ_w) so:

$$\left(\frac{\sigma_{x,Ed}}{f_y/\gamma_{M1}}\right)^2 + \left(\frac{\sigma_{z,Ed}}{f_y/\gamma_{M1}}\right)^2 - \left(\frac{\sigma_{x,Ed}}{f_y/\gamma_{M1}}\right)\left(\frac{\sigma_{z,Ed}}{f_y/\gamma_{M1}}\right) + 3\left(\frac{\tau_{Ed}}{f_y/\gamma_{M1}}\right)^2 \le \rho^2 \qquad \text{3-1-5/(10.4)}$$

Alternatively, and less conservatively, the reduction factors for each effect can be applied separately to the relevant stresses:

$$\left(\frac{\sigma_{x,Ed}}{\rho_x f_y/\gamma_{M1}}\right)^2 + \left(\frac{\sigma_{z,Ed}}{\rho_z f_y/\gamma_{M1}}\right)^2 - \left(\frac{\sigma_{x,Ed}}{\rho_x f_y/\gamma_{M1}}\right)\left(\frac{\sigma_{z,Ed}}{\rho_z f_y/\gamma_{M1}}\right) + 3\left(\frac{\tau_{Ed}}{\chi_w f_y/\gamma_{M1}}\right)^2 \le 1.0$$

3-1-5/(10.5)

The use of γ_{M1} for local buckling under direct stress is inconsistent with the use of γ_{M0} everywhere else in EN 1993 but it is considered to be necessary to give adequate reliabilty when using this method. If none of the calculated reduction factors are less than 1.0 after applying the rules, it would be reasonable to use γ_{M0} throughout expressions 3-1-5/(10.4) and 3-1-5/(10.5) to avoid a discontinuity with EN 1993-1-1 expression (6.1).

(ii) Hand calculations

Hand methods of calculation can still handle non-uniform panels by squaring them off based on the largest panel dimension. For stiffened plates, the lowest critical mode may be global plate buckling, local sub-panel buckling or a coupled mode at a lower load factor where global and local modes occur close together. However, if the slenderness in equation (D6.2-5) is determined by hand, coupled modes cannot be determined and a slenderness can then only be determined separately for global plate buckling and for buckling of each sub-panel as described below. The resistance is then checked separately for each case as allowed by ***Note 2*** of ***3-1-5/clause 10(3)***. This in itself can be slightly unconservative, but other aspects of this calculation are conservative. (It should be noted that the effective section method does combine the effects of sub-panel and overall buckling.) The following discussion therefore applies to the separate checks of both the overall panel and sub-panels. *3-1-5/clause 10(3)*

The first stage of calculation again requires the determination of $\alpha_{\text{ult,k}}$. The criterion for reaching the characteristic resistance is taken as the Von Mises yield criterion such that:

$$\frac{1}{\alpha_{\text{ult,k}}^2} = \left(\frac{\sigma_{\text{x,Ed}}}{f_y}\right)^2 + \left(\frac{\sigma_{\text{z,Ed}}}{f_y}\right)^2 - \left(\frac{\sigma_{\text{x,Ed}}}{f_y}\right)\left(\frac{\sigma_{\text{z,Ed}}}{f_y}\right) + 3\left(\frac{\tau_{\text{Ed}}}{f_y}\right)^2 \qquad \text{3-1-5/(10.3)}$$

where $\sigma_{\text{x,Ed}}$, $\sigma_{\text{z,Ed}}$ and τ_{Ed} are the direct stress in the longitudinal direction, the direct stress in the transverse direction and the shear stress respectively at a point in the plate which minimizes $\alpha_{\text{ult,k}}$. Where there is stress reversal across a plate, the check needs to be applied separately for the peak compressive and tensile regions of the plate for the reasons discussed below. Transverse stresses can be conservatively taken as the peak value in the panel being considered, allowing for the dispersal discussed in section 6.2.2.3 of this guide.

The second stage is to determine the lowest load factor α_{cr} to give elastic critical buckling under all the stresses combined. This lowest load factor in general needs to be determined for both plate-like and column-like behaviour. To determine these factors under the combined stress field would require a finite-element analysis as discussed above. This will not normally be very practical for bridges where there may be many panels to design and many load cases for each panel.

Without finite-element analysis, load factors for buckling will only be available for each stress component acting independently as these can be obtained from standard texts or other parts of EN 1993-1-5. For example, the load factor for buckling under $\sigma_{\text{x,Ed}}$ alone would be $\alpha_{\text{cr,x}} = \sigma_{\text{cr,x}}/\sigma_{\text{x,Ed}}$. In this situation, ***3-1-5/clause 10(6)*** gives a useful formula for combining these individual factors into one load factor for all effects acting together: *3-1-5/clause 10(6)*

$$\frac{1}{\alpha_{\text{cr}}} = \frac{1+\psi_x}{4\alpha_{\text{cr,x}}} + \frac{1+\psi_z}{4\alpha_{\text{cr,z}}} + \left[\left(\frac{1+\psi_x}{4\alpha_{\text{cr,x}}} + \frac{1+\psi_z}{4\alpha_{\text{cr,z}}}\right)^2 + \frac{1-\psi_x}{2\alpha_{\text{cr,x}}^2} + \frac{1-\psi_z}{2\alpha_{\text{cr,z}}^2} + \frac{1}{\alpha_{\text{cr},\tau}^2}\right]^{1/2} \qquad \text{3-1-5/(10.6)}$$

where ψ_x is the longitudinal direct stress ratio σ_2/σ_1 across the plate for either a sub-panel or overall stiffened panel as shown in Fig. 6.2-25. ψ_z has the same meaning for the transverse direct stresses. For compressive longitudinal direct stress, for example, $\alpha_{\text{cr,x}}$ should be calculated taking $\sigma_{\text{x,Ed}}$ as the greatest compressive stress in the sub-panel or overall plate as appropriate to the check being performed.

An alternative simpler and more conservative interaction is:

$$\frac{1}{\alpha_{\text{cr}}} = \frac{1}{\alpha_{\text{cr,x}}} + \frac{1}{\alpha_{\text{cr,z}}} + \frac{1}{\alpha_{\text{cr},\tau}} \qquad \text{(D6.2-6)}$$

If $\sigma_{\text{x,Ed}}$ is tensile throughout the panel, $\alpha_{\text{cr,x}}$ will need to be taken as infinity (∞). This appears to be slightly conservative as it ignores any benefit of 'straightening out the panel'

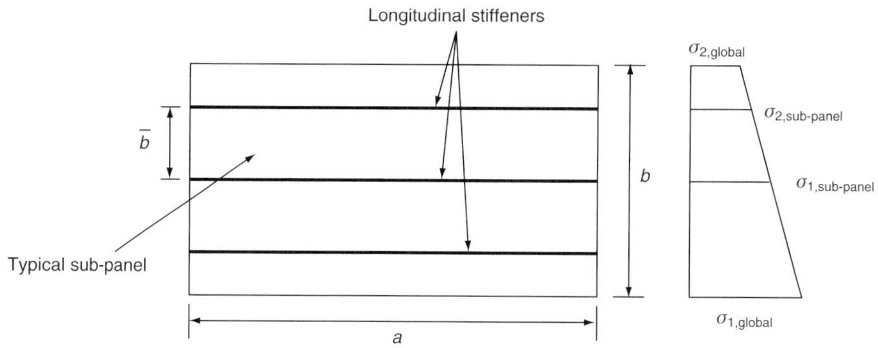

Fig. 6.2-25. Stresses for stress ratio calculation in a stiffened panel

for other buckling modes. However, this is indirectly accounted for in $\alpha_{ult,k}$ as its value is reduced by the tension which in turn reduces the slenderness and hence the reduction factor. The use of a negative value is not appropriate as expression 3-1-5/(10.6) breaks down for negative values of $\alpha_{cr,x}$, e.g. α_{cr} does not then equal $\alpha_{cr,x}$ with only longitudinal stress applied due to the square root of the square in the equation. It should be noted that one must still check panels which are wholly in tension for buckling, as shear buckling may still be significant.

If a flange is being checked, shear stresses for global and sub-panel buckling can be included in the same way as in the interaction check in 3-1-5/clause 7. This is discussed in section 6.2.9.2.3 of this guide.

When α_{cr} has been determined for both plate-like buckling ($\alpha_{p,cr}$) and column-like buckling ($\alpha_{c,cr}$), a slenderness is determined for each type of behaviour from equation (D6.2-5) and reduction factors are determined for each stress component. Calculation of the reduction factors for the two respective types of buckling are discussed in section 6.2.2.5 of this guide. In deriving $\alpha_{p,cr}$ and $\alpha_{c,cr}$, the critical load factor for shear acting alone, $\alpha_{cr,\tau} = \tau_{cr}/\tau_{Ed}$ will be the same in each case. An interaction is then performed between plate-like and column-like buckling to determine the final reduction factors for direct stress. This process is illustrated in Fig. 6.2-26; it is much simpler for cases without transverse stress, as in Worked Example 6.2-5. The final reduction factors are then:

- ρ_x for longitudinal direct stress, determined by interpolation between plate-like and column-like reductions according to 3-1-5/clause 4.5.4(1);
- ρ_z for transverse direct stress, determined by interpolation between plate-like and column-like reductions according to 3-1-5/clause 4.5.4(1);
- χ_w for shear stress, determined according to 3-1-5/clause 5.2(1) using the slenderness for plate-like behaviour.

For sub-panel buckling, these reduction factors are determined based on the stress distribution in the sub-panel. The interpolation between reduction factors for plate-like and column-like behaviour in an unstiffened panel is only necessary according to expression 3-1-5/(4.13) if benefit has been taken in deriving a critical stress for a given panel that is greater than that for a long panel of the same width. If the critical stress is determined using 3-1-5/Tables 4.1 and 4.2, long panel geometry is assumed and only plate-like behaviour need be considered in deriving ρ_x and ρ_z. If column-like behaviour is considered, the final reduction factors ρ_x and ρ_z should not be taken as less than those obtained by deriving the individual buckling factors for infinitely long plates in the direction of the applied stress considered.

For overall buckling, the critical direct stress for plate-like and column-like behaviour will often be very similar. It will frequently therefore not be worth the extra effort of calculating a load factor for plate-like behaviour; the factor for column-like behaviour can conservatively be used to determine reduction factors. This is illustrated in Worked Example 6.2-5 below. In this case, the reduction factor χ_w for shear stress is also determined using the overall slenderness derived considering column-like behaviour under direct stress which is slightly

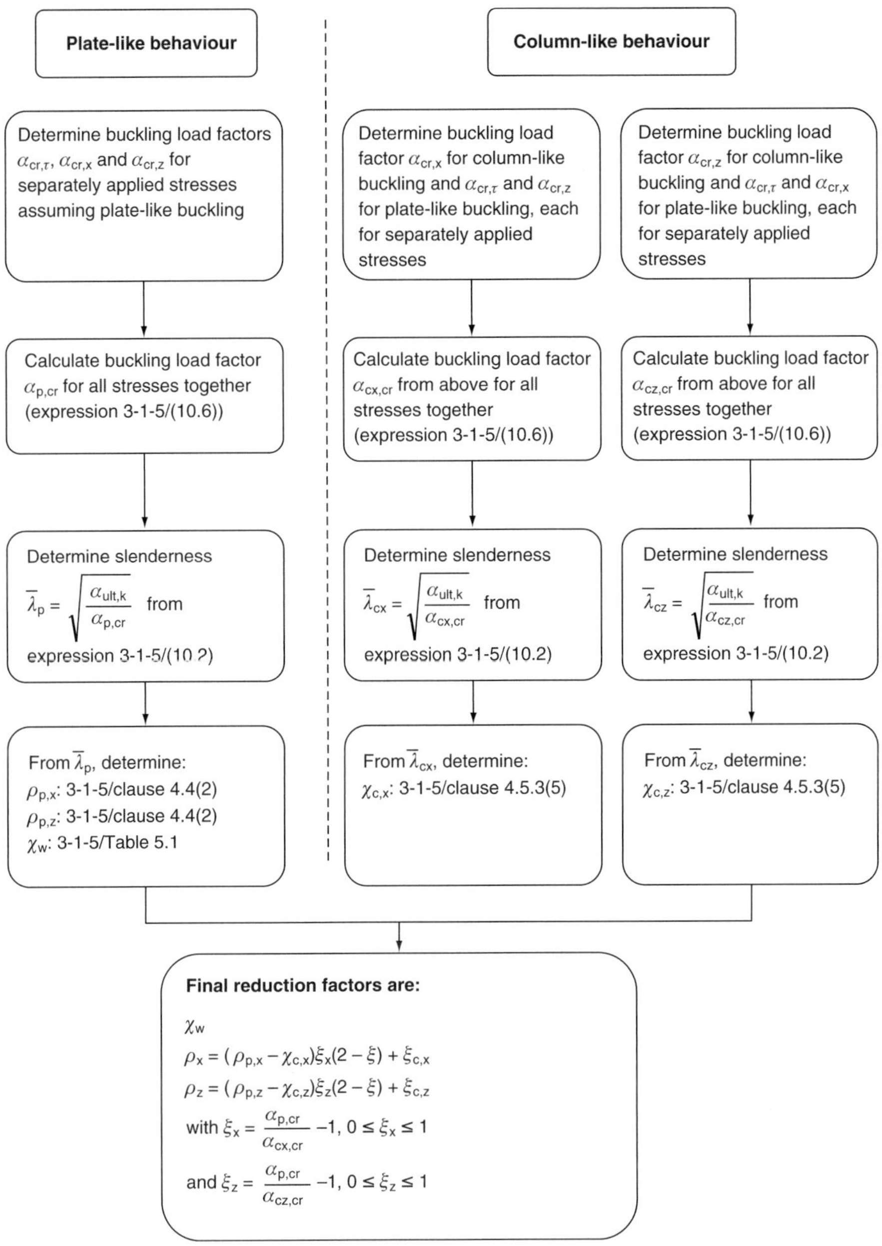

Fig. 6.2-26. Procedure for determining buckling reduction factors for expression 3-1-5/(10.5)

conservative; strictly, the slenderness for plate-like buckling should be used. When deriving the critical stress for overall shear buckling, the reduction factor of 3 on stiffener inertia implicit in the 3-1-5/clause A.3 formula discussed in section 6.2.6 should be removed as required by Note 1 of 3-1-5/clause 10(3).

The overall reduction factor for use in expression 3-1-5/(10.1) again depends on whether the mode of buckling is predominantly due to direct stresses or shear stresses as the reduction factor curves differ for each. The reduction factors for direct stresses and shear are applied to the cross-section check performed in the first stage, but this time using design values of the material properties:

$$\left(\frac{\sigma_{x,Ed}}{\rho_x f_y/\gamma_{M1}}\right)^2 + \left(\frac{\sigma_{z,Ed}}{\rho_z f_y/\gamma_{M1}}\right)^2 - \left(\frac{\sigma_{x,Ed}}{\rho_x f_y/\gamma_{M1}}\right)\left(\frac{\sigma_{z,Ed}}{\rho_z f_y/\gamma_{M1}}\right) + 3\left(\frac{\tau_{Ed}}{\chi_v f_y/\gamma_{M1}}\right)^2 \le 1.0$$

3-1-5/(10.5)

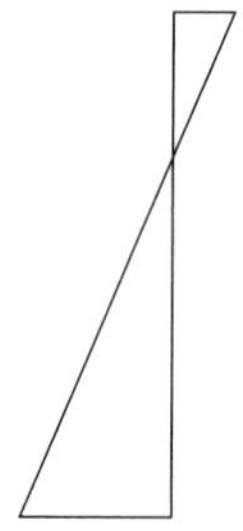

Fig. 6.2-27. Case of stress reversal where tensile stress exceeds the compressive stress

A problem arises with the use of expression 3-1-5/(10.5) in panels where the stress is tensile throughout or where there is stress reversal such that the compressive stress at one fibre is less in magnitude than the tensile stress at the opposite fibre. In the latter case, the greater tensile stress potentially ends up being magnified by the reduction factor determined using the critical stress for the compression zone if $\alpha_{ult,k}$ and the check in expression 3-1-5/(10.5) are evaluated using a tensile value of $\sigma_{x,Ed}$. This would be very conservative. In response to this problem, ***Note 2*** of ***3-1-5/clause 10(5)b)*** recommends that the check is only applied to the compressive part of the plate. There is logic for applying the method to the compressive parts. For direct stress alone, but with stress reversal as shown in Fig. 6.2-27, the slenderness according to 3-1-5/clause 4.4(2) is given by:

3-1-5/clause 10(5)b)

$$\bar{\lambda}_p = \sqrt{\frac{f_y}{\sigma_{cr}}}$$

and since $\sigma_{cr} = \alpha_{cr}\sigma_{comp}$

$$\bar{\lambda}_p = \sqrt{\frac{f_y/\sigma_{comp}}{\alpha_{cr}}} = \sqrt{\frac{\alpha_{ult,k}}{\alpha_{cr}}} \text{ with } \alpha_{ult,k} = f_y/\sigma_{comp}$$

Clearly in this case the slenderness is based on the compression fibre, even though the tensile stress is greater in magnitude. If the effective section method of 3-1-5/clause 4 was used, the tension zone would still however be checked for yielding but there would be no reduction to its effectiveness. The stress in it would rise slightly however from the gross-section value due to the loss of section in the compression zone.

Despite the recommendation of Note 2 of 3-1-5/clause 10(5)b), a check on the tensile zone should still however be made as the tensile stress in conjunction with the shear stress may cause yielding before yielding due to buckling occurs in the compression zone. There are several options for such a check and these are illustrated in Worked Example 6.2-5. Method (d) in the example is recommended for its greater compatibility with the results using the effective section method, but there is no directly equivalent method. If $\sigma_{x,Ed}$ is tensile throughout the panel being checked, the reduction factor ρ_x could be taken as 1.0, although this is not explicitly covered by EN 1993-1-5. The same applies to $\sigma_{z,Ed}$. Worked Example 6.2-6 illustrates a case of biaxial compression.

Although not explicitly stated, if the stress varies along the length of the panel, the verification of expression 3-1-5/(10.5) could be performed at a distance of $0.4a$ or $0.5b$, whichever is smaller, from the most highly stressed end of the panel. This is consistent with the approach allowed in the effective area method in 3-1-5/clause 4.6(3). If this is done, the yield check then needs to be repeated without reduction factors at the end of the panel. The comments on the use of γ_{M1} made in the discussion of the finite-element method above apply here also.

Worked Example 6.2-5: Footbridge

A steel footbridge has the section shown in Fig. 6.2-28. Cross-girders to flanges and transverse stiffeners to webs are provided at 2000 mm centres. The web is checked for the direct stresses shown in Fig. 6.2-29 and for a coexisting shear stress $\tau_{Ed} = 100$ MPa.

A computer elastic critical buckling analysis is not available to determine the load amplification factor for buckling with all stresses acting together, so amplification factors are determined for each stress component separately and then combined. The check must also be done separately for sub-panel and overall buckling of the web.

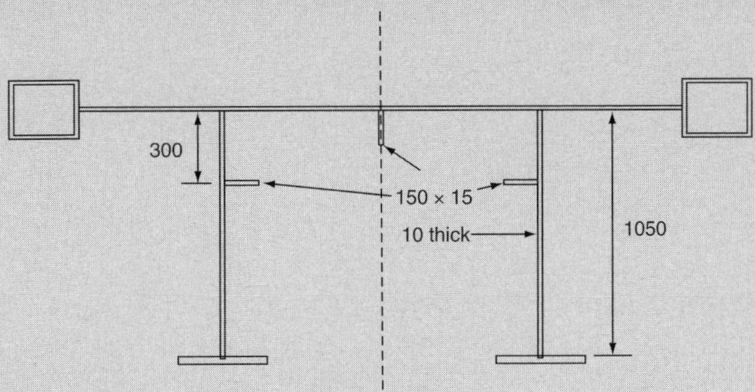

Fig. 6.2-28. Steel footbridge for Worked Example 6.2-5

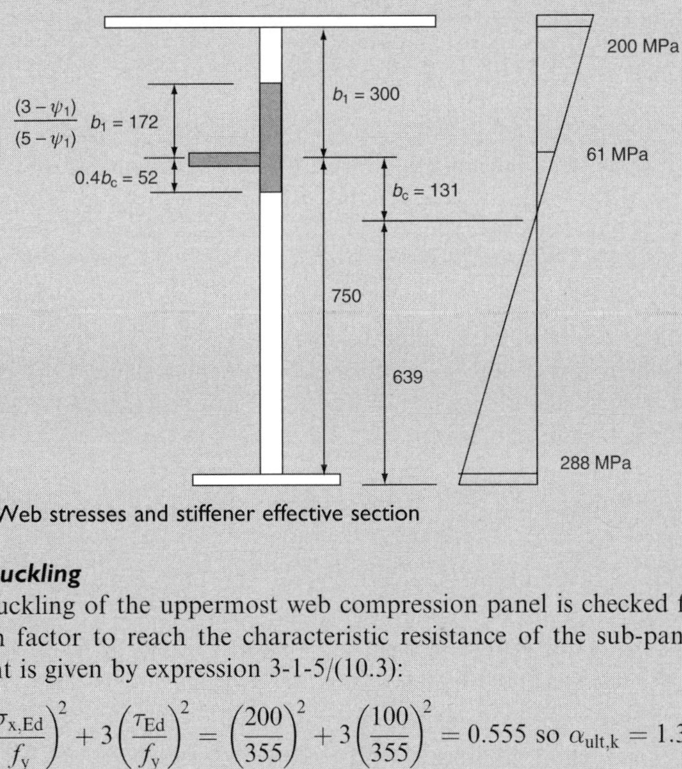

Fig. 6.2-29. Web stresses and stiffener effective section

Sub-panel buckling

Sub-panel buckling of the uppermost web compression panel is checked first. The load amplification factor to reach the characteristic resistance of the sub-panel at its most stressed point is given by expression 3-1-5/(10.3):

$$\frac{1}{\alpha_{ult,k}^2} = \left(\frac{\sigma_{x,Ed}}{f_y}\right)^2 + 3\left(\frac{\tau_{Ed}}{f_y}\right)^2 = \left(\frac{200}{355}\right)^2 + 3\left(\frac{100}{355}\right)^2 = 0.555 \text{ so } \alpha_{ult,k} = 1.342$$

Load factors for buckling are next calculated. By inspection, as the panels are long, there will be no need to consider column-like buckling as discussed in the main text.

Direct stresses:

For the top panel, $\psi = 131/431 = 0.30$ so from 3-1-5/clause 4.4:

$$\sigma_{cr,x} = \frac{k_\sigma \pi^2 E t^2}{12(1-\upsilon^2)\bar{b}^2} = \frac{6.07 \times \pi^2 \times 210 \times 10^3 \times 10^2}{12(1-0.3^2) \times 300^2} = 1280 \text{ MPa}$$

where $k_\sigma = 8.2/(1.05 + \psi) = 8.2/(1.05 + 0.3) = 6.07$ from 3-1-5/Table 4.1, conservatively assuming a long panel.

Shear stresses:

$$\tau_{cr} = \frac{k_\tau \pi^2 E t^2}{12(1-\nu^2)b^2} = \frac{5.43 \times \pi^2 \times 210 \times 10^3 \times 10^2}{12(1-0.3^2) \times 300^2} = 1145\,\text{MPa from 3-1-5/clause 5.3}$$

where $k_\tau = 5.34 + 4.00\left(\frac{\bar{b}}{a}\right)^2 = 5.34 + 4.00\left(\frac{300}{2000}\right)^2 = 5.43$

For separately applied stresses:

$$\alpha_{cr,x} = \frac{\sigma_{cr,x}}{\sigma_{Ed,x}} = \frac{1280}{200} = 6.40$$

$$\alpha_{cr,\tau} = \frac{\tau_{cr}}{\tau_{Ed}} = \frac{1145}{100} = 11.45$$

For stresses applied together, the critical load factor is given by expression 3-1-5/(10.6):

$$\frac{1}{\alpha_{cr}} = \frac{1+0.3}{4 \times 6.4} + \left[\left(\frac{1+0.3}{4 \times 6.4}\right)^2 + \frac{1-0.3}{2 \times 6.4^2} + \frac{1}{11.45^2}\right]^{1/2} = 0.188 \text{ so } \alpha_{cr} = 5.327$$

From equation (D6.2-5), the slenderness for sub-panel buckling is:

$$\bar{\lambda}_p = \sqrt{\frac{\alpha_{ult,k}}{\alpha_{cr}}} = \sqrt{\frac{1.342}{5.327}} = 0.502$$

From 3-1-5/clause 4.4(2), the reduction factor for longitudinal direct stress is $\rho_x = 1.00$. The reduction factor for shear stress at this slenderness is, from 3-1-5/Table 5.1:

$$\chi_w = \frac{0.83}{\bar{\lambda}_w} = \frac{0.83}{0.502} = 1.65 > \eta = 1.2 \text{ so } \chi_w = 1.2$$

The verification is therefore essentially just one of yielding as there are no reduction factors <1.00. Consequently the Von Mises check in 3-1-1/clause 6.2.1 could be used with the lower material factor $\gamma_{M0} = 1.0$. The check is, however, performed here as set out in 3-1-5/clause 10 using $\gamma_{M1} = 1.1$.

$$\left(\frac{200}{1.0 \times 355/1.1}\right)^2 + 3\left(\frac{100}{1.2 \times 355/1.1}\right)^2 = 0.58 \le 1.0 \text{ **so the upper panel is adequate**}$$

The lower sub-panel should next be checked as this will clearly now be more critical than the upper sub-panel on a yielding basis alone.

The load amplification factor to reach the characteristic resistance of the lower sub-panel at its most stressed point *in the compression zone* is given by:

$$\frac{1}{\alpha_{ult,k}^2} = \left(\frac{\sigma_{x,Ed}}{f_y}\right)^2 + 3\left(\frac{\tau_{Ed}}{f_y}\right)^2 = \left(\frac{61}{355}\right)^2 + 3\left(\frac{100}{355}\right)^2 = 0.268 \text{ so } \alpha_{ult,k} = 1.933$$

Direct stresses:
For the lower panel, $\psi = -619/131 = -4.73$ so from 3-1-5/clause 4.4:

$$\sigma_{cr,x} = \frac{k_\sigma \pi^2 E t^2}{12(1-\upsilon^2)\bar{b}^2} = \frac{95.68 \times \pi^2 \times 210 \times 10^3 \times 10^2}{12(1-0.3^2) \times 750^2} = 3228\,\text{MPa}$$

where, limiting ψ to -3.0 as in 3-1-5/Table 4.1, $k_\sigma = 5.98(1-\psi)^2 = 5.98(1+3)^2 = 95.68$

Shear stresses:

$$\tau_{cr} = \frac{k_\tau \pi^2 E t^2}{12(1-\nu^2)b^2} = \frac{5.90 \times \pi^2 \times 210 \times 10^3 \times 10^2}{12(1-0.3^2) \times 750^2} = 199 \text{ MPa from 3-1-5/clause 5.3:}$$

$$\text{where } k_\tau = 5.34 + 4.00\left(\frac{b}{a}\right)^2 = 5.34 + 4.00\left(\frac{750}{2000}\right)^2 = 5.90$$

For separately applied stresses:

$$\alpha_{cr,x} = \frac{\sigma_{cr,x}}{\sigma_{Ed,x}} = \frac{3228}{61} = 52.9$$

$$\alpha_{cr,\tau} = \frac{\tau_{cr}}{\tau_{Ed}} = \frac{199}{100} = 1.99$$

Since $\alpha_{cr,x}$ is so large, α_{cr} will tend to $\alpha_{cr,\tau}$ so $\alpha_{cr} \approx 1.99$

From equation (D6.2-5), the slenderness for sub-panel buckling is:

$$\bar{\lambda}_p = \sqrt{\frac{\alpha_{ult,k}}{\alpha_{cr}}} = \sqrt{\frac{1.933}{1.99}} = 0.99$$

The reduction factor for longitudinal direct stress is therefore from 3-1-5/clause 4.4:

$$\rho_x = \frac{\bar{\lambda}_p - 0.055(3+\psi)}{\bar{\lambda}_p^2} = \frac{0.99 - 0.055(3-4.73)}{0.99^2} = 1.11 > 1.0 \text{ so } \rho_x = 1.00$$

The reduction factor for shear stress at this slenderness is:

$$\chi_w = \frac{0.83}{\bar{\lambda}_w} = \frac{0.83}{0.99} = 0.84$$

The check is performed as set out in 3-1-5/clause 10 using $\gamma_{M1} = 1.1$:

$$\left(\frac{61}{1.0 \times 355/1.1}\right)^2 + 3\left(\frac{100}{0.84 \times 355/1.1}\right)^2 = 0.44 < 1.0 \textbf{ which is therefore adequate}$$

Although EN 1993-1-5 recommends that this check is only done for the compressive parts of panels, some check must still be performed for the tension part as the stresses (even ignoring buckling) could exceed yield. Some options for checking the tension zone are as follows.

(a) Perform a Von Mises check without reduction factors:

$$\left(\frac{288}{355/1.1}\right)^2 + 3\left(\frac{100}{355/1.1}\right)^2 = 1.08 < 1.0 \text{ so inadequate}$$

Although the Von Mises check is itself conservative for combinations of stresses, this method would not allow for shear buckling effects and could therefore become unconservative.

(b) Calculate the reduction factor for shear based on shear acting alone and apply expression 3-1-5/(10.5) with no reduction factor on direct stress:

$$\bar{\lambda}_w = 0.76\sqrt{\frac{f_{yw}}{\tau_{cr}}} = 0.76\sqrt{\frac{355}{199}} = 1.02$$

$$\chi_w = \frac{0.83}{1.02} = 0.81$$

$$\left(\frac{288}{1.0 \times 355/1.1}\right)^2 + 3\left(\frac{100}{0.81 \times 355/1.1}\right)^2 = 1.24 > 1.0 \text{ so inadequate}$$

This is more compatible with the approach in the effective section-based check in 3-1-5/clause 7 but is more conservative due to the conservative nature of the Von Mises check, which reduces the allowable direct stress in the presence of any shear.

(c) Repeat the check of expression 3-1-5/(10.5) using the same reduction factor for shear as calculated for the compressive side of the panel but again with no reduction factor on longitudinal direct stress:

$$\left(\frac{288}{1.0 \times 355/1.1}\right)^2 + 3\left(\frac{100}{0.84 \times 355/1.1}\right)^2 = 1.20 > 1.0 \text{ so inadequate}$$

This is not very logical and has the same conservatism as above.

(d) Recalculate the slenderness using $\alpha_{\text{ult,k}}$ for the tension side and take α_{cr} as calculated above for the whole stress field:

$$\frac{1}{\alpha_{\text{ult,k}}^2} = \left(\frac{\sigma_{\text{x,Ed}}}{f_y}\right)^2 + 3\left(\frac{\tau_{\text{Ed}}}{f_y}\right)^2 = \left(\frac{288}{355}\right)^2 + 3\left(\frac{100}{355}\right)^2 = 0.896 \text{ so } \alpha_{\text{ult,k}} = 1.056$$

From before, $\alpha_{\text{cr}} = 1.99$

From equation (D6.2-5), the slenderness for sub-panel buckling is:

$$\bar{\lambda}_p = \sqrt{\frac{\alpha_{\text{ult,k}}}{\alpha_{\text{cr}}}} = \sqrt{\frac{1.056}{1.99}} = 0.73$$

The reduction factor for shear stress at this slenderness is:

$$\chi_w = \frac{0.83}{\bar{\lambda}_w} = \frac{0.83}{0.73} = 1.14$$

The check is then performed using expression 3-1-5/(10.5) but with no reduction factors for the tensile stress:

$$\left(\frac{288}{1.0 \times 355/1.1}\right)^2 + 3\left(\frac{100}{1.14 \times 355/1.1}\right)^2 = 1.02 < 1.0 \ \textbf{which is almost adequate}$$

Method (d) is recommended here as it gives the best agreement with the interaction in 3-1-5/clause 7. If 3-1-5/clause 7 is applied to this panel in isolation, but using $\gamma_{M1} = 1.1$ for direct stress and elastic stresses for η_1 to further facilitate comparison:

$$\eta_1 + \left[1 - \frac{M_{\text{f,Rd}}}{M_{\text{pl,Rd}}}\right](2\bar{\eta}_3 - 1)^2 = \eta_1 + (2\bar{\eta}_3 - 1)^2 = \frac{288}{355/1.1} + \left(\frac{2 \times 100}{0.81 \times 355/\sqrt{3}} - 1\right)^2$$

$$= 1.00$$

compared with 1.02 from method (d).

Global plate buckling

Direct stresses:

The column buckling load only will be considered as, by inspection, there will be little benefit from considering stiffened plate behaviour. This avoids the need to consider interaction with plate-like behaviour as discussed in the main text. The effective section for the column is shown in Fig. 6.2-29.

From 3-1-5/Fig A.1, the upper attached width is:

$$\frac{(3 - \psi)}{(5 - \psi)} b_1 = \frac{(3 - 0.30)}{(5 - 0.30)} \times 300 = 172\,\text{mm}$$

The lower attached width $= 0.4b_c = 0.4 \times 131 = 52\,\text{mm}$

Therefore $A_{\text{sl,1}} = (172 + 52) \times 10 + 150 \times 15 = 4490\,\text{mm}^2$ and $I_{\text{sl,1}} = 1.142 \times 10^7\,\text{mm}^4$

From 3-1-5/clause 4.5.3:

$$\sigma_{cr,sl} = \frac{\pi^2 E I_{sl,1}}{A_{sl,1} a^2} = \frac{\pi^2 \times 210 \times 10^3 \times 1.142 \times 10^7}{4490 \times 2000^2} = 1318\,\text{MPa}$$

The critical stress based on the extreme compression fibre is therefore

$$\sigma_{cr,c} = \sigma_{cr,sl} \frac{b_c}{b_{sl,1}} = 1318 \times 431/131 = 4336\,\text{MPa}$$

from 3-1-5/clause 4.5.3(3) noting the unfortunate different definition of b_c in that clause to that in Fig. 6.2-29; the former defines b_c as the distance from neutral axis to extreme compression fibre of the web while the latter defines b_c as the distance from neutral axis to extreme compression fibre of the sub-panel bounded by the stiffener under consideration. (Calculation of critical stresses is discussed in section 6.2.2.5.)

Shear stresses:
$\alpha = a/b = 2000/1050 = 1.90 < 3$, so the shear buckling coefficient is obtained from expression 3-1-5/(A.6) but the reduction factor of 3 on stiffener second moment of area implicit in the formula should be removed as required by Note 1 to 3-1-5/clause 10(3). From 3-1-5/Fig. 5.3, each longitudinal stiffener has an attached piece of web of $30\varepsilon t$ plus the thickness of the stiffener $= 30 \times 0.81 \times 10 + 10 = 253$ mm. This is a slightly different effective section than that for direct stresses. The effective section therefore has inertia $= 1.186 \times 10^7\,\text{mm}^4$. For the purpose of these calculations, this inertia must be increased by a factor of 3.0 as stated above. From 3-1-5/Annex A.3:

$$k_\tau = 4.1 + \frac{6.3 + 0.18 \dfrac{I_{sl}}{t^3 b}}{\alpha^2} + 2.2 \sqrt[3]{\frac{I_{sl}}{t^3 b}}$$

$$= 4.1 + \frac{6.3 + 0.18 \times \dfrac{3 \times 1.186 \times 10^7}{10^3 \times 1050}}{1.90^2} + 2.2 \sqrt[3]{\frac{3 \times 1.186 \times 10^7}{10^3 \times 1050}} = 14.65$$

$$\tau_{cr} = \frac{k_\tau \pi^2 E t^2}{12(1 - \nu^2) b^2} = \frac{14.65 \times \pi^2 \times 210 \times 10^3 \times 10^2}{12(1 - 0.3^2) \times 1050^2} = 252\,\text{MPa}$$

$$\alpha_{cr,\tau} = \frac{\tau_{cr}}{\tau_{Ed}} = \frac{252}{100} = 2.52$$

$$\alpha_{cr,x} = \frac{\sigma_{cr,x}}{\sigma_{Ed,x}} = \frac{4336}{200} = 21.68$$

The stress ratio for the whole panel is:

$$\psi = \frac{200}{-288} = -0.694$$

From expression 3-1-5/(10.6):

$$\frac{1}{\alpha_{cr}} = \frac{1 - 0.694}{4 \times 21.68} + \left[\left(\frac{1 - 0.694}{4 \times 21.68} \right)^2 + \frac{1 - 0.3}{2 \times 21.68^2} + \frac{1}{2.52^2} \right]^{1/2} = 0.401$$

so $\alpha_{cr} = 2.492$, which is close to that for shear acting alone which clearly dominates.

The load amplification factor for the cross-section resistance is first derived for the compressive part of the plate:

$$\frac{1}{\alpha_{ult,k}^2} = \left(\frac{\sigma_{x,Ed}}{f_y} \right)^2 + 3 \left(\frac{\tau_{Ed}}{f_y} \right)^2 = \left(\frac{200}{355} \right)^2 + 3 \left(\frac{100}{355} \right)^2 = 0.555$$

$$\alpha_{ult,k} = 1.342$$

From equation (D6.2-5), the slenderness for overall buckling is:

$$\bar{\lambda}_c = \sqrt{\frac{\alpha_{ult,k}}{\alpha_{cr}}} = \sqrt{\frac{1.342}{2.490}} = 0.734$$

The reduction factor for longitudinal direct stress is then calculated from the column buckling curves using an imperfection:

$$\alpha_e = \alpha + \frac{0.09}{i/e} = 0.49 + \frac{0.09}{50.5/40} = 0.56$$

$$i = \sqrt{\frac{I_{st}}{A_{st}}} = \sqrt{\frac{1.142 \times 10^7}{4490}} = 50.5\,\text{mm}$$

$e = 150/2 + 10 - 45.1 = 40\,\text{mm}$ and $\alpha = 0.49$ for open stiffeners.

$$\Phi = 0.5[1 + \alpha_e(\bar{\lambda} - 0.2) + \bar{\lambda}^2] = 0.5[1 + 0.56(0.734 - 0.2) + 0.734^2] = 0.919$$

$$\rho_x = \chi = \frac{1}{\Phi + \sqrt{\Phi^2 - \bar{\lambda}^2}} = \frac{1}{0.919 + \sqrt{0.919^2 - 0.734^2}} = \mathbf{0.68}$$

The reduction factor for shear stress at this slenderness is:

$$\chi_w = \frac{0.83}{\bar{\lambda}_w} = \frac{0.83}{0.734} = 1.13$$

This is slightly conservative as the slenderness for shear should have been determined from the load amplifier considering plate-like buckling under direct stress. It makes virtually no difference in this case as the column- and plate-like buckling loads are virtually the same and shear buckling dominates the load amplifier in any case.

The final verification for overall behaviour is then from expression 3-1-5/(10.5):

$$\left(\frac{200}{0.68 \times 355/1.1}\right)^2 + 3\left(\frac{100}{1.13 \times 355/1.1}\right)^2 = 1.056 > 1.0$$

so the overall web is just inadequate

The actual usage factor is $\sqrt{1.056} = \mathbf{1.03}$

Although EN 1993-1-5 recommends that this check is only done for the compressive parts of panels, some check must still be performed for the tension part as the stresses (even ignoring buckling) could exceed yield. Option (d) above is used:

The slenderness is recalculated using $\alpha_{ult,k}$ for the tension side with α_{cr} as before:

$$\frac{1}{\alpha_{ult,k}^2} = \left(\frac{\sigma_{x,Ed}}{f_y}\right)^2 + 3\left(\frac{\tau_{Ed}}{f_y}\right)^2 = \left(\frac{288}{355}\right)^2 + 3\left(\frac{100}{355}\right)^2 = 0.896 \text{ so } \alpha_{ult,k} = 1.056$$

From above, $\alpha_{cr} = 2.49$

From equation (D6.2-5), the slenderness for buckling is:

$$\bar{\lambda}_c = \sqrt{\frac{\alpha_{ult,k}}{\alpha_{cr}}} = \sqrt{\frac{1.056}{2.49}} = 0.65$$

The reduction factor for shear stress at this slenderness is:

$$\chi_w = \frac{0.83}{\bar{\lambda}_w} = \frac{0.83}{0.65} = 1.28 > 1.2 \text{ so } \chi_w = 1.2$$

The check is performed as set out in 3-1-5/clause 10(5) with no reduction factor for the tensile stress:

$$\left(\frac{288}{1.0 \times 355/1.1}\right)^2 + 3\left(\frac{100}{1.2 \times 355/1.1}\right)^2 = 1.00 \text{ which is just adequate}$$

Worked Example 6.2-6: square panel under biaxial compression and shear
An unstiffened panel has dimensions 1000 mm by 1000 mm and 10 mm thick. It has $\sigma_{x,Ed} = \sigma_{z,Ed} = 100$ MPa and $\tau_{Ed} = 100$ MPa. Usage factors are determined for the panel under:

(i) $\sigma_{x,Ed}$ only,
(ii) $\sigma_{x,Ed}$ and $\sigma_{z,Ed}$ only, and
(iii) $\sigma_{x,Ed}$, $\sigma_{z,Ed}$ and τ_{Ed}.

The critical stresses for each stress acting separately are first calculated. The interaction with column-like buckling is not relevant for a square panel, but would be for other aspect ratios.

Direct stresses (3-1-5/clause 4.4):

$$\sigma_{cr,x} = \sigma_{cr,y} = \frac{k_\sigma \pi^2 E t^2}{12(1-\upsilon^2)\overline{b}^2} = \frac{4 \times \pi^2 \times 210 \times 10^3 \times 10^2}{12(1-0.3^2) \times 1000^2} = 75.9\,\text{MPa}$$

Shear stresses (3-1-5/clause 5.3):

$$\tau_{cr} = \frac{k_\tau \pi^2 E t^2}{12(1-\nu^2)b^2} = \frac{9.43 \times \pi^2 \times 210 \times 10^3 \times 10^2}{12(1-0.3^2) \times 1000^2} = 179\,\text{MPa}$$

$$\text{where } k_\tau = 5.34 + 4.00\left(\frac{b}{a}\right)^2 = 5.34 + 4.00\left(\frac{1000}{1000}\right)^2 = 9.43$$

(i) $\sigma_{x,Ed}$ only:
The load amplification factor to reach the characteristic resistance is from expression 3-1-5/(10.3):

$$\frac{1}{\alpha_{ult,k}^2} = \left(\frac{\sigma_{x,Ed}}{f_y}\right)^2 + \left(\frac{\sigma_{z,Ed}}{f_y}\right)^2 - \left(\frac{\sigma_{x,Ed}}{f_y}\right)\left(\frac{\sigma_{z,Ed}}{f_y}\right) + 3\left(\frac{\tau_{Ed}}{f_y}\right)^2 \text{ so } \alpha_{ult,k} = \left(\frac{355}{100}\right)$$
$$= 3.55$$

For separately applied stresses:

$\alpha_{cr,x} = 0.76$ so $\alpha_{cr} = 0.76$

From equation (D6.2-5), the slenderness for sub-panel buckling is:

$$\overline{\lambda}_p = \sqrt{\frac{\alpha_{ult,k}}{\alpha_{cr}}} = \sqrt{\frac{3.55}{0.76}} = 2.16$$

The reduction factor for longitudinal direct stress is therefore from expression 3-1-5/(4.2):

$$\rho_x = \frac{\overline{\lambda}_p - 0.055(3+\psi)}{\overline{\lambda}_p^2} = \frac{2.16 - 0.055(3+1)}{2.16^2} = 0.416$$

so from expression 3-1-5/(10.5):

$$\left(\frac{\sigma_{x,Ed}}{\rho_x f_y/\gamma_{M1}}\right)^2 + \left(\frac{\sigma_{z,Ed}}{\rho_z f_y/\gamma_{M1}}\right)^2 - \left(\frac{\sigma_{x,Ed}}{\rho_x f_y/\gamma_{M1}}\right)\left(\frac{\sigma_{z,Ed}}{\rho_z f_y/\gamma_{M1}}\right) + 3\left(\frac{\tau_{Ed}}{\chi_v f_y/\gamma_{M1}}\right)^2$$
$$= \left(\frac{100}{0.416 \times 355/1.1}\right)^2 = 0.55 < 1.0$$

The actual usage factor is $\sqrt{0.55} = \mathbf{0.74}$

(ii)* $\sigma_{x,Ed}$ *and* $\sigma_{z,Ed}$ *only:
The load amplification factor to reach the characteristic resistance is given by:

$$\frac{1}{\alpha_{ult,k}^2} = \left(\frac{\sigma_{x,Ed}}{f_y}\right)^2 + \left(\frac{\sigma_{z,Ed}}{f_y}\right)^2 - \left(\frac{\sigma_{x,Ed}}{f_y}\right)\left(\frac{\sigma_{z,Ed}}{f_y}\right) + 3\left(\frac{\tau_{Ed}}{f_y}\right)^2$$

$$= \left(\frac{100}{355}\right)^2 + \left(\frac{100}{355}\right)^2 - \left(\frac{100}{355}\right)\left(\frac{100}{355}\right) = 0.079 \text{ so } \alpha_{ult,k} = 3.55$$

the same as for uniaxial compression.

For separately applied stresses:

$$\alpha_{cr,x} = \alpha_{cr,z} = \frac{\sigma_{cr,x}}{\sigma_{Ed,x}} = \frac{75.9}{100} = 0.76$$

For stresses applied together, the critical load factor is given by expression 3-1-5/(10.6):

$$\frac{1}{\alpha_{cr}} = \frac{1+1}{4\times 0.76} + \frac{1+1}{4\times 0.76} + \left[\left(\frac{1+1}{4\times 0.76} + \frac{1+1}{4\times 0.76}\right)^2\right]^{1/2} = 2.632 \text{ so } \alpha_{cr} = 0.380$$

From equation (D6.2-5), the slenderness for sub-panel buckling is:

$$\bar{\lambda}_p = \sqrt{\frac{\alpha_{ult,k}}{\alpha_{cr}}} = \sqrt{\frac{3.55}{0.380}} = 3.056$$

The reduction factor for longitudinal and transverse direct stress is therefore:

$$\rho_x = \frac{\bar{\lambda}_p - 0.055(3+\psi)}{\bar{\lambda}_p^2} = \frac{3.056 - 0.055(3+1)}{3.056^2} = 0.304$$

so from expression 3-1-5/(10.5):

$$\left(\frac{\sigma_{x,Ed}}{\rho_x f_y/\gamma_{M1}}\right)^2 + \left(\frac{\sigma_{z,Ed}}{\rho_z f_y/\gamma_{M1}}\right)^2 - \left(\frac{\sigma_{x,Ed}}{\rho_x f_y/\gamma_{M1}}\right)\left(\frac{\sigma_{z,Ed}}{\rho_z f_y/\gamma_{M1}}\right) + 3\left(\frac{\tau_{Ed}}{\chi_v f_y/\gamma_{M1}}\right)^2$$

$$= \left(\frac{100}{0.304\times 355/1.1}\right)^2 + \left(\frac{100}{0.304\times 355/1.1}\right)^2$$

$$- \left(\frac{100}{0.304\times 355/1.1}\right)\left(\frac{100}{0.304\times 355/1.1}\right) = 1.04 > 1.0$$

The actual usage factor is $\sqrt{1.04} = \mathbf{1.02}$

(iii)* $\sigma_{x,Ed}$*,* $\sigma_{z,Ed}$ *and* τ_{Ed}*:
The load amplification factor to reach the characteristic resistance is given by:

$$\frac{1}{\alpha_{ult,k}^2} = \left(\frac{\sigma_{x,Ed}}{f_y}\right)^2 + \left(\frac{\sigma_{z,Ed}}{f_y}\right)^2 - \left(\frac{\sigma_{x,Ed}}{f_y}\right)\left(\frac{\sigma_{z,Ed}}{f_y}\right) + 3\left(\frac{\tau_{Ed}}{f_y}\right)^2$$

$$= \left(\frac{100}{355}\right)^2 + \left(\frac{100}{355}\right)^2 - \left(\frac{100}{355}\right)\left(\frac{100}{355}\right) + 3\left(\frac{100}{355}\right)^2 = 0.31 \text{ so } \alpha_{ult,k} = 1.775$$

For separately applied stresses:

$$\alpha_{cr,x} = \alpha_{cr,z} = \frac{\sigma_{cr,x}}{\sigma_{Ed,x}} = \frac{75.9}{100} = 0.76$$

$$\alpha_{cr,\tau} = \frac{\tau_{cr}}{\tau_{Ed}} = \frac{179}{100} = 1.79$$

For stresses applied together, the critical load factor is given by expression 3-1-5/(10.6):

$$\frac{1}{\alpha_{cr}} = \frac{1+1}{4 \times 0.76} + \frac{1+1}{4 \times 0.76} + \left[\left(\frac{1+1}{4 \times 0.76} + \frac{1+1}{4 \times 0.76}\right)^2 + 0 + 0 + \frac{1}{1.79^2}\right]^{1/2} = 2.745$$

so $\alpha_{cr} = 0.364$.

From equation (D6.2-5), the slenderness for sub-panel buckling is:

$$\overline{\lambda}_p = \sqrt{\frac{\alpha_{ult,k}}{\alpha_{cr}}} = \sqrt{\frac{1.775}{0.364}} = 2.207$$

The reduction factor for longitudinal and transverse direct stress is therefore:

$$\rho_x = \frac{\overline{\lambda}_p - 0.055(3+\psi)}{\overline{\lambda}_p^2} = \frac{2.207 - 0.055(3+1)}{2.207^2} = 0.408$$

The reduction factor for shear stress at this slenderness, assuming rigid end-post boundary conditions, is:

$$\chi_w = \frac{1.37}{(0.7 + \overline{\lambda}_w)} = \frac{1.37}{(0.7 + 2.207)} = 0.471, \text{ so from expression 3-1-5/(10.5):}$$

$$\left(\frac{\sigma_{x,Ed}}{\rho_x f_y/\gamma_{M1}}\right)^2 + \left(\frac{\sigma_{z,Ed}}{\rho_z f_y/\gamma_{M1}}\right)^2 - \left(\frac{\sigma_{x,Ed}}{\rho_x f_y/\gamma_{M1}}\right)\left(\frac{\sigma_{z,Ed}}{\rho_z f_y/\gamma_{M1}}\right) + 3\left(\frac{\tau_{Ed}}{\chi_v f_y/\gamma_{M1}}\right)^2$$

$$= \left(\frac{100}{0.41 \times 355/1.1}\right)^2 + \left(\frac{100}{0.41 \times 355/1.1}\right)^2$$

$$- \left(\frac{100}{0.41 \times 355/1.1}\right)\left(\frac{100}{0.41 \times 355/1.1}\right) + 3\left(\frac{100}{0.471 \times 355/1.1}\right)^2 = 1.86 > 1.0$$

The actual usage factor is $\sqrt{1.86} = \mathbf{1.36}$

6.2.2.6.3. Comparison with effective section method of EN 1993-1-5 clause 4

Despite apparent conservatism in the reduced stress method (such as no consideration of load shedding, higher value of material factor and shear moment interaction for all magnitudes of shear and moment), it can be *less* conservative than the effective section method:

(i) For unstiffened panels in isolation, the methods are equivalent for uniform compression. 3-1-5/clause 10 is however most conservative because of the higher material factor.

(ii) For stiffened panels, unless finite-element analysis is done, coupled modes of sub-panel and overall plate buckling cannot be checked, whereas this is considered in 3-1-5/clause 4. 3-1-5/clause 10 is therefore not always more conservative than 3-1-5/clause 4 for a single stiffened panel in isolation, despite using a lower material factor.

(iii) For stiffened panels with uniform compression where there is no reduction in strength due to overall plate buckling, 3-1-5/clause 10 is more conservative as the reduction for sub-panel buckling is effectively applied to the stiffener outstands as well, since no load shedding is possible and stresses develop uniformly across the gross cross-section.

(iv) For stiffened panels with uniform compression and with some overall reduction in strength due to buckling, the most conservative method varies depending on the relative area of sub-panels and stiffener outstands.

(v) For stiffened panels with very high slenderness for overall plate buckling modes, the methods are again equivalent as $\chi \rightarrow 1/\overline{\lambda}^2$ as $\overline{\lambda} \rightarrow \infty$.

For 3-1-5/clause 4, for overall buckling:

$$\overline{\lambda}_c = \sqrt{\frac{A_{eff} f_y}{A\sigma_{cr,c}}} \text{ so } \chi = \frac{1}{\overline{\lambda}^2} = \frac{A\sigma_{cr,c}}{A_{eff} f_y} \text{ and panel resistance} = \chi A_{eff} f_y = A\sigma_{cr,c}$$

For 3-1-5/clause 10, for overall buckling:

$$\bar{\lambda}_c = \sqrt{\frac{f_y}{\sigma_{cr,c}}} \text{ so } \chi = \frac{1}{\bar{\lambda}^2} = \frac{\sigma_{cr,c}}{f_y} \text{ and panel resistance} = \chi A f_y = A\sigma_{cr,c}$$

This is like the Perry-Robertson result for struts where very slender members fail at the Euler load, which does not depend on the cross-section area.

(vi) For flanges, which are essentially like the isolated panel cases above, neither 3-1-5/clause 4 nor 3-1-5/clause 10 is always the more conservative for the reasons above.

(vii) For web panels, 3-1-5/clause 10 will generally be most conservative despite the above since in the effective section method of 3-1-5/clause 4, the web can shed most of its direct stress to the flanges without the overall flange direct stress increasing much. In 3-1-5/clause 10, a single overstressed web panel can govern the design.

3-1-1/clause 6.2.3(1)
3-1-1/clause 6.2.3(2)

6.2.3. Tension members

3-1-1/clause 6.2.3(1) and ***3-1-1/clause 6.2.3(2)*** give the basic requirement for the cross-section resistance as follows:

$$\frac{N_{Ed}}{N_{t,Rd}} \leq 1.0 \qquad \text{3-1-1/(6.5)}$$

where N_{Ed} is the applied design tension force and $N_{t,Rd}$ is the tension resistance taken as the lesser of:

(a) The design plastic resistance of the gross cross-section $N_{pl,Rd}$:

$$N_{pl,Rd} = \frac{Af_y}{\gamma_{M0}} \qquad \text{3-1-1/(6.6)}$$

where A is the gross area of the steel component and f_y is the yield stress of the steel component.

(b) The design ultimate resistance of the net cross-section $N_{u,Rd}$ (fastener holes deducted):

$$N_{u,Rd} = \frac{0.9A_{net}f_u}{\gamma_{M2}} \qquad \text{3-1-1/(6.7)}$$

where A_{net} is the net area determined in accordance with 3-1-1/clause 6.2.2.2 and f_u is the ultimate tensile stress of the steel component. The 0.9 factor on A_{net} allows for a non-uniform distribution of stress across the net section arising from stress concentrations or minor eccentricities.

Tension members are deemed to 'fail' when their increase in length becomes unacceptable or a section ruptures. Expression 3-1-1/(6.7) allows the ultimate tensile stress to be taken in conjunction with the net cross-section because the length of the connection is usually small compared to the total length of the steel component. The resulting increase in length caused by the plastic strain of the connection zone will generally be minimal compared to the increase in length of the rest of the member. Use of the ultimate tensile stress is not allowed, however, in conjunction with category C connections (which are non-slip at ultimate) according to ***3-1-1/clause 6.2.3(4)***. This is because large plastic strains in the material adjacent to the bolt would result in a reduction of thickness of the plate and a consequent reduction in bolt clamping force. In this case, yield has to be checked on the net section as follows:

3-1-1/clause 6.2.3(4)

$$N_{net,Rd} \leq \frac{A_{net}f_y}{\gamma_{M0}} \qquad \text{3-1-1/(6.8)}$$

Situations where category C connections might be required for bridges are discussed in section 5.2.1.2 of this guide.

Expressions 3-1-1/(6.6) and 3-1-1/(6.7) can be combined to give allowable ratios of A_{net}/A for which the effect of bolt holes on the section resistance can be neglected as follows:

$$\frac{A_{net}}{A} \geq \frac{f_y \gamma_{M2}}{0.9 f_u \gamma_{M0}} \tag{D6.2-6}$$

If the recommended partial material factors are accepted, then the minimum allowable ratio A_{net}/A for S355 steel is 0.97 (using $\gamma_{M2} = 1.25$, $\gamma_{M0} = 1.00$, $f_y = 355$ MPa and $f_u = 510$ MPa from 3-1-1/Table 3.1). This shows that for tension in an S355 steel, the presence of even small bolt holes may reduce the section resistance. If the ultimate strength is taken from EN 10025, then $f_u = 490$ MPa and the minimum ratio A_{net}/A becomes 1.0. This means that the check in expression 3-1-1/(6.7) would always govern. There is advantage therefore in using material properties from 3-1-1/Table 3.1, but the UK National Annex requires properties to be taken from EN 10025. For S275 steel the minimum ratio A_{net}/A is either 0.89 or 0.93 depending on whether the ultimate tensile strength is taken from 3-1-1/Table 3.1 or EN 10025 respectively.

A further restriction on $N_{t,Rd}$, imposed by ***3-1-1/clause 6.2.3(3)***, occurs when a structure is required to have ductile behaviour for seismic design in accordance with EN 1998. This means that the gross-section should fail by yielding rather than by rupture of the net section. To achieve this, the resistance from expression 3-1-1/(6.7) must exceed that from expression 3-1-1/(6.6) and the limiting ratio A_{net}/A should be as calculated above from equation (D6.2-6). *3-1-1/clause 6.2.3(3)*

Where sections have eccentric end connections (either due to member asymmetry or asymmetric connections), this eccentricity should be allowed for. The check of net section for angles connected through a single leg is explicitly covered in ***3-1-1/clause 6.2.3(5)*** by reference to 3-1-8/clause 3.10.3. These resistances allow for the eccentricity by placing a reduction factor on the net area used. Where an unequal angle is connected via holes on its smaller leg only, the net area for use with EN 1993-1-8 is based on a fictitious equal angle with leg size based on the smaller of those for the real unequal angle. Worked Example 6.2-7 illustrates the use of the formulae in EN 1993-1-8 (which are not otherwise reproduced here). Welded connections can similarly be treated with A_{net} calculated for a section with no holes as recommended in 3-1-8/clause 4.13. Other section types are not explicitly covered. For single T-sections connected through the flange and channel sections connected through the web, A_{net} could reasonably be taken as the effective net area of the connected part of the cross-section plus half the area of the outstand parts. This net area could then be used with expression 3-1-1/(6.7). It should also be ensured that the gross area of the same section satisfies the yield check of expression 3-1-1/(6.6). *3-1-1/clause 6.2.3(5)*

Worked Example 6.2-7: Angle in tension

A 100 × 100 × 12 Rolled Steel Angle (Grade S355) contains 4 No. 26 mm diameter holes for fasteners at either end. The connection detail is category B (slip allowed at ULS) with geometry as follows.

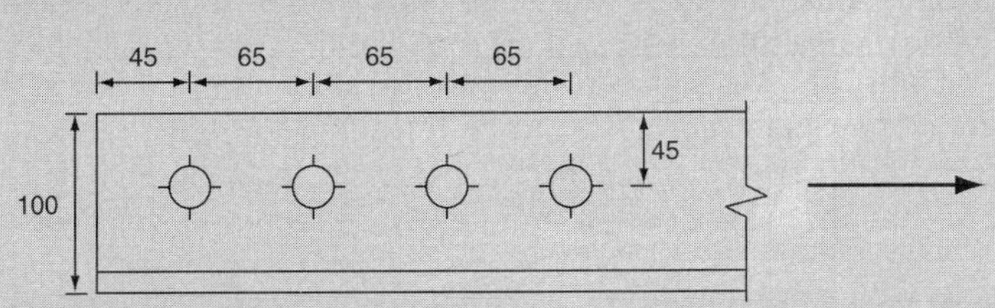

The maximum tensile resistance of the angle at the connection is calculated.

Gross area of angle = 2270 mm^2.

From 3-1-1/Table 3.1: $f_y = 355$ MPa, $f_u = 510$ MPa for 12 mm plate. (Note that a National Annex may require properties to be derived from EN 10025; the UK National Annex does.)

The design plastic resistance of the gross cross-section is given by:

$$N_{\text{pl,Rd}} = \frac{Af_y}{\gamma_{M0}} = \frac{2270 \times 355}{1.0} = 805.9\,\text{kN}$$

As the connection involves a single angle connected by one leg, the eccentricity must be considered and the rules for net area in 3-1-8/clause 3.10.3 apply. The net area is modified by a factor β_3 to allow for the eccentricity of the bolts. As the angle is an equal angle, A_{net} may be determined from the actual gross area.

$$N_{\text{u,Rd}} = \frac{\beta_3 A_{\text{net}} f_u}{\gamma_{M2}} \quad \text{where } \beta_3 = 0.5 \text{ for bolt pitch} \leq 2.5 \times \text{hole diameter}$$

$$N_{\text{u,Rd}} = \frac{0.5 \times (2270 - 26 \times 12) \times 510}{1.25} = 399.4\,\text{kN}$$

Therefore, $N_{\text{t,Rd}}$ is governed by the net area. $N_{\text{t,Rd}} = \mathbf{399.4\,kN}$

6.2.4. Compression members

3-1-1/clause 6.2.4(1)

3-1-1/clause 6.2.4(1) gives the basic requirement for the cross-section resistance as follows:

$$\frac{N_{\text{Ed}}}{N_{\text{c,Rd}}} \leq 1.0 \qquad \text{3-1-1/(6.9)}$$

where N_{Ed} is the applied design compressive force and $N_{\text{c,Rd}}$ is the design resistance for uniform compression.

The checks in 3-2/clause 6.2.4 relate only to local cross-sections. If the overall member is prone to flexural or flexural–torsional buckling, then this mode of failure must also be checked as discussed in section 6.3.1 of this guide.

3-2/clause 6.2.4(2)

3-2/clause 6.2.4(2) requires different calculation approaches for members which are in Class 1, 2 or 3 and those which are in Class 4.

Provided that the cross-section is either Class 1, 2 or 3, it will be able to achieve the yield stress in compression without local buckling occurring. The cross-section resistance will then simply be the product of the area of the section and the yield stress as follows:

$$N_{\text{c,Rd}} = \frac{Af_y}{\gamma_{M0}} \qquad \text{3-2/(6.1)}$$

Class 4 cross-sections are susceptible to local buckling at a stress lower than the yield stress. Two options are provided for calculating the cross-section compression resistance.

The first option is to use the effective section method, discussed in detail in section 6.2.2.5 of this guide. For bisymmetric sections, the compression resistance is as follows:

$$N_{\text{c,Rd}} = \frac{A_{\text{eff}} f_y}{\gamma_{M0}} \qquad \text{3-2/(6.2)}$$

The effective area, A_{eff}, is calculated as discussed in section 6.2.2.5 of this guide.

When the effective section for axial load is calculated for an asymmetric section, the neutral axis will shift an amount e_N from its original position on the gross cross-section. This shift produces a moment of the axial force on the cross-section about the new neutral axis position. The additional bending stresses must be included in the check of cross-section resistance as discussed in more detail in section 6.2.10.3 of this guide which deals with bending and axial force. ***3-1-1/clause 6.2.4(4)*** is a reminder that this effect must be considered. It does not apply if the second method below is used to design the Class 4 cross-section.

3-1-1/clause 6.2.4(4)

The second option is to use the gross cross-section area but limit the axial stresses to some derived value less than the yield strength as follows:

$$N_{\text{c,Rd}} = \frac{A\sigma_{\text{limit}}}{\gamma_{M0}} \qquad \text{3-2/(6.3)}$$

where σ_{limit} is the limiting compressive stress of the weakest part of the cross-section as determined by the method of 3-1-5/clause 10 which is discussed in detail in section 6.2.2.6 of this guide. This method is also discussed in more detail in the section on bending and axial force in section 6.2.10 of this guide.

3-1-1/clause 6.2.4(3)

3-1-1/clause 6.2.4(3) states that fastener holes do not need to be deducted from the area provided that they are 'filled' by fasteners and are not oversize or slotted. Previous practice in the UK was similar, although some reduction was made for holes containing black bolts (which would be Category A connections in 3-1-8/clause 3.4.1(1)), because black bolts were not deemed to 'fill' the holes. A similar distinction could be made here, but as 3-2/clause 2.1.3.3(4) requires bolted connections in bridges to be Category B or C or alternatively to use closely fitted bolts, it will always be possible to consider holes to be filled in accordance with this clause.

Worked Example 6.2-8: universal column in compression

A 152 × 152 × 37 Universal Column (Grade S355) is fully restrained with regard to flexural buckling. Calculate the maximum compression force that can be withstood by the Universal Column. All holes in the section are filled with preloaded bolts.

The first check is to determine the cross-section classification of the section to see whether local buckling is possible.

From section tables:
Area of UC = 4740 mm^2
Flange outstand aspect ratio $(c/t) = 6.36$ (conservatively taken to face of web)
Web aspect ratio $(c/t) = 17.1$ (conservatively taken between faces of flanges)

From 3-1-1/Table 5.2:
Flange is Class 1 $(c/t \leq 9\varepsilon = 9 \times 0.81 = 7.29)$
Web is Class 1 $(c/t \leq 33\varepsilon = 33 \times 0.81 = 40.7)$
Therefore section is Class 1 – local buckling will not occur.

From expression 3-2/(6.1):

$$N_{c,Rd} = \frac{Af_y}{\gamma_{M0}} = \frac{4740 \times 355}{1.0} = 1682.7\,\text{kN}$$

Therefore, compression resistance of UC = **1682.7 kN**

6.2.5. Bending moment

3-1-1/clause 6.2.5(1)

3-2/clause 6.2.5(1) refers to ***3-1-1/clause 6.2.5(1)*** for cross-section bending resistance. The basic requirement is as follows:

$$\frac{M_{Ed}}{M_{c,Rd}} \leq 1.0 \qquad \text{3-1-1/(6.12)}$$

where M_{Ed} is the applied design bending moment and $M_{c,Rd}$ is the design resistance for bending of the steel beam.

3-2/clause 6.2.5(2)

The checks in this section relate only to local cross-sections. If the overall member is prone to lateral torsional buckling, then this mode of failure must also be checked as discussed in section 6.3.2 of this guide. ***3-2/clause 6.2.5(2)*** requires different approaches for cross-section design depending on section Class.

Class 1 cross-sections

As discussed in section 5.5 of this guide, Class 1 cross-sections can develop a full plastic hinge. The design resistance of the beam corresponds to a fully plastic internal stress distribution as shown in Fig. 6.2-30.

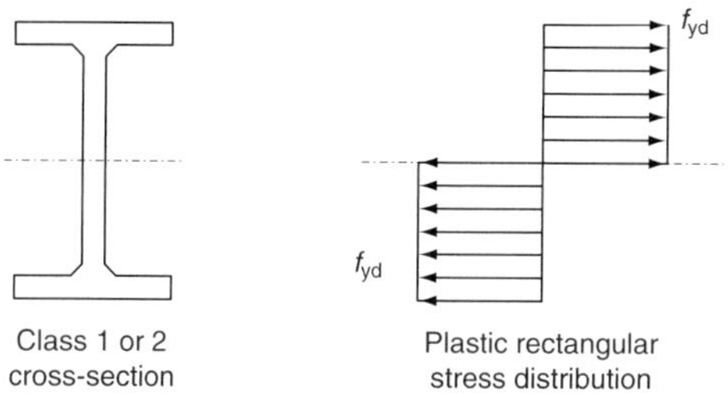

Fig. 6.2-30. Stress block for Class 1 or 2 cross-sections

The resistance moment is given by:

$$M_{c,Rd} = \frac{W_{pl} f_y}{\gamma_{M0}} \qquad 3\text{-}2/(6.4)$$

If the yield strength is not constant throughout the cross-section, then the plastic modulus, W_{pl}, cannot be used and the resistance needs to be computed directly from the plastic stress block. This will frequently be the case and the form of expression 3-2/(6.4) is not very useful.

Class 2 cross-sections

Class 2 cross-sections can also develop a full plastic resistance but have limited rotation capacity. The design resistance again corresponds to the plastic stress distribution as shown in Fig. 6.2-30, with resistance according to expression 3-2/(6.4).

If all sections in a bridge of continuous construction are not in either Class 1 or Class 2, then some care should be used with mixing classes in cross-section design throughout the bridge when elastic global analysis has been used. This is because when the yield point of a Class 1 or 2 cross-section is reached, its stiffness will be reduced for further increments of load, even though it may be some way off its final full plastic resistance. This loss of stiffness means that the moment attracted to adjacent unyielded areas with bending moment of the opposite sign will be greater than that predicted by elastic analysis. If these areas have Class 3 or Class 4 cross-sections, failure at these sections will be by local buckling with limited rotation capacity. This shedding of moment to a Class 3 or 4 section must be checked such that its resistance is not exceeded. If mixed class section design is to be used, the checks suggested in section 5.4.2 of this guide (where the problem is discussed in more detail) should be made.

Class 3 cross-sections

Class 3 cross-sections can develop compressive yield at their extreme fibres but will fail by local buckling if this yielding starts to spread further into the cross-section. The maximum resistance is therefore reached when the extreme compression fibre reaches yield. Generally, partial plastification of the tension zone is not considered in design and the resistance is considered to be reached when the stress from an elastic stress distribution reaches yield at either fibre, whether compressive or tensile, as shown in Fig. 6.2-31. Note that an extreme fibre is defined in 3-1-1/clause 6.2.1(9) as being at the mid-plane of a flange rather than its outer surface.

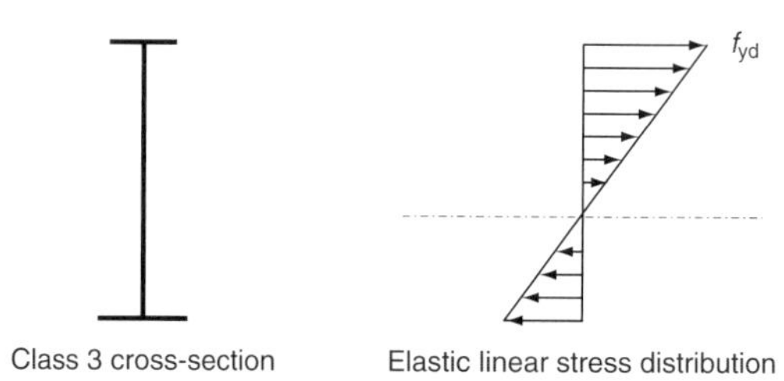

Fig. 6.2-31. Elastic stress distribution for Class 3 cross-sections

Fig. 6.2-32. Partially plastic stress distribution for Class 3 sections

The resistance moment is then given by:

$$M_{c,Rd} = \frac{W_{el,min} f_y}{\gamma_{M0}} \qquad \text{3-1-1/(6.5)}$$

where $W_{el,min}$ is the section modulus at the fibre with maximum stress for the reason given above.

Partial plastification of the tension zone may however be considered in accordance with 3-1-1/clause 6.2.1(10) if yielding first occurs on the tension side of a Class 3 cross-section, as a plastic stress block can develop in the tension zone until yield is reached at the extreme compression fibre. This could occur where the compression flange is the larger flange, as illustrated in Fig. 6.2-32. The resistance moment is then determined for this situation by assuming plane sections remain plane, a bilinear stress–strain curve and by balancing forces in tension and compression zones. The neutral axis will move as plasticity spreads throughout the tension zone and this can then affect the section classification. This complexity is one reason for the usual simplification of restriction to elastic behaviour.

Class 4 cross-sections

Class 4 cross-sections fail by local buckling before they reach yield. 3-2/clause 6.2.5(2)b) allows two methods to be used to calculate the bending resistance; the effective area method and the limiting stress method. These methods are explained in detail in sections 6.2.2.5 and 6.2.2.6 of this guide respectively. The latter method can be conservative as it does not allow shedding of load between panels.

For the effective area method, the resistance moment is obtained when yield is reached at an extreme fibre of the effective section as illustrated in Fig. 6.2-33.

$$M_{c,Rd} = \frac{W_{eff,min} f_y}{\gamma_{M0}} \qquad \text{3-2/(6.6)}$$

where $W_{eff,min}$ is the smallest elastic section modulus of the effective cross-section determined as discussed in section 6.2.2.5.

For the limiting stress method, the gross cross-section is used but the resistance moment is deemed to be obtained when the weakest panel in compression fails by local buckling. This leads to the use of a limiting stress, σ_{limit}, less than the yield stress as shown in Fig. 6.2-34.

$$M_{c,Rd} = \frac{W_{el,min} \sigma_{limit}}{\gamma_{M0}} \qquad \text{3-2/(6.7)}$$

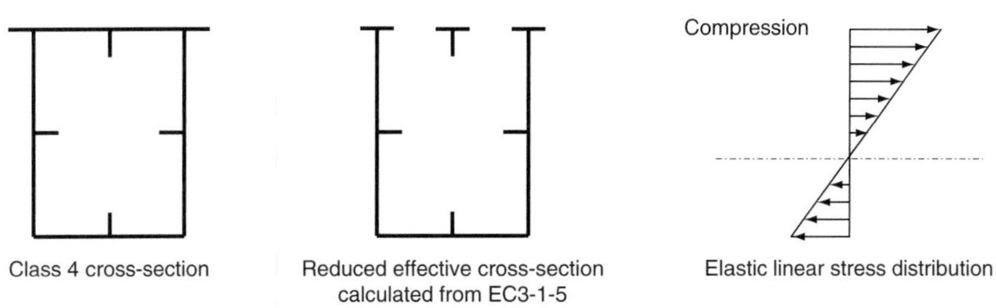

Fig. 6.2-33. Elastic stress distribution for Class 4 sections

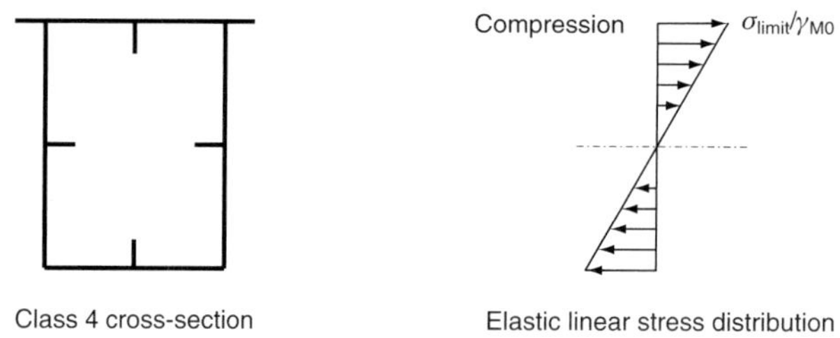

Fig. 6.2-34. Elastic stress distribution for Class 4 equivalent Class 3 section

The concept of σ_{limit} is a slightly strange one for cross-section checks as, in order to determine σ_{limit}, the section must first be checked under bending stresses alone according to the method of 3-1-5/clause 10. This involves checking all the constituent parts of the cross-section, which may have different allowable stresses, and verifying that they are all satisfactory. The verification of 3-1-5/clause 10 is thus itself a check of the cross-section and there is no real need to determine σ_{limit} itself for cross-section checks.

The definition of σ_{limit} in 3-2/clause 6.2.5 as 'the limiting stress of the weakest part of the cross-section in compression' is very conservative where the panel which buckles first is not at the extreme fibre and is consequently not subject to the maximum stress. σ_{limit} could therefore be determined as the peak compressive bending stress at an extreme fibre such that failure occurs by local buckling somewhere within the cross-section, not necessarily at the most stressed fibre where σ_{limit} is attained. The value of σ_{limit} should obviously not exceed f_y. In expression 3-2/(6.7), the value of $W_{el,min}$ can conservatively be taken as the minimum for either compression or tension fibre as is currently stated in EN 1993-2. However, it is more logical to check the compression fibre for a stress of σ_{limit} and the tension fibre for a stress of f_y so the moment resistance is the minimum value of:

$$M_{c,Rd} = \frac{W_{el,comp}\sigma_{limit}}{\gamma_{M0}}$$

$$M_{c,Rd} = \frac{W_{el,ten}f_y}{\gamma_{M0}}$$

A full check to 3-1-5/clause 10 requires shear, axial force, bending moment and transverse load to be considered at the same time. When this full check is carried out, a check under bending moment on its own becomes redundant (unless the other effects are zero); the full check will be more critical. Consequently, it is recommended here that if Class 4 cross-sections are to be treated as Class 3, the entire check should be performed using 3-1-5/clause 10, as discussed in section 6.2.2.6 of this guide, without reference to σ_{limit}. There is an inconsistency with expression 3-2/(6.7) in that the material factor γ_{M1} is used in 3-1-5/clause 10.

It should be noted however that σ_{limit} will still be needed for member buckling checks for Class 4 members if they are treated as Class 3, as discussed in section 6.2.2.6 of this guide.

Fastener holes

Fastener holes in the beam cross-section tension zone need to be considered when calculating the relevant section properties. ***3-1-1/clause 6.2.5(4)*** allows fastener holes in the tension flange to be neglected provided the following equation is met:

3-1-1/clause 6.2.5(4)

$$\frac{A_{f,net}0.9f_u}{\gamma_{M2}} \geq \frac{A_f f_y}{\gamma_{M0}} \qquad \text{3-1-1/(6.16)}$$

where $A_{f,net}$ is the net area of the tension flange. This is the same as equation (D6.2-6) derived in section 6.2.3 for tension members. The area of the tension flange used in the bending check will need to be reduced if the above equation cannot be met. Either the net area could be used (which would be very conservative) or it would be possible to reduce the flange area to an effective value, A'_f, such that expression 3-1-1/(6.16) is satisfied. Consequently, the reduced flange area is given by:

$$A'_{\mathrm{f}} = \frac{A_{\mathrm{f,net}} 0.9 f_{\mathrm{u}} / \gamma_{\mathrm{M2}}}{f_{\mathrm{y}} / \gamma_{\mathrm{M0}}} \qquad \text{(D6.2-7)}$$

An exception to the use of equation (D6.2-7) is where a bridge is required to have ductile behaviour for seismic design in accordance with EN 1998, or where plastic global analysis is to be used. In these cases, the gross-section should fail by yielding rather than by rupture of the net section. To achieve this, the criterion in expression 3-1-1/(6.16) needs to be met.

3-1-1/clause 6.2.5(5) allows fastener holes in the web tension zone to be neglected if expression 3-1-1/(6.16) is satisfied for the entire tension zone comprising tension flange and the part of the web in tension where there are holes. In this case, the relevant areas are those for the entire tension zone. *3-1-1/clause 6.2.5(5)*

Fastener holes in the compression zone need not be allowed for, according to ***3-1-1/clause 6.2.5(6)***, providing they are filled by fasteners and are not oversize or slotted holes. This requirement is discussed in section 6.2.4 of this guide. *3-1-1/clause 6.2.5(6)*

6.2.6. Shear

This sub-section of EN 1993-2 has been split into two further sub-sections in this guide which deal with the plastic shear resistance and the shear buckling resistance respectively.

6.2.6.1. Shear resistance without shear buckling

A feature of shear design to EN 1993 that will be unfamiliar to UK designers is that the shear resistance of a stocky web may exceed its resistance based on the Von Mises yield stress in shear, $f_{\mathrm{y}}/\sqrt{3}$. This is because tests have shown that strain hardening allows a higher resistance to be mobilized without excessive deformation occurring. Both 3-1-1/clause 6.2.6 and 3-1-5/clause 5 include a factor, η, to take this into account. This factor is defined in ***3-1-5/clause 5.1(2)*** but its numerical value is subject to national choice. For steel grades up to S460, $\eta = 1.2$ is recommended, which is equivalent to an average web shear stress of $0.7 f_{\mathrm{y}}$. For grades above S460, $\eta = 1.0$ is recommended since strain hardening is less significant with higher steel grades. However, a background paper by Johansson *et al.*[12] recommended that the value of $\eta = 1.2$ should only be used for steels up to Grade S355 due to a lack of test results for higher grades. The typical ratios $f_{\mathrm{u}}/f_{\mathrm{y}}$ for S460 steels suggest that $\eta = 1.2$ might be acceptable for these also, but the recommendation of this guide is to take $\eta = 1.0$ for steels of Grade S460 and above in the absence of test evidence. In EN 1993-1-1, the factor η is included in the shear area (3-1-1/clause 6.2.6(3)) but in EN 1993-1-5 it appears directly in the resistance formula (3-1-5/clause 5.2(1)), so care must be taken not to include the effect twice when switching between parts of EN 1993. *3-1-5/clause 5.1(2)*

In the absence of shear buckling, the shear resistance is based on the plastic resistance from ***3-1-1/clause 6.2.6(2)***: *3-1-1/clause 6.2.6(2)*

$$V_{\mathrm{pl,Rd}} = \frac{A_{\mathrm{v}} f_{\mathrm{y}}}{\sqrt{3}\gamma_{\mathrm{M0}}} \qquad \text{3-1-1/(6.18)}$$

The shear area makes allowance for the effects of strain hardening as discussed above and values of A_{v} are given in ***3-1-1/clause 6.2.6(3)***. The shear area for the web of a fabricated I-girder being sheared parallel to the web is $\eta h_{\mathrm{w}} t_{\mathrm{w}}$, where h_{w} and t_{w} are the height and thickness of the web respectively. η is obtained from EN 1993-1-5 as discussed above. If the web thickness is not constant, either the minimum thickness should be used in expression 3-1-1/(6.18) or the resistance based on the elastic shear flow distribution described below could be used. *3-1-1/clause 6.2.6(3)*

In situations where there is no interaction formula given in the Eurocodes for combinations of shear and other internal effects, it will be necessary to apply the Von Mises yield criterion, discussed in section 6.2.1 of this guide, to all points of the web. This is conservative as it ignores the plastic redistribution that is assumed in other interaction formulae. The elastic shear stress at a point, when the section properties are constant along the

3-1-1/clause 6.2.6(4)

member, is given in ***3-1-1/clause 6.2.6(4)***:

$$\tau_{Ed} = \frac{V_{Ed}S}{It} = \frac{V_{Ed}Az}{It} \qquad \text{3-1-1/(6.20)}$$

where:

A is the area above the plane being checked;
z is the distance from the member neutral axis to the centroid of area A;
I is the second moment of area of the whole cross-section; and
t is the thickness of the section at the point being checked.

Where the section properties vary along the beam, expression 3-1-1/(6.20) is no longer correct and the shear stress is given by:

$$\tau_{Ed} = \frac{V_{Ed}Az}{It} + \frac{M_{Ed}}{t}\frac{d}{dx}\left(\frac{Az}{I}\right) \qquad \text{(D6.2-8)}$$

In hogging zones where the depth increases towards the support, the second term of equation (D6.2-8) reduces the shear flow so can conservatively be ignored. In sagging zones of beams with a parabolic soffit, the second term can increase the shear flow. The average shear stress, $\tau_{Ed} = V_{Ed}/h_w t_w$, may however be used for I and H sections in accordance with ***3-1-1/clause 6.2.6(5)*** if $A_f/h_w t_w \geq 0.6$, where A_f is the area of one flange.

3-1-1/clause 6.2.6(5)
3-1-1/clause 6.2.6(7)

3-1-1/clause 6.2.6(7) states that fastener holes need not be allowed for in shear design other than in verifying the design shear resistance at connections according to EN 1993-1-8 as discussed in section 8 of this guide. A check of a splice is presented there together with interpretation of the section properties used for shear and bending in the cover plates. Since shear comprises bands of principal tension and compression, it is at first difficult to see why no reduction should be made for bolt holes in shear design when a reduction is made for tension, particularly as the plastic shear resistance already considers allowance for strain hardening. The latter is significant as strain hardening is the justification for allowing some holes in tension members without reducing the yield resistance of the gross cross-section – see section 6.2.3.

Since the tension from shear is inclined, the bolt holes are effectively staggered in the direction of tension and the deduction to area for holes according to 3-1-1/clause 6.2.2 is therefore correspondingly less. However, in view of the inclusion of strain hardening in the shear resistance, it is recommended that caution should be exercised in designing webs up to their full plastic resistance when there are bolt holes, particularly when there are multiple lines of bolts (where the staggering effect is less). A conservative approach would be to fully deduct holes from the shear area when evaluating the full plastic resistance with $\eta = 1.2$.

No guidance is given in EN 1993 on the design of webs with larger holes, such as may be provided for access or services. It is suggested in this guide that the height of holes should simply be deducted from the web height when applying expression 3-1-1/(6.18) if the hole diameter does not exceed 5% of the height of the web. (This is consistent with limitations on the use of rules for shear buckling in 3-1-5/clause 5.1(1).) For greater hole dimensions, the hole should be framed by stiffeners and the stiffened sections designed for the local distribution of shear above and below the hole, together with the secondary bending (Vierendeel action) induced around the hole.

6.2.6.2. Shear buckling

3-1-1/clause 6.2.6(6)
3-1-5/clause 5.1(2)

The resistance of plate girders to shear buckling in 3-1-5/clause 5 is based on the rotated stress field theory proposed by Höglund.[13] Webs become susceptible to shear buckling when the height to thickness ratio, h_w/t, exceeds certain limits. Both ***3-1-1/clause 6.2.6(6)*** and ***3-1-5/clause 5.1(2)*** give such limits. The latter clause gives the following limits beyond which buckling must be checked:

$$h_w/t > \frac{72}{\eta}\varepsilon \text{ for webs without longitudinal stiffeners}$$

$h_w/t > \frac{31}{\eta}\varepsilon\sqrt{k_\tau}$ for webs with longitudinal stiffeners

where:

$\varepsilon = \sqrt{\frac{235}{f_y}}$ and k_τ is a shear buckling coefficient discussed later.

The derivation of the shear buckling rules for the case of widely spaced transverse stiffeners and no longitudinal stiffeners is presented below. It is based on that presented in Reference 12, but includes some minor corrections and extensions to it.

For low shear in the absence of direct stresses, a state of pure shear exists and the principal stresses occur at 45° to the horizontal. For increasing shear, elastic critical buckling occurs and the major principal stress rotates to form an angle of less than 45° to the horizontal due to the formation of tensile horizontal membrane stresses, σ_H. The stress state remains near to pure shear near the flanges however. The state of stress in the web is such that there is no vertical direct stress on the plate edges. The situation is shown in Fig. 6.2-35. The rotated principal tensile stress is at an angle of ϕ to the horizontal and the principal stresses, with tension positive, are therefore:

$$\sigma_1 = \tau/\tan\phi \qquad \text{(D6.2-9)}$$

$$\sigma_2 = -\tau\tan\phi \qquad \text{(D6.2-10)}$$

The angle ϕ is obviously required to proceed and this has to be derived from test observations which suggest that the principal compressive stress remains approximately equal to the elastic critical stress for shear buckling, despite the stress field rotation with increasing shear. Therefore:

$$\sigma_2 = -\tau_{cr} \qquad \text{(D6.2-11)}$$

From equations (D6.2-10) and (D6.2-11), $\tan\phi = \tau_{cr}/\tau$ so equation (D6.2-9) gives:

$$\sigma_1 = \frac{\tau^2}{\tau_{cr}} \qquad \text{(D6.2-12)}$$

The ultimate strength of the web is then assumed to be reached when the equivalent stress, using the Von Mises criterion, reaches yield:

$$\sigma_1^2 + \sigma_2^2 - \sigma_1\sigma_2 = f_y^2 \qquad \text{(D6.2-13)}$$

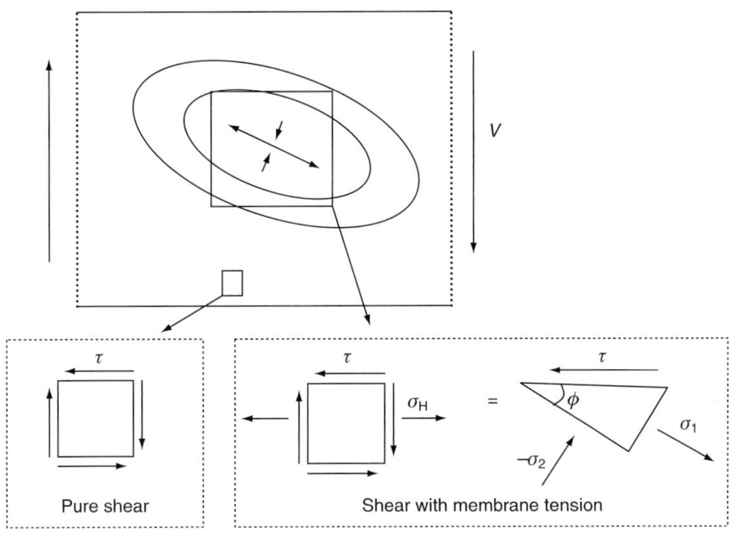

Fig. 6.2-35. Stress field for web after initial elastic buckling load reached

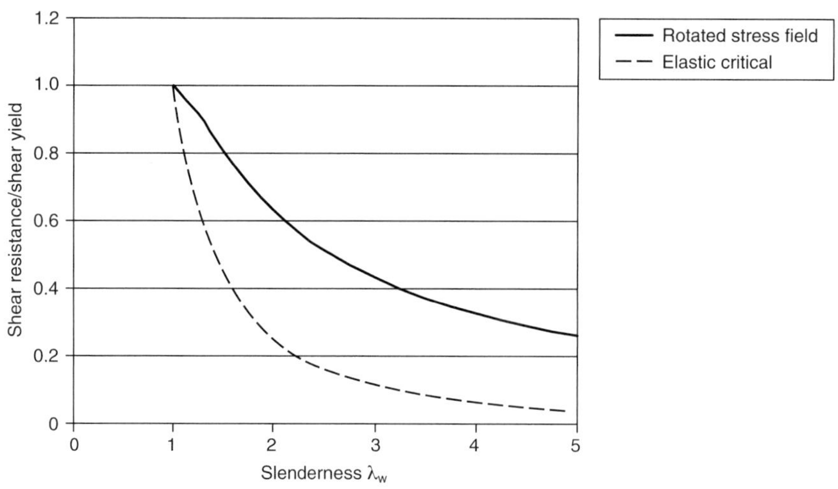

Fig. 6.2-36. Comparison of theoretical rotated stress theory resistance with elastic critical resistance for web without longitudinal stiffeners

Substituting equations (D6.2-11) and (D6.2-12) into equation (D6.2-13) gives the following shear resistance:

$$\frac{\tau_u}{f_v} = \frac{\sqrt[4]{3}}{\lambda_w}\sqrt{\sqrt{1 - \frac{1}{4\lambda_w^4}} - \frac{1}{2\sqrt{3}\lambda_w^2}} \quad \text{(D6.2-14)}$$

with $\lambda_w = \sqrt{f_v/\tau_{cr}}$ and $f_v = f_y/\sqrt{3}$.

The resistance from equation (D6.2-14) is shown in Fig. 6.2-36. It matches reasonably well with test results for cases of shear with rigid end-posts (which can resist the resulting membrane tension assumed above at the beam ends) but is an overestimate for cases where there are no rigid end-posts. Tests, however, show that the longitudinal tension field still develops in girders with non-rigid end-posts, but to a lesser extent. The rigid end-post theory is also adequate for shear at internal supports which are therefore well away from the beam ends. Rigid end posts are discussed in section 6.7 of this guide under the topic of bearing stiffeners. They have to be designed as two double-sided stiffeners to resist the membrane tension acting as a beam spanning between flanges.

It follows from above that the membrane tension to be carried by a rigid end-post can be taken as a force N_H based on the stress σ_H which is assumed to be uniformly distributed over the web depth. This is conservative since the stress state near to the flanges is closer to pure shear as shown in Fig. 6.2-35. The force can be derived as follows.

From equations (D6.2-9) and (D6.2-10) the maximum principal tension is:

$$\sigma_1 = \tau^2/\tau_{cr} \quad \text{(D6.2-15)}$$

From the Mohr's circle of stress in Fig. 6.2-37, the maximum principal tension is related to σ_H and τ by:

$$\sigma_1 = \frac{\sigma_H}{2} + \sqrt{\left(\frac{\sigma_H}{2}\right)^2 + \tau^2} \quad \text{(D6.2-16)}$$

From equations (D6.2-15) and (D6.2-16), the horizontal membrane stress is found to be:

$$\sigma_H = \frac{\tau^2}{\tau_{cr}} - \tau_{cr} \quad \text{(D6.2-17)}$$

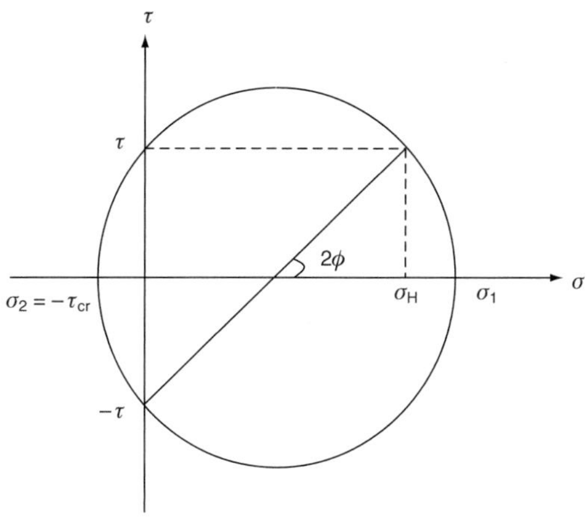

Fig. 6.2-37. Mohr's circle for web element undergoing tension field action

Conservatively assuming that the membrane stress is uniform over the height of the web, the membrane force for a perfectly flat plate is then given by:

$$N_H = h_w t_w \left(\frac{\tau^2}{\tau_{cr}} - \tau_{cr} \right) \quad \text{(D6.2-18)}$$

where h_w and t_w are the height and thickness of the web panel respectively. An expression for τ_{cr} is given in 3-1-5/clause 5.2(3). Equation (D6.2-18) is not strictly valid for real plates with imperfections, but it is used in section 6.7.2.3 of this guide to develop a design equation.

The above method for calculating shear resistance was also shown to be adequate where vertical stiffeners are present by simply including their contribution in the calculation of τ_{cr}.[14] For webs with longitudinal stiffeners however, test results indicate that if the full theoretical elastic critical stress is used to calculate the slenderness, the results are unsafe. This is because a longitudinally stiffened web possesses less post-buckling strength than an unstiffened web. Better agreement with tests on girders with open stiffeners is obtained when the critical stress τ_{cr} is derived using one-third of the longitudinal stiffener second moment of area and this reduction is included in the formulae in 3-1-5/Annex A.3. If formulae are derived indepen dently for the critical stress of stiffened panels, it is essential that a similar reduction to stiffener second moment of area is made before calculating the slenderness – ***3-1-5/clause 5.3(4)*** refers. It is also essential to consider hinged supports at the panel boundaries when deriving the critical stress for compatibility with the resistance curves used in EN 1993-1-5. *3-1-5/clause 5.3(4)*

It should be noted that vertical stiffeners generally have to be designed to be 'rigid' in EN 1993-1-5 if the formulae for shear are to be used, as they assume rigid support along these transverse boundaries. It should be noted that the rotated stress field theory above does not assume any vertical force to be developed in these stiffeners unless the flange can anchor off some additional tension field (the $V_{bf,Rd}$ term) as discussed below. This has led to some considerable debate on the applicable design loads for the stiffeners themselves as discussed in section 6.6 of this guide.

Design resistance to shear buckling – 3-1-5/clause 5.2

The shear buckling resistance in ***3-1-5/clause 5.2(1)*** is given as: *3-1-5/clause 5.2(1)*

$$V_{b,Rd} = V_{bw,Rd} + V_{bf,Rd} \leq \frac{\eta f_{yw} h_w t}{\sqrt{3}\gamma_{M1}} \quad \text{3-1-5/(5.1)}$$

where $V_{bw,Rd}$ is the contribution from the web and $V_{bf,Rd}$ is the contribution from the flange. The background to the web contribution is as discussed above. If a web is inclined, as in a large box girder, the design should be done in the plane of the web, taking h_w as the depth of the web in its plane, and the vertical shear force should be accordingly increased to account for shear acting in this plane. The geometric limitations given on the use of this

Table 6.2-2. EN 1993-1-5 Table 5.1 for contribution of the web, χ_w

	Rigid end post	Non-rigid end post
$\bar{\lambda}_w < 0.83/\eta$	η	η
$0.83/\eta \leq \bar{\lambda}_w < 1.08$	$0.83/\bar{\lambda}_w$	$0.83/\bar{\lambda}_w$
$\bar{\lambda}_w \geq 1.08$	$1.37/(0.7 + \bar{\lambda}_w)$	$0.83/\bar{\lambda}_w$

method are the same as those for the use of the rules on Class 4 effective sections discussed in section 6.4.4.2 of this guide.

Contribution from the web – 3-1-5/clause 5.3

3-1-5/clause 5.3(1)

The contribution from the web given in ***3-1-5/clause 5.3(1)***:

$$V_{\mathrm{bw,Rd}} = \frac{\chi_w f_{yw} h_w t}{\sqrt{3}\gamma_{M1}}$$

is determined from 3-1-5/Table 5.1 or 3-1-5/Fig. 5.2 and depends on web slenderness. The final resistances in EN 1993-1-5 Table 5.1 and Fig. 5.2 are slightly lower than those from the theory above in Fig. 6.2-36 to allow for test result scatter, and a lower branch is added to cover cases with no rigid end-posts. Tests show that the longitudinal tension field still develops in girders with non-rigid end-posts, but to a lesser extent. The reduction factor, χ_w, ignores any contribution from the flanges, which is discussed later. EN 1993-1-5 Table 5.1 is reproduced above as Table 6.2-2.

The reduction factor, χ_w, for an η value of 1.2 derived from 3-1-5/Table 5.1 is shown plotted against slenderness in Fig. 6.2-38 for both rigid and non-rigid end-posts.

The general expression for slenderness takes the usual Eurocode form:

$$\bar{\lambda}_w = \sqrt{\frac{f_v}{\tau_{cr}}} = 0.76\sqrt{\frac{f_{yw}}{\tau_{cr}}} \qquad \text{3-1-5/(5.3)}$$

$$\text{with } f_v = \frac{f_{yw}}{\sqrt{3}} \text{ and } \tau_{cr} = k_\tau \sigma_E = \frac{k_\tau \pi^2 E t^2}{12(1-\nu^2)b^2} \qquad \text{3-1-5/(5.4)}$$

where k_τ is the buckling coefficient which will vary depending on whether sub-panel or overall stiffened panel buckling is being checked. σ_E is defined in 3-1-5/Annex A. For a web within 3-1-5/clause 5, b is the overall web depth, h_w, for overall buckling or is the sub-panel depth, h_{wi}, for sub-panel buckling. The term b is used more generally in

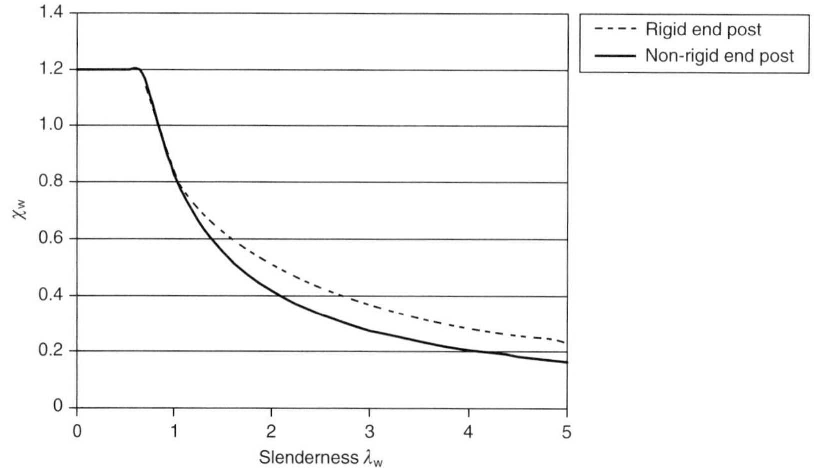

Fig. 6.2-38. χ_w against slenderness for $\eta = 1.2$

EN 1993-1-5 for the overall width or depth of a panel because the provisions on buckling apply equally to webs and flanges. $\bar{b}$ is similarly used in place of h_{wi} for the width of a sub-panel elsewhere in EN 1993-1-5. The designer must think carefully what the appropriate value of b is for each check. The lowest value of τ_{cr} from overall or sub-panel buckling is used to determine the slenderness. Values of k_τ are presented in 3-1-5/Annex A.3 as follows, but h_w has been replaced by b below in line with the above discussion.

Where there are no longitudinal stiffeners, three or more longitudinal stiffeners or all cases with $a/b \geq 3$:

$$k_\tau = 5.34 + 4.00\left(\frac{b}{a}\right)^2 + k_{\tau sl} \text{ when } a/b \geq 1 \qquad \text{(D6.2-19)}$$

$$k_\tau = 4.00 + 5.34\left(\frac{b}{a}\right)^2 + k_{\tau sl} \text{ when } a/b < 1 \qquad \text{(D6.2-20)}$$

$$\text{with } k_{\tau sl} = 9\left(\frac{b}{a}\right)^2 \sqrt[4]{\left(\frac{I_{sl}}{t^3 b}\right)^3} \text{ but not less than } k_{\tau sl} = \frac{2.1}{t}\sqrt[3]{\frac{I_{sl}}{b}}$$

where b is the overall web depth, h_w, for overall buckling or is the sub-panel depth, h_{wi}, for sub-panel buckling. For sub-panels, $k_{\tau sl}$ is taken as zero.

These formulae include the necessary reduction in contribution from the longitudinal stiffeners as discussed above. I_{sl} refers to the total second moment of area of all the longitudinal stiffeners, calculated assuming an attached width of web of $15\varepsilon t$ each side of the stiffener. The inclusion of the yield ratio, ε, is difficult to explain in this context as it has nothing to do with elastic stiffness.

Where there are fewer than three longitudinal stiffeners and $a/b < 3$, an alternative formula is required to account for the discrete nature of the stiffeners, since the above formulae were found to overestimate the resistance in this case. This is provided by expression 3-1-5/(A.6):

$$k_\tau = 4.1 + \frac{6.3 + 0.18\dfrac{I_{sl}}{t^3 b}}{\alpha^2} + 2.2\sqrt[3]{\frac{I_{sl}}{t^3 b}} \qquad \text{(D6.2-21)}$$

Equation (D6.2-21) was derived from Kloppel charts[15] for various stiffener positions and either one or two stiffeners. In the case of one stiffener, the stiffener was not considered to be closer to the flange than $0.2b$ in the derivation. Moving it closer would make equation (D6.2-21) unsafe in itself but it is likely that the check of the large remaining sub-panel would then govern in any case. It can be noted that unfortunately there is a discontinuity in the values calculated according to equations (D6.2-19) and (D6.2-21) at $a/b = 3$.

Substitution of the expression for τ_{cr} in expression 3-1-5/(5.4) into expression 3-1-5/(5.3) gives the general expression for overall slenderness for webs with transverse stiffeners and/or longitudinal stiffeners in ***3-1-5/clause 5.3(3)b)***: *3-1-5/clause 5.3(3)b)*

$$\bar{\lambda}_w = \frac{h_w}{37.4 t\varepsilon\sqrt{k_\tau}} \qquad \text{3-1-5/(5.6)}$$

For members without longitudinal stiffeners and transverse stiffeners at supports only, taking $b/a = 0$ in equation (D6.2-19) gives the expression in ***3-1-5/clause 5.3(3)a)***: *3-1-5/clause 5.3(3)a)*

$$\bar{\lambda}_w = \frac{h_w}{86.4 t\varepsilon} \qquad \text{3-1-5/(5.5)}$$

EN 1993-1-5 assumes that all transverse stiffeners are rigid and design in accordance with 3-1-5/clause 9 is intended to ensure this. It is, in principle, still possible to improve shear resistance by adding flexible transverse stiffeners to the web in a similar way to the inclusion of flexible longitudinal stiffeners, but no formulae are given in EN 1993-1-5 to

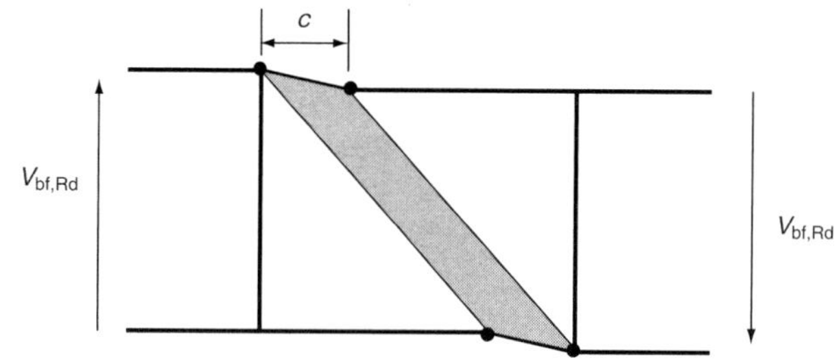

Fig. 6.2-39. Origin of flange component, $V_{bf,Rd}$

include the effect of flexible transverse stiffeners in τ_{cr}. If it is desired to do this, reference should be made to standard texts such as Bulson.[9]

3-1-5/clause 5.3(5)

Where there are longitudinal stiffeners, a check on the most slender sub-panel must also be made to prevent local buckling according to ***3-1-5/clause 5.3(5)***:

$$\bar{\lambda}_w = \frac{h_{wi}}{37.4t\varepsilon\sqrt{k_{\tau i}}} \qquad \text{3-1-5/(5.7)}$$

where h_{wi} is the depth of the sub-panel and $k_{\tau i}$ is the buckling coefficient for the sub-panel from equation (D6.2-19) or (D6.2-20), ignoring the longitudinal stiffeners other than in their function of providing a rigid boundary to the sub-panel.

Contribution from the flanges – 3-1-5/clause 5.4

If there are intermediate transverse stiffeners at reasonably close centres, an additional tension field mechanism can also be mobilized. This occurs because the flanges can span between stiffeners and give restraint to the web pulling in vertically over a length c as shown in Fig. 6.2-39. The predicted magnitude of this tension field which has to be supported by the stiffeners is less than in previous UK practice because the rotated stress field in the web provides the post-buckling web resistance for cases with weak flanges. This is discussed further in section 6.6 of this guide.

By considering the flange collapse mechanism in Fig. 6.2-39, the shear supported by the bending flange can be shown from energy considerations to be:

$$V_{bf,Rd} = \frac{b_f t_f^2 f_{yf}}{c\gamma_{M1}}$$

3-1-5/clause 5.4(1)

When the coexisting longitudinal stresses in the flange from global action are included, the contribution according to ***3-1-5/clause 5.4(1)*** is as follows:

$$V_{bf,Rd} = \frac{b_f t_f^2 f_{yf}}{c\gamma_{M1}}\left[1 - \left(\frac{M_{Ed}}{M_{f,Rd}}\right)^2\right] \qquad \text{3-1-5/(5.8)}$$

where $M_{f,Rd}$ is the design bending resistance of the section based on the effective flanges only. It needs to be reduced in the presence of axial load according to expression 3-1-5/(5.9). The width of the tension band is given by:

$$c = a\left(0.25 + \frac{1.6b_f t_f^2 f_{yf}}{t h_w^2 f_{yw}}\right)$$

It can be seen that the flange contribution contains an interaction with bending moment and this is illustrated in section 6.2.9.2.1 of this guide. Expression 3-1-5/(5.8) should be applied to both flanges separately and the lowest calculated contribution, $V_{bf,Rd}$, taken. b_f should not include a greater width of flange on each side of the web than $15\varepsilon t_f$. Where a flange is present on only one side of the web, it is advisable to take $b_f = 0$ to avoid considerations of torsion in the flange.

The flange contribution can always be conservatively ignored to avoid the additional calculation effort. Often it will be small in any case.

Worked Example 6.2-9: Girder without longitudinal stiffeners
A continuous girder in S355 steel has plate sizes as shown in Fig. 6.2-40(a). All bearing stiffeners comprise single double-sided stiffeners only and there are no intermediate transverse stiffeners. The shear resistance is calculated at an internal support and at an end support.

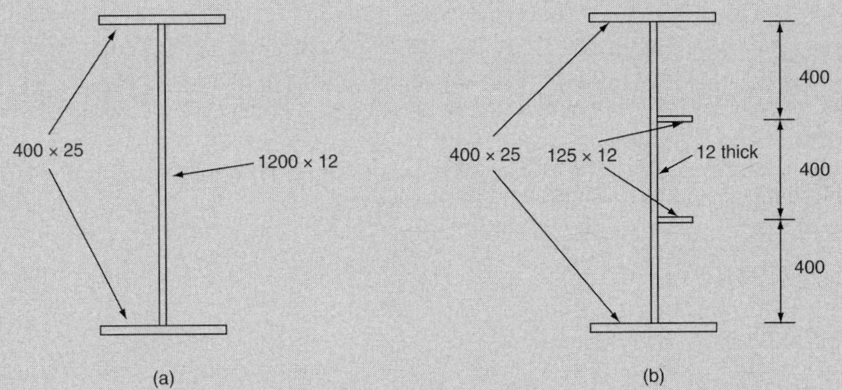

Fig. 6.2-40. Girders for (a) Worked Example 6.2-9; and (b) Worked Example 6.2-10

Consider first an internal support.
For no intermediate stiffeners, the slenderness is obtained from expression 3-1-5/(5.5):

$$\bar{\lambda}_w = \frac{h_w}{86.4t\varepsilon} = \frac{1200}{86.4 \times 12 \times 0.81} = 1.429$$

At an internal support, the rigid end-post case applies, so from 3-1-5/Table 5.1:

$$\chi_w = \frac{1.37}{0.7 + \bar{\lambda}_w} = \frac{1.37}{0.7 + 1.429} = 0.64$$

Any contribution from the flanges will be negligible as the transverse stiffeners are far apart, so the resistance is therefore:

$$V_{bw,Rd} = \frac{\chi_w f_{yw} h_w t}{\sqrt{3}\gamma_{M1}} = \frac{0.64 \times 355 \times 1200 \times 12}{\sqrt{3} \times 1.1} = \mathbf{1727\,kN}$$

Considering now an end support, the slenderness is again obtained from expression 3-1-5/(5.5):

$$\bar{\lambda}_w = \frac{h_w}{86.4t\varepsilon} = \frac{1200}{86.4 \times 12 \times 0.81} = 1.429$$

At an end support with single bearing stiffener, the non-rigid end-post case applies, so from 3-1-5/Table 5.1:

$$\chi_w = \frac{0.83}{\bar{\lambda}_w} = \frac{0.83}{1.429} = 0.58$$

If any small contribution from the flanges is conservatively ignored, the resistance is therefore:

$$V_{bw,Rd} = \frac{\chi_w f_{yw} h_w t}{\sqrt{3}\gamma_{M1}} = \frac{0.58 \times 355 \times 1200 \times 12}{\sqrt{3} \times 1.1} = \mathbf{1558\,kN}$$

Worked Example 6.2-10: Girder with longitudinal stiffeners

A continuous girder in S355 steel has the same plate sizes as in Worked Example 6.2-9, but incorporates longitudinal stiffeners as shown in Fig. 6.2-40(b). All bearing stiffeners comprise single double-sided stiffeners only and there are intermediate transverse stiffeners at 4000 mm centres. The shear resistance is calculated at an internal support.

Slenderness for overall shear buckling of the stiffened panel is checked first. $a/b = 4000/1200 = 3.33 > 3$, so the shear buckling coefficient is obtained from equation (D6.2-19). From 3-1-5/Fig. 5.3, each longitudinal stiffener has an attached piece of web of $30\varepsilon t$ plus the thickness of the stiffener $= 30 \times 0.81 \times 12 + 12 = 304\,\text{mm} < 400\,\text{mm}$ available. Each effective section therefore has second moment of area $= 6.982 \times 10^6\,\text{mm}^4$ so $I_{sl} = 2 \times 6.982 \times 10^6 = 1.396 \times 10^7\,\text{mm}^4$. From equation (D6.2-19):

$$k_{\tau sl} = 9\left(\frac{b}{a}\right)^2 \sqrt[4]{\left(\frac{I_{sl}}{t^3 b}\right)^3} = 9\left(\frac{1200}{4000}\right)^2 \sqrt[4]{\left(\frac{2.095 \times 10^7}{12^3 \times 1200}\right)^3} = 3.386$$

but not less than

$$k_{\tau sl} = \frac{2.1}{t}\sqrt[3]{\frac{I_{sl}}{b}} = \frac{2.1}{12}\sqrt[3]{\frac{2.095 \times 10^7}{1200}} = 4.540$$

$$k_\tau = 5.34 + 4.00\left(\frac{b}{a}\right)^2 + k_{\tau sl} = 5.34 + 4.00\left(\frac{1200}{4000}\right)^2 + 4.540 = 10.24$$

The slenderness for overall buckling is obtained from expression 3-1-5/(5.6):

$$\bar{\lambda}_w = \frac{h_w}{37.4 t\varepsilon\sqrt{k_\tau}} = \frac{1200}{37.4 \times 12 \times 0.81 \times \sqrt{10.24}} = 1.032$$

Next, the slenderness for sub-panel buckling is calculated.

For sub-panel buckling, $a = 4000\,\text{mm}$ and $\bar{b} = 400\,\text{mm}$ and from equation (D6.2-19):

$$k_{\tau i} = 5.34 + 4.00\left(\frac{\bar{b}}{a}\right)^2 = 5.34 + 4.00\left(\frac{400}{4000}\right)^2 = 5.38$$

The slenderness for sub-panel buckling is obtained from expression 3-1-5/(5.7):

$$\bar{\lambda}_w = \frac{h_{wi}}{37.4 t\varepsilon\sqrt{k_{\tau i}}} = \frac{400}{37.4 \times 12 \times 0.81 \times \sqrt{5.38}} = 0.474 < 1.032$$

for overall buckling so sub-panel buckling does not govern.

At an internal support, the rigid end-post case applies but since $\bar{\lambda}_w < 1.08$, from 3-1-5/Table 5.1 it does not matter whether or not there is a rigid end-post.

$$\chi_w = \frac{0.83}{\bar{\lambda}_w} = \frac{0.83}{1.024} = 0.80$$

If the contribution from the flanges is conservatively ignored, the resistance is therefore:

$$V_{bw,Rd} = \frac{\chi_w f_{yw} h_w t}{\sqrt{3}\gamma_{M1}} = \frac{0.80 \times 355 \times 1200 \times 12}{\sqrt{3} \times 1.1} = \mathbf{2159\,kN}$$

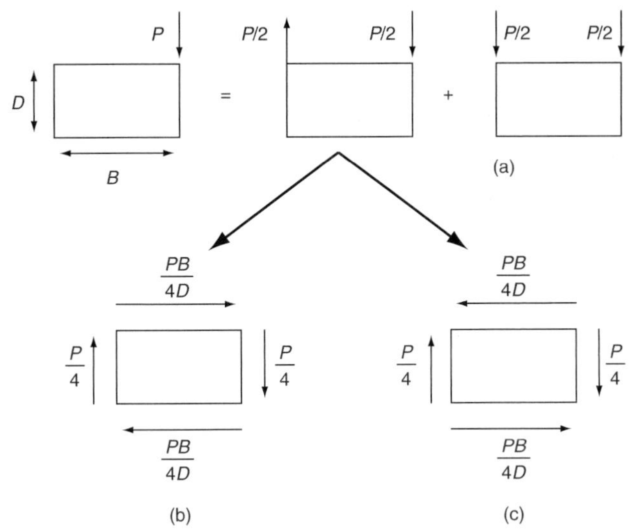

Fig. 6.2-41. Distortion from eccentric load on a box girder: (a) symmetrical component; (b) torsional component; (c) distortional component

6.2.7. Torsion

6.2.7.1. General

Torsion and distortion

3-2/clause 6.2.7.1 is primarily concerned with box girders. If torsional loading is applied to a box section by forces with the same distribution as the St Venant shear flow around the box due to pure torsion, the cross-section will not distort and the section may be analysed for torsion in accordance with section 6.2.7.2 below. However, if this is not the case, the section will distort. The effect is illustrated in Fig. 6.2-41 for a simple rectangular hollow cross-section with an eccentric load.

The eccentric load can be split into symmetric and anti-symmetric loadings. The latter case can be further split into a torsional component (where the shear flows can be derived from the expected St Venant torsional shear flow as discussed in section 6.2.7.2) and a distortional component. The distortional component leads to a distortion of the cross section which is illustrated in Fig. 6.2-42.

From Fig. 6.2-42 it can be seen that the distortional component leads to both a transverse bending of the box walls (transverse distortional bending) and an in-plane bending of the box walls (distortional warping) between the points where distortion of the cross-section is restrained. These are the distortional effects to which ***3-2/clause 6.2.7.1(1)*** refers. The magnitude of the stresses obtained by each mechanism can be seen to depend on the relative stiffness of the plates acting transversely and longitudinally. Distortional restraint can be provided by diaphragms, ring frames or cross-bracing. Generally, diaphragms and cross-bracings will be sufficiently stiff to act as a fully rigid restraint to distortion whereas ring frames may not be, as they themselves resist the distortion by frame bending. To be effective against distortion, restraints clearly need to have both adequate stiffness and

3-2/clause 6.2.7.1(1)

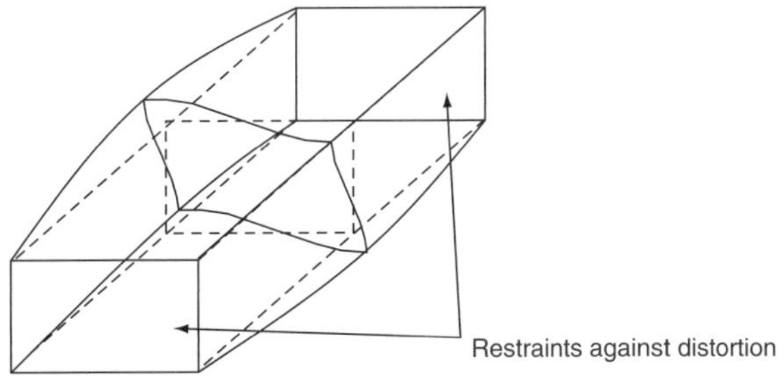

Fig. 6.2-42. Effect of distortional component

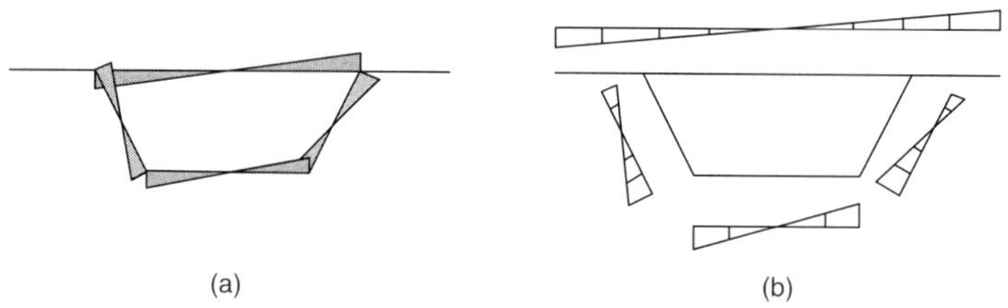

Fig. 6.2-43. Distribution of (a) transverse distortional moments and (b) longitudinal warping stresses

strength. If the torsional load is actually applied at a 'rigid' restraint, the distortional forces in Fig. 6.2-41(c) are taken directly by the restraint, and the box itself acts only in torsion.

The distribution of longitudinal stresses due to distortional warping and transverse distortional moments are shown in Fig. 6.2-43.

Methods of designing for distortion

3-2/clause 6.2.7.1(2)

3-2/clause 6.2.7.1(2) requires an appropriate elastic model to be used to assess distorsional effects. The most accurate method of modelling distortional effects is through the use of a finite-element shell model of the whole box structure. Elastic finite-element (FE) modelling is becoming increasingly quick to carry out and is an excellent predictor of behaviour at serviceability and fatigue limit states, but can be somewhat conservative for ultimate limit states. This is because EN 1993 permits certain effects to be neglected at ultimate limit states (such as those from torsional warping – see section 6.2.7.2) and others to be modified for plasticity (such as the effects of shear lag) and these cannot be dissociated from the overall results of an elastic FE model. Several simpler models are therefore possible which allow these individual effects to be separated. These include:

- beam on elastic foundations (BEF) analogy
- shear flexible grillage
- space frame.

A detailed discussion of these design methods is beyond the scope of this guide as no one method is prescribed by the Eurocode. All these methods are discussed in *Bridge Deck Behaviour*.[16] Non-linear FE modelling is a further alternative for making allowance for plastic redistribution at ULS, but it is unlikely to be feasible for most day-to-day design because of the analysis time involved and because superposition of loadings cannot be performed.

The BEF analogy, as illustrated in Fig. 6.2-44, is a commonly used method. In this analogy, the beam inertia represents the in-plane bending (warping) stiffness of the plates and the elastic foundation springs represent the transverse distortional bending stiffness of the box. Concentrated torsional loads are modelled as point loads on the beam. Warping stresses are proportional to the bending moment in the beam and the forces in the springs are proportional to the distortional forces carried by the local cross-section and hence to the transverse distortional bending moments. Flexible restraints (typically ring frames and some bracing systems) can be modelled as discrete additional springs. The 'spring' stiffness can be obtained from a plane frame model and care should be taken to include

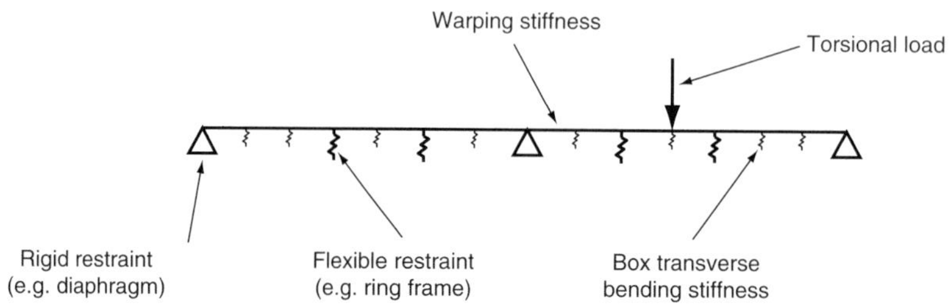

Fig. 6.2-44. BEF analogy for the effects of distortional load

the effects of non-noding of bracing members which can increase flexibility under distortional loading. Rigid diaphragms (permitting warping) are modelled as fixed supports. A support preventing warping would be modelled as a built-in support but such a support is unlikely to be achievable in practice. This analogy shows, for example, that if a load is applied between restraints which are a long way apart and the box is relatively stiff transversely compared to its longitudinal warping stiffness, then most of this distortional load will be carried in transverse distortional bending. This result is intuitively correct. More detail on the use of this method is provided in the original paper by Wright *et al.*[17]

The bridge code BS 5400: Part 3[4] used the BEF analogy to derive design equations. However the application of these is somewhat limited as they assume restraints are extremely stiff. Most diaphragms will comply with the stiffness requirements but other forms of restraint, such as ring frames, are unlikely to comply. It is therefore better to make reference to the original paper by Wright *et al.*[17] when using the BEF.

Distortion of the cross-section leads to an apparent softening of the torsional stiffness. For multi-beam decks, where less torsional load is attracted if the torsional stiffness is reduced, there may be some benefit in 'softening' the torsional constant in analysis to allow for the distortion which will occur. This can reduce the torsional load attracted and thus also the distortional stresses. An effective reduced torsional inertia can be derived using References 16 or 17. It should be recognized that such a method is approximate as the distortional displacements and hence modified torsional stiffness are dependent on load configuration and therefore would vary with each load case. This is not made clear in Reference 16.

When combining distortional effects with those from bending, shear and axial load, it is simplest to use elastic cross-section analysis. Warping stresses should be added to other direct stresses. Distortional bending stresses can be combined with other stresses using the Von Mises equivalent stress criterion. This can be done in the same manner as the combination of local and global effects discussed in section 6.5.2 of this guide.

Design of distortional restraints

The reference in ***3-2/clause 6.2.7.1(4)*** to the need to design diaphragms for the actions resulting from their 'load distributing effect' includes the effects arising from resisting distortion. This applies to restraints in general. The distortional forces acting on a restraint in Fig. 6.2-45(a) are found from the applied distortional torque at the restraint, T, as described above. If a torque is not applied directly to a restraint or the restraint is very flexible, the BEF model can be used to determine the share of the distortional torque applied to the restraint from the reaction developed at the restraint. These forces can be represented by the equivalent diagonal forces shown in Fig. 6.2-45(b). If the torque is applied in the manner of Fig. 6.2-41 then the restraints can be designed as follows. *3-2/clause 6.2.7.1(4)*

A plate diaphragm can be designed for a shear stress according to:

$$\tau = \frac{T}{D(B_T + B_B)t_D} \qquad \text{(D6.2-22)}$$

where t_D is the thickness of the diaphragm plate. If part of the distortional torque is applied between restraints, the torque can be apportioned to the restraints by statics.

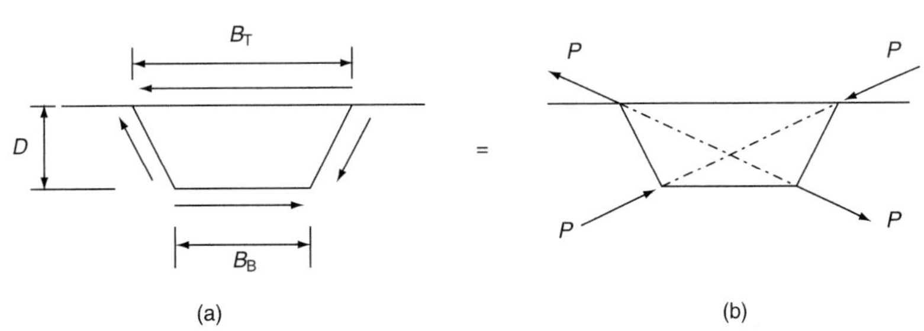

Fig. 6.2-45. Distribution of forces for design of distortional restraints: (a) distortional shear flow; (b) equivalent forces for design of restraints

A ring frame or cross-bracing can be designed using a plane frame model resisting the forces shown in Fig. 6.2-45(b) where the forces are as follows:

$$P = \frac{T\sqrt{1+\left(\frac{B_T+B_B}{2D}\right)^2}}{2B_T\left(1+\frac{B_T}{B_B}\right)} \qquad \text{(D6.2-23)}$$

Plane frame models are useful because they can also pick up additional moments caused in bracing systems from non-noding of the members with the box walls. As discussed above, restraints also need to have adequate stiffness to limit the distortional stresses in the main box.

In ring frame details, it is particularly important to ensure continuity of the web and flange transverse members. As seen in Fig. 6.2-43(a), the frame moments are a maximum in the box corners and non-continuous transverse members can lead to very large stresses being developed in the box web and flange plates and the weld between them. The latter is usually particularly prone to fatigue damage in such a situation.

6.2.7.2. Torsion for which distortional effects may be neglected

Where distortional effects can be neglected, sections may be designed for torsion alone in accordance with 3-1-1/clause 6.2.7. The basic design requirement for sections in torsion is

3-1-1/clause 6.2.7(1)

given in ***3-1-1/clause 6.2.7(1)*** as:

$$\frac{T_{Ed}}{T_{Rd}} \le 1.0 \qquad \text{3-1-1/(6.23)}$$

where T_{Ed} is the applied torsional moment and T_{Rd} is the design torsional resistance of the section.

3-1-1/clause 6.2.7(2)

In general, torsion can be resisted by two mechanisms such that the applied torque can be split into two components according to ***3-1-1/clause 6.2.7(2)*** thus:

$$T_{Ed} = T_{t,Ed} + T_{w,Ed} \qquad \text{3-1-1/(6.24)}$$

where $T_{t,Ed}$ is the St Venant torsion involving a closed flow of shear around the section perimeter and $T_{w,Ed}$ is the warping torsion involving transverse bending of the constituent plates of the section.

These two types of torsion are quite different in both their mechanism and the way they interact with other internal actions such as bending and shear. The concept of an overall torsional resistance T_{Rd} in expression 3-1-1/(6.23) is therefore not a particularly useful one and is not used anywhere else in EN 1993. The behaviours under these two different types of torsion are also very different for open and closed sections and they are therefore dealt with separately below.

6.2.7.2.1. Open sections (additional sub-section)

An 'open section' is any section that does not form a hollow section. Examples of open sections are fabricated I-girders, universal beams, universal columns and channels. Where torsion arises due to eccentric loading on the section, the relevant eccentricity is that from the section's shear centre. Loads applied through the shear centre will not give rise to any twist. The location of the shear centre for some commonly used sections is given in section 6.3.1.4 of this guide. Open sections resist torsion via two mechanisms: St Venant

3-1-1/clause 6.2.7(3)
3-1-1/clause 6.2.7(4)

torsion and warping torsion. The share of torsion between these mechanisms can be determined by elastic analysis in accordance with ***3-1-1/clause 6.2.7(3)*** as described below. The effects referred to in ***3-1-1/clause 6.2.7(4)*** are also discussed.

(i) St Venant torsion

St Venant torsion involves a closed flow of shear stress around the section perimeter, as illustrated in Fig. 6.2-46. In this case the shear stress is given by:

$$\tau_{T,Ed} = tG\frac{d\theta}{dx} \qquad \text{(D6.2-24)}$$

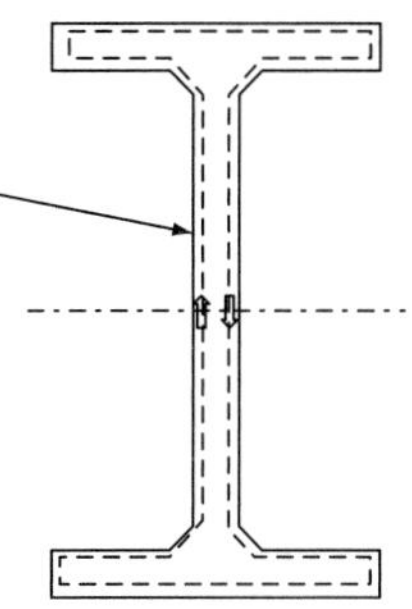

Fig. 6.2-46. Open section resisting torsion through St Venant shear flow

where:

G is the shear modulus of the steel component;
t is the thickness of constituent part being considered;
$\frac{d\theta}{dx}$ is the rate of twist of the open section with length along the member.

The torque resisted by St Venant shear flow is given by:

$$T_{t,Ed} = GI_T \frac{d\theta}{dx} \qquad \text{(D6.2-25)}$$

Therefore if the section is free to warp or (warping is neglected) so that $T_{Ed} = T_{t,Ed}$, the shear stress is given by:

$$\tau_{t,Ed} = \frac{T_{Ed} t}{I_T} \qquad \text{(D6.2-26)}$$

Since the resistance of a section in torsion based on St Venant shear flow is usually very small, it is common to neglect St Venant torsion and carry the entire torque by warping where this mechanism is possible. ***3-1-1/clause 6.2.7(7)*** allows designers to neglect the effect of St Venant torsion in open sections, but it is essential that an imposed torsion is then fully resisted by another mechanism as discussed below. *3-1-1/clause 6.2.7(7)*

(ii) Warping torsion

As illustrated in Fig. 6.2-47 below, the torque can also be resisted in an I-beam by in-plane shear and bending of the flanges. The opposing transverse moments produced in the flanges by the action of the opposing shearing forces T_{Ed}/h resisting torsion is referred to as the 'bi-moment', B_{Ed}. This resistance mechanism is referred to as the 'torsional warping resistance'. As both flanges will bend in different directions, the section will change shape or 'warp' as it resists the torsion. A similar mechanism occurs in other flanged members such as channels, but in the case of channels there is also in-plane vertical bending of the web plate due to the compatibility of longitudinal bending stresses which has to be maintained at the junctions between web and flanges. The transverse shear force and bi-moment produce transverse

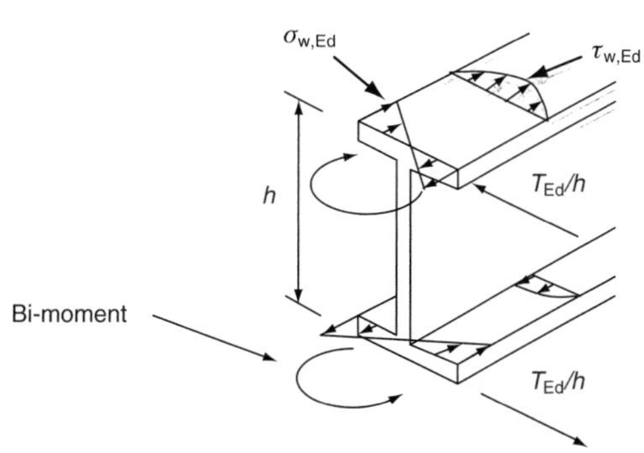

Fig. 6.2-47. Open section resisting torsion by warping

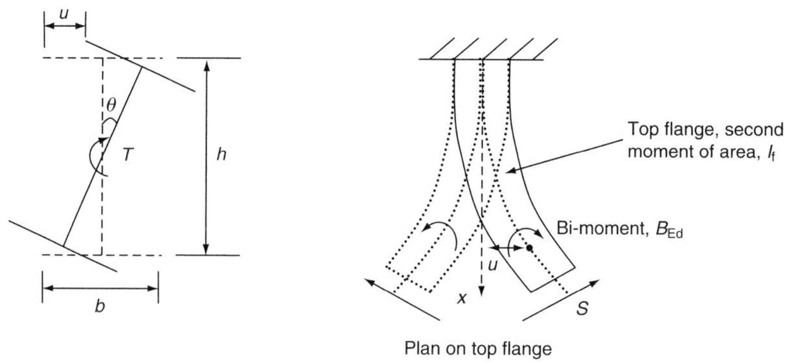

Fig. 6.2-48. Bisymmetric I-beam resisting torsion by warping

shear stresses $\tau_{w,Ed}$ and bending stresses $\sigma_{w,Ed}$ respectively as shown in Fig. 6.2-47. These stresses together with $\tau_{t,Ed}$ from St Venant torsion must be considered in accordance with ***3-1-1/clause 6.2.7(4)***.

3-1-1/clause 6.2.7(4)

The torque resisted by warping is as follows:

$$T_{w,Ed} = -EI_W \frac{d^3\theta}{dx^3} \tag{D6.2-27}$$

A simple derivation of this formula helps to illustrate the behaviour. Considering the cantilevered bisymmetric I-beam in Fig. 6.2-48 under the action of an end torque, the moment in each flange, B_{Ed}, is obtained from the curvature of the top flange:

$$B_{Ed} = -EI_f \frac{d^2u}{dx^2} \tag{D6.2-28}$$

The flange shear force is given by:

$$S = \frac{dB_{Ed}}{dx} = -EI_f \frac{d^3u}{dx^3} \tag{D6.2-29}$$

Noting that $u = h\theta/2$, the applied torque is given by:

$$T_{Ed} = Sh = -EI_f h \frac{d^3u}{dx^3} = -EI_f \frac{h^2}{2} \frac{d^3\theta}{dx^3} \tag{D6.2-30}$$

The term $I_f(h^2/2)$ is thus the warping constant I_w for a symmetric I-beam with equal flanges so that $T_{w,Ed} = -EI_W(d^3\theta/dx^3)$ as in equation (D6.2-27).

The maximum transverse shear stresses $\tau_{w,Ed_{max}}$ and bending stresses $\sigma_{w,Ed_{max}}$ can be shown to be:

$$\tau_{w,Ed_{max}} = Ek_{1,max} \frac{d^3\theta}{dx^3} \tag{D6.2-31}$$

$$\sigma_{w,Ed_{max}} = Ek_{2,max} \frac{d^2\theta}{dx^2} \tag{D6.2-32}$$

where $k_{1,max}$ is the torsional warping shear constant appropriate to point of maximum shear stress and $k_{2,max}$ is the torsional warping bending constant appropriate to point of maximum direct stress. Equations (D6.2-31) and (D6.2-32) can also be rewritten more generally as

$$\tau_{w,Ed} = Ek_1 \frac{d^3\theta}{dx^3} \text{ and } \sigma_{w,Ed} = Ek_2 \frac{d^2\theta}{dx^2}$$

in which case k_1 and k_2 relate to whichever point in the cross-section is being checked. Solutions for k_1, k_2, θ, $d^2\theta/dz^2$ and $d^3\theta/dz^3$ for thin-walled open sections under torsion in

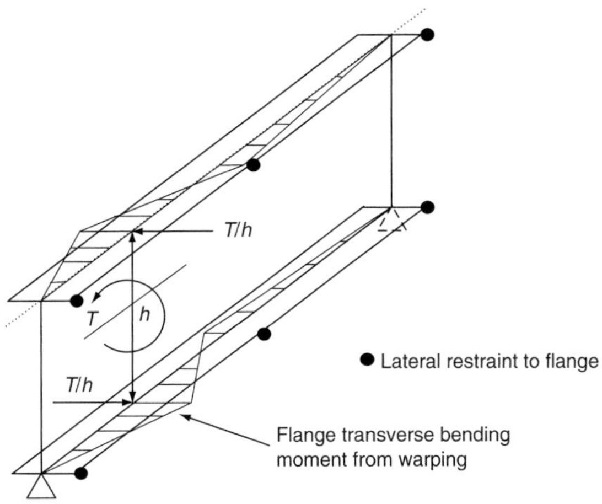

Fig. 6.2-49. Simple model for determining warping stresses in an I-beam (ignoring St Venant torsion)

a variety of different load configurations are provided in Reference 18. For a bisymmetric I-beam, $k_{1,\max} = hb^2/16$ and $k_{2,\max} = hb/4$.

However, as discussed above, the elastic torque will be carried by a combination of warping and St Venant torsion and the relative contributions of the two are determined from considerations of compatibility from elastic analysis according to the following differential equation which combines equations (D6.2-25) and (D6.2-27):

$$T_{\mathrm{Ed}} = GI_{\mathrm{T}} \frac{\mathrm{d}\theta}{\mathrm{d}x} - EI_{\mathrm{W}} \frac{\mathrm{d}^3\theta}{\mathrm{d}x^3} \qquad \text{(D6.2-33)}$$

If the effects of St Venant torsion are to be neglected as allowed by 3-1-1/clause 6.2.7(7), the calculated stresses from warping torsion obtained from equation (D6.2-33) need to be increased accordingly so that the full applied torsion (including the redistributed St Venant component) is still resisted. For serviceability limit states and for fatigue calculations, the torsional stresses should, however, be determined from the actual contributions of St Venant and warping torsion.

Since 3-1-1/clause 6.2.7(7) permits St Venant torsion to be ignored at ULS and warping is often the most efficient means of carrying torsion, it will frequently be simpler to consider the torque to be resisted by opposing bending in the flanges, rather than to struggle with the solution of differential equations. A simple case of carrying torsion in this manner is illustrated in Fig. 6.2-49 for a length of I-beam between rigid restraints provided by bracing. If the length between restraints becomes very long, then the warping bending stresses would become very large and the section would try to resist the torsion predominantly through St Venant shear flow. In this case it may be better to derive the actual contributions from St Venant and warping torsion. (A shell finite-element model can be used to determine these combined stresses directly.) It should also be noted that there needs to be a mechanism for introducing the torsional load into the flanges as in Fig. 6.2-49. If there is a stiffener at the point of application of the eccentric load, the rigidity of the stiffener provides this mechanism. Without a stiffener, an eccentrically applied vertical load will bend the web out-of-plane and this local bending also needs to be considered.

Shear and torsion

The shear stresses induced by the torsion will have a detrimental effect on the shear resistance and ***3-1-1/clause 6.2.7(9)*** requires the plastic shear resistance of steel components to be modified to account for torsion. The reduced plastic shear resistance in the presence of torsion is denoted $V_{\mathrm{pl,T,Rd}}$ and is used in subsequent interactions between shear, bending and axial force in place of $V_{\mathrm{pl,Rd}}$. *3-1-1/clause 6.2.7(9)*

For an I or H section:

$$V_{\mathrm{pl,T,Rd}} = \sqrt{1 - \frac{\tau_{\mathrm{t,Ed}}}{1.25(f_y/\sqrt{3})/\gamma_{\mathrm{M0}}}}\,V_{\mathrm{pl,Rd}} \qquad \text{3-1-1/(6.26)}$$

where $V_{\mathrm{pl,Rd}}$ is as given in 3-1-1/clause 6.2.6.

Warping stresses here do not reduce the shear strength as the warping action only involves transverse shear stresses in the flanges and not the webs. If St Venant torsion is ignored and all the torque is carried by warping then there is no reduction to make to the plastic shear resistance.

For a channel section:

$$V_{\mathrm{pl,T,Rd}} = \left[\sqrt{1 - \frac{\tau_{\mathrm{t,Ed}}}{1.25(f_y/\sqrt{3})/\gamma_{\mathrm{M0}}}} - \frac{\tau_{\mathrm{w,Ed}}}{(f_y/\sqrt{3})/\gamma_{\mathrm{M0}}}\right] V_{\mathrm{pl,Rd}} \qquad \text{3-1-1/(6.27)}$$

Warping stresses here do reduce the shear strength as the elastic warping action in a channel involves transverse shear stresses in both flanges and the webs. If at the ultimate limit state the flanges are able to resist the torque in opposing transverse bending without any contribution from the web, it may be possible to consider this simplified mechanism for resisting the torsion without reducing the resistance of the web to vertical shear.

No guidance is given for the effect of torsion on shear buckling resistance. Where warping shear stresses are developed in a web, such as occurs in channel sections, these warping shear stresses should also be added to the vertical shear stresses when verifying the shear buckling resistance. The treatment of St Venant torsional shear stress in an open section is less straightforward. As there is no net vertical shear produced in a web from the circulatory St Venant torsional shear flow, it does not actually promote an overall shear buckling mode, so fully adding this stress to that from vertical shear in a buckling check would be very conservative. It can however lead to premature failure by causing yielding. This effect is more akin to an increased equivalent geometric imperfection in the plate, which would reduce the shear buckling resistance itself. A possibility would be to add an additional term, $\tau_{\mathrm{t,Ed}}/\tau_{\mathrm{yd}}$, to η_3 in the shear buckling interaction of 3-1-5/clause 7.1 (see section 6.2.9.2 of this guide). To avoid this problem it is simplest to ensure all the torsion is carried in a warping mode where possible.

Shear, torsion and bending

If a member is subjected to a major axis bending moment and a torque, the twist from the torque will induce a bending moment component about the minor axis as illustrated in Fig. 6.2-50. This induced moment is not specifically mentioned in EN 1993 but will usually be small. It gives rise to an additional flange transverse bending stress of $\sigma_{\mathrm{Mz,Ed}}$.

Both the warping bi-moment and this additional minor axis moment give rise to transverse curvature in the flanges. This curvature and hence the moments will be magnified by the presence of axial load in the flange from bending. One way of allowing for this in the member buckling check is to multiply these bending stresses by a magnifier in the following

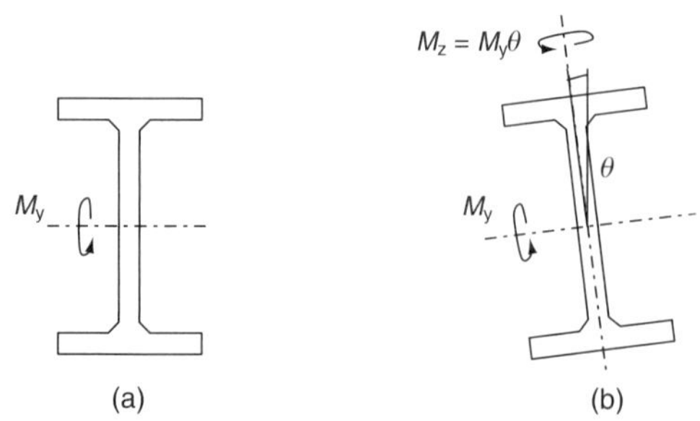

Fig. 6.2-50. (a) Section under bending moment; (b) minor axis moment induced by twisting of an open section

interaction:

$$\frac{M_{y,Ed}}{\chi_{LT} M_{y,Rk}/\gamma_{M1}} + \frac{(\sigma_{w,Ed} + \sigma_{Mz,Ed})}{f_{yd}} \left(\frac{1}{1 - \sigma_{Ed}/\sigma_{cr}} \right) \leq 1.0 \qquad \text{(D6.2-34)}$$

where σ_{cr} is the critical buckling stress for the compression flange which could be determined from section 6.3.4.2 of this guide and σ_{Ed} is the stress in the flange.

Beams with bending, shear and torsion should also be checked for cross-section resistance. If warping torsion is considered, the shear–moment interaction check (see section 6.2.9 of this guide) needs to account for the reduction in beam bending resistance due to the flange warping stresses. This can conservatively be achieved by reducing the effective yield stress of each flange, by an amount equal to the warping stress, when calculating global bending resistance. ***3-1-1/clause 6.2.7(6)*** refers to such a consideration. Where there is no shear buckling, the Von Mises yield criterion could alternatively be applied to all parts of the beam, as allowed by ***3-1-1/clause 6.2.7(5)***, but this will be conservative compared to the use of a modified interaction equation as suggested above. Where shear buckling can occur, the Von Mises check alone will not suffice. *3-1-1/clause 6.2.7(6)* *3-1-1/clause 6.2.7(5)*

6.2.7.2.2. Closed sections (additional sub-section)

Examples of closed sections include fabricated box girders, rectangular hollow sections, square hollow sections and circular hollow sections. Closed sections resist torsion predominately by St Venant circulatory shear flow around the hollow section as illustrated in Fig. 6.2-51. The treatment of torsion in closed sections is therefore quite different to that of open sections as St Venant torsion is a very efficient mechanism for carrying the torsion. As discussed in section 6.2.7.1, the designer must also consider distortional effects if the torque is not applied uniformly around the walls of the box.

(i) St Venant torsion

The shear stress $\tau_{t,Ed}$ due to torque in a thin-walled section is given by:

$$\tau_{t,Ed} = \frac{T_{Ed}}{2A_0 t} \qquad \text{(D6.2-35)}$$

where A_0 is the area enclosed by a perimeter running through the centre of the walls and t is the wall thickness of plate considered.

The rotation per unit length is given by:

$$\frac{d\theta}{dx} = \frac{T}{GI_T} \qquad \text{(D6.2-36)}$$

where $I_T = \dfrac{4A_0}{\oint \frac{ds}{t}}$

The St Venant shear stresses will, however, reduce the plastic shear resistance of the webs. ***3-1-1/clause 6.2.7(9)*** provides the following formula for the reduced shear resistance of a *3-1-1/clause 6.2.7(9)*

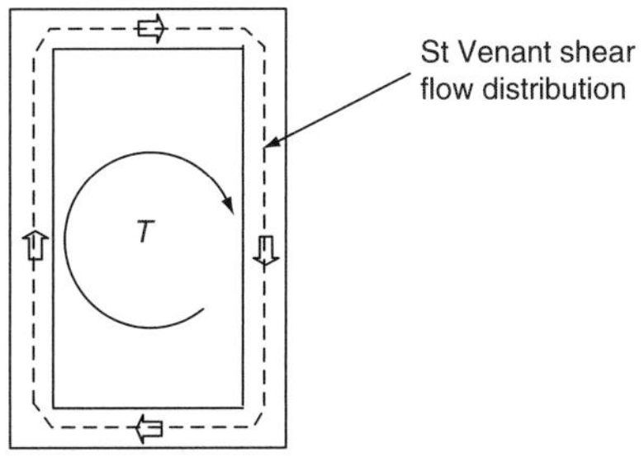

Fig. 6.2-51. St Venant shear flow in a closed section

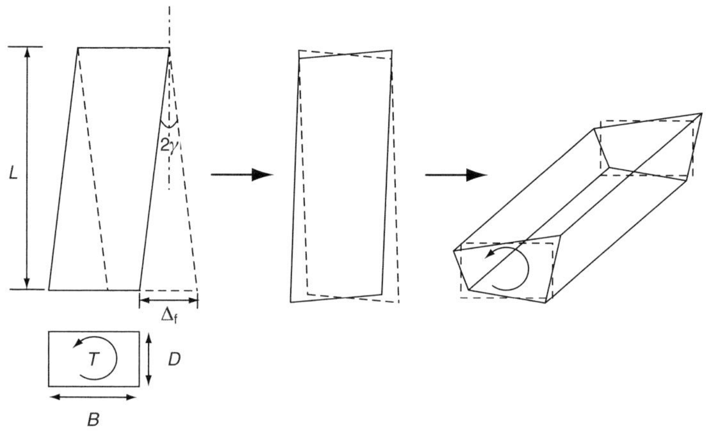

Fig. 6.2-52. Origin of torsional warping in box girders: (a) shear displacement of flanges (ends remaining plane); (b) reduction of displacement by warping of ends; (c) torsional warping of cross-section

web, $V_{pl,T,Rd}$, in a closed section:

$$V_{pl,T,Rd} = \left[1 - \frac{\tau_{t,Ed}}{(f_y/\sqrt{3})/\gamma_{M0}}\right] V_{pl,Rd} \qquad \text{3-1-1/(6.28)}$$

No guidance is given in EN 1993-1-1 for the effect of torsion on shear buckling resistance but where St Venant shear stresses are developed in a web, these shear stresses should be added to the vertical shear stresses when verifying the shear buckling resistance of a web in accordance with 3-1-5/clause 5. ***3-1-1/clause 6.2.7(8)*** gives a similar requirement but does not cover the interaction with shear; it refers only indirectly to considering shear buckling of individual web and flange plates in the derivation of the torsion resistance.

3-1-1/clause 6.2.7(8)

(ii) Torsional warping

The strains accompanying the St Venant shear stresses around the plates of hollow sections may cause the closed cross-section to change its shape and warp. If the simple rectangular cross-section in Fig. 6.2-52 is considered (with constant thickness for simplicity), the St Venant shear stress in the plates is everywhere $\tau = T/2BDt$. If twisting (but not warping) is prevented at one end and the flanges are assumed to remain plane as shown in Fig. 6.2-52(a), the shear displacement at the other end is then:

$$\Delta_f = 2\gamma L = \frac{2\tau L}{G} = \frac{TL}{GBDt}$$

and the apparent rotation of the two flanges is:

$$\theta_f = \frac{\Delta_f}{D} = \frac{TL}{GBD^2 t}$$

If the same calculation is applied to the web plates then the apparent angle of web rotation is:

$$\theta_w = \frac{TL}{GB^2 Dt}$$

The St Venant torsional rotation from equation (D6.2-36) is calculated to be:

$$\theta = \frac{T(B+D)L}{2GB^2 D^2 t}$$

For a square cross-section $\theta_f = \theta_w = \theta$ and therefore the ends of the cross-section remain plane as assumed. For the rectangular cross-section with $B > D$ however, $\theta_f > \theta > \theta_w$ from the above analysis but since the actual rotation must be equal to a unique value θ for the whole cross-section, to achieve this the flanges and webs must warp. Figure 6.2-52(b)

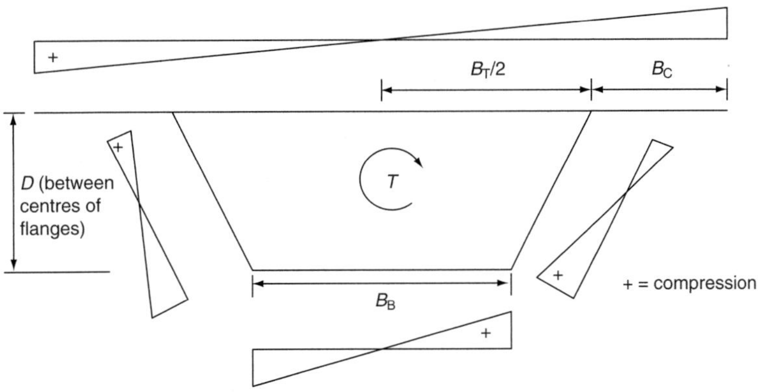

Fig. 6.2-53. Distribution of torsional warping stresses

shows how the flanges must warp to *reduce* their apparent rotation and the webs will warp in a similar way so as to *increase* their apparent rotation. The final cross-sectional warping is therefore shown in Fig. 6.2-52(c). For perfectly circular or square sections no warping of the cross-section will occur.

If the in-plane warping deformation is prevented by a rigid diaphragm (at free ends) or by an adjacent span (in the case of a continuous beam) or by symmetry (as at mid-span with symmetric loading and support conditions) then longitudinal stresses will be induced. These stresses are referred to as being due to 'restraint of torsional warping'. In reality, the out-of-plane stiffness of steel diaphragms normally found in steel box girder bridges is insufficient to generate warping restraint, although concrete diaphragms might generate such restraint. For the reasons above, no warping stresses develop in perfectly circular or square sections.

As an approximation, when an increment of torque, ΔT, is applied at a section of a box girder such as that in Fig. 6.2-53 (other than at a free end where there cannot be any longitudinal warping stresses), the resulting maximum longitudinal stress at this section due to restraint of torsional warping at the junction between the bottom flange and the web is given by:

$$\sigma_{\mathrm{TWB}} = \frac{D\Delta T}{I_{\mathrm{T}}} \qquad \text{(D6.2-37)}$$

The stress at the junction between the top flange and the web is:

$$\sigma_{\mathrm{TWT}} = \left(\frac{B_{\mathrm{B}}}{B_{\mathrm{T}}}\right)^2 \frac{D\Delta T}{I_{\mathrm{T}}\left(1 + \dfrac{2B_{\mathrm{c}}}{B_{\mathrm{T}}}\right)^3} \qquad \text{(D6.2-38)}$$

The stresses decay away quickly remote from the section where the torque is applied so that at a distance x away, the above stresses are reduced exponentially according to equation (D6.2-39):

$$\sigma_{\mathrm{TW}} = \sigma_{\mathrm{TW}}\, e^{-(2x/B_{\mathrm{B}})} \qquad \text{(D6.2-39)}$$

These formulae (which are given in BS 5400: Part 3[4]) are only approximate and, despite the discussions above, would predict a torsional warping stress for square and circular sections. For real boxes, the estimate of stress produced is reasonable however. The distribution across the section of the longitudinal stress due to restraint of torsional warping can be assumed to be as shown in Fig. 6.2-53.

Torsional warping longitudinal restraint stresses can be safely neglected at the ultimate limit state as they do not contribute to the carrying of the torsion and can therefore be relieved by plastic redistribution. This is stated in ***3-1-1/clause 6.2.7(7)***. They should however be considered for serviceability and fatigue stress checks as they do increase stresses in the corners of the box. *3-1-1/clause 6.2.7(7)*

6.2.8. Bending, axial load, shear and transverse loads

This section of the guide is split into four sub-sections as follows:

- Resistance to transverse loads *Section 6.2.8.1*
- Interaction of transverse loads with other effects *Section 6.2.8.2*
- Bending, axial load and transverse loads *Section 6.2.8.3*
- Bending, axial load, shear and transverse loads *Section 6.2.8.4*

6.2.8.1. Resistance to transverse loads

Large local transverse loads are relatively uncommon in bridge design other than during launching operations, from special vehicles or from heavy construction loads, such as from a crane outrigger. Strictly, patch loading from local wheel loads should be checked but is unlikely to be significant. Neither EN 1993-1-1 nor EN 1993-2 deal with patch loads on beams. A method of calculation of resistance is presented in 3-1-5/clause 6 which is intended for use in subsequent interaction equations. This will be discussed here. An alternative method, based on individual panel checks and the Von Mises yield criterion, can also be used as discussed in section 6.2.2.6 of this guide. That method, however, is more conservative as it does not make the same allowances for plasticity that are implicit in the empirical interaction-based methods.

The patch loading rules given in 3-1-5/clause 6 make allowance for failure by either plastic failure of the web, with associated plastic bending deformation of the flange, or by buckling of the web. In ENV 1993-1-1,[19] the latter failure mode was separated into web crippling (where the buckling failure was local to the flange) or web buckling (where most of the depth of the web buckled). The rules for patch loading can only be used if the geometric conditions discussed in section 6.2.2.5 of this guide are met, otherwise the method discussed in section 6.2.2.6 should be used. ***3-1-5/clause 6.1(1)*** also requires that the compression flange is 'adequately restrained' laterally. This restraint requirement is not defined, but it should be satisfied if the flange is continuously braced by, for example, a deck slab or if there are sufficient restraints to prevent lateral torsional buckling.

3-1-5/clause 6.1(1)

The slenderness for buckling failure under patch loading follows the usual Eurocode format. The slenderness is the square root of a plastic resistance divided by an elastic critical force according to ***3-1-5/clause 6.4(1)***:

3-1-5/clause 6.4(1)

$$\bar{\lambda}_F = \sqrt{\frac{F_y}{F_{cr}}} = \sqrt{\frac{l_y t_w f_{yw}}{F_{cr}}} \qquad \text{3-1-5/(6.4)}$$

Yielding

The plastic resistance, $F_y = l_y t_w f_{yw}$, depends on the length l_y which is the effective loaded length acting on the top of the web resulting from the distributing effect of the loaded flange. This length, l_y, depends on the loading configuration shown in Fig. 6.2-55. For cases (a) and (b), it is assumed that four plastic hinges form in the flange as shown in Fig. 6.2-54 and the spread length is calculated such that the flange system fully mobilizes the four hinges. For stocky webs, the plastic resistance of the hinges is based on that of the top flange alone. Using a work equation for this situation and equating work done externally to work done internally gives:

$$[l_y - (s_s + 2t_f) - s_y/2] f_{yw} t_w \Delta = 4M_p \theta \qquad \text{(D6.2-40)}$$

Since the plastic moment resistance of flange alone $= M_p = b_f t_f^2 f_{yf}/4$ and $\theta = 2\Delta/s_y$, equation (D6.2-40) becomes:

$$[l_y - (s_s + 2t_f) - s_y/2] f_{yw} t_w = 2b_f t_f^2 f_{yf}/s_y \qquad \text{(D6.2-41)}$$

The width of flange b_f should be limited to $15\varepsilon t_f$ on each side of the web as elsewhere for attached widths of plate acting with stiffeners – ***3-1-5/clause 6.5(1)*** refers. The effective bearing length is given by:

3-1-5/clause 6.5(1)

$$l_y = (s_s + 2t_f) + s_y \qquad \text{(D6.2-42)}$$

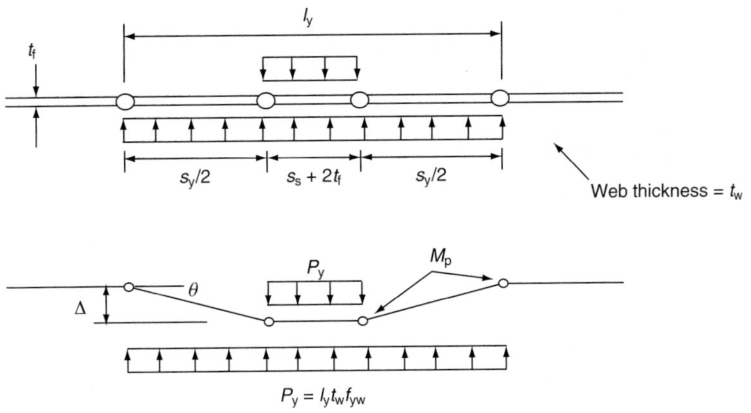

Fig. 6.2-54. Flange collapse mechanism for determination of bearing length l_y

Substitution into equation (D6.2-41) gives:

$$s_y/2 = \frac{2b_f t_f^2 f_{yf}}{t_w f_{yw} s_y} \qquad \text{(D6.2-43)}$$

and hence

$$s_y = 2t_f\sqrt{\frac{b_f f_{yf}}{t_w f_{yw}}} \qquad \text{(D6.2-44)}$$

From equation (D6.2-42) the effective bearing length is:

$$l_y = (s_s + 2t_f) + 2t_f\sqrt{\frac{b_f f_{yf}}{t_w f_{yw}}} \qquad \text{(D6.2-45)}$$

If the parameter m_1 is introduced so $m_1 = b_f f_{yf}/(t_w f_{yw})$ then equation (D6.2-45) becomes:

$$l_y = s_s + 2t_f(1 + \sqrt{m_1}) \qquad \text{(D6.2-46)}$$

The mechanism in Fig. 6.2-54 uses a distance of $s_s + 2t_f$ between inner hinges to allow for the spread of load through the flange so that the effective loaded length on the web is at least the stiff loaded length plus the spread through the flange. If the top flange is composite with a concrete deck it will be conservative to ignore the contribution of the reinforced concrete to the plastic bending resistance of the flange. No testing is available to validate inclusion of any contribution.

The expression in ***3-1-5/clause 6.5(2)*** is similar to equation (D6.2-46) but there is an additional term, m_2: *3-1-5/clause 6.5(2)*

$$l_y = s_s + 2t_f(1 + \sqrt{m_1 + m_2}) \qquad \text{3-1-5/(6.10)}$$

For webs that are slender, such that the full yield force cannot be reached, it is assumed that a part of the web plate acts with the flange (forming a T section) at the outer hinges and increases the plastic moment M_p at those locations. This is based on test observations that suggested the depth of web acting increased with depth of section for slender members. This leads to the introduction of the parameter

$$m_2 = 0.02\left(\frac{h_w}{t_f}\right)^2$$

to represent the increasing outer hinge resistance with web depth. If $\overline{\lambda}_F \le 0.5$, such that the web is stocky and the full web yield force can be reached, then $m_2 = 0$ and the web contribution to hinge resistance is ignored. This is done to avoid overestimating the resistance of stocky webs as found by testing. This may lead to an iteration being needed

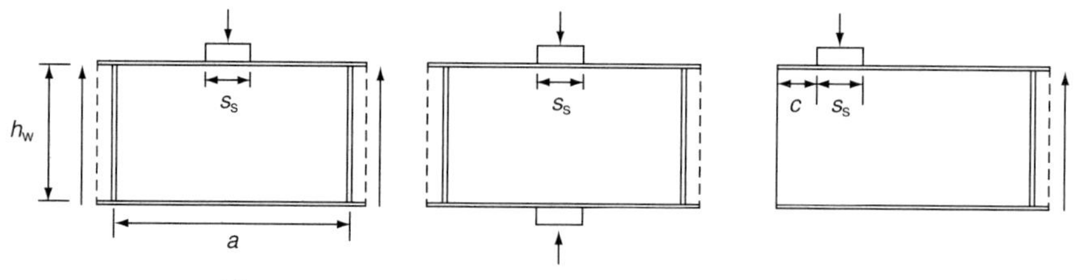

Fig. 6.2-55. Buckling coefficient k_F in various loading situations for type (a), type (b) and type (c)

to see whether m_2 may be taken greater than 0. It also leads to an unfortunate discontinuity in resistance at $\overline{\lambda}_F = 0.5$ such that the resistance of a web with $\overline{\lambda}_F$ just greater than 0.5 may have a resistance which is greater than a more stocky web with $\overline{\lambda}_F$ just less than 0.5. A typical procedure might be to first calculate $\overline{\lambda}_F$ assuming $m_2 = 0$. If $\overline{\lambda}_F > 0.5$, the calculation of $\overline{\lambda}_F$ can be repeated with a non-zero value of m_2, but it must then be checked that $\overline{\lambda}_F$ is still greater than 0.5. If it is not, the original slenderness value based on $m_2 = 0$ must be used.

For Fig. 6.2-55 case (c), the analysis above has to be modified slightly due to the different support conditions. In this case, the length l_y is taken as the lower of that calculated above and the following further two equations:

$$l_y = l_e + t_f\sqrt{\frac{m_1}{2} + \left(\frac{l_e}{t_f}\right)^2 + m_2} \qquad \text{3-1-5/(6.11)}$$

$$l_y = l_e + t_f\sqrt{m_1 + m_2} \qquad \text{3-1-5/(6.12)}$$

where $l_e = k_F E t_w^2/(2 f_{yw} h_w) \leq s_s + c$

Buckling

3-1-5/clause 6.4(1)

The elastic critical buckling load in ***3-1-5/clause 6.4(1)*** follows a standard format from elastic theory:

$$F_{cr} = k_F \frac{\pi^2 E}{12(1-\upsilon^2)} \frac{t_w^3}{h_w} = 0.9 k_F E \frac{t_w^3}{h_w} \qquad \text{3-1-5/(6.5)}$$

The buckling coefficient k_F depends on the load application type as shown in Fig. 6.2-55. Simple values of k_F are not readily available from elastic theory and the following results were based on finite-element studies[20] for webs without longitudinal stiffeners. The coefficients (given in 3-1-5/Fig. 6.1) do not allow for variations in the length of the load in cases (a) and (b) and therefore may become conservative for long loaded length:

$$\text{Type (a): } k_F = 6 + 2\left(\frac{h_w}{a}\right)^2 \qquad \text{(D6.2-47)}$$

$$\text{Type (b): } k_F = 3.5 + 2\left(\frac{h_w}{a}\right)^2 \qquad \text{(D6.2-48)}$$

$$\text{Type (c): } k_F = 2 + 6\left(\frac{s_s + c}{h_w}\right) \leq 6 \qquad \text{(D6.2-49)}$$

3-1-5/clause 6.4(2)

For webs with longitudinal stiffeners, the National Annex can give values of k_F. ***3-1-5/clause 6.4(2)*** gives one solution for the most commonly encountered case of type (a) loading:

$$k_F = 6 + 2\left(\frac{h_w}{a}\right)^2 + \left[5.44\frac{b_1}{a} - 0.21\right]\sqrt{\gamma_s} \qquad \text{3-1-5/(6.6)}$$

where

$$\gamma_s = 10.9\left(\frac{I_{sl,1}}{h_w t_w^3}\right) \leq 13\left[\frac{a}{h_w}\right]^3 + 210\left[0.3 - \frac{b_1}{a}\right]$$

with $I_{sl,1}$ equal to the second moment of area of the effective section (comprising the stiffener outstand and an attached width of web of up to $15\varepsilon t_w$ on each side) of the longitudinal stiffener nearest the loaded flange. This is based on the resistance of an unstiffened panel in equation (D6.2-47) with some additional resistance arising from the restraint to the web panel offered by the longitudinal stiffeners. The equation is only valid for $0.05 \leq b_1/a \leq 0.3$ and $b_1/h_w \leq 0.3$ where b_1 is the depth of the sub-panel adjacent to the loaded flange.

Expression 3-1-5/(6.6) leads to a resistance lower than for an unstiffened panel if $b_1/a < 0.039$. In this case, the value for an unstiffened panel can conservatively be used. It will be found that, for b_1 within the above limits, the resistance actually increases with increasing b_1, sometimes up to a maximum before reducing again with further increases in b_1. This occurs because the analysis assumes that there is only one stiffener on the web. For small b_1, this stiffener is close to the loaded flange and is not very effective in stabilising the web; buckling then occurs in the sub-panel below the stiffener. For some geometries, no maximum is reached within the limits of application and the resistance simply rises with increasing b_1. Clearly this is incorrect for several equally spaced longitudinal stiffeners down the web as it suggests that the web is weakened by adding more stiffeners. In situations such as these (or for cases outside the limits of applications), it is possible to determine the elastic critical patch load from a finite-element analysis. If this is done, the plate boundaries should be modelled with hinged edges in order to be compatible with the analysis behind the derivation of the reduction factor curve in 3-1-5/clause 6.4(1). Alternatively, non-linear analysis with imperfections could be used.

Reduction factor

The reduction factor is calculated as follows from ***3-1-5/clause 6.4(1)***: *3-1-5/clause 6.4(1)*

$$\chi_F = \frac{0.5}{\bar{\lambda}_F} \leq 1.0 \qquad \text{3-1-5/(6.3)}$$

The design resistance is then:

$$F_{Rd} = \chi_F \frac{f_{yw} l_y t_w}{\gamma_{M1}} \qquad \text{(D6.2-50)}$$

Note that ***3-1-5/clause 6.2(1)*** presents equation (D6.2-50) as: *3-1-5/clause 6.2(1)*

$$F_{Rd} = \frac{f_{yw} L_{eff} t_w}{\gamma_{M1}}$$

with $L_{eff} = \chi_F l_y$ which introduces another effective length, L_{eff}, and thus some possibility for confusion. It also has little physical significance.

6.2.8.2. Interaction of transverse loads with other effects

For the interaction-based approach, EN 1993-1-5 section 7 is used as discussed in section 6.2.8.3 below. Transverse patch loads only need to be combined with bending and axial load when using this method. Some concerns have been expressed by the authors and others that no interaction with shear is required in EN 1993-1-5, even at very high shear, as this case has been less well investigated. Some tests have been examined with coexisting shear up to about 75% of the shear resistance[20] and these gave little influence of the shear. The high shear was however mostly produced by the patch load itself. The latter point means that, if shear were included in an interaction, there would be some degree of double-counting as the transverse load is transformed into shear in the vicinity of the load. The test results would therefore be applicable to the check of a bridge girder web during launching, but would not cover the case of a large concentrated patch load on a web, where the patch load itself was a small proportion of the total load.

The results of a study by Kuhlman and Seitz[21] suggested that shear could have an influence on the resistance to patch loads of longitudinally stiffened webs. However, in that study the limiting patch load was still well in excess of the predictions of EN 1993-1-5. From the limited tests, it would therefore appear that the interaction presented in 3-1-5/clause 7.2 is generally safe but some caution is advised when:

- the coexisting shear exceeds 75% of the shear resistance;
- the patch load itself produces only a relatively small amount of the total shear at the location of the load.

If the designer is concerned about the interaction with shear in a particular situation, the combination of effects could be considered by performing panel checks in accordance with 3-1-5/clause 10 as discussed in section 6.2.2.6 of this guide. This is however much more conservative. As a final point, it should be noted that the rules for patch loads given in BS 5400: Part 3[4] also did not include any interaction with shear and panel checks had to be performed if the designer wanted to include it. The situation in EN 1993 is therefore not very different!

6.2.8.3. Bending, axial load and transverse loads

3-1-5/clause 7.2(1)

If transverse load is present, its interaction with bending and axial force can be checked according to the interaction given in ***3-1-5/clause 7.2(1)*** as follows:

$$\eta_2 + 0.8\eta_1 = 1.4 \qquad \text{3-1-5/(7.2)}$$

where:

$$\eta_2 = \frac{\sigma_{z,Ed}}{f_{yw}/\gamma_{M1}} = \frac{F_{Ed}}{f_{yw}L_{eff}t_w/\gamma_{M1}} = \frac{F_{Ed}}{F_{Rd}}$$

is the usage factor for transverse load acting alone. $\sigma_{z,Ed}$ has little real physical significance in this case.

$$\eta_1 = \frac{\sigma_{x,Ed}}{f_y/\gamma_{M0}} = \frac{N_{Ed}}{f_yA_{eff}/\gamma_{M0}} + \frac{M_{Ed} + N_{Ed}e_N}{f_yW_{eff}/\gamma_{M0}}$$

is the usage factor for direct stress alone, calculated elastically. The calculation of η_1 is discussed in section 6.2.10 of this guide.

It can be seen that this interaction in expression 3-1-5/(7.2) does not allow for a plastic distribution of stress under bending and axial force. If the section is Class 1 or 2, this may initially seem an unreasonable penalty simply because a transverse load (which could be very small) has been applied to the section. However, it will not lead to any discontinuity with the plastic interaction between bending and axial load (or with shear) as only 80% of the elastic bending stress has to be considered and the limiting value of the interaction is 1.4. Since the ratio between a plastic and elastic moment resistance for beams is typically less than 1.2, it can be seen that the interaction will not lead to any discontinuity when a small transverse load is applied.

This interaction has been produced assuming that the patch load is applied to the compression flange. If the load is applied to the tension flange, the Von Mises yield criterion in section 6.2.1 of this guide should be satisfied. The transverse stress should be based on the distribution discussed in section 6.2.2.3.2 of this guide, rather than on the effective length l_y derived from the mechanism approach in section 6.2.8.2.

The need to only consider transverse load with longitudinal direct stresses and not shear is discussed in section 6.2.8.2. In theory, since the patch load rules of EN 1993-1-5 can only be used if certain geometric constraints are met as discussed in section 6.2.5.1 of this guide, it may, however, sometimes be necessary to use the Von Mises yield criterion and panel buckling checks discussed in section 6.2.2.6 of this guide. This will be more conservative and does then include shear stresses.

Limited comparison has been carried out by the authors with Annex D of BS 5400: Part 3: 2000.[4] This suggests that results are quite similar for long panels, typical bridge girder cross-sections and flange stresses of about 50% of the yield stress. For short panels, shallow girders and higher flange stresses, EN 1993-1-5 becomes considerably less conservative.

Worked Example 6.2-11: Patch load on bridge beam

A crane outrigger applies a 500 kN ultimate limit state load over a square area of 300 mm side to the web of the steel and concrete composite beam shown in Fig. 6.2-56. The top flange can be assumed to be restrained from lateral movement. The yield strength is taken as 355 MPa for all plate sizes in accordance with 3-1-1/Table 3.1 (but note that EN 10025 specifies 345 MPa for the 40 mm thick plate and the UK National Annex requires this to be used). The coexistent compressive bending stress in the top flange is 300 MPa. The beam is checked for the combined effect of bending and transverse load.

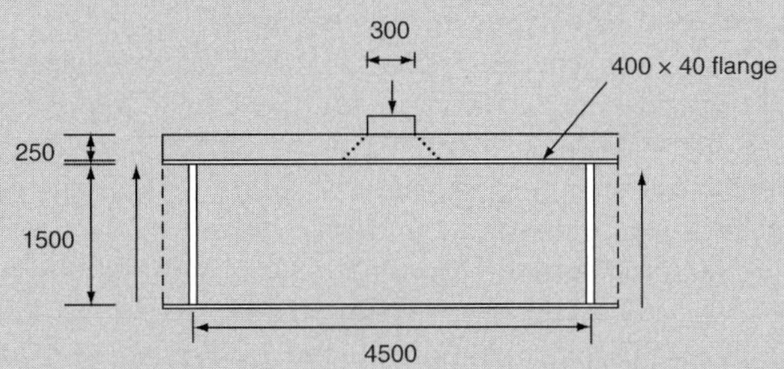

Fig. 6.2-56. Elevation of composite beam in Worked Example 6.2-11

Ignoring any contribution from the concrete to the top flange plastic moment resistance, from 3-1-5/clause 6.5(1):

$$m_1 = \frac{b_f f_{yf}}{t_w f_{yw}} = \frac{400 \times 355}{15 \times 355} = 26.67$$

$$m_2 = 0.02\left(\frac{h_w}{t_f}\right)^2 = 0.02\left(\frac{1500}{40}\right)^2 = 28.1$$

assuming that the slenderness exceeds 0.5, which is found to be the case below.

The stiff bearing length can include a spread through the concrete (taken as 1:1 here) so $s_s = 300 + 2 \times 250 = 800$ mm.

For a patch load applied to the top flange between stiffeners, from 3-1-5/clause 6.5(2): $l_y = s_s + 2t_f(1 + \sqrt{m_1 + m_2}) = 800 + 2 \times 40(1 + \sqrt{26.67 + 28.1}) = 1472$ mm.

From 3-1-5/Fig. 6.1:

$$k_F = 6 + 2\left(\frac{h_w}{a}\right)^2 = 6 + 2\left(\frac{1500}{4500}\right)^2 = 6.22$$

From 3-1-5/clause 6.4(1):

$$F_{cr} = 0.9 k_F E \frac{t_w^3}{h_w} = 0.9 \times 6.22 \times 210 \times 10^3 \frac{15^3}{1500} = 2.582 \times 10^6\ \text{N}$$

The slenderness is therefore:

$$\bar{\lambda}_F = \sqrt{\frac{F_y}{F_{cr}}} = \sqrt{\frac{l_y t_w f_{yw}}{F_{cr}}} = \sqrt{\frac{1472 \times 15 \times 355}{2.5862 \times 10^6}} = 1.74$$

$$\chi_F = \frac{0.5}{\bar{\lambda}_F} = \frac{0.5}{1.74} = 0.287$$

The resistance to patch load is therefore given by 3-1-5/clause 6.2(1):

$$F_{Rd} = \chi_F \frac{f_{yw} l_y t_w}{\gamma_{M1}} = 0.287 \times \frac{355 \times 1472 \times 15}{1.1} = 2047\ \text{kN}$$

The interaction between bending and transverse load is then as follows:

$\eta_2 + 0.8\eta_1 = 0.244 + 0.8 \times 0.845 = \mathbf{0.92} < \mathbf{1.4}$ so the web is adequate.

In the above:

$$\eta_2 = \frac{F_{Ed}}{F_{Rd}} = \frac{500}{2047} = 0.244 \text{ and } \eta_1 = \frac{\sigma_{x,Ed}}{f_y/\gamma_{M0}} = \frac{300}{355/1.0} = 0.845$$

6.2.8.4. Bending, axial load, shear and transverse loads

The method to be used for checking combinations of bending, axial force, shear and transverse load depends on whether or not the steel component is susceptible to local buckling. 'Local buckling' in this context means either buckling under longitudinal direct stresses (i.e. the cross-section is classified as Class 4 as discussed in section 5.5 of this guide), or shear buckling as discussed in section 6.2.6. As discussed in section 6.2.8.2 above, there is no requirement to combine shear with transverse loads but the designer may choose to do so by performing panel checks as discussed in section

3-2/clause 6.2.8(1)

6.2.2.6 of this guide. The different methods allowed by ***3-2/clause 6.2.8(1)*** are discussed below.

(i) Section not susceptible to local buckling

Method 1

Interaction methods given in EN 1993-2 clauses 6.2.9, 6.2.10 and 6.2.11 can be used to check combinations of bending, axial load and shear. These interactions are discussed in sections 6.2.9 to 6.2.11 of this guide and will generally give the most economic design as they allow some plastic stress redistribution in the steel component after yielding. This method is therefore recommended here. If transverse load is present, its interaction with bending and axial force must be checked according to the interaction discussed in section 6.2.8.3 above.

Method 2

The combined stress field can be considered using the Von Mises yield criterion discussed in section 6.2.1. This will generally be conservative as the method does not allow any plastic redistribution of stress after yield. If transverse load is present, its interaction with other effects can be included by using the panel buckling check method of 3-1-5/clause 10 as discussed in section 6.2.2.6 of this guide.

(ii) Section is susceptible to local buckling

Method 1

Interaction methods given in EN 1993-1-5 have to be used to check combinations of bending, axial load and shear as the methods allow for local buckling. This method is recommended here as it allows for shedding of load from overstressed plate panels such that the section resistance is not necessarily limited by initial buckling of the weakest sub-panel. Certain geometric criteria as discussed in section 6.2.2.5.1 have to be met to use this method, otherwise method 2 has to be used. If transverse load is present, its interaction with bending and axial force must be checked according to the interaction discussed in section 6.2.8.3 above.

Method 2

The combined stress field can be considered using the Von Mises yield criterion and panel buckling checks in 3-1-5/clause 10. This method can be conservative for the reasons discussed in section 6.2.2.6 of this guide.

6.2.9. Bending and shear

Shear forces acting on a steel section may reduce the bending resistance of the section if the shear force is sufficiently large and hence it is necessary to consider an interaction between the two internal actions. The method of calculation depends on the classification of the cross-section for longitudinal direct stresses and also whether or not the steel section is prone to shear buckling before reaching the plastic shear resistance.

This section of the guide is split into two sub-sections as follows:

- Sections not susceptible to shear buckling *Section 6.2.9.1*
- Sections susceptible to shear buckling *Section 6.2.9.2*

6.2.9.1. Sections not susceptible to shear buckling

6.2.9.1.1. Class 1 and 2 cross-sections

Theoretical interactions between bending and shear can be derived by reducing the effectiveness of the web in resisting direct stresses to allow for the effects of shear. A reduced effectiveness can be derived based on the Von Mises yield criterion, but this implies a reduced bending resistance in the presence of any shear. Testing has shown that, even for relatively high values of shear force, the reduction due to shear on the plastic moment resistance is negligible. This can be explained by the strain hardening of the steel. ***3-1-1/clause 6.2.8(2)*** allows the interaction between shear and bending to be ignored when the design shear force is less than 50% of the plastic shear resistance. Where the design shear force is greater than 50% of the plastic shear resistance, ***3-1-1/clause 6.2.8(3)*** requires shear to be taken into account by reducing the yield stress in the 'shear area' for bending calculation by a factor $(1 - \rho)$, such that:

3-1-1/clause 6.2.8(2)
3-1-1/clause 6.2.8(3)

Allowable flexural stress in the 'shear area' $= (1 - \rho)f_y$ 3-1-1/(6.29)

where

$$\rho = \left(\frac{2V_{Ed}}{V_{pl,Rd}} - 1\right)^2$$

V_{Ed} is the applied shear force and $V_{pl,Rd}$ is the plastic shear resistance determined as discussed in section 6.2.6 of this guide. As an alternative to reducing the web strength, the thickness of the web could be reduced by the same factor. This reduced web thickness should not of course be used to reclassify the web as a higher class for direct stresses.

'Shear area' has been placed in inverted commas above because the same term is used in 3-1-1/clause 6.2.6 to describe a different parameter, A_v, which is the numerical area used to calculate the shear resistance. The shear area referred to in expression 3-1-1/(6.29) is only intended to be the web area $A_w = h_w t_w$. A_v for a plate girder, however, may be up to 1.2 times the web area as discussed in section 6.2.6 of this guide. This definition of shear area for use in expression 3-1-1/(6.29) is illustrated by the derivation of expression 3-1-1/(6.30) below for I-beams.

The formula for ρ is modified in ***3-1-1/clause 6.2.8(4)*** for components resisting combined shear and torsion by using

3-1-1/clause 6.2.8(4)

$$\rho = \left(\frac{2V_{Ed}}{V_{pl,T,Rd}} - 1\right)^2$$

where the derivation of $V_{pl,T,Rd}$ is discussed in section 6.2.7 of this guide.

The relationship produced between shear and bending is illustrated in Fig. 6.2-57.

For Class 1 and 2 symmetrical I-sections, the contribution of the web to the full plastic bending resistance is:

$$M_{pl,web} = \frac{h_w^2 t_w f_{yw}}{4} = \frac{A_w^2 f_{yw}}{4t_w} \qquad \text{(D6.2-51)}$$

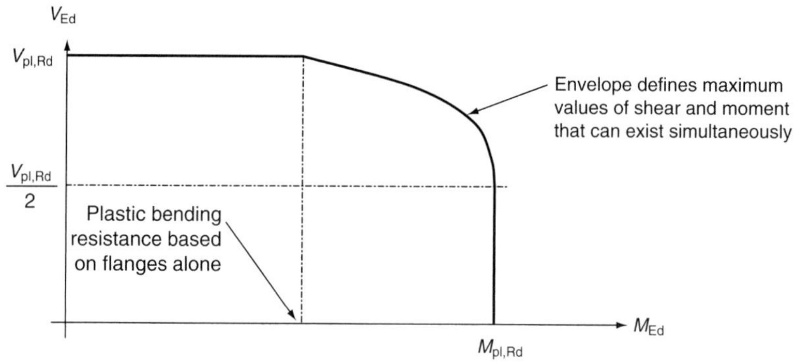

Fig. 6.2-57. Shear–moment interaction for Class 1 and 2 cross-sections

The reduction to web plastic resistance moment caused by shear is thus

$$\frac{\rho A_w^2 f_{yw}}{4t_w}$$

3-1-1/clause 6.2.8(5)

and therefore if web and flange have the same yield strength, the expression for moment resistance for symmetrical beams in ***3-1-1/clause 6.2.8(5)*** is obtained:

$$M_{y,V,Rd} = \frac{\left(W_{pl,y} - \frac{\rho A_w^2}{4t_w}\right) f_y}{\gamma_{M0}} \qquad \text{3-1-1/(6.30)}$$

Expression 3-1-1/(6.30) cannot be used for non-symmetric sections as any reduction of f_{yw} in the web will cause a shift in the plastic neutral axis location. As a result expression 3-1-1/(6.30) will rarely be of use in bridge design, where beams are generally non-symmetric and may have different yield strengths for flanges and webs. In this case, the revised plastic resistance moment must be found by first determining the new plastic neutral axis and calculating the moment resistance about this axis. This is illustrated in Worked Example 6.2-12.

It should be noted that there is no requirement in EN 1993 for the section classification, which was established with the gross-section, to be rechecked for the shift in plastic neutral axis produced by reducing the web strength. The main argument for not reclassifying the section is that the method of determining the reduced moment resistance is not intended to be a true model of the girder behaviour, only a means to produce resistances that lie safely within those from tests. A very similar interaction could have been achieved through the provision of an interaction equation (as in 3-1-5/clause 7.1 discussed below, which could be used in this case also), whereupon the issue of reclassification would not arise. EN 1994-2 clause 6.2.2.5(4) specifically clarifies this because a number of misinterpretations arose with earlier drafts.

6.2.9.1.2. Class 3 cross-sections

The depth-to-thickness ratio required to prevent shear buckling in beams with widely spaced transverse stiffeners (and no longitudinal stiffeners) is given as:

$$\frac{h_w}{t_w} \leq \frac{72}{\eta}\varepsilon$$

in 3-1-5/clause 5.1(2). For beams with widely spaced stiffeners, it is unlikely that a beam could be Class 3 in bending without being susceptible to shear buckling. This situation can however occur where transverse stiffening is provided to the web to allow the plastic shear resistance to be reached.

Some interpretation of the interaction method is necessary for Class 3 cross-sections. As written, 3-1-1/clause 6.2.8(3) requires the 'design resistance of the cross-section' to be

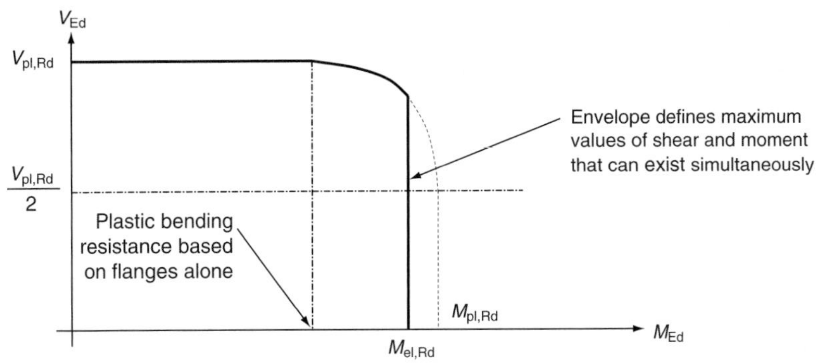

Fig. 6.2-58. Bending and shear interaction envelope for Class 3 cross-sections

determined using the reduced web strength in expression 3-1-1/(6.29). Since Class 3 section design requires elastic stresses to be limited to yield, the bending resistance at full shear would be zero, governed by web yield. This is clearly incorrect. Another interpretation would be to reduce the web thickness, rather than the web yield strength. This leads to more credible results such that the resistance moment equals that due to the flanges alone when the web is fully stressed in shear and the elastic moment resistance is reduced when the shear exceeds 50% of the plastic shear resistance.

A further interpretation arises from considering 3-1-5/clause 7.1 and the ENV version of EN 1993-1-1. In both of these, the interaction is conducted using the plastic bending resistance, but the moment resistance calculated is limited to the elastic resistance moment in the absence of shear. The reason for applying the interaction in this way is that test results on symmetric beams with Class 3 and Class 4 webs (Reference 22) and computer simulations on composite bridge girders (therefore with unequal flanges) (Reference 23) showed very weak interaction with shear. The former physical tests showed virtually no interaction at all and the latter typically showed some minor interaction only after 80% of the shear resistance had been reached. The use of a plastic resistance moment in the interaction helps to force this behaviour as seen below. The formula in 3-1-1/clause 6.2.8(5) implicitly allows this same approach for I-beams with equal flanges but appears to be deliberately non-committal for the case of beams with unequal flanges. This reflects the fact that most available testing relates to symmetric beams where the web has no net compressive force. The literal interpretation of 3-1-1/clause 6.2.8 seems to be that the elastic bending resistance should be used in the interaction, which puts it at odds with EN 1993-1-5.

If the interpretation of using plastic moment resistances in the interaction is used (following the EN 1993-1-5 method), the procedure for treating Class 3 cross-sections is then identical to that for Class 1 and 2 sections above, except that the reduced moment $M_{y,V,Rd}$ should not be allowed to exceed the elastic design moment $M_{y,c,Rd} = M_{el,Rd}$ calculated in accordance with 3-1-1/clause 6.2.5. However, before $M_{el,Rd}$ is reached, the shear reduction to bending resistance is still derived from the plastic resistance moment of the Class 3 section. The shear–moment interaction diagram for a typical Class 3 section is then as illustrated in Fig. 6.2-58. It is effectively the same curve as in Fig. 6.2-57 but it is truncated by limiting the resistance moment to the elastic value. This ensures that shear forces well in excess of 50% of the plastic shear resistance can be achieved without affecting the bending resistance in line with the test results. The comments made above for Class 1 and 2 cross-sections, regarding not reclassifying the beam for bending in the presence of shear, also apply to Class 3 cross-sections.

6.2.9.1.3. Class 4 cross-sections

Class 4 sections have to be dealt with using one of two possible methods given in EN 1993-1-5. These are discussed in section 6.2.9.2.3 below as the procedure is the same whether or not there is shear buckling.

Worked Example 6.2-12: Shear–moment interaction for Class 2 plate girder cross-section without shear buckling

A plate girder in S355 steel is shown in Fig. 6.2-59. The thickness-dependent yield stresses are taken from 3-1-1/Table 3.1 which gives a constant yield stress of 355 MPa throughout. (The UK National Annex to EN 1993-2 requires the values in EN 10025 to be used.) The girder is a Class 2 cross-section for hogging moment and has the following properties in the absence of shear:

- Plastic neutral axis = 525 mm from bottom flange
- Plastic resistance moment = 7834 kNm

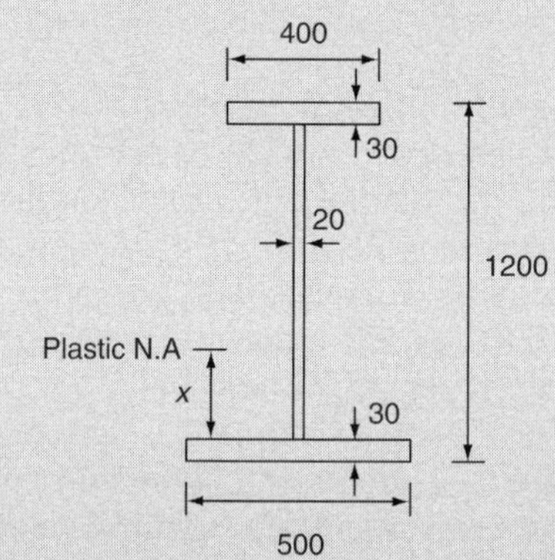

Fig. 6.2-59. Plate girder for Worked Example 6.2-12

The girder is restrained against lateral torsional buckling and stable against shear buckling. The maximum bending moment that the section can withstand is calculated in conjunction with a shear force of 4486 kN.

Web area $A_w = h_w t_w = (1200 - 60) \times 20 = 22\,800\,\text{mm}^2$

Plastic shear resistance:

$$V_{pl,Rd} = \frac{\eta A_w (f_y/\sqrt{3})}{\gamma_{M0}} = \frac{1.2 \times 22\,800(355/\sqrt{3})}{1.00} = 5608\,\text{kN}$$

where $A_v = \eta A_w$ and η is taken as 1.2 as recommended in 3-1-5/clause 5.1(2). V_{Ed} is greater than $0.5 \times V_{pl,Rd}$ so shear will reduce the resistance moment to $M_{y,V,Rd}$.

From expression 3-1-1/(6.29):

$$\rho = \left(\frac{2V_{Ed}}{V_{pl,Rd}} - 1\right)^2 = \left(\frac{2 \times 4486}{5608} - 1\right)^2 = 0.360$$

Allowable stress in web = $(1 - \rho)f_y = (1 - 0.36) \times 355 = 227.2\,\text{MPa}$

The plastic moment of resistance with the reduced web allowable stress is calculated next. The plastic neutral axis will shift from its position without shear, caused by the reduction in web strength. The new plastic neutral axis at height x from the top of the bottom flange is found by balancing forces:

$$(500 \times 30 \times 355) + (20 \times 227.2 \times x) = (400 \times 30 \times 355) + [20 \times 227.2 \times (1140 - x)]$$

for which $x = 452.8\,\text{mm}$.

The moment resistance in the presence of shear is found by taking moments about the plastic neutral axis:

$$M_{y,V,Rd} = \frac{(500 \times 30 \times 467.8 + 400 \times 30 \times 702.2) \times 355}{1.00}$$
$$+ \frac{[452.8^2 \times 20 \times 0.5 + (1140 - 452.8)^2 \times 20 \times 0.5] \times 227.2}{1.00} = \mathbf{7021\,kNm}$$

The plastic moment resistance of the section reduces to 7021 kNm from 7834 kNm with a coexistent shear force of 4486 kN.

Worked Example 6.2-13: Shear–moment interaction for Class 3 plate girder without shear buckling

The steel plate girder shown in Fig. 6.2-60 is on the upper limit for a Class 3 cross-section. It has the following properties in the absence of shear:

- Plastic section modulus of girder, $W_{pl,y} = 4.436 \times 10^7\,mm^3$
- Elastic section modulus of girder, $W_{el,min} = 3.750 \times 10^7\,mm^3$(based on the mid-plane of the flange as allowed by 3-1-1/clause 6.2.1(9).

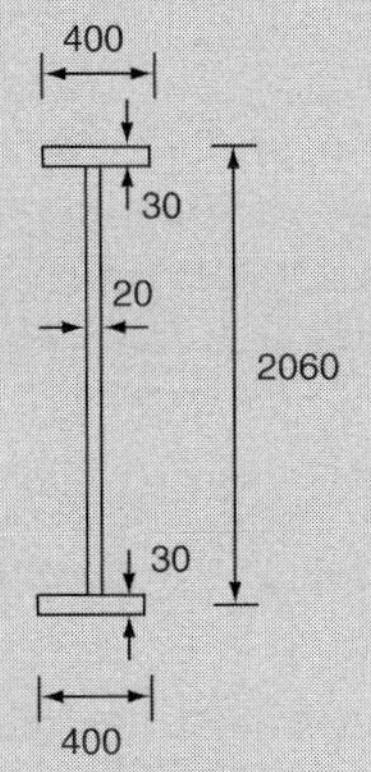

Fig. 6.2-60. Plate girder for Worked Example 6.2-13

All plates are Grade S355 to EN 10025 and the girder is restrained against lateral torsional buckling and stable against shear buckling due to the presence of closely spaced transverse stiffeners. The thickness-dependent yield stresses are taken from 3-1-1/Table 3.1 which gives a constant yield stress of 355 MPa throughout. (The UK National Annex to EN 1993-2 requires the values in EN 10025 to be used.) The maximum bending moment that the section can withstand in conjunction with a shear force of 7871 kN is calculated.

Web area $A_w = h_w t_w = (2060 - 60) \times 20 = 40\,000\,mm^2$

Plastic shear resistance:

$$V_{pl,Rd} = \frac{\eta A_w (f_y/\sqrt{3})}{\gamma_{M0}} = \frac{1.2 \times 40\,000(355/\sqrt{3})}{1.00} = 9838\,kN$$

with η taken as 1.2 as recommended in 3-1-5/clause 5.1(2). V_{Ed} is greater than $0.5 \times V_{pl,Rd}$ so shear will reduce moment resistance to $M_{y,V,Rd}$.

From expression 3-1-1/(6.29):

$$\rho = \left(\frac{2V_{Ed}}{V_{pl,Rd}} - 1\right)^2 = \left(\frac{2 \times 7871}{9838} - 1\right)^2 = 0.360$$

As the beam is symmetric and the yield strength is the same everywhere, expression 3-1-1/(6.30) can be used.

$$M_{y,V,Rd} = \frac{\left[W_{pl,y} - \frac{\rho A_w^2}{4t_w}\right] f_y}{\gamma_{M0}} = \frac{\left[4.436 \times 10^7 - \frac{0.36 \times (2000 \times 20)^2}{4 \times 20}\right] \times 355}{1.00}$$

$$= 13\,192\,\text{kNm}$$

but not greater than:

$$M_{c,Rd} = \frac{W_{el,min} f_y}{\gamma_{M0}} = \frac{3.750 \times 10^7 \times 355}{1.0} = 13\,317\,\text{kNm} > M_{y,V,Rd}$$

just. Therefore, the moment resistance of the section reduces from 13 317 kNm to **13 192 kNm** with a coexistent shear force of 7871 kN.

6.2.9.2. Sections susceptible to shear buckling

3-1-1/clause 6.2.8(2)

If the section's shear resistance is limited by shear buckling as discussed in section 6.2.6 of this guide, ***3-1-1/clause 6.2.8(2)*** effectively requires section 7 of EN 1993-1-5 to be used to perform the interaction between shear and bending.

6.2.9.2.1. Class 1 and 2 cross-sections

3-1-5/clause 7.1(1)

The approach is similar to that for no shear buckling. ***3-1-5/clause 7.1(1)*** allows the designer to neglect the interaction between shear and bending moment when the design shear force is less than 50% of the shear buckling resistance based on the web contribution alone. Where the design shear force exceeds this value, the following interaction has to be satisfied:

$$\bar{\eta}_1 + \left(1 - \frac{M_{f,Rd}}{M_{pl,Rd}}\right)(2\bar{\eta}_3 - 1)^2 \le 1.0 \qquad \text{3-1-5/(7.1)}$$

3-1-5/clause 7.1(3)

where $\bar{\eta}_3$ is the ratio $V_{Ed}/V_{bw,Rd}$ and $\bar{\eta}_1$ is the usage factor for bending, $M_{Ed}/M_{pl,Rd}$, based on the plastic moment resistance of the section. $M_{f,Rd}$ is the design plastic bending resistance based on a section comprising the flanges only. For unequal flanges, this may, for simplicity, be taken as the smaller plastic resistance of the two flanges multiplied by the distance between the centroids of the flanges according to ***3-1-5/clause 7.1(3)***. The interaction produced is illustrated in Fig. 6.2-61. The full web shear resistance contribution $V_{bw,Rd}$ is obtained at a moment of $M_{f,Rd}$. For smaller moments, the coexisting shear can increase further due to the additional flange shear contribution, $V_{bf,Rd}$, from 3-1-5/clause 5.4.

The expression $(2\bar{\eta}_3 - 1)^2$ can be rewritten as:

$$\left(\frac{2V_{Ed}}{V_{bw,Rd}} - 1\right)^2$$

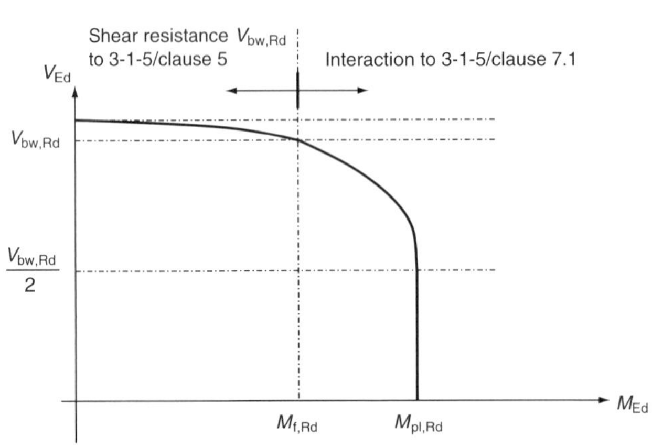

Fig. 6.2-61. Interaction to 3-1-5/clause 7.1 for Class 1 and 2 cross-sections

which is the same form as the web strength reduction factor:

$$\rho = \left(\frac{2V_{\mathrm{Ed}}}{V_{\mathrm{pl,Rd}}} - 1\right)^2$$

which is used when there is no shear buckling. For no shear buckling and symmetrical sections, expression 3-1-5/(7.1) would therefore give the same result as the method in section 6.2.9.1.1 above. For sections with no shear buckling and unequal flanges, expression 3-1-5/(7.1) would give a slightly more conservative result than the method in section 6.2.9.1.1 above. It is worth noting that expression 3-1-5/(7.1) is not used for Class 1 and 2 cross-sections in EN 1994 when there is shear buckling. Instead, the web strength is reduced by the factor $(1 - \rho)$ where:

$$\rho = \left(\frac{2V_{\mathrm{Ed}}}{V_{\mathrm{b,Rd}}} - 1\right)^2$$

and the plastic moment resistance is recalculated. Unfortunately, the two Eurocode parts have not been reconciled but interchanging methods will generally have little practical consequence.

3-1-5/clause 7.1(4) requires that where axial force is present such that the whole web is in compression, $M_{\mathrm{f,Rd}}$ should be taken as zero in accordance with 3-1-5/clause 7.1(5). It is unclear what to do if there is no external axial force but the whole web is still in compression, as could occur with an asymmetric beam. A safe interpretation, given the relatively small amount of testing on asymmetric sections, would be to take $M_{\mathrm{f,Rd}}$ as zero in this case also. This is likely to be conservative at high shear, given the weak interaction between bending and shear found in the tests on composite beams discussed in section 6.2.9.1.2 above. *3-1-5/clause 7.1(4)*

3-1-5/clause 7.1(2) does not require the interaction in 3-1-5/clause 7.1(1) to be verified at sections nearer than $h_{\mathrm{w}}/2$ to a support, where it is assumed that there is a bearing stiffener present. This is because the effect of buckling is small adjacent to a stiffener. However, the cross-section resistance should still be verified at the support. It is therefore recommended here that 3-1-5/clause 7.1(1) should be applied at the support, but using the plastic shear resistance in place of the shear buckling resistance. *3-1-5/clause 7.1(2)*

6.2.9.2.2. Class 3 cross-sections

The approach is identical to that above for Class 1 and 2 sections, except that the resulting bending resistance must additionally not exceed the elastic bending resistance. This effectively truncates the interaction diagram in Fig. 6.2-61 in the same way as in Fig. 6.2-58. The plastic bending resistance is again used in the interaction because of the weakness of interaction between bending and shear found in the studies identified in section 6.2.9.1.2 above. This ensures that shears well in excess of 50% of the web contribution to shear resistance can be accommodated before any reduction is made to the elastic bending resistance. In earlier drafts of EN 1993-1-5, $\bar{\eta}_1$ in expression 3-1-5/(7.1) was taken as $M_{\mathrm{Ed}}/M_{\mathrm{el,Rd}}$, based on the elastic bending resistance. This had the disadvantage that the bending resistance predicted was less than that of the flanges alone when the shear force was equal to $V_{\mathrm{bw,Rd}}$.

For composite beams where the cross-section is built up in stages, the same interaction can be applied and guidance on the relevant value of M_{Ed} to use is given in the *Designers' Guide to EN 1994-2*.[7] A separate check must be made of the accumulated elastic stresses, via η_1 from 3-1-5/clause 4.6. In general, it will always be conservative to base $\bar{\eta}_1$ on the ratio of accumulated stress to the allowable stress i.e. η_1.

The comments made above for Class 1 and 2 sections regarding the use of 3-1-5/clause 7.1(4) for asymmetric sections and on 3-1-5/clause 7.1(2) for sections close to supports also apply to Class 3 sections.

6.2.9.2.3. Class 4 cross-sections, including beams with longitudinal stiffeners

Two methods are possible for Class 4 cross-sections. If the required geometric constraints are met as discussed in section 6.2.2.5.1 of this guide, it will usually be most economic to use the same interaction method as above for Class 1, 2 and 3 sections. Expression 3-1-5/(7.1) again applies but the calculation of $M_{f,Rd}$ and $M_{pl,Rd}$ must consider effective widths for flanges, allowing for plate buckling. $M_{pl,Rd}$ is, however, calculated using the gross web, regardless of any reduction that might be required for local buckling under direct stress. The reason for allowing plastic properties to be used in the interaction is again due to the weakness of interaction found in the tests on beams with Class 4 webs identified in section 6.2.9.1.2 above. It is still necessary to verify the girder under direct stresses alone to 3-1-5/clause 4.6, using elastic design and appropriate effective sections for flanges and webs. This again truncates the interaction.

While the interaction of expression 3-1-5/(7.1) applies to beams with longitudinally stiffened webs, the authors are not aware of similar test justification to support the use of plastic properties in the interaction. Such webs have less post-buckling strength when overall web buckling is critical, but the approach once again leads to an interaction with shear only at very high percentages of the web shear resistance. A safer option is to replace $\overline{\eta}_1$ by η_1 in the interaction until such time as there have been further studies to confirm this to be unnecessary. In these cases, if the section is built up in stages, η_1 is the usage factor based on accumulated stress.

The comments made above for Class 1 and 2 cross-sections regarding the use of 3-1-5/clause 7.1(4) for asymmetric sections and on 3-1-5/clause 7.1(2) for sections close to supports also apply to Class 4 cross-sections. In the latter case, some interpretation is required for longitudinally stiffened webs. It is suggested here that the distance $h_w/2$ be replaced by $b_{max}/2$ (where b_{max} is the height of the largest sub-panel) when checking buckling of sub-panels.

Expression 3-1-5/(7.1) should also be used to verify flanges in box girders. However, in this case, $M_{f,Rd}$ is taken equal to zero according to 3-1-5/clause 7.1(5), $\overline{\eta}_1$ is replaced by η_1 and $\overline{\eta}_3$ is determined as the greater value obtained for overall flange shear buckling (based on the average shear stress in the flange but not less than half the maximum flange shear stress) and for sub-panel buckling (based on the average shear stress in the most critical sub-panel, determined from the elastic shear flow distribution).

For a single-cell box girder with vertical shear only, the flange shear stress varies linearly from a maximum positive value τ_{shear} at one web to a negative value $-\tau_{shear}$ at the other web. The average shear stress is therefore zero. The relevant shear stress to use for overall flange buckling is then governed by the requirement to be not less than half the maximum value, which occurs at a web junction, i.e. $0.5\tau_{shear}$. It is not entirely clear if this sign change is to be considered. If the sign is not considered, only the magnitude, the average shear stress is equal to half the maximum value (i.e. $0.5\tau_{shear}$) and the two requirements are the same.

When torsional shear stress τ_{tor}, which is uniform throughout the flange, is included, consideration of sign of the shear stress does make a difference. If it is considered, the average stress is τ_{tor} and half the maximum is $0.5\tau_{shear} + 0.5\tau_{tor}$. This is probably the intended interpretation. If it is not considered, the average stress is $0.5\tau_{shear} + \tau_{tor}$ and half the maximum is $0.5\tau_{shear} + 0.5\tau_{tor}$. This is more conservative, whereupon the shear stress is 50% of the shear stress at the web–flange junction due to the beam vertical shear force plus 100% of the torsional shear stress, which was the requirement in BS 5400: Part 3.[4] This latter interpretation has been conservatively used in Worked Example 6.2-15 but it was probably *not* the drafters' intended interpretation. If shear stress from distortional warping or transverse loading on the box is present, this must also be included.

The interaction for a box girder flange becomes:

$$\eta_1 + (2\overline{\eta}_3 - 1)^2 \leq 1.0 \qquad \text{(D6.2-52)}$$

This means that there is no interaction between direct stress and shear in the flange when $\overline{\eta}_3 \leq 0.5$ but that no direct stress can be carried when $\overline{\eta}_3 = 1.0$ as shown in Fig. 6.2-62.

Worked Example 6.2-15 illustrates the check of a box flange. It is noted that closed stiffeners are not explicitly covered in 3-1-5/Annex A.3 when determining shear buckling resistance. If closed stiffeners are provided on the flange, it is suggested here that the effective stiffener second moment of area is derived for a section which comprises:

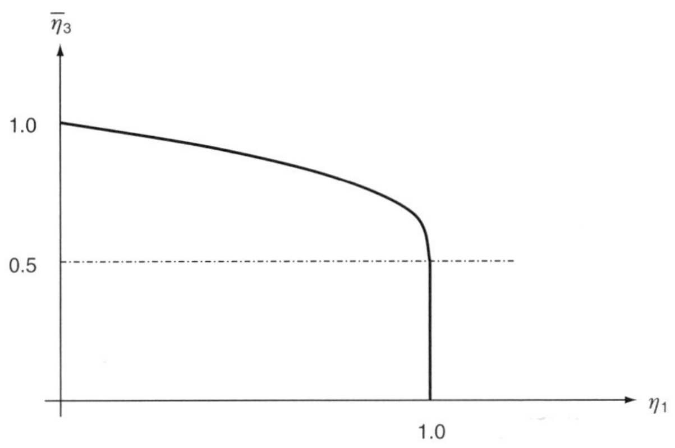

Fig. 6.2-62. Interaction to 3-1-5/clause 7.1 for flanges

- the stiffener itself, with reduced area derived in accordance with 3-1-5/clause 4.4 if necessary;
- an attached width of flange plate at each connection to the stiffener of $15\varepsilon t$ each side of the connecting stiffener leg (or half the distance to an adjacent stiffener leg if smaller) plus the thickness of the stiffener leg as provided in 3-1-5/Fig. 5.3.

The formulae in 3-1-5/Annex A.3 are very conservative for closed stiffeners as they do not allow for their significant torsional stiffness.

Where the geometric constraints discussed in section 6.2.2.5.1 are not met, the method of 3-1-5/clause 10 as discussed in section 6.2.2.6 of this guide may be used. This will, however, be much more conservative as there is no allowance made for plastic redistribution, and shear stresses reduce the allowable resistance moment, whatever their magnitude.

Worked Example 6.2-14: Shear–moment interaction for Class 3 plate girder with shear buckling

The steel plate girder in Worked Example 6.2-13 is modified to have no web transverse stiffeners except at supports. The girder is checked for a moment of 10 000 kNm and a coexisting shear force of 4000 kN.

The shear buckling resistance is first determined.
For no intermediate stiffeners, the slenderness is obtained from expression 3-1-5/(5.5):

$$\bar{\lambda}_w = \frac{h_w}{86.4t\varepsilon} = \frac{2000}{86.4 \times 20 \times 0.81} = 1.429$$

At an internal support, the rigid end-post case applies, so from 3-1-5/Table 5.1:

$$\chi_w = \frac{1.37}{0.7 + \bar{\lambda}_w} = \frac{1.37}{0.7 + 1.429} = 0.64$$

Any contribution from the flanges will be negligible as the transverse stiffeners are far apart, so the resistance is therefore:

$$V_{bw,Rd} = \frac{\chi_w f_{yw} h_w t}{\sqrt{3}\gamma_{M1}} = \frac{0.64 \times 355 \times 2000 \times 20}{\sqrt{3} \times 1.1} = 4770\,\text{kN}$$

The shear ratio $\bar{\eta}_3 = 4000/4770 = 0.839$, which exceeds 0.5 so the interaction with bending moment must be performed using expression 3-1-5/(7.1). Section moduli are taken from Worked Example 6.2-13 as follows.

The elastic bending resistance equals:

$$M_{c,Rd} = \frac{W_{el,min} f_y}{\gamma_{M0}} = \frac{3.750 \times 10^7 \times 355}{1.0} = 13\,317\,\text{kNm} > 10\,000\,\text{kNm applied}$$

The plastic bending resistance:

$$M_{\text{pl,Rd}} = \frac{W_{\text{pl}} f_{\text{y}}}{\gamma_{\text{M0}}} = \frac{4.436 \times 10^7 \times 355}{1.0} = 15\,748\,\text{kNm}$$

The bending ratio:

$$\bar{\eta}_1 = \frac{10\,000}{15\,748} = 0.635$$

The plastic bending resistance ignoring the web is:

$$M_{\text{f,Rd}} = \frac{\left[W_{\text{pl,y}} - \dfrac{A_{\text{w}}^2}{4t_{\text{w}}}\right] f_{\text{y}}}{\gamma_{\text{M0}}} = \frac{\left[4.436 \times 10^7 - \dfrac{(2000 \times 20)^2}{4 \times 20}\right] \times 355}{1.00} = 8648\,\text{kNm}$$

From expression 3-1-5/(7.1):

$$\bar{\eta}_1 + \left[1 - \frac{M_{\text{f,Rd}}}{M_{\text{pl,Rd}}}\right](2\bar{\eta}_3 - 1)^2 = 0.635 + \left[1 - \frac{8648}{15\,748}\right](2 \times 0.839 - 1)^2 = 0.842 < 1.0$$

The plate girder is therefore adequate.

Worked Example 6.2-15: Box girder flange with longitudinal stiffeners

A continuous girder in S355 steel has a 10 000 mm wide and 10 mm thick bottom flange with 24 No. angle stiffeners at 400 mm centres (as illustrated in Fig. 6.2-63). Diaphragms are provided at 4000 mm centres along the bridge. Each angle stiffener together with attached parent plate in accordance with 3-1-5/Fig. 5.3 has a second moment of area $= 3.621 \times 10^7\,\text{mm}^4$. The shear stress due to vertical shear at the junction between web and bottom flange is 130 MPa and the torsional shear stress in the bottom flange is 10 MPa. The direct stress in the bottom flange, calculated from an effective section determined in accordance with 3-1-5/clause 4.5, is 250 MPa. Check the bottom flange for combined bending and shear.

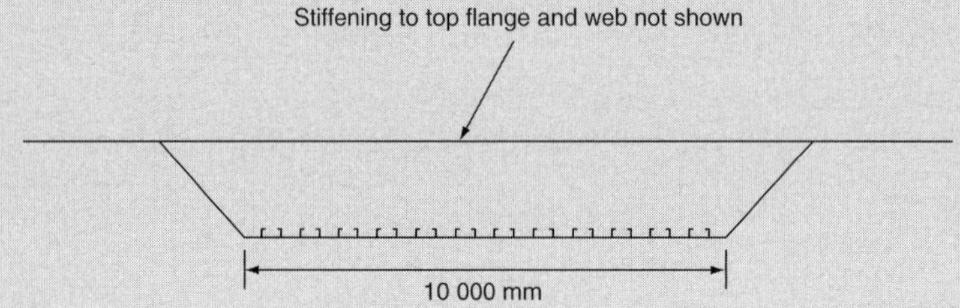

Fig. 6.2-63. Box girder for Worked Example 6.2-15

The slenderness for overall shear buckling of the stiffened panel is calculated first using 3-1-5/Annex A.3:

$$k_{\tau\text{sl}} = 9\left(\frac{b}{a}\right)^2 \sqrt[4]{\left(\frac{I_{\text{sl}}}{t^3 b}\right)^3} = 9\left(\frac{10\,000}{4000}\right)^2 \sqrt[4]{\left(\frac{8.690 \times 10^8}{10^3 \times 10\,000}\right)^3} = 1601$$

with $I_{\text{sl}} = 24 \times 3.621 \times 10^7 = 8.690 \times 10^8\,\text{mm}^4$ but not less than:

$$k_{\tau\text{sl}} = \frac{2.1}{t}\sqrt[3]{\frac{I_{\text{sl}}}{b}} = \frac{2.1}{10}\sqrt[3]{\frac{8.690 \times 10^8}{10\,000}} = 9.3. \text{ Since } a/b = 4000/10\,000 = 0.4 < 1.0:$$

$$k_\tau = 4.00 + 5.34\left(\frac{b}{a}\right)^2 + k_{\tau\text{sl}} = 4.00 + 5.34\left(\frac{10\,000}{4000}\right)^2 + 1601 = 1638.4$$

The slenderness for overall buckling is obtained from expression 3-1-5/(5.6):

$$\bar{\lambda}_w = \frac{b}{37.4t\varepsilon\sqrt{k_\tau}} = \frac{10\,000}{37.4 \times 10 \times 0.81 \times \sqrt{1638.4}} = 0.816$$

At an internal support, the rigid end-post case applies and from 3-1-5/Table 5.1:

$$\chi_w = \frac{0.83}{\bar{\lambda}_w} = \frac{0.83}{0.816} = 1.02$$

$\bar{\eta}_3$ is calculated using the conservative interpretation of flange shear stress calculation discussed in the main text above, such that the total shear stress used is $0.5\tau_{shear} + \tau_{tor}$:

$$\bar{\eta}_3 = \frac{130/2 + 10}{1.02 \times \dfrac{355}{\sqrt{3} \times 1.1}} = 0.39 < 0.5$$

so there is no interaction with direct stress in an overall buckling mode.

Next, the slenderness for sub-panel buckling is calculated. For sub-panel buckling, $a = 4000$ mm and $\bar{b} = 400$ mm so $a/\bar{b} = 10 > 1.0$ and

$$k_{\tau i} = 5.34 + 4.00\left(\frac{\bar{b}}{a}\right)^2 = 5.34 + 4.00\left(\frac{400}{4000}\right)^2 = 5.38$$

The slenderness for sub-panel buckling is obtained from expression 3-1-5/(5.7):

$$\bar{\lambda}_w = \frac{\bar{b}}{37.4t\varepsilon\sqrt{k_{\tau i}}} = \frac{400}{37.4 \times 10 \times 0.81 \times \sqrt{5.38}} = 0.569$$

From 3-1-5/Table 5.1, $\chi_w = 1.2$. $\bar{\eta}_3$ is calculated using the average shear stress in the flange sub-panel as discussed in the main text:

$$\bar{\eta}_3 = \frac{130 + 10}{1.20 \times \dfrac{355}{\sqrt{3} \times 1.1}} = 0.63 > 0.5$$

so there is interaction with direct stress in a local buckling mode. The interaction is checked using equation (D6.2-52):

$$\eta_1 + (2\bar{\eta}_3 - 1)^2 = \frac{250}{355/1.0} + (2 \times 0.63 - 1)^3 = 0.77 \le 1.0$$

so the flange is adequate in combined bending and shear.

6.2.10. Bending and axial force

Axial force will reduce the ultimate bending resistance of a cross-section where parts that contribute to the bending resistance are also required to resist the axial force. When axial force is present, it is vitally important to be consistent between global analysis and cross-section design with respect to the height at which the axial force is assumed to act. For elastic analysis, if an axial force is applied anywhere other than at the elastic centroid of the cross-section it will generate bending stresses. Consequently, if the axial force does not act at the section centroid, it is usual to refer the force to the centroid and add the resulting moment produced by the eccentricity of the force from the centroid to any other moment that is applied. Where plastic design is used to derive an overall stress block under axial force and bending moment, it may sometimes be preferable to refer the axial force to the plastic neutral axis (for bending alone) as discussed below.

6.2.10.1. Class 1 and 2 cross-sections

Class 1 and 2 cross-sections will not be very common in bridge design for members acting in combined bending and axial force. This is because the web in typical deep bridge members is

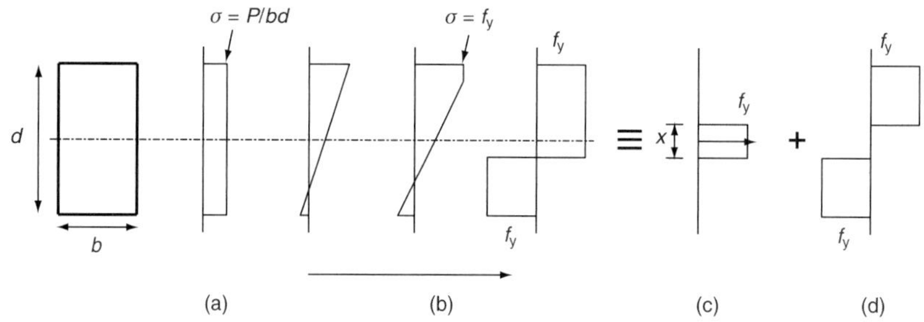

Fig. 6.2-64. Effect of axial force on plastic moment resistance: for a rectangular section; (a) stress due to axial force P; (b) stresses due to increasing moment acting with P; (c) component resisting axial force; (d) component resisting bending moment

often Class 3 even under bending alone, due to the large depth of web in compression; the axial force typically increases this depth.

Class 1 and 2 cross-sections can develop full plasticity throughout the entire depth of the section. This complicates the check of the cross-section as the stresses from bending and axial force cannot simply be superposed if advantage is to be taken of this plasticity. ***3-1-1/clause 6.2.9.1(1)*** gives recommendations for assessing the reductions to bending resistance due to axial force. The general requirement is as follows:

3-1-1/clause 6.2.9.1(1)

$$M_{Ed} \leq M_{N,Rd} \quad \text{3-1-1/(6.31)}$$

where M_{Ed} is the applied moment and $M_{N,Rd}$ is the reduced plastic bending resistance of the section in the presence of an axial force N_{Ed}.

Plastic stress blocks are illustrated most simply (and with least direct practicality) by reference to a rectangular solid section. The procedure for calculating $M_{N,Rd}$ in this simple case is as illustrated in Fig. 6.2-64.

As illustrated in Fig. 6.2-64, the bending resistance of the beam, $M_{N,Rd}$, will be reduced because the depth x of the section is required to resist the axial force as follows:

$$M_{N,Rd} = M_{pl,Rd} - M_{pl,x} \quad \text{(D6.2-53)}$$

where $M_{pl,Rd}$ is the plastic moment resistance of the full section and $M_{pl,x}$ is the plastic moment resistance of the section component resisting axial force.

If the section is subjected to an axial force, N_{Ed}, equal to its design plastic resistance, $N_{pl,Rd}$, then the dimension x defined in Fig. 6.2-64 will be equal to the depth d of the gross-section. For lesser values of N_{Ed}, $x = (N_{Ed}/N_{pl,Rd})d$ and therefore:

$$M_{pl,x} = W_{plx}f_y = \left(\frac{bx^2}{4}f_y\right) = \left[\frac{bd^2}{4}\left(\frac{N_{Ed}}{N_{pl,Rd}}\right)^2 f_y\right] \quad \text{(D6.2-54)}$$

Substituting equation (D6.2-54) into equation (D6.2-53):

$$M_{N,Rd} = M_{pl,Rd} - M_{pl,Rd}\left(\frac{N_{Ed}}{N_{pl,Rd}}\right)^2 = M_{pl,Rd}\left[1 - \left(\frac{N_{Ed}}{N_{pl,Rd}}\right)^2\right]$$

3-1-1/clause 6.2.9.1(3)

as given in ***3-1-1/clause 6.2.9.1(3)*** for rectangular sections.

This same basic procedure can be used for general symmetrical flanged beams but account has to be taken in deriving formulae of whether the zone required to resist axial force extends into the flanges or not. The procedure is more complicated for asymmetric sections. Even though the above analysis indicates that any magnitude of axial force will have a detrimental effect on bending resistance, ***3-1-1/clause 6.2.9.1(4)*** allows the designer to neglect the effect when the following criteria are satisfied.

3-1-1/clause 6.2.9.1(4)

(a) For doubly symmetric flanged sections resisting moment about the y–y axis:

$$N_{Ed} \leq 0.25N_{pl,Rd} \text{ and } N_{Ed} \leq \frac{0.5h_w t_w f_y}{\gamma_{M0}}$$

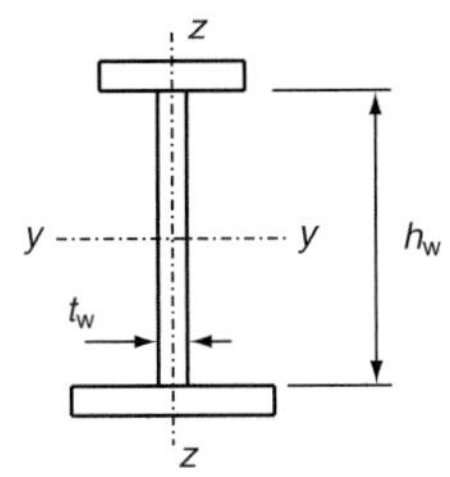

Fig. 6.2-65. Sign convention for axes

The calculated reduction under these assumptions would in any case be very small, which is the justification for the simplification.

(b) For I and H sections symmetrical about the z–z axis and resisting moment about the z–z axis:

$$N_{Ed} \leq \frac{h_w t_w f_y}{\gamma_{M0}}$$

This is a fairly obvious simplification as the web provides virtually no contribution to the bending resistance for z–z bending.

The sign convention for axes is the same as in 3-1-1/Table 6.2, reproduced here in Fig. 6.2-65 for convenience.

In order to facilitate the calculation process, ***3-1-1/clause 6.2.9.1(5)*** provides various approximations for estimating $M_{N,Rd}$ for symmetrical sections with equal flange widths. However, as most steel components used in bridge engineering do not have symmetrical flanges, these formulae are of limited applicability. A general method is therefore given here in Fig. 6.2-66 for calculating $M_{N,Rd}$ for non-symmetrical Class 1 and 2 cross-sections. Worked Example 6.2-16 illustrates the method. *3-1-1/clause 6.2.9.1(5)*

Where sections are symmetrical, the axial force derived from analysis is usually acting at the middle of the section, where both the elastic and plastic bending neutral axes coincide. Where cross-sections are not symmetrical, it is vital to carefully consider where the axial force determined from global analysis is assumed to act. This is particularly important as the elastic and plastic bending neutral axes will no longer be at the same location. In the method in Fig. 6.2-66 it is assumed that the axial force acts at the **plastic** neutral axis for bending, so if the axial force from global analysis is assumed to act at the elastic neutral axis, the axial force needs to be referred to the plastic neutral axis and an additional moment added to the section to account for this shift.

From Fig. 6.2-66:

$$M_{N,Rd} = M_{pl,Rd} - M_{2fyd} \quad \text{(D6.2-55)}$$

Depth a is determined such that N_{Ed} = area in height $a \times 2f_{yd}$ and $f_{yd} = f_y/\gamma_{M0}$ with f_y appropriate to the thickness of the parts within the depth a.

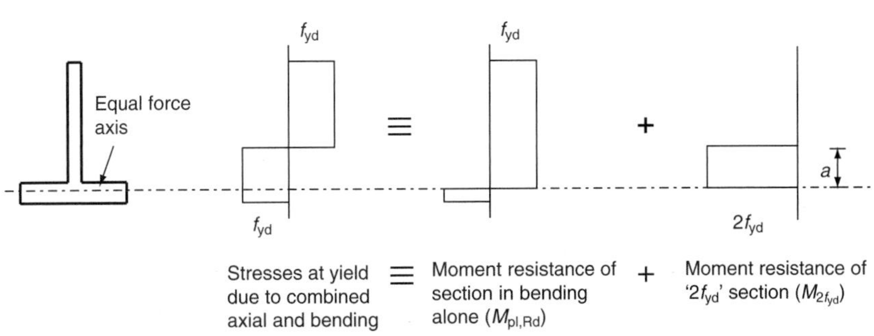

Fig. 6.2-66. Procedure for calculating $M_{N,Rd}$ for non-symmetrical sections

It is important to check that the web is indeed Class 1 or 2 when the above plastic stress block has been determined as the axial force may increase the section class from that obtained for bending alone.

Biaxial bending

In many practical situations, steel sections will be subjected to axial force as well as bending about both axes of the cross-section. Calculation of the collapse load is further complicated by the addition of moments about both axes of the section. Using the same principles as above of reducing the moment resistance by removing components to resist both

3-1-1/clause 6.2.9.1(6)

axial force and biaxial moment, a solution can be found. ***3-1-1/clause 6.2.9.1(6)*** provides an approximate failure criterion for biaxial bending of Class 1 and 2 cross-sections as follows:

$$\left[\frac{M_{y,Ed}}{M_{N,y,Rd}}\right]^{\alpha} + \left[\frac{M_{z,Ed}}{M_{N,z,Rd}}\right]^{\beta} \leq 1.0 \qquad \text{3-1-1/(6.41)}$$

where α and β are constants which may be conservatively taken as 1.0 or as follows.

I and H sections:

$$\alpha = 2;\ \beta = 5n \quad \text{but} \geq 1.0 \qquad \text{where } n = \frac{N_{Ed}}{N_{pl,Rd}}$$

Circular hollow sections:

$$\alpha = 2;\ \beta = 2$$

Rectangular hollow sections:

$$\alpha = \beta = \frac{1.66}{1 - 1.13n^2} \quad \text{but } \alpha = \beta \leq 6 \qquad \text{where } n = \frac{N_{Ed}}{N_{pl,Rd}}$$

Simple linear interaction

To avoid the complexity of calculating plastic bending resistances with coexisting axial force, it is possible to use a simplified conservative linear interaction according to 3-1-1/clause 6.2.1 as follows:

$$\frac{N_{Ed}}{N_{Rd}} + \frac{M_{y,Ed}}{M_{pl,y,Rd}} + \frac{M_{z,Ed}}{M_{pl,z,Rd}} \leq 1.0 \qquad \text{(D6.2-56)}$$

This simplified form of interaction will be familiar to UK bridge designers. It also has the advantage of being simple to use to cater for high shear also, as the resistance moments in the interaction can simply be reduced for the presence of shear as discussed in section 6.2.9 of this guide. For uniaxial bending, the above relationship is compared qualitatively in Fig. 6.2-67 with the more exact interaction obtained from the use of the resultant

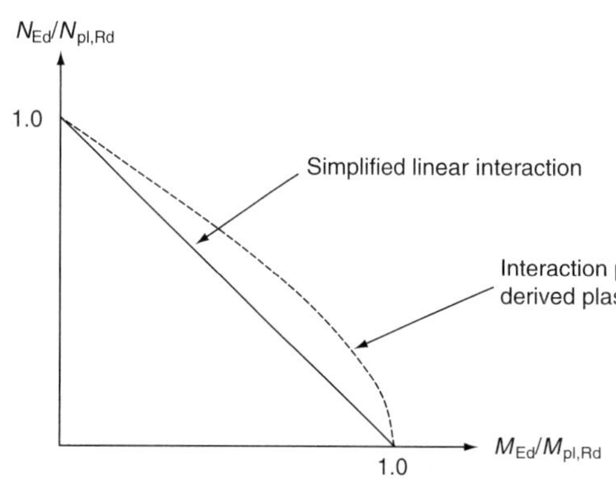

Fig. 6.2-67. Uniaxial bending and compression

plastic stress block. One problem with the use of equation (D6.2-56) is that it is still necessary to classify the cross-section to decide whether or not the use of the plastic bending resistance is appropriate. If separate section classification is performed to avoid the need to determine the plastic stress block, it is possible (indeed likely) that the beam will be Class 1 or 2 for bending but Class 3 or even 4 for axial force. In this case, the safe approach is to calculate the moment resistance based on the class obtained for axial force alone. Unfortunately, for typical bridge beams, this is likely to lead to a classification other than Class 1 or 2 and the combined stress block might then need to be investigated.

6.2.10.2. Class 3 cross-sections

As Class 3 cross-sections become susceptible to local buckling when the yield point is reached in compression, the plastic interactions of axial force and bending moment discussed above cannot be applied. Instead, a simple superposition of elastic stresses is performed and the resulting stress limited to the design yield strength at all locations within the beam according to ***3-1-1/clause 6.2.9.2(1)***: *3-1-1/clause 6.2.9.2(1)*

$$\sigma_{x,Ed} \leq \frac{f_y}{\gamma_{M0}} \qquad \text{3-1-1/(6.41)}$$

For components subjected to both axial force and biaxial bending, the value of $\sigma_{x,Ed}$ at an extreme fibre is therefore:

$$\sigma_{x,Ed} = \frac{N_{Ed}}{A} + \frac{M_{y,Ed}}{W_{el,y}} + \frac{M_{z,Rd}}{W_{el,z}} \qquad \text{(D6.2-57)}$$

where:

N_{Ed} is the applied axial force;
A is the gross area in accordance with 3-1-1/clause 6.2.3 or 6.2.4. This may need to be modified to the net area for fastener holes as appropriate;
$W_{el,y}$ is the elastic section modulus about the Y–Y axis;
$W_{el,z}$ is the elastic section modulus about the Z–Z axis.

If this approach is used for biaxial bending, each check should correspond to a unique point on the cross-section to avoid excessive conservatism. Cruciform sections would, for example, be checked too conservatively by equation (D6.2-57) if the Y and Z section moduli used in a single check referred to the extreme Y and Z fibres, as the stresses at these points do not coexist. For an I-beam however, these stresses usually do coexist.

Class 4 sections may also be treated as Class 3 sections, in accordance with ***3-2/clause 6.2.10.2(2)***, by using cross-section properties and limiting the direct stress to σ_{limit} thus: *3-2/clause 6.2.10.2(2)*

$$\sigma_{x,Ed} \leq \frac{\sigma_{limit}}{\gamma_{M0}} \qquad \text{3-2/(6.8)}$$

where σ_{limit} is defined in 3-2/clause 6.2.4(2) as 'the limiting stress of the weakest part of the cross-section in compression'. The concept of σ_{limit} is a slightly strange one for cross-section checks as, in order to determine σ_{limit}, the section must first be checked under the combined stress field according to the method of 3-1-5/clause 10, which is discussed in detail in section 6.2.2.6 of this guide. This involves checking all the constituent parts of the cross-section, which may have different allowable stresses, and verifying that they are all satisfactory. The verification of 3-1-5/clause 10 is thus itself a check of the cross-section and there is no real need to determine σ_{limit} itself for cross-section checks. Additionally, while a unique value of σ_{limit} can be determined under axial force alone such that failure was determined by the weakest part of the cross-section, the same is not true for bending and compression. In this latter case, the weakest and governing part of the cross-section may not experience the greatest applied stress. σ_{limit} is therefore more appropriately defined as the peak compressive stress, under bending and axial force, at an extreme fibre such that failure occurs by local buckling somewhere within the cross-section. This location need not necessarily be at the most stressed fibre where σ_{limit} is attained.

A full check to 3-1-5/clause 10 requires shear, axial force, bending moment and transverse force to be considered at the same time. When this full check is carried out, a check under bending moment and axial force on its own becomes redundant (unless the other effects are zero); the full check will be more critical. Consequently, it is recommended here that if Class 4 cross-sections are to be treated as Class 3, the entire check should be performed using 3-1-5/clause 10, as discussed in section 6.2.2.6 of this guide, without reference to σ_{limit}. Additional comments on the σ_{limit} method are made under the heading Class 4 cross-sections in section 6.2.5 of this guide. In general, the method is more conservative than the use of effective sections for Class 4 members as discussed below.

6.2.10.3. Class 4 cross-sections

3-1-1/clause 6.2.9.3(1)

Class 4 cross-sections are treated in a similar way as Class 3 cross-sections, but an effective section to allow for plate buckling is used in calculating the maximum stress $\sigma_{x,Ed}$. ***3-1-1/clause 6.2.9.3(1)*** gives the same criterion as for Class 3 cross-sections:

$$\sigma_{x,Ed} \leq \frac{f_y}{\gamma_{M0}} \qquad \text{3-1-1/(6.43)}$$

where $\sigma_{x,Ed}$ is the maximum longitudinal stress in the section, taking account of bolt holes where necessary and the effects of local buckling. This requirement is stated in a more useful way in expression 3-1-5/(4.15), reproduced below.

The derivation of effective section properties is discussed at length in section 6.2.2.5 of this guide. The section properties A_{eff} and W_{eff} may be derived either separately for axial force and bending moment or may be derived from the combined stress distribution from axial force and moment acting together. The latter will typically not be very practical as it will require recalculation of the section properties for every load case and requires an iterative approach as discussed below. It may however give some increase in economy, so could be tried if a section is just failing its stress check.

Usually, the axial force will have been referred to the centroid of the gross cross-section and any moments arising from eccentricity of its actual position added to other imposed moments. When the effective section for axial force alone is calculated for an asymmetric section, the neutral axis will shift an amount e_N as illustrated in Fig. 6.2-68. This shift produces an additional moment of the axial force from its old assumed neutral axis position on the gross cross-section to the new neutral axis position on the effective cross-section. This must be included when calculating bending stresses. If the axial force was originally derived from a global analysis already using effective cross-section properties, and was assumed to act at the centroid of this effective section, no further shift would be necessary. These additional moments are accounted for in the following interaction given in both 3-1-5/clause 4.6 and ***3-1-1/clause 6.2.9.3(2)***:

3-1-1/clause 6.2.9.3(2)

$$\eta_1 = \frac{N_{Ed}}{A_{eff} f_y/\gamma_{M0}} + \frac{M_{y,Ed} + N_{Ed} e_{Ny}}{W_{y,eff} f_y/\gamma_{M0}} + \frac{M_{z,Ed} + N_{Ed} e_{Nz}}{W_{z,eff} f_y/\gamma_{M0}} \leq 1.0 \qquad \text{3-1-5/(4.15)}$$

where:

A_{eff} is the effective area of the cross-section when subjected to uniform compression;
W_{eff} is the effective elastic section modulus of the cross-section when subjected to moment about the relevant axis;
e_N is the shift of the relevant centroidal axis when the cross-section is subjected to compression only.

3-1-5/clause 4.6(3)

If the stress varies along the length of a panel, ***3-1-5/clause 4.6(3)*** permits the verification according to expression 3-1-5/(4.15) to be performed at a distance of 0.4a or 0.5b, whichever is smaller, from the most highly stressed end of the panel. This is because failure is most influenced by stresses within the middle portion of the buckling waveform, rather than at its boundaries; for individual plate panels, although the longitudinal membrane stresses are greatest at longitudinal supported edges of the plate, the longitudinal and transverse bending stresses are greatest at the centre of the buckle which leads to a greater effective

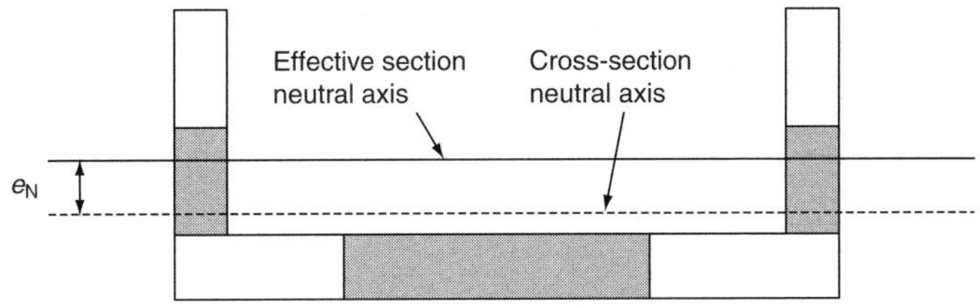

Fig. 6.2-68. Shift in neutral axis for Class 4 section under axial force

stress. If advantage is taken of 3-1-5/clause 4.6(3), the stress check needs to be repeated at the end of the panel using gross section properties. For longitudinally stiffened beams, some care is necessary with the definition of '*b*'. For example, if the reduction in effective section is dominated by sub-panel buckling under near uniform compressive stress, '*b*' should relate to the sub-panel dimension, rather than the overall width of the stiffened plate.

If a unique effective section is derived for bending and compression together, the shift in neutral axis will lead to a change in the applied moment which will in turn lead to a change in stress distribution and hence effective section again. The procedure therefore becomes iterative. The final stress check, when convergence has been obtained, can then be performed using expression 3-1-5/(4.15) and e_N will be the final shift from the gross section to the final unique effective section. This adds to the impracticality of this approach.

As an alternative to using the effective section approach, gross section properties may be used and the check based on the method of 3-1-5/clause 10 as discussed above in the section on Class 3 gross sections. This can be considerably more conservative as discussed in section 6.2.2.6 of this guide.

Worked Example 6.2-16: Calculation of the reduced resistance moment of a steel plate girder with Class 2 cross-section under combined moment and axial force

The steel plate girder shown in Fig. 6.2-69 is restrained against lateral torsional buckling and is initially assumed to be a Class 2 cross-section under bending and axial force. The girder is part of a single-span integral bridge and receives a compressive thrust from the abutments of 10 600 kN *applied at the level of the plastic neutral axis for bending moment alone*. The maximum sagging bending moment that the section can withstand in conjunction with the axial force is calculated and a check is made to ensure that the cross-section remains Class 2. All plates are grade S355 to EN 10025 and the yield strengths for different plate thicknesses are to be taken from 3-1-1/Table 3.1. (Note that the UK National Annex requires the values from EN 10025 to be used.)

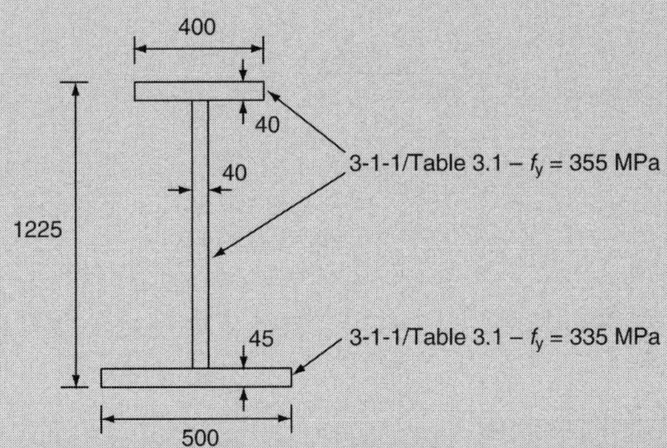

Fig. 6.2-69. Plate girder for Worked Example 6.2-16

The compression flange is first classified using 3-1-1/Table 5.2. Conservatively ignoring the web-to-flange welds, the flange outstand $c = (400 - 40)/2 = 180$ mm. $c/t = 180/40 = 4.5$. $9\varepsilon = 7.3 \geq 4.5$, so the flange is Class 1.

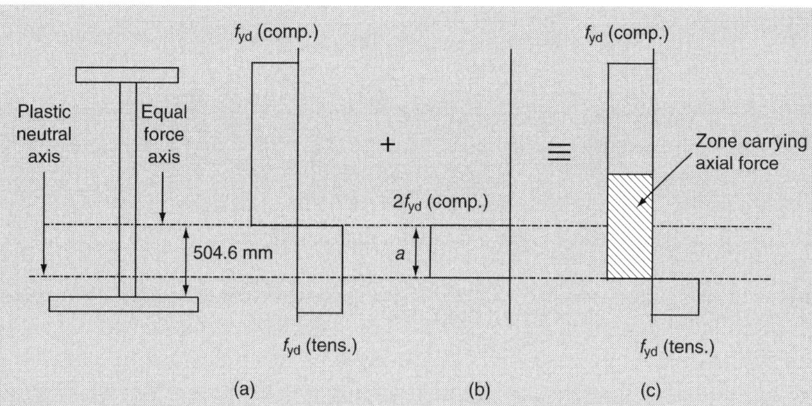

Fig. 6.2-70. Stress block for Worked Example 6.2-16: (a) stress block for bending; (b) stress due to axial force; (c) final stress block

The plastic section properties of the girder are found to be as follows.

Equal force axis = 504.6 mm from bottom of web. This is the location of the plastic neutral axis for bending alone.

Plastic moment of resistance ($M_{pl,Rd}$) = 12 370 kNm.

Figure 6.2-70 shows the stress distribution under combined bending and axial force.

The depth 'a' is first calculated.

Assuming the plastic neutral axis occurs in the web, force balance gives:

$10\,600 \times 10^3 = a \times 40 \times 2 \times 355/1.0$ so $a = \mathbf{373.2\,mm} < 504.6\,\text{mm}$

The assumption is therefore correct – the plastic neutral axis occurs in the web.

Therefore, the plastic moment of resistance about the equal force axis of the section resisting axial force (M_{2fyd}) = $373.2 \times 40 \times 2 \times 355 \times 373.2/2 \times 1 \times 10^{-6} = 1977$ kNm.

Therefore the resulting plastic moment of resistance in the presence of axial force $M_{N,Rd} = M_{pl,Rd} - M_{2fyd} = 12\,370 - 1977 = 10\,393$ kNm.

The section can withstand a maximum sagging moment of **10 393 kNm** in the presence of a 10 600 kN axial force (applied at the level of the plastic neutral axis for bending moment alone). This method would need modification if the yield stress was different in web and flange and the neutral axis was located in the flange. It would be simplest to use the smallest value of yield stress throughout.

It is now checked that the cross-section is still Class 2 in the presence of the axial force. 3-1-1/Table 5.2 – Web is 'Part subject to bending and compression.'

$\alpha > 0.5$ (by inspection)

c = depth of web = 1140 mm

αc = depth of web in compression = $1140 - 504.6 + 373.2 = 1008.6$ mm

Therefore, $\alpha = 1008.6/1140 = 0.885$

For the web to be classified as Class 2, $c/t \le 456\varepsilon/(13\alpha - 1)$ where:

t = thickness of web = 40 mm

$\varepsilon = 0.81$ (3-1-1/Table 5.2)

$$\frac{c}{t} = \frac{1140}{40} = 28.5 \text{ and } \frac{456\varepsilon}{13\alpha - 1} = \frac{456 \times 0.81}{13 \times 0.885 - 1} = 35.2 > 28.5$$

Therefore the web is still Class 2 despite the compression forces. It will be further noted that this section would still be compact if the whole web depth was in compression.

6.2.11. Bending, shear and axial force

The bending resistance of cross-sections resisting combined bending, shear and axial force may be reduced by both the axial force and shear components of the loading. ***3-1-1/clause 6.2.10(1)*** requires this effect to be considered. The method of interaction again depends on the class of the cross-section and whether or not the shear resistance is limited by shear buckling. *3-1-1/clause 6.2.10(1)*

This section of the guide is split into two sub-sections as follows:

- Sections not susceptible to shear buckling *Section 6.2.11.1*
- Sections susceptible to shear buckling *Section 6.2.11.2*

6.2.11.1. Sections not susceptible to shear buckling

6.2.11.1.1. Class 1 and 2 cross-sections

Where there is no shear buckling, the effect of shear on cross-section resistance need only be considered if the shear force exceeds 50% of the design plastic resistance – ***3-1-1/clause 6.2.10(2)*** refers. The first step in checking the cross-section is to establish the reduced web yield strength, or thickness, caused by shear as discussed in section 6.2.9.1.1 of this guide – ***3-1-1/clause 6.2.10(3)*** refers. The resulting reduced section is then checked for combined bending and axial force using plastic section design in accordance with section 6.2.10.1 of this guide. If the section is not symmetric, the plastic neutral axis will shift when the reduction in web strength is made. The comments made in section 6.2.9.1.1 regarding not reclassifying the web to 3-1-1/Table 5.2 after modifying the cross-section for shear apply here also; the section classification is checked first under the bending moment and axial force before any reduction is made to the web strength for shear. *3-1-1/clause 6.2.10(2)* *3-1-1/clause 6.2.10(3)*

An alternative simpler and more conservative approach is to use a linear interaction as permitted by 3-1-1/clause 6.2.1(7):

$$\frac{N_{\mathrm{Ed}}}{N_{\mathrm{v,Rd}}} + \frac{M_{\mathrm{y,Ed}}}{M_{\mathrm{v,y,Rd}}} + \frac{M_{\mathrm{z,Ed}}}{M_{\mathrm{v,z,Rd}}} \leq 1.0 \qquad \text{(D6.2-58)}$$

where $M_{\mathrm{v,y,Rd}}$ and $M_{\mathrm{v,z,Rd}}$ are the reduced resistance moments allowing for shear, but not axial force, about the y–y and z–z axes respectively. $N_{\mathrm{v,Rd}}$ is the axial resistance based on a cross-section with reductions for shear. The comments on section classification under axial force and bending made in section 6.2.10.1 of this guide apply when using this linear interaction.

6.2.11.1.2. Class 3 cross-sections

The procedure for treating Class 3 cross-sections is slightly different to that for Class 1 and 2 cross-sections since elastic section design has to be used for combinations of bending and axial force. There are three possibilities for doing the check:

(i) Establish the reduced web yield strength or thickness caused by shear (if the shear force exceeds 50% of the plastic resistance) as for Class 1 and 2 cross-sections. The resulting reduced cross-section is then checked for combined bending and axial force using elastic section design in accordance with section 6.2.10.2 of this guide. It makes most sense to reduce the web thickness rather than its yield strength or the beam resistance will be governed by yielding of the web at this reduced yield stress.

(ii) Use the interaction given in 3-1-5/clause 7.1 which is intended for cases with shear buckling as discussed in section 6.2.11.2.2 of this guide.

(iii) Use the interaction of equation (D6.2-58). The use of plastic section properties in determining $M_{\mathrm{v,Rd}}$ (as long as the resulting resistance does not exceed the elastic bending resistance) is discussed in section 6.2.9.1.2. It is necessary to use this method to avoid a discontinuity with the moment resistance with shear alone if this has been determined to EC3-1-1 clause 6.2.8 as discussed in section 6.2.9.1.2 of this guide.

The three methods are illustrated in Worked Example 6.2-18.

Once again, if the section is not symmetric, the neutral axis will shift when the reduction in web thickness for shear is made. There is no need for the section classification to be rechecked for this shift. Methods (ii) and (iii) are significantly more economical.

6.2.11.1.3. Class 4 cross-sections
Class 4 cross-sections have to be dealt with using one of two possible methods given in EN 1993-1-5. These are discussed in section 6.2.11.2.3 below as the procedure is the same whether or not there is shear buckling.

Worked Example 6.2-17: Calculation of the moment resistance of a plate girder with Class 2 cross-section subjected to combined moment, shear and axial force
The steel plate girder shown in Fig. 6.2-71 is formed from S355 steel and is initially assumed to be a Class 2 cross-section. The section is located at the central support of a two-span integral bridge. The beam is restrained laterally and the web is stable against shear buckling. As all plate thicknesses are less than 40 mm, the yield stresses of all plates can be taken as 355 MPa according to 3-1-1/Table 3.1. (Note that the UK National Annex to EN 1993-2 requires the yield stress variation with thickness to be taken from EN 10025.) The design plastic bending resistance in the absence of shear and axial force, $M_{pl,Rd}$, is 7874 kNm and the height of the plastic neutral axis for bending alone is 495 mm from the upper surface of the bottom flange.

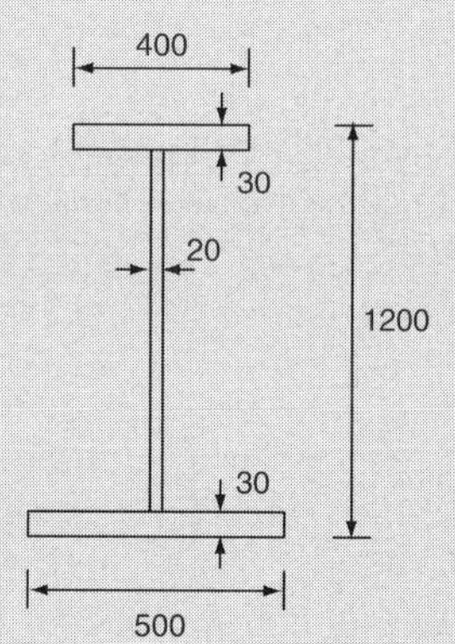

Fig. 6.2-71. Class 2 beam cross-section for Worked Example 6.2-17

The maximum hogging bending moment that the section can withstand in conjunction with a shear force (V_{Ed}) of 4486 kN and a compressive axial force (N_{Ed}) of 2200 kN, *applied at the height of the plastic neutral axis for bending alone*, is calculated. (The height of the axial force is therefore 495 mm from the upper surface of the bottom flange.) It is also verified that the cross-section remains Class 2 under combined bending and axial force.

The compression flange is first classified using 3-1-1/Table 5.2. Conservatively ignoring the web-to-flange welds, the flange outstand $c = (500 - 20)/2 = 240$ mm. $c/t = 240/30 = 8.0$. $10\varepsilon = 8.1 \geq 8.0$, so the flange is just Class 2.

Next it is necessary to check compactness of the web under bending and axial force alone. Following the calculation method in Worked Example 6.2-16 of section 6.2.10 of this guide, the plastic neutral axis for maximum bending resistance with a coexisting axial force of 2200 kN is:

$$495 + \frac{2200 \times 10^3}{2 \times 20 \times 355} = 650\,\text{mm}$$

above the upper surface of the bottom flange.

From 3-1-1/Table 5.2:

$$\alpha = \frac{650}{1140} = 0.570$$

For the cross-section to be Class 2:

$$\frac{c}{t} \leq \frac{456\varepsilon}{13\alpha - 1} = \frac{456 \times 0.81}{13 \times 0.570 - 1} = 57.6$$

The actual $c/t = 1140/20 = 57.0$ so the cross-section is just Class 2 for the combination of bending and axial force.

Next the shear resistance is calculated from 3-1-1/clause 6.2.6:

$$V_{\text{pl,Rd}} = \frac{\eta A_{\text{w}}(f_{\text{y}}/\sqrt{3})}{\gamma_{\text{M0}}} = \frac{1.2 \times 1140 \times 20(355\sqrt{3})}{1.00} = 5608\,\text{kN}$$

$V_{\text{Ed}} = 4486\,\text{kN}$ is greater than $0.5 \times V_{\text{pl,Rd}}$, so shear will reduce the moment resistance to $M_{\text{y,V,Rd}}$.

From 3-1-1/clause 6.2.8(3):

$$\rho = \left(\frac{2V_{\text{Ed}}}{V_{\text{pl,Rd}}} - 1\right)^2 = \left(\frac{2 \times 4486}{5608} - 1\right)^2 = 0.360$$

The allowable stress in the web $= (1 - \rho)f_{\text{y}} = (1 - 0.36) \times 355 = 227.2\,\text{MPa}$

It is simplest to reduce the web thickness rather than its yield stress. Therefore, the reduced thickness of web equals:

$$20 \times \frac{227.2}{355} = 12.8\,\text{mm}$$

The revised cross-section allowing for shear is shown in Fig. 6.2-72.

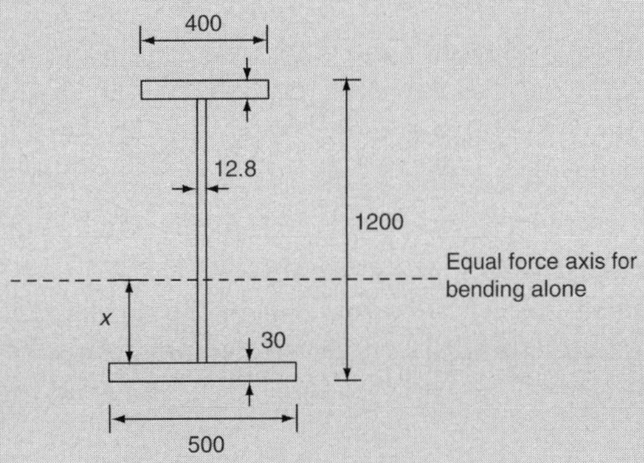

Fig. 6.2-72. Effective section with effective web thickness reduced for shear

The height of the equal force axis x in Fig. 6.2-72 is found from force balance:

$$(500 \times 30) + (12.8 \times x) = (400 \times 30) + [12.8 \times (1140 - x)]$$

for which $x = 452.8\,\text{mm}$.

The bending resistance in the presence of shear but without axial force is therefore:

$$M_{\text{y,V,Rd}} = \frac{(500 \times 30 \times 467.8 + 400 \times 30 \times 702.2) \times 355}{1.00}$$

$$+ \frac{(452.8^2 \times 20 \times 0.5 + (1140 - 452.8)^2 \times 20 \times 0.5) \times 227.2}{1.00} = 7021\,\text{kNm}$$

Now the effect of axial force is added in.

From Fig. 6.2-73, the stress distribution under combined bending and axial force will be as follows.

To calculate the depth 'a' it is initially assumed that the plastic neutral axis occurs in the web, therefore: $2200 \times 10^3 = a \times 12.8 \times 2 \times 355$ and $a = 242.1\,\text{mm}$ so the plastic neutral

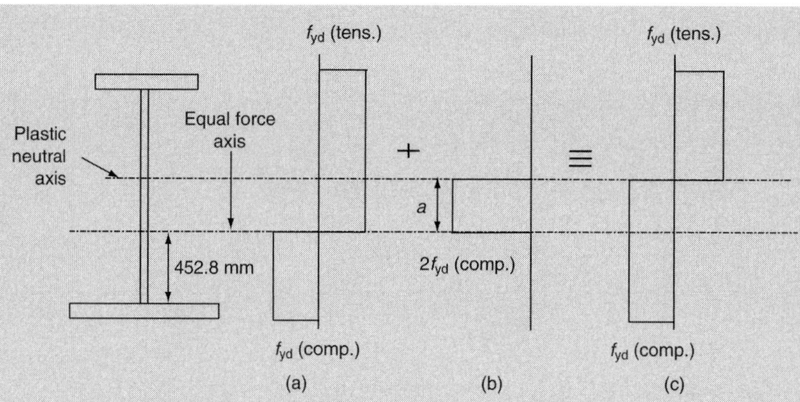

Fig. 6.2-73. Stresses due to combined bending and axial force on the cross-section with web reduced for shear: (a) stress block for bending alone; (b) stress due to axial force; (c) final stress block

axis does occur in the web at (452.8 + 242.1) mm = 694.9 mm from the top surface of the bottom flange.

The bending resistance of the axial force component about the equal force axis is:

$$M_{2\text{fyd}} = \frac{242.1^2 \times 12.8}{2} \times \frac{2 \times 355}{1.0} = 266.3\,\text{kNm}$$

The resulting plastic moment of resistance in the presence of shear and axial force is then $M_{\text{N,v,Rd}} = 7021 - 266.3 = 6754.7$ kNm. However, the axial force was applied at the level of the plastic neutral axis for bending alone which was at a height of 495 mm from the top of the bottom flange. When shear is taken into account, this axis shifts down by $495 - 452.8 = 42.2$ mm. The axial force therefore produces a sagging moment of $2200 \times 0.0422 = 92.8$ kNm about this new axis, so the hogging moment that can be applied together with an axial force of 2200 kN at 495 mm above the top of the bottom flange is $M_{\text{y,Ed}} = 6754.7 + 92.8 =$ **6848 kNm**.

6.2.11.2. Sections susceptible to shear buckling

3-1-1/clause 6.2.10(2)

If the section's shear resistance is limited by shear buckling as discussed in section 6.2.6 of this guide, then ***3-1-1/clause 6.2.10(2)*** effectively requires clause 7 of EN 1993-1-5 to be used to perform the interaction between bending, shear and axial force.

6.2.11.2.1. Class 1 and 2 cross-sections

3-1-5/clause 7.1(1)
3-1-5/clause 7.1(4)

The approach is similar to that above where there is no shear buckling. 3-1-5/clause 7.1 allows the interaction with shear to be neglected when the design shear force is less than 50% of the shear buckling resistance. Where the design shear force exceeds 50% of the shear buckling resistance, the following interaction has to be satisfied, which is the one given in ***3-1-5/clause 7.1(1)*** with the modifications required by ***3-1-5/clause 7.1(4)***:

$$\bar{\eta}_1 + \left[1 - \frac{\alpha_{\text{N}} M_{\text{f,Rd}}}{M_{\text{N,Rd}}}\right](2\bar{\eta}_3 - 1)^2 \leq 1.0 \qquad \text{(D6.2-59)}$$

where:

$\bar{\eta}_3$ is the ratio $V_{\text{Ed}}/V_{\text{bw,Rd}}$;

$\bar{\eta}_1$ is the usage factor for bending and axial force, $M_{\text{Ed}}/M_{\text{N,Rd}}$, determined as discussed in section 6.2.10.1 of this guide. ($M_{\text{N,Rd}}$ is the reduced resistance moment in the presence of axial force);

$M_{\text{f,Rd}}$ is the design plastic bending resistance based on a section comprising the flanges only; its definition is discussed in section 6.2.9.2.1 of this guide. If the whole web is in compression, 3-1-5/clause 7.1(4) requires $M_{\text{f,Rd}}$ to be taken as zero, which can lead to a discontinuity in resistance;

α_N is a factor from ***3-1-5/5.4(2)*** to allow for the effect of axial force on the effectiveness of the flanges: *3-1-5/clause 5.4(2)*

$$\alpha_N = \left[1 - \frac{N_{Ed}}{(A_{f1} + A_{f2})f_{yf}/\gamma_{M0}}\right]$$

where A_{f1} and A_{f2} are the areas of top and bottom flanges. This factor has been added into equation (D6.2-59) for clarity. In 3-1-5/clause 7.1, it is dealt only within the text of clause 7.1(4).

The comments made in section 6.2.9.2.1 of this guide regarding the use of 3-1-5/clause 7.1(2) for sections close to supports also apply here.

6.2.11.2.2. Class 3 cross-sections

The approach is identical to that above for Class 1 and 2 cross-sections, except that the elastic stresses from bending and axial force should also be checked according to 3-1-5/clause 4.6 as discussed in section 6.2.10.2 of this guide. The use of plastic resistances for bending and axial force is again used in the interaction in 3-1-5/clause 7.1(1) because of the weakness of interaction between bending and shear found in the studies identified in section 6.2.9.1.2 of this guide. The authors are not however aware of similar test justification covering cases where there is significant axial force present. The requirement in 3-1-5/clause 7.1(4) to reduce $M_{f,Rd}$ to zero and to replace $\overline{\eta}_1$ by η_1 where the whole web is in compression was introduced to cover this uncertainty. In general, it will always be conservative to base $\overline{\eta}_1$ on η_1 in the interaction.

The comments made in 6.2.9.2.1 of this guide regarding the use of 3-1-5/clause 7.1(2) for sections close to supports also apply to Class 3 cross-sections.

6.2.11.2.3. Class 4 cross-sections

Two methods are possible for Class 4 cross-sections. If the required geometric constraints on the section are met as discussed in section 6.2.2.5.1 of this guide, it will usually be most economic to use the same interaction method as above for Class 1, 2 and 3 cross-sections. Equation (D6.2-59) again applies but the calculation of α_N, $M_{f,Rd}$ and $M_{N,Rd}$ must consider effective widths for flanges, allowing for plate buckling. The gross web section may however be considered. The reason for allowing plastic properties to be used in the interaction is again due to the weakness of shear–moment interaction found in the tests on beams with Class 4 webs identified in section 6.2.9.1.2 of this guide. The comments made above for Class 3 cross-sections with significant axial force also apply to Class 4 cross-sections; a more cautious approach would therefore be to replace $\overline{\eta}_1$ in the interaction by the elastic parameter η_1 from 3-1-5/clause 4.6, which uses effective sections throughout.

While the interaction of expression 3-1-5/(7.1) applies to beams with longitudinally stiffened webs, the authors are not aware of similar test justification to support the use of plastic properties in the interaction for such beams. Such webs have less post-buckling strength when overall web buckling is critical, but the approach leads to an interaction with shear only at very high percentages of the web shear resistance. A safer option is to replace $\overline{\eta}_1$ by η_1 in the interaction until such time as further studies are available to show this to be unnecessary. Where η_1 is used, if the cross-section is built up in stages, η_1 is the usage factor based on accumulated stress.

The interpretation of 3-1-5/clause 7.1(2) for sections close to supports is discussed in section 6.2.9.2.3 of this guide.

Stiffened flanges must be checked separately for the interaction of shear, bending and axial force in accordance with ***3-1-5/clause 7.1(5)***. This is described in section 6.2.9.2.3 of this guide and a worked example is provided. *3-1-5/clause 7.1(5)*

Where the geometric constraints discussed in section 6.2.2.5.1 are not met, the method of 3-1-5/clause 10 may be used. This will however be much more conservative as there is no allowance made for plasticity and shear stresses reduce the allowable resistance to other effects whatever their magnitude.

Worked Example 6.2-18: Calculation of the moment resistance of a plate girder with Class 3 cross-section subjected to combined moment, shear and axial force

The steel plate girder shown in Fig. 6.2-74 is initially assumed to be a Class 3 cross-section. The maximum bending moment that the cross-section can withstand in conjunction with a shear force (V_{Ed}) of 9600 kN and axial force of 500 kN (acting at the centroid of the gross cross-section) is calculated and a check is made of the compactness of the section under this bending moment and axial force. All plates are grade S355 to EN 10025 and the girder is restrained laterally and is stable against shear buckling. The thickness-dependent yield strengths are taken from 3-1-1/Table 3.1 which gives a constant yield stress of 355 MPa throughout. (The UK National Annex to EN 1993-2 requires the values in EN 10025 to be used.)

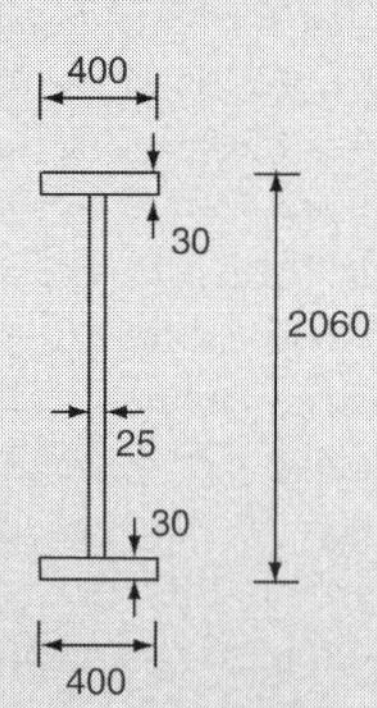

Fig. 6.2-74. Plate girder section for Worked Example 6.2-18

The main girder properties are as follows:

I_{yy} of girder $= 4.139 \times 10^{10}\,\text{mm}^4$
Elastic section modulus, $W_{el,min} = 4.078 \times 10^7\,\text{mm}^3$ (based on centres of the flanges)
Area of girder $= 74\,000\,\text{mm}^2$
Design plastic resistance moment $M_{pl,Rd} = 17\,523\,\text{kNm}$

The plastic shear resistance is calculated from 3-1-1/clause 6.2.6:

$$V_{pl,Rd} = \frac{\eta A_w (f_y/\sqrt{3})}{\gamma_{M0}} = \frac{1.2 \times 2000 \times 25 \times (355/\sqrt{3})}{1.00} = 12\,298\,\text{kN}$$

V_{Ed} is greater than $0.5 \times V_{pl,Rd}$, so shear will reduce the resistance moment to $M_{y,V,Rd}$.

$$\rho = \left(\frac{2V_{Ed}}{V_{pl,Rd}} - 1\right)^2 = \left(\frac{2 \times 9600}{12\,298} - 1\right)^2 = 0.315$$

Therefore, the allowable stress in web $= (1-\rho)f_y = (1-0.315) \times 355 = 243.2\,\text{MPa}$

There are three possibilities, as identified in section 6.2.11.1.2 above, for calculating the maximum allowable moment as follows.

1. Limitation of direct stresses to first yield on cross-section with reduced web thickness

Reduced thickness of web $= 25 \times 243.2/355 = 17.1\,\text{mm}$
Revised reduced elastic properties are therefore:

$$I_{yy} = \frac{400 \times 2060^3}{12} - \frac{(400 - 17.1) \times 2000^3}{12} = 3.613 \times 10^{10}\,\text{mm}^4$$

$$W_y = \frac{3.613 \times 10^{10}}{1015} = 3.560 \times 10^7\,\text{mm}^3 \text{ (based on centres of the flanges)}$$

Area = (2060 × 400) − [2000 × (400 − 17.1)] = 58 200 mm^2

Longitudinal stress in member $= \frac{P}{A} + \frac{M}{W_y} = \frac{500 \times 10^3}{58\,200} + \frac{M_{y,Ed}}{3.560 \times 10^7} \leq \frac{355}{1.00}$

Therefore $M_{y,Ed}$ = maximum allowable bending moment = **12 331 kNm**

It is next checked that the cross-section is still classified as Class 3 under moment and axial force, based on gross section properties without reduction for shear.

Stress in gross member at centroid of flanges equals:

$$\frac{P}{A} \pm \frac{M_{y,Ed}}{W_{el,min}} = \frac{500 \times 10^3}{74\,000} \pm \frac{12\,331 \times 10^6}{4.078 \times 10^7} = +309\,\text{MPa} - 296\,\text{MPa}$$

From 3-1-1/Table 5.2, $\psi = -296/309 = -0.958$ based on stress variations between flanges. (The variation over the web height should strictly be used but this is approximately the same.)

Therefore

$$\frac{c}{t} \leq \frac{42\varepsilon}{0.67 + 0.33\psi} = \frac{42 \times 0.81}{(0.67 + 0.33 \times -0.958)} = 96.1$$

The actual $c/t = 2000/25 = 80 < 96.1$, so the section is still Class 3 with axial force.

2. Linear interaction of equation (D6.2-58)

The above check was conservative because for moment and shear alone, 3-1-1/clause 6.2.8 allows the moment resistance for symmetric Class 3 cross-sections (and arguably for asymmetric cross-sections – see section 6.2.9.1.2 of this guide) to be based on plastic section properties as long as the resulting resistance does not exceed the elastic bending resistance. The check in method 1 above reduces the bending strength for any shear in excess of 50% of the shear resistance.

The plastic resistance in the presence of shear, $M_{y,V,Rd}$, using the reduced yield strength above, is found to be 14 718 kNm but the bending resistance should not be taken as greater than:

$$M_{c,Rd} = \frac{W_{el,min} f_y}{\gamma_{M0}} = \frac{4.078 \times 10^7 \times 355}{1.0} = 14\,477\,\text{kNm}$$

so $M_{y,V,Rd}$ is taken as the elastic resistance = 14 477 kNm. There is therefore effectively no interaction between bending and shear in this method for this loading situation. There will however be interaction between axial force and shear. The area of the section with reduction to the web width for shear is 58 200 mm^{2}as calculated in method 1 above.

The interaction with axial force is then performed according to the linear interaction of equation (D6.2-58):

$$\frac{N_{Ed}}{N_{v,Rd}} + \frac{M_{y,Ed}}{M_{v,y,Rd}} + \frac{M_{z,Ed}}{M_{v,z,Rd}} = \frac{500 \times 10^3}{58\,200 \times 355/1.0} + \frac{M_{y,Ed}}{14\,477} = 1.0$$

Therefore $M_{y,Ed}$ = maximum allowable bending moment = **14 127 kNm**

This is significantly greater than the value above in method 1 above. A similar calculation to that in method 1 shows that the cross-section remains Class 3 when the axial force is applied.

3. Method of EN 1993-1-5 clause 7.1

The interaction in EN 1993-1-5 in the presence of axial force is from equation (D6.2-59):

$$\bar{\eta}_1 + \left[1 - \frac{\alpha_N M_{f,Rd}}{M_{N,Rd}}\right](2\bar{\eta}_3 - 1)^2 \leq 1.0$$

where:

$$(2\bar{\eta}_3 - 1)^2 = \left(\frac{2 \times 9600}{12\,298} - 1\right)^2 = 0.315$$

$$\alpha_N = \left[1 - \frac{N_{Ed}}{(A_{f1} + A_{f2})f_{yf}/\gamma_{M0}}\right] = \left[1 - \frac{500 \times 10^3}{(400 \times 30 + 400 \times 30) \times 355/1.0}\right] = 0.941$$

$$M_{f,Rd} = 400 \times 30 \times \frac{355}{1.0} \times 2030 = 8648\,\text{kNm}$$

The web depth required to resist axial force $= \dfrac{500 \times 10^3}{25 \times 355/1.0} = 56\,\text{mm}$

Therefore $M_{N,Rd} = 17\,523 \times 10^6 - 25 \times 56^2/4 \times \dfrac{355}{1.0} = 17\,516\,\text{kNm}$

$$\bar{\eta}_1 = 1.0 - \left[1 - \frac{\alpha_N M_{f,Rd}}{M_{N,Rd}}\right](2\bar{\eta}_3 - 1)^2 = 1.0 - \left[1 - \frac{0.941 \times 8648}{17\,516}\right] \times 0.315 = 0.83$$

so $M_{y,Ed} = 0.83 \times 17\,516 = 14\,538\,\text{kNm}$. Clearly this check does not govern as the bending resistance produced exceeds the elastic bending resistance of 14 477 kNm from above. It is also necessary to verify axial force and bending without shear, using 3-1-5/clause 4.6:

$$\eta_1 = \frac{P}{A} + \frac{M_{y,Ed}}{W_{el,min}} = \frac{500 \times 10^3}{74\,000} + \frac{M_{y,Ed}}{4.078 \times 10^7} \le 355/1.0 \text{ and hence } M_{y,Ed} = \mathbf{14\,201\,kNm}$$

There is therefore no interaction with shear according to this method for this loading situation. A similar calculation to that in method 1 above shows that the cross-section remains Class 3 when the axial force is applied.

6.3. Buckling resistance of members

6.3.1. Uniform members in compression

6.3.1.1. Buckling resistance

In addition to cross-section checks discussed in section 6.2, compression members need to be checked for buckling resistance. The basic requirement in ***3-1-1/clause 6.3.1.1(1)*** is as follows:

3-1-1/clause 6.3.1.1(1)

$$\frac{N_{Ed}}{N_{b,Rd}} \le 1.0 \qquad \text{3-1-1/(6.46)}$$

where N_{Ed} is the design value of the compression force and $N_{b,Rd}$ is the design buckling resistance of the compression member.

Three modes of buckling must be checked: flexural buckling (upon which the derivation of the buckling curves is based), torsional buckling and flexural–torsional buckling. $N_{b,Rd}$ is derived from the following equations in ***3-1-1/clause 6.3.1.1(3)***:

3-1-1/clause 6.3.1.1(3)

$$N_{b,Rd} = \frac{\chi A f_y}{\gamma_{M1}} \quad \text{for Class 1, 2 and 3 cross-sections} \qquad \text{3-1-1/(6.47)}$$

$$N_{b,Rd} = \frac{\chi A_{eff} f_y}{\gamma_{M1}} \quad \text{for Class 4 cross-sections} \qquad \text{3-1-1/(6.48)}$$

where χ is the reduction factor for the relevant buckling mode, which is determined from the buckling curves in 3-1-1/clause 6.3.1.2. A_{eff} is the effective area allowing for local buckling. The cross-sectional areas need not allow for holes at the end connections of pin-jointed members where the flexural stresses from buckling are very small. ***3-1-1/clause 6.3.1.1(4)*** provides a similar relaxation, but does not restrict it to pin-jointed ends. If the end connections are designed to carry moment and to provide an effective length shorter than

3-1-1/clause 6.3.1.1(4)

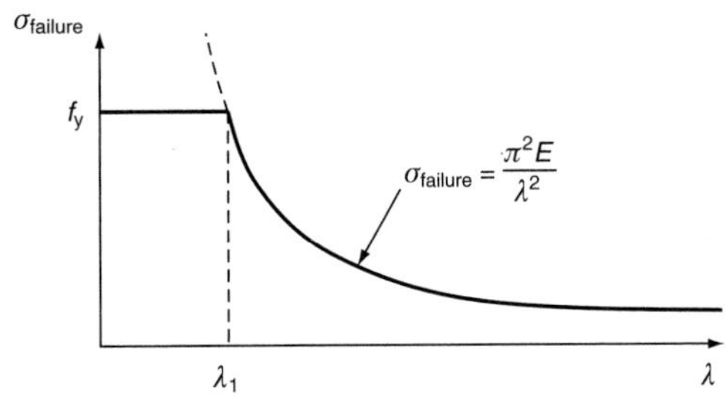

Fig. 6.3-1. Relationship between Euler strut failure load and slenderness

the member length, it would be necessary to make some allowance for the holes. For holes in other locations, judgement is needed as the flexural stresses may similarly be considerably less than the peak values within the middle third of each half-wavelength of buckling. Holes can always be conservatively included.

3-1-1/clause 6.3.1.1(2)

3-1-1/clause 6.3.1.1(2) is a reminder that for asymmetric Class 4 cross-sections, an additional moment may arise due to the eccentricity between the gross cross-section centroid and that of the effective cross-section – see section 6.2.10.3 of this guide. This requires a check of buckling under combined bending and axial force to 3-2/clause 6.3.3 or 3-2/clause 6.3.4.

6.3.1.2. Buckling curves

Euler first derived the now well-known equation for the flexural buckling load, N_{cr}, of a pin-ended strut of length L_{cr}:

$$N_{cr} = \frac{\pi^2 EI}{L_{cr}^2} \qquad \text{(D6.3-1)}$$

The axial stress, σ_{cr}, in the strut when elastic critical buckling occurs is thus:

$$\sigma_{cr} = \frac{\pi^2 E}{\lambda^2} \qquad \text{(D6.3-2)}$$

where λ is the slenderness of the strut equal to L_{cr}/i and i is the radius of gyration for the plane of buckling. From equation (D6.3-2), if σ_{cr} exceeds the yield stress of the strut, f_y, the column might be expected to fail by yielding in compression as opposed to buckling. Figure 6.3-1 shows the relationship between the stress at which an initially perfectly straight strut fails and the ratio L_{cr}/i.

An important value in Fig. 6.3-1 is λ_1 which corresponds to the limiting slenderness for yielding to occur, above which the initially perfectly straight strut would fail by buckling. For this condition:

$$\sigma_{failure} = f_y = \frac{\pi^2 E}{\lambda_1^2} \text{ so } \lambda_1 = \sqrt{\frac{\pi^2 E}{f_y}} \qquad \text{(D6.3-3)}$$

By changing the axes in Fig. 6.3-1 to χ ($= \sigma_{failure}/f_y$) and $\bar{\lambda}(= \lambda/\lambda_1)$ respectively, the same curve can be plotted non-dimensionally as shown in Fig. 6.3-2.

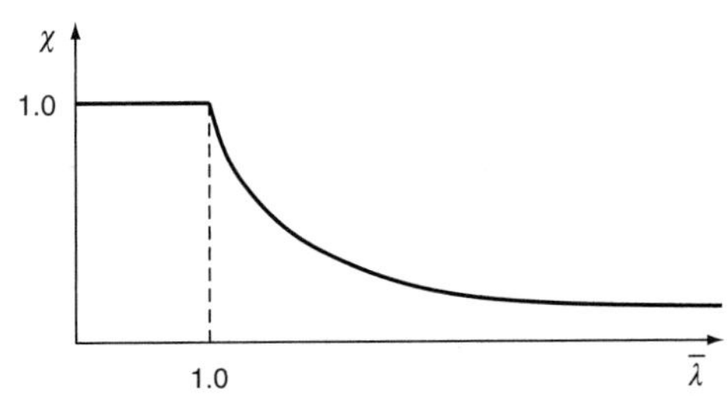

Fig. 6.3-2. Non-dimensional relationship between Euler strut buckling load and slenderness

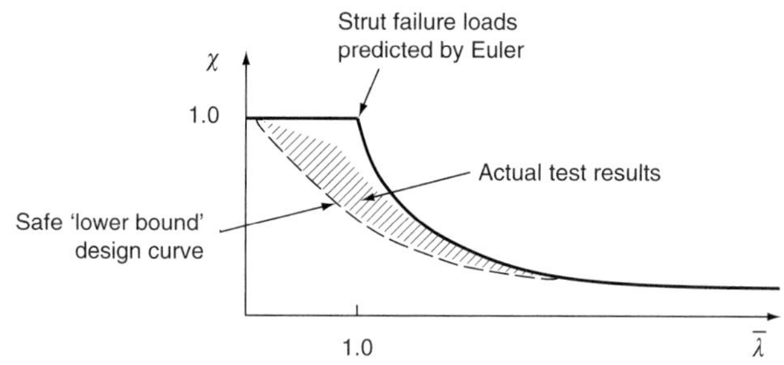

Fig. 6.3-3. Relationship between actual column failure loads and those predicted by Euler

If the actual failure loads of a range of steel struts tested in a laboratory are plotted against the failure loads predicted above by Euler, as on Fig. 6.3-3, a problem with the Euler theory becomes apparent. The Euler collapse load correlates well with actual failure loads at high slenderness values but significantly overestimates the actual failure loads at intermediate slenderness values. However, at very low slenderness, the test results show that the strut resistances are unaffected by buckling and the failure loads reach the yield load.

The difference arises because Euler's derivation of N_{cr} assumed a perfectly straight, linear elastic strut. 'Real' columns however contain imperfections as discussed in section 5.3 of this guide. These significantly modify the behaviour assumed above. Imperfections include:

- **Initial out of straightness.** In reality, all struts will have some degree of initial curvature. This induces bending in the strut which reduces the failure load.
- **Eccentricity of loading.** A strut nominally loaded through its centroidal axis will usually have some bending moment induced by unavoidable minor eccentricities. These additional moments will reduce the resistance.
- **Residual stresses due to welding and rolling.** Struts that have not been stress-relieved will invariably have self-equilibrating residual stresses, caused by welding and rolling procedures, locked into them. These residual stresses cause premature yielding and reduce the stiffness and buckling resistance of a strut.
- **Lack of a clearly defined yield point.** Some steels do not exhibit a sharply defined yield point but show a gradual transition from elastic to plastic behaviour. This can reduce the buckling resistance of struts with intermediate slenderness.

In order to provide a safe lower bound to test results, most design codes have derived design curves by modifying the Euler theory to allow for an initial lack of straightness in the column. The remaining sources of imperfection are taken into account by adjusting the shape of each design curve by effectively increasing the initial bows to provide equivalent geometric imperfections. The design curves in EN 1993-1-1 use this approach and the analysis is presented later in this section.

A single lower bound strut design resistance curve, as illustrated in Fig. 6.3-3, would always give a safe resistance but would also give an unnecessarily conservative answer in certain scenarios. For example, rolled sections will have a higher buckling load than equivalent welded members because welding leads to significantly greater residual stresses. For an I-section, a lower resistance curve is required for buckling about the minor axis compared to the major axis because the y/i ratio will be higher about the minor axis. The importance of this ratio can be seen in the derivation of the imperfection parameter below. Different strut design curves are therefore given for different situations as schematically illustrated in Fig. 6.3-4.

Five design curves are given in 3-1-1/Fig. 6.4. The relevant curve depends on the method of manufacture of the section, the shape of the section, the axis of buckling and the yield strength as determined from 3-1-1/Table 6.2. Each buckling curve is also represented mathematically in ***3-1-1/clause 6.3.1.2(1)*** as follows:

3-1-1/clause 6.3.1.2(1)

$$\chi = \frac{1}{\Phi + \sqrt{\Phi^2 - \bar{\lambda}^2}} \text{ but } \chi \leq 1.0 \qquad \text{3-1-1/(6.49)}$$

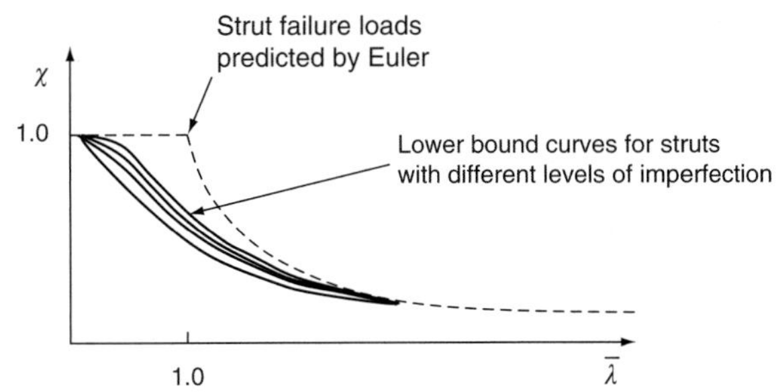

Fig. 6.3-4. Lower bound strut buckling curves representing struts with different levels of imperfection

where:

$$\Phi = 0.5[1 + \alpha(\bar{\lambda} - 0.2) + \bar{\lambda}^2]$$

$$\bar{\lambda} = \sqrt{\frac{Af_y}{N_{cr}}} \text{ for Class 1, 2 and 3 cross-sections}$$

Similarly for Class 4 cross-sections:

$$\bar{\lambda} = \sqrt{\frac{A_{eff}f_y}{N_{cr}}}$$

α is an imperfection factor derived from 3-1-1/Table 6.1, reproduced below as Table 6.3-1. The relevant buckling curve is selected from 3-1-1/Table 6.2 and depends on the factors discussed above, including the y/i ratio discussed below which generally differs for different axes of buckling.

Expression 3-1-1/(6.49) is derived from the Perry–Robertson theory which considers an initial sinusoidal bow imperfection of e_0 in the strut and which predicts failure to occur when the most critical compression fibre reaches the yield stress. The moment from the initial imperfection is:

$$M_{Ed} = N_{Ed}\left[\frac{e_0}{1 - (N_{Ed}/N_{cr})}\right]$$

from section 5.2 of this guide. Equating the stress from this moment plus the stress from the axial force to the yield strength leads to the following failure criterion:

$$(\sigma_a - f_y)(\sigma_a - \sigma_{cr}) = \eta\sigma_{cr}\sigma_a \qquad \text{(D6.3-4)}$$

where:

σ_a is the axial stress when the yield stress is reached at an extreme fibre
σ_{cr} is $\pi^2 Ei^2/L_{cr}^2$
η is an imperfection parameter which is equal to ye_0/i^2 from the above analysis, where y is the maximum distance from the cross-section centroidal axis to an extreme fibre in the plane of bending.

The larger the imperfection parameter, the smaller the allowable compressive stress becomes. It can therefore be seen that increasing the ratio y/i reduces buckling resistance. As discussed above, the equivalent geometric imperfection e_0 includes not only geometric imperfections (which are length dependent) but also the effects of residual stresses.

Table 6.3-1. Imperfection factors for buckling curves

Buckling curve	a_0	a	b	c	d
Imperfection factor α	0.13	0.21	0.34	0.49	0.76

Consequently the imperfection parameter in EN 1993 is taken as:

$$\eta = \alpha(\bar{\lambda} - 0.2) \qquad \text{(D6.3-5)}$$

This imperfection parameter reduces to zero at low slenderness, which reflects observed behaviour that stocky struts can reach the full squash load. Solution of equation (D6.3-4) leads to:

$$\sigma_a/f_y = \chi = 0.5\left\{[1 + (1+\eta)/\bar{\lambda}^2] - \sqrt{[1 + (1+\eta)/\bar{\lambda}^2]^2 - 4/\bar{\lambda}^2}\right\} \qquad \text{(D6.3-6)}$$

3-1-1/clause 6.3.1.2(3)

This can alternatively be presented in the convenient form of expression 3-1-1/(6.49) or graphically as in 3-1-1/Fig. 6.4 – ***3-1-1/clause 6.3.1.2(3)*** refers. The same resistance formula is also applied to torsional and flexural–torsional buckling by analogy.

3-1-1/clause 6.3.1.2(4)

It will be seen from 3-1-1/Fig. 6.4 that there is a plateau of resistance for slenderness up to $\bar{\lambda} = 0.2$. For $\bar{\lambda} \le 0.2$, the full squash load can be obtained and buckling need not be checked – ***3-1-1/clause 6.3.1.2(4)*** refers. If it is not intended to load the cross-section up to its full squash load, the same clause allows a higher slenderness ratio to be attained before buckling need be considered. This is achieved by exchanging the actual design axial force for the squash load in the non-dimensional slenderness and checking that $\bar{\lambda} = \sqrt{N_{Ed}/N_{cr}} \le 0.2$ and hence that $N_{Ed}/N_{cr} \le 0.04$.

6.3.1.3. Slenderness for flexural buckling

3-1-1/clause 6.3.1.3(1)

To determine the flexural buckling load for a strut from expression 3-1-1/(6.49) or 3-1-1/Fig. 6.4, $\bar{\lambda}$ must first be calculated from ***3-1-1/clause 6.3.1.3(1)*** as follows:

$$\bar{\lambda} = \sqrt{\frac{Af_y}{N_{cr}}} = \frac{L_{cr}}{i\lambda_1} \text{ for Class 1, 2 and 3 cross-sections} \qquad \text{3-1-1/(6.50)}$$

$$\bar{\lambda} = \sqrt{\frac{A_{eff}f_y}{N_{cr}}} = \frac{L_{cr}\sqrt{\dfrac{A_{eff}}{A}}}{i\lambda_1} \text{ for Class 4 cross-sections} \qquad \text{3-1-1/(6.51)}$$

where:

L_{cr} is the buckling length in the buckling plane considered, i.e. the effective length;
i is the radius of gyration about the relevant axis, determined using the properties of the gross section;

$$\lambda_1 = \pi\sqrt{\frac{E}{f_y}} = 96.9\varepsilon \text{ with } \varepsilon = \sqrt{\frac{235}{f_y}}$$

The background to expression 3-1-1/(6.50) is discussed in section 6.3.1.2. It should be noted that, for Class 4 cross-sections, the effective area A_{eff} allowing for plate buckling is used in the numerator of expression 3-1-1/(6.51). However, a similar reduction for plate buckling is not made when determining N_{cr}. This is because the loss of strength due to plate buckling is much more severe than the loss of stiffness.

Effective length for flexural buckling, L_{cr}

The elastic critical buckling load, N_{cr}, is given for a range of struts with different end restraint conditions in Fig. 6.3-5. Using an effective length, L_{cr}, the equation for N_{cr} is:

$$N_{cr} = \frac{\pi^2 EI}{L_{cr}^2} \qquad \text{(D6.3-7)}$$

The theoretical values of L_{cr} for each set of end restraints are also shown in Fig. 6.3-5. Fully rigid end rotational restraints will never actually exist in practice so the theoretical effective length for rigid cases should generally be increased to allow for this flexibility. If the restraint rotational stiffnesses are known, the effective length can be calculated using the method in section 5.2.2.3 of this guide. If the real stiffness cannot be obtained, the

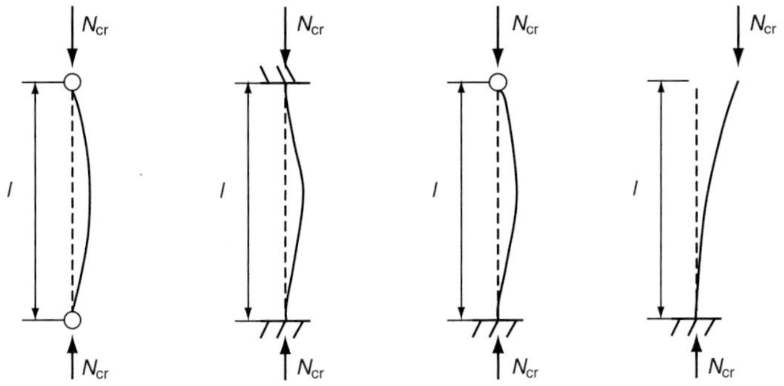

Elastic critical buckling load, N_{cr}	$N_{cr} = \frac{\pi^2 EI}{l^2}$	$N_{cr} = \frac{4\pi^2 EI}{l^2}$	$N_{cr} = \frac{2.04\pi^2 EI}{l^2}$	$N_{cr} = \frac{0.25\pi^2 EI}{l^2}$
Theoretical values of L_{cr}	$1.0l$	$0.5l$	$0.7l$	$2.0l$
Recommended values of L_{cr} from ref. 4	$1.0l$	$0.7l$	$0.85l$	$2.0l$

Fig. 6.3-5. Elastic critical buckling loads for struts with different end restraints

flexibility can be taken into account approximately by using the recommended increased effective lengths given in Fig. 6.3-5, which were taken from BS 5400: Part 3.[4] Care is required with the use of the cantilever effective length; the method in section 5.2.2.3 of this guide should be used where there are concerns over end rotational flexibility as the value in BS 5400 made no such allowance in this case.

For more complex load restraint conditions, N_{cr} can be calculated directly from a computer elastic critical buckling analysis as discussed in section 5.2.2 of this guide. N_{cr} can then be used to determine slenderness directly from expression 3-1-1/(6.50) or (6.51) as appropriate. This procedure can also be used for members with varying section or varying compression.

3-2/Annex D gives methods of calculating effective lengths for isolated bridge members in trusses and for buckling of arch bridges. It also gives imperfections for arches where second-order analysis is to be carried out.

Worked Example 6.3-1: Calculation of buckling resistance for a column

A 355.6 × 12.5 circular hollow section in S355 steel cantilevers 7.5 m from a rigid foundation. The flexural buckling resistance, $N_{b,Rd}$, is calculated.

Area of CHS = 135 cm^2

i = radius of gyration of CHS = 121 mm

From 3-1-1/clause 6.3.1.3(1):

$$\bar{\lambda} = \sqrt{\frac{Af_y}{N_{cr}}} = \frac{L_{cr}}{i\lambda_1}$$

From Fig. 6.3-5:

$$L_{cr} = 2.0l = 2.0 \times 7.5 = 15\,\text{m}$$

From 3-1-1/clause 6.3.1.3:

$$\lambda_1 = \pi\sqrt{\frac{E}{f_y}} = \pi\sqrt{\frac{205 \times 10^3}{355}} = 75.5$$

Therefore:

$$\bar{\lambda} = \frac{L_{cr}}{i\lambda_1} = \frac{15\,000}{121 \times 75.5} = 1.64$$

3-1-1/Table 6.2: For S355 hot rolled CHS, use buckling curve a
3-1-1/Fig. 6.4: For $\bar{\lambda} = 1.64$, reduction factor $\chi = 0.32$
3-1-1/clause 6.3.1.1(3):

$$N_{b,Rd} = \frac{\chi A f_y}{\gamma_{M1}} = \frac{0.32 \times 13\,500 \times 355}{1.10} = 1394\,\text{kN}$$

Therefore the flexural buckling resistance of the CHS is **1394 kN**.

6.3.1.4. Slenderness for torsional and flexural–torsional buckling

3-1-1/clause 6.3.1.4(1)
3-1-1/clause 6.3.1.4(2)

It is possible for sections to fail in overall buckling under axial load at a lower load than that from flexural buckling by either a torsional or flexural–torsional mode. ***3-1-1/clause 6.3.1.4(1)*** therefore requires these modes to be checked. The slenderness for Class 1, 2 and 3 cross-sections is determined from ***3-1-1/clause 6.3.1.4(2)***:

$$\bar{\lambda}_T = \sqrt{\frac{Af_y}{N_{cr}}} \qquad \text{3-1-1/(6.52)}$$

3-1-1/clause 6.3.1.4(3)

where A is the area of the section (with allowance for holes if necessary) and N_{cr} is the lowest critical buckling load from flexural–torsional buckling or torsional buckling modes. The slenderness for Class 4 cross-sections is similar but the effective area A_{eff} is used in place of the gross area in the numerator. N_{cr} is still calculated based on the gross cross-section as plate buckling has little influence on member stiffness. No guidance is given on the determination of this buckling load in EN 1993-1-1; some is provided below. When determining the reduction factor for this slenderness, ***3-1-1/clause 6.3.1.4(3)*** permits the curve appropriate to the z–z axis to be determined from 3-1-1/Table 6.2.

Torsional buckling (bisymmetric sections)

Bisymmetric sections alone may buckle in a purely torsional mode as shown in Fig. 6.3-6(a) at an axial load less than either of the principal axis flexural buckling loads. For the special case of bisymmetric sections, torsional buckling occurs without interaction with the two flexural modes so there is no flexural–torsional mode. The elastic critical torsional buckling load may be calculated as follows:

$$N_{cr,T} = [GI_T + \pi^2 EI_w / L^2] / i_g^2 \qquad \text{(D6.3-8)}$$

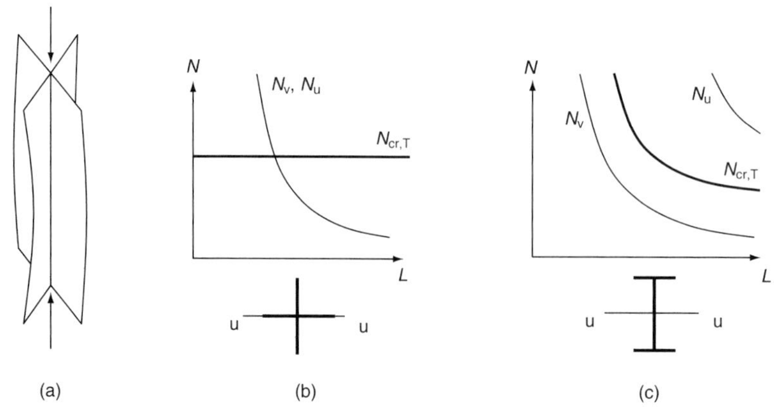

Fig. 6.3-6. Torsional buckling of bisymmetric sections: (a) torsional mode; (b) cruciforms; (c) symmetric I-beams

where:

- L is the effective length between points where rotation is prevented about the axis of the member, or a shorter length if warping is also prevented;
- I_T is the St Venant torsional inertia;
- I_w is the warping constant;
- i_g is the radius of gyration about the centre of gravity $= \sqrt{(I_u + I_v)/A}$;
- I_u, I_v are the second moment of area about the major and minor principal axes respectively.

It can be seen from equation (D6.3-8) that the resistance is independent of length when the warping constant is small. Consequently, torsional buckling is likely to govern the resistance of sections with small warping constant, such as cruciform sections, at short effective lengths as shown in Fig. 6.3-6(b). N_u and N_v are the flexural buckling loads about the major and minor principal axes respectively. Sections with appreciable warping constant, such as I-beams, are unlikely to be governed by torsional buckling rather than flexural buckling as the torsional buckling load increases with reducing length in the same way as for flexural buckling as seen in Fig. 6.3-6(c). They should nevertheless be checked, as some section geometries (such as small height-to-width ratio) can lead to torsional buckling becoming critical.

For a cruciform section without warping resistance and outstands of thickness t and width b, the elastic torsional buckling resistance is $4Gt^3/b$. This is the same as the sum of the elastic critical plate buckling loads of the four outstands. In UK practice, it has often been assumed therefore that torsional buckling is not a problem where the outstand shape limits have been observed (such that yield can occur without buckling of the outstands). However, in EN 1993-1-1, the reduction factor for member buckling is greater than that for plate buckling, so compliance with the plate outstand limits in 3-1-1/Table 5.2 will not necessarily prevent buckling from being predicted in an overall torsional mode in preference to a flexural mode and a check has to be made. If the outstand shape limits are met, the torsional buckling check should never be significantly more onerous than the flexural buckling check.

Flexural–torsional buckling (monosymmetric and asymmetric sections)

For monosymmetric and asymmetric sections, the buckling modes are interdependent and a flexural–torsional mode becomes possible. This can, in principle, govern the design but it will have little influence on most bridge members such as the slender bracing member in Worked Example 6.3-2. Problems may arise where the member has very low warping resistance and has been designed as a stocky column for flexural buckling design as again illustrated at the end of Worked Example 6.3-2.

For asymmetric sections, flexural–torsional buckling always occurs at a lower load ($N_{cr,TF}$, involving torsion and flexure about both principal axes) than for either of the principal axis flexural buckling loads (N_u and N_v) or the torsional buckling load ($N_{cr,T}$). However, where either the minor principal axis flexural buckling load or the torsional buckling load is much smaller than the other, the buckling load will tend to this smaller value. This is illustrated in Fig. 6.3-7(a) for an asymmetric angle, which may have a flexural torsional buckling load much lower than the minor axis flexural buckling load at short length.

For monosymmetric sections, the section buckles at the lower of the minor axis flexural buckling load or a combined flexural–torsional mode involving torsion and flexure about the major axis as shown for a channel in Fig. 6.3-7(b). Where there is only small warping resistance, such as for equal angles and T-beams, behaviour is similar to that in Fig. 6.3-7(a) but the curve for $N_{cr,TF}$ actually meets that of N_v at high length, rather than tending towards it, so N_v may become the lower buckling load.

Typically, the lowest critical buckling load tends to the torsional load at short effective length and the minor axis flexural buckling load at greater effective length as illustrated in Fig. 6.3-7(a). Channel sections buckle slightly below the torsional load at short effective lengths and will achieve the minor axis flexural buckling load at longer length as shown in

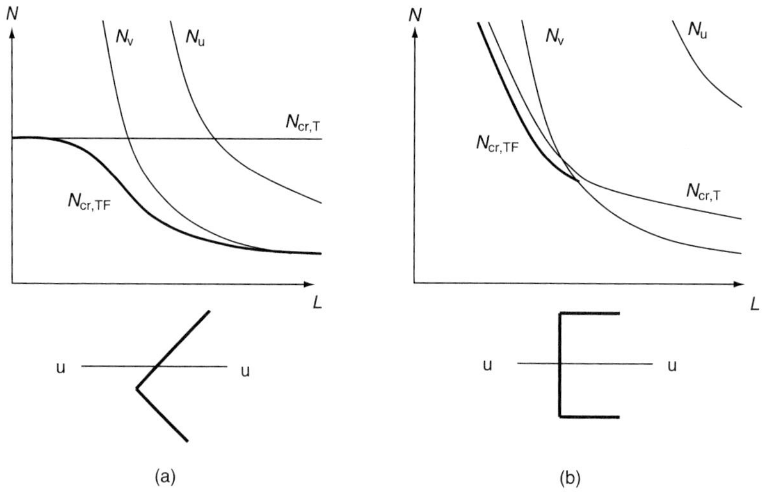

Fig. 6.3-7. Torsional buckling of asymmetric and monosymmetric sections: (a) asymmetric angle; (b) monosymmetric channel

Fig. 6.3-7(b). If the channel was given a lip to increase the minor axis inertia, the torsional buckling load can become relatively small compared to the minor axis flexural load at all lengths and then buckling will occur at a load near to, but lower than, the torsional value at all lengths.

The flexural–torsional buckling load may generally be obtained as the lowest root of the following equation:

$$N_{cr,TF}^3(i_s^2 - u_s^2 - v_s^2) - N_{cr,TF}^2[(N_u + N_v + N_{cr,T})i_s^2 - N_v u_s^2 - N_u v_s^2] + N_{cr,TF} i_s^2(N_u N_v + N_v N_{cr,T} + N_{cr,T} N_u) - N_u N_v N_{cr,T} i_s^2 = 0 \quad \text{(D6.3-9)}$$

where:

$N_u = \pi^2 EI_u / L_u^2$ (major axis flexural buckling);
$N_v = \pi^2 EI_v / L_v^2$ (minor axis flexural buckling);
$N_{cr,T} = (GI_T + \pi^2 EI_w / L_x^2)/i_s^2$ (torsional buckling);
L_u, L_v, L_x are the effective lengths for the relevant buckling mode (see discussion later);
I_T is the St Venant torsional inertia;
I_u, I_v are the second moment of area about the major and minor principal axes respectively;
I_w is the warping constant;
i_s^2 is the square of the radius of gyration about the shear centre $= (I_u + I_v)/A + u_s^2 + v_s^2$;
u_s is the distance from the centre of gravity to the shear centre in the u direction;
v_s is the distance from the centre of gravity to the shear centre in the v direction.

Where a section has one axis of symmetry about the u–u axis, such as for the channel in Fig. 6.3-7(b), equation (D6.3-9) simplifies to:

$$N_{cr,TF} = \frac{(N_u + N_{cr,T}) - \sqrt{(N_u + N_{cr,T})^2 - \dfrac{4N_u N_{cr,T} i_g^2}{i_g^2 + u_s^2}}}{2i_g^2/(i_g^2 + u_s^2)} \quad \text{(D6.3-10)}$$

where the notations have their meanings above.

Effective length

For calculations of torsional buckling and flexural–torsional buckling resistance, the effective length for torsional buckling can conservatively be taken as the member length

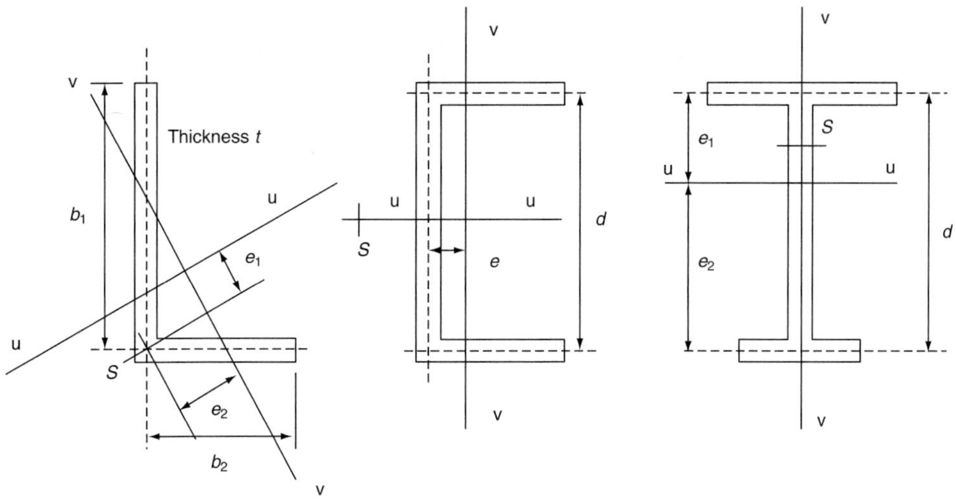

Fig. 6.3-8. Definitions for some commonly used bridge sections

between points at which rotation about the member axis is effectively restrained. A lower value could be considered where warping is effectively restrained. Theoretically, a value of effective length equal to 0.5 times the distance between points of full warping restraint could be used in the case of warping restraint at both ends, but it will not be possible to provide full rigidity in practice, so a value of 0.7 times the distance between points of warping restraint might be more appropriate in this case. Effective lengths for flexural buckling are discussed in section 6.3.1.3 of this guide.

For columns with continuous restraint, the resistance in flexural–torsional buckling would have to be obtained from first principles and this is beyond the scope of this guide.

Formula for warping constant and shear centre

Some formulae are given below for warping constant and shear centre for the cross-section shapes in Fig. 6.3-8.

Angle:

$$u_s = e_2, v_s = e_1 \text{ and } I_w = \frac{t^3}{36}(b_1^3 + b_2^3)$$

Channel:

$$u_s = -e\left(1 + \frac{d^2 A}{4I_u}\right), v_s = 0 \text{ and } I_w = \frac{d^2}{4}\left[I_v + e^2 A\left(1 - \frac{d^2 A}{4I_u}\right)\right]$$

where A is the area of the section and I_u and I_v are the principal second moments of area.

I-beam:

$$u_s = 0, v_s = \frac{e_2 I_2 - e_1 I_1}{I_1 + I_2} \text{ and } I_w = \frac{d^2 I_1 I_2}{I_1 + I_2}$$

where I_1 and I_2 are the second moments of area of the top and bottom flange respectively acting alone about the v–v axis.

Worked Example 6.3-2: Main beam angle bracing member

A 150 × 150 × 12 horizontal bracing angle in S275 steel has an effective length for flexural and torsional buckling of 3.2 m (taken as the distance between end connections) and is used to brace a pair of beams in part of a multi-beam deck. No resistance to warping is provided at the angle end connections. The reduction factor for buckling is determined under axial load.

Section classification first has to be carried out. The angle meets the limit for outstands for angles but does not meet the criterion for angle perimeter slenderness

$$\frac{b+h}{2t} \leq 11.5\varepsilon \text{ in 3-1-1/Table 5.2}$$

since

$$\frac{150+150}{2\times 12} = 12.5 > 11.5\varepsilon = 11.5\times 0.92 = 10.6$$

The section is therefore Class 4. However, when EN 1993-1-5 is used to determine the effective outstands, it is found that the full section area is available, which is not surprising as the individual outstands were Class 3 to EN 1993-1-1. The above perimeter limit was required in previous drafts of EN 1993-1-1 because no explicit checks on torsional and flexural–torsional buckling were made; it now appears to be redundant.

The section is monosymmetric so the buckling load is expected to be the lower of the minor axis flexural load or a flexural–torsional mode. From section tables:

$$u_s = e_2 = 49.8\,\text{mm}, v_s = e_1 = 0\,\text{mm}$$

The warping constant is small and could be neglected but it is calculated here.

$$I_w = \frac{t^3}{36}(b_1^3 + b_2^3) = \frac{12^3}{36}(144^3 + 144^3) = 2.867\times 10^8\,\text{mm}^6$$

$$J = \frac{12^3}{3}(144+144) = 1.659\times 10^5\,\text{mm}^4$$

$$i_s^2 = (I_u + I_v)/A + u_s^2 + v_s^2 = (1170\times 10^4 + 303\times 10^4)/34.8\times 10^2 + 49.8^2 = 6713\,\text{mm}^2$$

$$i_g^2 = (I_u + I_v)/A = (1170\times 10^4 + 303\times 10^4)/34.8\times 10^2 = 4233\,\text{mm}^2$$

$$N_u = \pi^2 E I_u / L_u^2 = \pi^2\times 210\times 10^3\times 1170\times 10^4/3200^2 = 2368\,\text{kN}$$

The torsional buckling load is:

$$N_{cr,T} = (GJ + \pi^2 E I_w / L_x^2)/i_s^2$$

$$= (81\times 10^3\times 1.659\times 10^5 + \pi^2\times 210\times 10^3\times 2.867\times 10^8/3200^2)/6713$$

$$= 2010\,\text{kN}$$

The warping resistance increases the resistance by less than 1% here so could have been neglected. The flexural torsional buckling load from equation (D6.3-10) is:

$$N_{cr,TF} = \frac{(N_u + N_{cr,T}) - \sqrt{(N_u + N_{cr,T})^2 - \dfrac{4N_u N_{cr,T} i_g^2}{i_g^2 + u_s^2}}}{2i_g^2/(i_g^2 + u_s^2)}$$

$$= \frac{(2368\times 10^3 + 2010\times 10^3) - \sqrt{(2368\times 10^3 + 2010\times 10^3)^2 - \dfrac{4\times 2368\times 10^3\times 2010\times 10^3\times 4233}{4233 + 49.8^2}}}{2\times 4233/(4233 + 49.8^2)}$$

$$= 1349\,\text{kN}$$

The minor axis flexural buckling load is:

$$N_v = \pi^2 E I_v / L_v^2 = \pi^2\times 210\times 10^3\times 303\times 10^4/3200^2 = 613\,\text{kN} < 1349\,\text{kN}$$

The minor axis buckling load therefore is lower than that for flexural torsional buckling, so neglecting flexural torsional buckling would have been safe here. The slenderness for flexural buckling is therefore given by:

$$\bar{\lambda} = \sqrt{\frac{Af_y}{N_{cr}}} = \sqrt{\frac{34.8 \times 10^2 \times 275}{613 \times 10^3}} = 1.25$$

The reduction factor for flexural buckling from curve b (chosen according to 3-1-1/ Table 6.2) is from 3-1-1/Fig. 6.4, $\chi = \mathbf{0.46}$.

If the length of the brace is now halved to 1600 mm, then:

$N_u = 2368 \times 4 = 9472\,\text{kN}$

$N_{cr,T}$ will essentially remain the same, as the warping contribution is still very small. Thus $N_{cr,T} = 2010\,\text{kN}$.

$$N_{cr,TF} = \frac{(N_u + N_{cr,T}) - \sqrt{(N_u + N_{cr,T})^2 - \dfrac{4N_u N_{cr,T} i_g^2}{i_g^2 + u_s^2}}}{2i_g^2/(i_g^2 + u_s^2)}$$

$$= \frac{(9472 \times 10^3 + 2010 \times 10^3) - \sqrt{(9472 \times 10^3 + 2010 \times 10^3)^2 - \dfrac{4 \times 9472 \times 10^3 \times 2010 \times 10^3 \times 4233}{4233 + 49.8^2}}}{2 \times 4233/(4233 + 49.8^2)}$$

$$= 1845\,\text{kN}$$

which is closer to the torsional buckling load as the major axis buckling load has increased considerably. The minor axis flexural buckling load is now:

$N_v = 4 \times 613\,\text{kN} = 2452\,\text{kN} > 1845\,\text{kN}$

so the minor axis buckling load now therefore exceeds that for flexural–torsional buckling.

For flexural buckling, $\bar{\lambda} = 0.62$ and the reduction factor from curve b is $\chi = 0.83$.

For flexural–torsional buckling:

$$\bar{\lambda}_T = \sqrt{\frac{34.8 \times 10^2 \times 275}{1845 \times 10^3}} = 0.72$$

and the reduction factor from curve b is $\chi = \mathbf{0.76} < 0.83$.

This illustrates the dangers of using open sections with low warping resistance designed to be stocky against flexural buckling (i.e. short), as flexural–torsional buckling may govern in such cases. If the calculation is repeated for a $150 \times 150 \times 15$ angle (which meets the second shape limit criterion discussed above), the results are similar except that flexural–torsional buckling becomes critical at a shorter length.

6.3.1.5. Use of Class 3 section properties with stress limits

3-2/clause 6.3.1.5(1) permits Class 4 cross-sections to be treated as Class 3 sections in the above buckling checks, provided a reduced stress is used in the calculation in accordance with 3-1-5/clause 10. Determination of this reduced stress is discussed in section 6.2.2.6 of this guide. *3-2/clause 6.3.1.5(1)*

6.3.2. Uniform members in bending

6.3.2.1. Buckling resistance

The bending resistance of steel members can be reduced to a value lower than the cross-section resistance by lateral torsional buckling (LTB) or similar mechanisms essentially involving lateral buckling of the compression flange under the action of moment. Figure 6.3-9 shows lateral torsional buckling of a beam under uniform moment with torsional

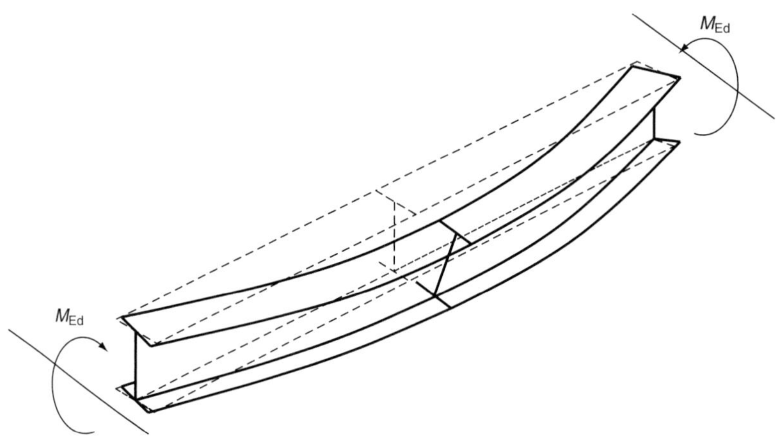

Fig. 6.3-9. Lateral torsional buckling of beam under end moments

rotation prevented at the ends but with the flanges allowed to rotate in plan, i.e. no warping restraint. (The lateral movement of the tension flange has been exaggerated here.) Both lateral and torsional movement of the beam can be observed at the centre of the beam. The tendency for lateral torsional buckling can therefore be reduced by bracing the compression flange against lateral movement or by torsional bracing to prevent rotation of the beam. Where beams are braced together in pairs to prevent LTB of individual beams, it is also necessary to consider the stability of the braced pair. This is particularly important for paired beams during construction prior to the addition of a decking system, but is rarely a problem once the decking system has been added.

For an initially straight beam with equal flanges and bisymmetric cross-section, the elastic critical moment to cause buckling into the above shape is conservatively given by:

$$M_{cr} = \frac{\pi^2 EI_z}{L^2}\left(\frac{I_w}{I_z} + \frac{L^2 GI_T}{\pi^2 EI_z}\right)^{0.5} \qquad \text{(D6.3-11)}$$

or written in another format:

$$M_{cr} = \left[\frac{\pi^2 EI_z}{L^2}\left(GI_T + \frac{\pi^2 EI_w}{L^2}\right)\right]^{0.5} \qquad \text{(D6.3-12)}$$

where:

I_w is the warping constant (formulae for certain sections are given in section 6.3.1.4 of this guide);
I_z is the minor axis second moment of area;
I_T is the St Venant torsional inertia; and
L is the length of the beam between points of restraint.

Equation (D6.3-12) contains terms relating to the transverse flexural inertia and the twisting stiffness (torsional and warping) as both lateral and torsional deformations occur in true lateral torsional buckling. The formulae ignore any pre-buckling deflections in the plane of bending. Where the stiffnesses EI_z and GI_T are comparable to or greater than the stiffness in the plane of bending, EI_y, equation (D6.3-12) becomes very conservative and does not predict, for example, the fact that circular hollow sections are stable against lateral torsional buckling. This is reflected in the wording of ***3-1-1/clause 6.3.2.1(2)***. In such circumstances, a more accurate equation is required, such as that found in Reference 24.

3-1-1/clause 6.3.2.1(2)

The load at which a beam buckles depends on a large number of factors including:

- section properties
- distribution of moment between restraints
- height of the loading above the shear centre

- support conditions (resistance is enhanced if warping restraint is also present in addition to full torsional restraint, but resistance is reduced if the torsional restraint is not rigid)
- stiffness and type of intermediate restraints.

The calculation becomes very much more complicated for monosymmetric and asymmetric beams.

Equation (D6.3-11) does not give the actual resistance moment of a 'real' beam, but it is a useful tool for calculating the real resistance. The slenderness of a beam in EN 1993 relates to its elastic critical buckling moment, M_{cr}, as discussed in the next section. True lateral torsional buckling is not very common in bridges because beams usually have either a deck slab, which offers continuous restraint to one flange as in composite construction, or have regularly spaced cross-girders carrying a decking system between beams (U-frame construction) which provides much stiffer support to one flange than the other. 'Lateral torsional buckling' is therefore often simplified to consider only buckling of the compression chord as a strut. This effectively ignores the torsional resistance of the section in equation (D6.3-11). This simplification is discussed in section 6.3.4.2 of this guide; it covers many of the real practical bridge cases. It also avoids the complexity of calculating M_{cr} as discussed here.

The design buckling resistance of a member is given in ***3-1-1/clause 6.3.2.1(3)***: *3-1-1/clause 6.3.2.1(3)*

$$M_{b,Rd} = \chi_{LT} W_y \frac{f_y}{\gamma_{M1}} \qquad \text{3-1-1/(6.55)}$$

where W_y is the plastic section modulus for members in Class 1 and 2, the elastic section modulus for members in Class 3 and the elastic effective section modulus for members in Class 4. χ_{LT} is the reduction factor for lateral torsional buckling.

6.3.2.2. *Lateral–torsional buckling curves – general case*

The form of the buckling resistance curves is the same as for flexural buckling. They have been produced by analogy with strut behaviour as discussed in section 6.3.1.2 of this guide and the slenderness in ***3-1-1/clause 6.3.2.2(1)*** is therefore taken as: *3-1-1/clause 6.3.2.2(1)*

$$\bar{\lambda}_{LT} = \sqrt{\frac{M_{Rk}}{M_{cr}}} = \sqrt{\frac{W_y f_y}{M_{cr}}}$$

by analogy with the slenderness for flexural buckling of:

$$\sqrt{\frac{N_{Rk}}{N_{cr}}} = \sqrt{\frac{A f_y}{N_{cr}}}$$

If the actual failure loads of a range of steel beams are plotted against the failure moments bounded by yield and elastic critical buckling, once again it can be seen that the actual failure moments at high slenderness values tend to the elastic critical moments but are significantly lower at low slenderness values (Fig. 6.3-10).

Once again, the difference between elastic critical and real behaviour is explained by the presence of imperfections as discussed in section 6.3.1.2 of this guide. However, for beams it is not easy to derive a simple criterion to allow for imperfections like the Perry–Robertson formula for struts so a criterion is made by analogy to that for struts. This leads to the following failure criterion which forms the basis of the EN 1993 design curves:

$$(M_{b,Rk} - M_{Rk})(M_{b,Rk} - M_{cr}) = \eta M_{cr} M_{b,Rk} \qquad \text{(D6.3-13)}$$

where:

$M_{b,Rk}$ is the characteristic buckling resistance for real beams;
M_{Rk} is the characteristic resistance of the beam cross-section ignoring buckling; and
η is an imperfection parameter which allows for similar imperfections to those discussed for struts.

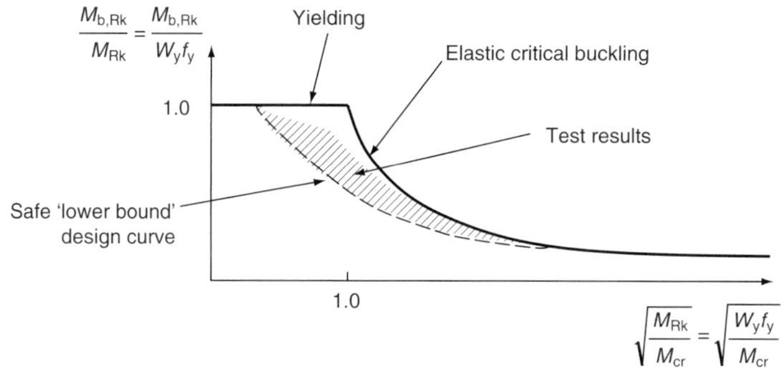

Fig. 6.3-10. Relationship between actual failure moment and elastic critical moment

In EN 1993, different curves are used for rolled and welded sections, as welding leads to significantly greater residual stresses. This is illustrated in Fig. 6.3-11.

The buckling curves are represented mathematically in 3-1-1/clause 6.3.2.2(1) as follows:

$$\chi_{LT} = \frac{1}{\Phi_{LT} + \sqrt{\Phi_{LT}^2 - \bar{\lambda}_{LT}^2}} \quad \text{but } \chi_{LT} \leq 1.0 \qquad \text{3-1-1/(6.56)}$$

where:

$\Phi_{LT} = 0.5[1 + \alpha_{LT}(\bar{\lambda}_{LT} - 0.2) + \bar{\lambda}_{LT}^2]$
α_{LT} is an imperfection factor from 3-1-1/Table 6.3 reproduced as Table 6.3-2 below; and
$\bar{\lambda}_{LT} = \sqrt{W_y f_y / M_{cr}}$, where W_y is either the elastic or plastic section modulus depending on the section classification.

3-1-1/clause 6.3.2.2(2)

For Class 4 cross-sections, the elastic section modulus is based on an effective section allowing for local plate buckling. ***3-1-1/clause 6.3.2.2(2)*** states that M_{cr} should always be based on gross cross-section properties. This applies even when a cross-section is Class 4 because the loss of strength due to local plate buckling is much more severe than the loss of stiffness it causes. It would therefore be too conservative to consider a reduction to M_{cr} in the slenderness calculation.

3-2/clause 6.3.2.2(4)

The buckling curves in 3-1-1/clause 6.3.2.2 have been conservatively taken to be the same as those for struts and therefore have a plateau length of 0.2 along the slenderness axis. For this reason, ***3-2/clause 6.3.2.2(4)*** permits lateral torsional buckling effects to be ignored where $\bar{\lambda}_{LT} \leq 0.2$. They may also be ignored where $M_{Ed}/M_{cr} \leq 0.04$ for the reasons discussed under the equivalent clause (3-1-1/clause 6.3.1.2(4)) for flexural buckling.

3-2/clause 6.3.2.3(1)

An alternative set of buckling curves is given in 3-1-1/clause 6.3.2.3, by way of ***3-2/clause 6.3.2.3(1)***, with a longer plateau length of 0.4 on the slenderness axis before a reduction for buckling occurs. These apply only to 'rolled sections or equivalent welded sections'. The reference to equivalent welded sections is intended to limit the use of the clause to members of the same size as available rolled sections. The drafters of EN 1993 considered there was insufficient test evidence available to support the use of a plateau length of 0.4

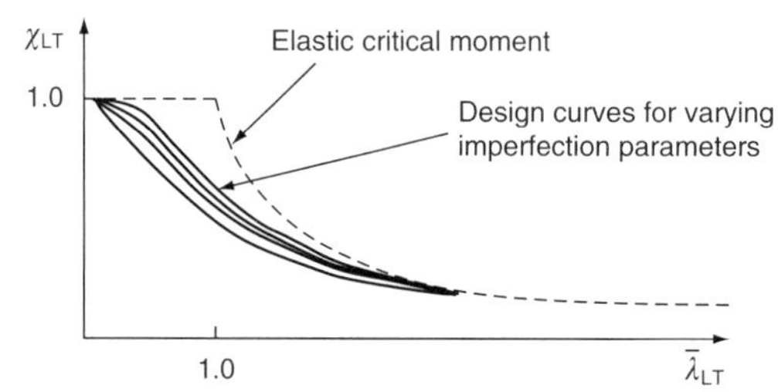

Fig. 6.3-11. Diagrammatic design curves for lateral torsional buckling resistance

Table 6.3-2. Imperfection factors for lateral torsional buckling

Buckling curve	*a*	*b*	*c*	*d*
Imperfection factor α_{LT}	0.21	0.34	0.49	0.76

for deeper members. However, a plateau length of 0.4 was used in previous UK practice to BS 5400: Part 3[4] for lengths of beam between rigid restraints, so the prohibition of the use of the longer plateau may lead to a loss of economy in some instances, or more closely spaced bracings.

3-1-1/clause 6.3.2.3(2)

Where 3-2/clause 6.3.2.3 is applied, ***3-1-1/clause 6.3.2.3(2)*** can be used to gain some additional benefit by way of the factor f. While this factor includes the shape of the moment diagram which is also included in the calculation of M_{cr}, it is not serving the same function and does not double-count the benefit. The peak benefit of the recommended expression for f occurs at a slenderness of 0.8, with the benefit reducing each side of this slenderness.

The main difficulty in the check of lateral torsional buckling according to this method is the determination of M_{cr}, as EN 1993 gives no formula for its calculation. Such a calculation becomes particularly complicated for monosymmetric or asymmetric beams. Previous UK codes have been based on the same theoretical buckling approach but with some simplifications made to reduce the complexity of the calculations. The next section discusses theoretical and computer-based calculations of M_{cr} while section 6.3.2.4 discusses a more empirical approach, based on the rules in BS 5400: Part 3.[4] A further alternative method is discussed in section 6.3.4.2 which covers most in-service cases for steel–concrete composite bridges.

6.3.2.3. Values of M_{cr} (additional sub-section)

Guidance on the use of 3-2/clause 6.3.2.3 is given in section 6.3.2.2 above. This sub-section focuses on calculation of M_{cr}.

Formulae for the elastic critical moment are not provided in EN 1993 so designers must find a way of determining this themselves. To do this, it is necessary either to refer to theoretical texts or to determine a value directly from an elastic finite-element model. It is becoming increasingly easy to calculate M_{cr} directly from a computer elastic critical buckling analysis, using a shell finite-element model, and many engineers will find this the quickest and most accurate method. Some experience is required, however, to determine M_{cr} from the output as often the first buckling mode observed does not correspond to the required global buckling mode; there may be many local plate buckling modes for the web and flanges before the first global mode is found.

The earlier ENV version of EN 1993-1-1[19] did provide formulae for M_{cr}, but agreement could not be reached on the values of accompanying coefficients and the majority of real bridge situations were not well covered. The complexity of calculating M_{cr} means it will often be preferable to use the simple compression chord model of 3-2/clause 6.3.4.2, described in section 6.3.4.2 of this guide. This is particularly applicable for U-frame bridges or steel and concrete composite bridges with a deck slab and with or without intermediate bracings in the span. Section 6.3.2 of this guide therefore only briefly discusses theoretical expressions for M_{cr}. A more empirical means of determining slenderness is discussed in section 6.3.2.4.

Bisymmetric sections

Bisymmetric sections, such as I-girders with equal flanges, are simplest to analyse. The elastic critical moment can be derived from the following formula:

$$M_{cr} = C_1 \frac{\pi^2 EI_z}{(kL)^2} \left\{ \left[\left(\frac{k}{k_w} \right)^2 \frac{I_w}{I_z} + \frac{(kL)^2 GI_T}{\pi^2 EI_z} + (C_2 z_g)^2 \right]^{0.5} - C_2 z_g \right\} \qquad \text{(D6.3-14)}$$

where the symbols have their definitions as in equation (D6.3-11) for the simple case of uniform bending together with the following additional definitions:

C_1 is a parameter that allows for the shape of the moment diagram between points of restraint;
C_2 is a parameter that allows for the destabilising or stabilising effects of loads applied to the beam between restraints;
z_g is the height of the load relative to the height of the shear centre with loads applied above the shear centre taken as positive;
k is an effective length factor with respect to minor axis buckling. For no restraint against rotation of the beam in plan, as is typical, $k = 1.0$. This assumes full torsional rotational restraint is provided;
k_w is an effective length factor with respect to warping of the beam at its ends. For no restraint against rotation of the beam in plan, as is typical, $k = 1.0$. This still assumes full torsional rotational restraint is provided.

The term C_2 can lead to an increase in resistance for stabilising load (applied below the shear centre) as well as destabilising load (applied above the shear centre). The additional complexity of trying to identify values of C_2 can be avoided by making an approximate modification to the effective length as was previous UK practice so that:

$$M_{cr} = C_1 \frac{\pi^2 EI_z}{(k_1 kL)^2} \left[\left(\frac{k}{k_w} \right)^2 \frac{I_w}{I_z} + \frac{(k_1 kL)^2 GI_T}{\pi^2 EI_z} \right]^{0.5} \tag{D6.3-15}$$

k_1 was taken as 1.2 for destabilising load or 1.0 otherwise, but this is not always conservative. However, as most real bridges do not have destabilising load, as the beams are either loaded below their shear centres (in half through bridges) or have a deck slab to prevent movement of the load, this approach is generally adequate.

The term C_1allows for the shape of the moment diagram. For bisymmetric flanges, for a given distribution of moments, reversing the sign of all the moments does not make any difference as both compression flanges have the same individual buckling resistance. Where the moment does not change sign and there is no restraint against rotation in plan at internal supports, C_1 is equivalent to m in section 6.3.4.2. However, where the moment does change sign between restraints, care must be taken with choosing a value of C_1. Where one flange is *not* continuously held by a deck, the equivalence of C_1 and m is lost as the values of m assume that the tension flange is restrained. It is then not always safe to use the value of m for $M_2 = 0$ (which is $m = 1.88$) when the moment at end two reverses as allowed in 3-2/clause 6.3.4.2 as illustrated in Fig. 6.3-12. This is because the opposite flange goes into compression and may become more critical. In case (c) of Fig. 6.3-12, the moment reversal leads to a length of top flange in compression that has a higher average compressive load than does the bottom flange in case (a). This is equivalent to saying that the greatest flange compressive load in the middle third is greater in case (c) than case (a). As a consequence, case (c) produces buckling at a lower value of M_1 than does case (a) and therefore C_1 is lower for case (c). In these cases, C_1 can either conservatively be taken as 1.0 or can be taken from text books. BS 5400: Part 3[4] contained values of

$$\eta = \frac{1}{\sqrt{C_1}}$$

for a wider variety of moment conditions and these could be used to obtain

$$C_1 = \frac{1}{\eta^2}$$

Values of η are reproduced in Fig. 6.3-13.

The equivalence with the simplified LTB model, which considers only buckling of the compression chord, can be shown by conservatively ignoring the torsional stiffness of

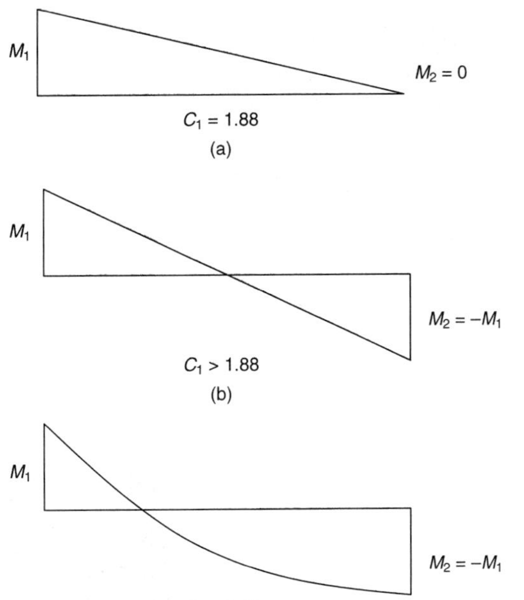

Fig. 6.3-12. Values of C_1 for beam with varying moment between restraints and with $k = 1.0$ and no continuous restraint to either flange (M_1 is hogging)

the beam in equation (D6.3-11) whereupon:

$$M_{cr} = \frac{\pi^2 EI_z}{L^2}\left[\frac{I_w}{I_z}\right]^{0.5} = \frac{\pi^2 EI_z}{L^2}\left[\frac{I_z d^2/4}{I_z}\right]^{0.5} = \frac{\pi^2 EI_z/2}{L^2} d \quad \text{(D6.3-16)}$$

Since $I_z/2$ is approximately the second moment of area of one flange about its stiff axis, the critical moment can be seen to be the flexural buckling load of one flange multiplied by the lever arm between flanges. The same analogy holds for any beams where there is an enforced centre of rotation at tension flange level, such as occurs with several beams connected by a composite deck slab.

Monosymmetric beams

The case of monosymmetric beams is very much more complicated as, in this case, it matters which way up the beam is when exposed to the given moment field. As a consequence, more parameters are necessary in addition to C_1 and C_2 to calculate the resistance. ENV 1993-1-1[19] gives the following formula:

$$M_{cr} = C_1 \frac{\pi^2 EI_z}{(kL)^2}\left\{\left[\left(\frac{k}{k_w}\right)^2 \frac{I_w}{I_z} + \frac{(kL)^2 GI_T}{\pi^2 EI_z} + (C_2 z_g - C_3 z_j)^2\right]^{0.5} - (C_2 z_g - C_3 z_j)\right\} \quad \text{(D6.3-17)}$$

where:

- C_3 is a parameter that accounts for the shape of the bending moment in conjunction with z_j;
- z_j is a measure of the asymmetry of the cross-section. It is zero for bisymmetric sections and positive where the compression flange with greatest second moment of area is in compression at the point of maximum moment. This reflects the intuitive fact that asymmetric beams are most stable when bent such that the larger flange is in compression.

Values of the various parameters can be obtained by reference to ENV 1993-1-1,[19] but the designer will find that the cases presented generally are inadequate for bridge design. There is also no general agreement over the appropriateness of the values given.

Cantilevers

Determination of the relevant parameters for M_{cr} for cantilever situations is difficult and is not attempted here. The value of M_{cr} is very sensitive to the location of load application and the restraints to the beam at the position of load, at cantilever tip and at the cantilever root. It illustrates that either some more pragmatic rules are required, as discussed in the next section, or a computer elastic critical buckling analysis is needed.

6.3.2.4. Determination of slenderness without explicit calculation of M_{cr}

Previous UK practice has been to determine slenderness using an effective length approach, analogous to that for flexural buckling. The effects of shape of moment diagram and beam asymmetry discussed above in section 6.3.2.3 are dealt with by factors within the basic expression for slenderness. In BS 5400: Part 3: 2000,[4] the slenderness, with a few minor changes to symbols to suit EN 1993 notation, is defined as follows:

$$\lambda_{LT} = \sqrt{\frac{\pi^2 E}{f_y}\frac{M_{pl,Rk}}{M_{cr}}} \qquad \text{(D6.3-18)}$$

where $M_{pl,Rk}$ is the characteristic plastic moment resistance of the section and f_y is the characteristic yield strength of the compression flange. This differs from the slenderness definition in EN 1993 where:

$$\bar{\lambda}_{LT} = \sqrt{\frac{W_y f_y}{M_{cr}}}$$

with W_y as either the elastic or plastic section modulus depending on the section classification. Adjusting for this different definition gives the following for the slenderness of a Class 1 or 2 section to EN 1993:

$$\bar{\lambda}_{LT} = \lambda_{LT}\sqrt{\frac{f_y}{\pi^2 E}} \qquad \text{(D6.3-19)}$$

For a beam with Class 3 or 4 cross-section, the slenderness in EN 1993 is:

$$\bar{\lambda}_{LT} = \lambda_{LT}\sqrt{\frac{f_y}{\pi^2 E}\frac{M_{el,Rk}}{M_{pl,Rk}}} \qquad \text{(D6.3-20)}$$

where $M_{el,Rk}$ is the characteristic elastic resistance moment of the section.

Uniform I, channel, tee and angle sections

A simplified version of equation (D6.3-18) is given in BS 5400: Part 3[4] for the case of uniform I, channel, tee and angle sections bending about the y–y axis and this can be used in conjunction with equations (D6.3-19) and (D6.3-20) to determine a slenderness to EN 1993:

$$\lambda_{LT} = \frac{l_e}{i_z} k_4 \eta \nu \qquad \text{(D6.3-21)}$$

where

l_e is the effective length. For lengths of girder between rigid restraints to the compression flange, the effective length is taken as the distance between restraints. etailed methods of calculation for l_e are provided in BS 5400: Part 3[4] for other situations, including cases where there is no plan bracing provided;

i_z is the radius of gyration of the gross cross-section of the beam about its z–z axis;

k_4 = 0.9 for rolled I or channel section beams or any I-section symmetrical about both axes with t_f not greater than twice the web thickness, or = 1.0 for all other beams;

η = 1.0, but where the bending moment varies substantially within the half-wavelength of buckling of the compression flange, advantage may be obtained by using η in Fig. 6.3-13, which has been reproduced from BS 5400:

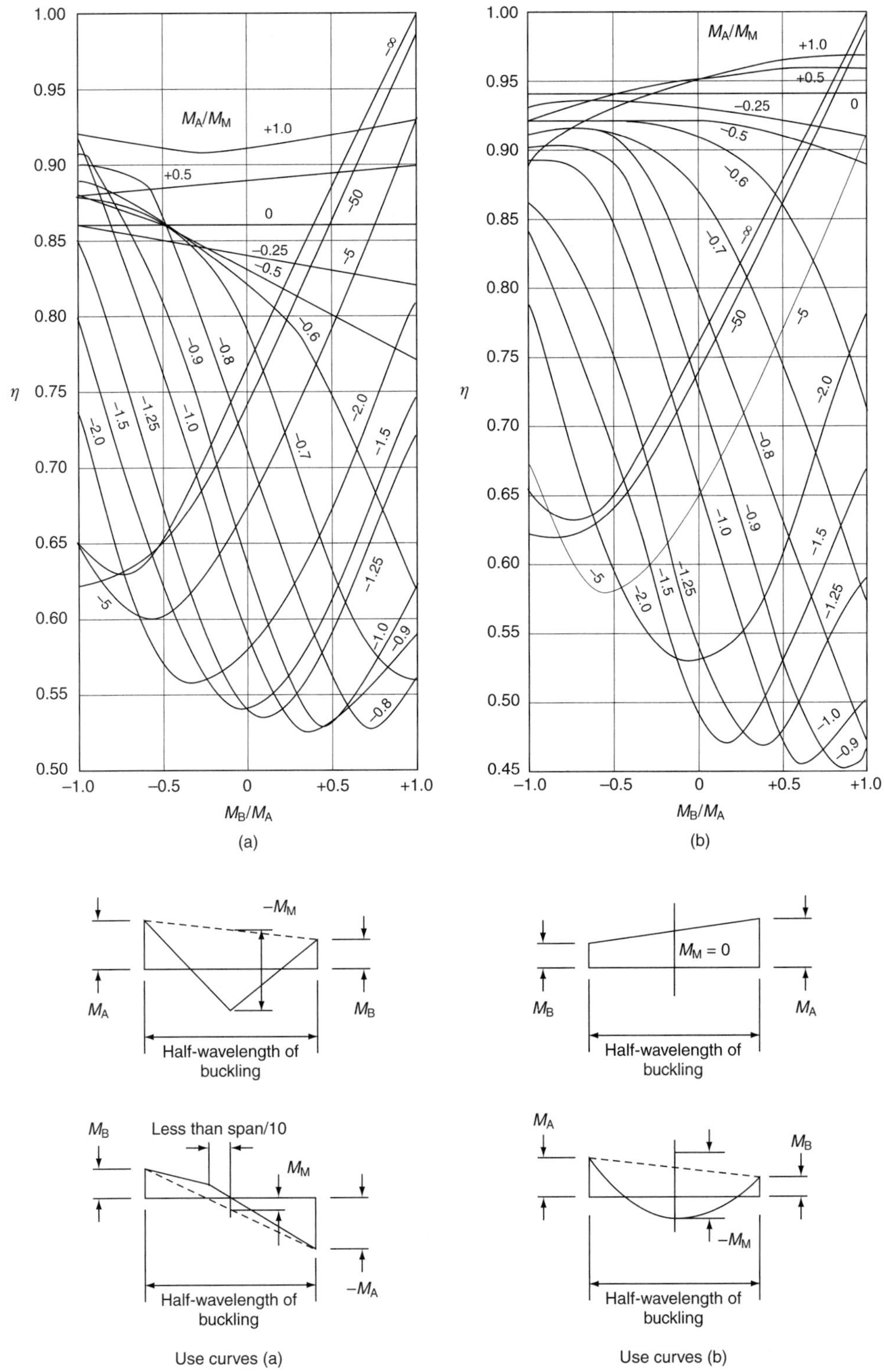

Fig. 6.3-13. Slenderness factor η for bending moment variation: (a) applied loading substantially concentrated within the middle fifth of the half-wavelength of buckling; (b) applied loading other than for (a)

Part 3/Fig. 10.[4] In using Fig. 6.3-13, hogging moments are positive and the ends A and B should be chosen so that $M_A \geq M_B$ regardless of sign;

ν is dependent on the shape of the beam, and may be obtained from Table 6.3-3, which has been reproduced from BS 5400: Part 3/Table 9,[4] using the parameters:

$$\lambda_F = \frac{l_e}{i_z}\left(\frac{t_f}{D}\right) \quad \text{and} \quad i = \frac{I_c}{I_c + I_t}$$

D is the depth of the cross-section;

t_f is the mean thickness of the two flanges of an I or channel section, or the mean thickness of the table of a tee or leg of an angle section;

I_c and I_t are the second moments of area of the compression and tension flange, respectively, about their z–z axes, at the section being checked. For beams with $I_c \geq I_t$ or with $\lambda_F \geq 8$, λ_{LT} may conservatively be taken as l_e/i_z.

When using Table 6.3-3, intermediate values to the right of the stepped line should be determined from the following formula, rather than from linear interpolation:

$$v = \{[4i(1-i) + 0.05\lambda_F^2 + \psi_i^2]^{0.5} + \psi_i\}^{-0.5}$$

with $\psi_i = 2i - 1$ when $I_c < I_t$ and $\psi_i = 0.8(2i - 1)$ when $I_c \geq I_t$.

This method can be applied to composite bridges also as an alternative to the 'continuous inverted U-frame' model of EN 1994-2 by conservatively ignoring the rotational restraint provided by the transverse continuity of the deck slab across the beams. Where a flange is common to two or more (n numbers) beams, the properties i_z, I_c or I_t may be calculated by attributing a fraction $1/n$ of the lateral second moment of area and of the area of the common flange to the section of each beam. In calculating t_f, I_c and I_t for composite beams, the equivalent thickness of the composite flange in compression should be based on the appropriate modular ratio. Concrete in tension should be ignored and the equivalent thickness of tension reinforcement should be taken as the area of reinforcement divided by the flange width over which it is placed. Methods of checking LTB for composite bridges are given in the *Designers' Guide to EN 1994-2*.[7]

Table 6.3-3. Slenderness factor ν for beams of uniform section

0	1.0	0.8	0.6	0.5	0.4	0.3	0.2	0.1	0
λ_F									
0.0	0.791	0.842	0.932	1.000	1.119	1.291	1.582	2.237	∞
1.0	0.784	0.834	0.922	0.988	1.102	1.266	1.535	2.110	6.364
2.0	0.764	0.813	0.895	0.956	1.057	1.200	1.421	1.840	3.237
3.0	0.737	0.784	0.859	0.912	0.998	1.116	1.287	1.573	2.214
4.0	0.708	0.752	0.818	0.864	0.936	1.031	1.162	1.359	1.711
5.0	0.679	0.719	0.778	0.817	0.878	0.954	1.055	1.196	1.415
6.0	0.651	0.688	0.740	0.774	0.824	0.887	0.966	1.071	1.219
7.0	0.626	0.660	0.705	0.734	0.777	0.829	0.892	0.973	1.080
8.0	0.602	0.633	0.674	0.699	0.736	0.779	0.831	0.895	0.977
9.0	0.581	0.609	0.645	0.668	0.699	0.786	0.780	0.832	0.896
10.0	0.562	0.587	0.620	0.639	0.667	0.699	0.736	0.779	0.831
11.0	0.544	0.567	0.597	0.614	0.639	0.666	0.698	0.735	0.778
12.0	0.528	0.549	0.576	0.591	0.613	0.638	0.665	0.697	0.733
13.0	0.512	0.533	0.557	0.571	0.590	0.612	0.636	0.664	0.695
14.0	0.499	0.517	0.539	0.552	0.570	0.589	0.611	0.635	0.662
15.0	0.486	0.503	0.523	0.535	0.551	0.568	0.588	0.609	0.633
16.0	0.474	0.490	0.509	0.519	0.534	0.550	0.567	0.586	0.607
17.0	0.463	0.478	0.495	0.505	0.518	0.533	0.548	0.566	0.585
18.0	0.452	0.466	0.482	0.492	0.504	0.517	0.531	0.547	0.564
19.0	0.442	0.456	0.471	0.479	0.491	0.503	0.516	0.530	0.546
20.0	0.433	0.446	0.460	0.468	0.478	0.489	0.502	0.515	0.529

Equation (D6.3-21) was not intended to be used for U-frame-type calculations where the intermediate restraints are not rigid enough to restrict the effective length to the distance between restraints. In this case, the method in section 6.3.4.2 of this guide is more appropriate. (It can be used for cases with rigid braces also.)

If equation (D6.3-21) is substituted into equation (D6.3-19), the following is obtained for the slenderness of a Class 1 or 2 cross-section to EN 1993:

$$\bar{\lambda}_{LT} = \lambda_{LT}\sqrt{\frac{f_y}{\pi^2 E}} = \frac{l_e}{i_z}k_4\eta\nu\sqrt{\frac{f_y}{\pi^2 E}} \quad \text{(D6.3-22)}$$

Similarly, for a beam with Class 3 or 4 cross-section, the slenderness in EN 1993 is:

$$\bar{\lambda}_{LT} = \lambda_{LT}\sqrt{\frac{f_y}{\pi^2 E}\frac{M_{el,Rk}}{M_{pl,Rk}}} = \frac{l_e}{i_z}k_4\eta\nu\sqrt{\frac{f_y}{\pi^2 E}\frac{M_{el,Rk}}{M_{pl,Rk}}} \quad \text{(D6.3-23)}$$

In both these equations, the symbols have their meanings defined above. An alternative to having two formats for slenderness depending on section classification is to define an equivalent elastic critical moment for use in expression 3-1-1/(6.56) as follows:

$$M_{cr} = \frac{M_{pl,Rk}\pi^2 E i_z^2}{l_e^2 k_4^2 \eta^2 \nu^2 f_y} \quad \text{(D6.3-24)}$$

The disadvantage of this presentation is that the real elastic critical is independent of any plastic properties.

The use of these equations is not discussed further here. The purpose of this section is merely to show how the slenderness in BS 5400: Part 3: 2000[4] can be converted into the Eurocode format. BS 5400: Part 3 gives extensive guidance on effective length calculation which allows most typical bridge situations to be covered fairly simply, including the temporary erection condition where there may be only torsional bracing and no deck or plan bracing system. Alternative methods of analysis for lateral torsional buckling are discussed in section 6.3.4.2 of this guide and increasingly designers will find the quickest and most economical way of checking buckling is with a computer elastic critical buckling analysis.

Uniform box sections

A similar conversion between BS 5400: Part 3 and EN 1993 slenderness definitions can be performed but this is not discussed further here.

6.3.3. Uniform members in bending and axial compression

3-2/clause 6.3.3 and 3-1-1/clause 6.3.3 provide rules for checking member stability under combinations of moments and axial force. The rules referenced in EN 1993-1-1 are only intended for use in checking bending and compression in uniform bisymmetric sections (***3-1-1/clause 6.3.3(1)***), so are somewhat limited in their application in bridge design. The general rules in 3-1-1/clause 6.3.4 can, however, be used for non-bisymmetric sections. Alternatively, it is possible to avoid a buckling interaction check if a second-order analysis has been used which considers all the relevant global and local imperfections and possible modes of buckling as discussed in section 5.2 of this guide. *3-1-1/clause 6.3.3(1)*

For steel and concrete composite beams, the simpler methods presented in the *Designers' Guide to EN 1994-2*[7] can be used. The methods therein can also be applied to all-steel bridges where one flange is continuously braced.

This section of the guide is split into two sub-sections as follows:

- Interaction in EN 1993-1-1 — *Section 6.3.3.1*
- Simplified interaction in 3-2/clause 6.3.3(1) for uniaxial bending — *Section 6.3.3.2*

6.3.3.1. Interaction in EN 1993-1-1

The simplest case to consider is axial force and bending where buckling is restricted to occurring in the plane of bending only. A reminder of the axes convention in EN 1993 is

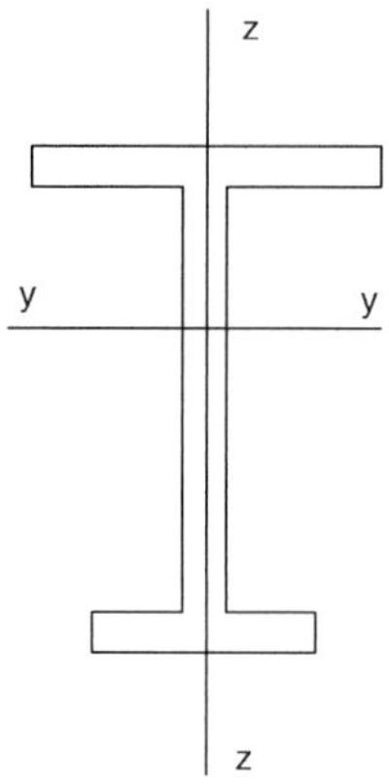

Fig. 6.3-14. Axes convention for I-beams

given in Fig. 6.3-14. Under axial load, additional moments are generated from the growth of initial imperfections, a_0, as discussed in section 5.2. For imperfections leading to bending about the y–y axis, the moment is given by:

$$M_{\mathrm{imp}} = N_{\mathrm{Ed}}\left[\frac{a_0}{1-(N_{\mathrm{Ed}}/N_{\mathrm{cr,y}})}\right] \qquad \text{(D6.3-25)}$$

This additional moment is included in the resistance formulae for flexural buckling so does not have to be included in the code interaction check. However, the applied moments $M^{\mathrm{I}}_{\mathrm{y,Ed}}$ are also magnified by the axial force, giving a second-order moment as follows:

$$M^{\mathrm{II}}_{\mathrm{y,Ed}} = M^{\mathrm{I}}_{\mathrm{y,Ed}}\left[\frac{1}{1-(N_{\mathrm{Ed}}/N_{\mathrm{cr,y}})}\right] \qquad \text{(D6.3-26)}$$

This increase in the applied in-plane moments due to second-order effects is not included in the resistance formulae for either axial force or bending and therefore needs to be included in the interaction between bending and axial force. If the interaction is performed on the basis of summing stresses, the following is obtained:

$$\left(\frac{N_{\mathrm{Ed}}}{Af_{\mathrm{yd}}}+\frac{M_{\mathrm{imp}}}{W_{\mathrm{el}}f_{\mathrm{yd}}}\right)+\frac{M^{\mathrm{II}}_{\mathrm{y,Ed}}}{W_{\mathrm{el,y}}f_{\mathrm{yd}}} \le 1.0 \qquad \text{(D6.3-27)}$$

where:

$$f_{\mathrm{yd}} = \frac{f_{\mathrm{y}}}{\gamma_{\mathrm{M1}}}$$

and $W_{\mathrm{el,y}}$ is the section modulus for the fibre considered. Since the effects of imperfections are included in the resistance formulae for flexural buckling as discussed above, equation (D6.3-27) can be re-expressed as a simple linear interaction:

$$\frac{N_{\mathrm{Ed}}}{\chi_{\mathrm{y}}N_{\mathrm{pl,Rd}}}+\frac{M^{\mathrm{II}}_{\mathrm{y,Ed}}}{M_{\mathrm{y,Rd}}} \le 1.0 \qquad \text{(D6.3-28)}$$

with

$$N_{\mathrm{pl,Rd}} = \frac{Af_{\mathrm{y}}}{\gamma_{\mathrm{M1}}} \text{ and } M_{\mathrm{y,Rd}} = \frac{W_{\mathrm{el,y}}f_{\mathrm{y}}}{\gamma_{\mathrm{M1}}}$$

This is similar to the simple interaction for cross-section design given in 3-1-1/clause 6.2.1, but the effect of moments from imperfections is included in the first term by way of the buckling reduction factor χ_{y}. It conservatively assumes that the peak applied moment and the peak second-order moment from imperfections and deflections both coexist at the same cross-section. Introducing equation (D6.3-26), equation (D6.3-28) then becomes:

$$\frac{N_{\mathrm{Ed}}}{\chi_{\mathrm{y}}N_{\mathrm{pl,Rd}}}+\frac{1}{1-(N_{\mathrm{Ed}}/N_{\mathrm{cr,y}})}\frac{M^{\mathrm{I}}_{\mathrm{y,Ed}}}{M_{\mathrm{y,Rd}}} \le 1.0 \qquad \text{(D6.3-29)}$$

Equation (D6.3-28) is not identical to equation (D6.3-27). The first term of equation (D6.3-27) will be smaller than that in equation (D6.3-28) since the maximum fibre stress produced under axial force does not increase linearly with the axial force because of the non-linear magnification of the moments from imperfections in equation (D6.3-25). This means that the ratio $N_{\mathrm{Ed}}/(\chi_{\mathrm{y}} N_{\mathrm{pl,Rd}})$ does not give an actual measure of the ratio of extreme fibre stress to yield stress under a given axial load, unless the reduction factor χ_{y} approaches 1.0 and the imperfections do not have any significant effect on axial resistance. This makes the interaction of equation (D6.3-29) conservative.

Equation (D6.3-29) may become more conservative where the applied bending moment is not uniform throughout the effective length and the peak applied moment does not occur at the same location as the peak moment from the second-order effects as discussed in section 5.2. To overcome the latter conservatism, a factor can be applied to the maximum moment to account for the distribution of moments, and this is done in the EN 1993-1-1 interaction equations discussed below.

3-1-1/clause 6.3.3(4)

In ***3-1-1/clause 6.3.3(4)***, two equations are presented for checking the interaction of bending and axial force for members prone to buckling. The first equation corresponds to the interaction discussed above:

$$\frac{N_{\mathrm{Ed}}}{\dfrac{\chi_{\mathrm{y}} N_{\mathrm{Rk}}}{\gamma_{\mathrm{M1}}}} + k_{\mathrm{yy}} \frac{M_{\mathrm{y,Ed}} + \Delta M_{\mathrm{y,Ed}}}{\chi_{\mathrm{LT}} \dfrac{M_{\mathrm{y,Rk}}}{\gamma_{\mathrm{M1}}}} + k_{\mathrm{yz}} \frac{M_{\mathrm{z,Ed}} + \Delta M_{\mathrm{z,Ed}}}{\dfrac{M_{\mathrm{z,Rk}}}{\gamma_{\mathrm{M1}}}} \leq 1.0 \qquad \text{3-1-1/(6.61)}$$

Expression 3-1-1/(6.61) introduces the possibility of biaxial bending and also the additional moment from the axial force due to the shift in neutral axis position for Class 4 sections. Considering first only uniaxial bending about the y–y axis, k_{yy} deals with, among other things, the amplification of moments by the axial load, the shape of the moment diagram and the ratio of elastic to plastic section resistance for Class 1 and 2 cross-sections. Two informative annexes are provided in EN 1993-1-1 (Annexes A and B) to determine values of k_{yy}. These are not reproduced here. In 3-1-1/Annex A, k_{yy} deals with the following:

- Amplification of the applied moment about the y–y axis by the factor:

  $$\frac{1}{1 - (N_{\mathrm{Ed}}/N_{\mathrm{cr,y}})}$$

  as discussed above.
- Magnification of the lateral and torsional displacements involved in lateral torsional buckling under axial force by an analogous factor:

  $$\frac{1}{\sqrt{\left(1 - \dfrac{N_{\mathrm{Ed}}}{N_{\mathrm{cr,z}}}\right)\left(1 - \dfrac{N_{\mathrm{Ed}}}{N_{\mathrm{cr,T}}}\right)}}$$

  within the term C_{mLT}. This term can be taken as unity if the slenderness for lateral torsional buckling is zero, but this will rarely occur in practice as continuous restraint would be required. It would not be unreasonable to modify this criterion so that $C_{\mathrm{mLT}} = 1.0$ if $\bar{\lambda}_{\mathrm{LT}} \leq 0.2$.
- Shape of the applied first-order moment diagram by way of the parameters C_{my} and C_{mLT}. Previous UK practice has been to use the maximum moment within the middle third of the buckling length to avoid the need for equivalent moment factors in the interaction. C_{my} is determined from 3-1-1/Table A.2 and relates to the shape of the bending moment about the y–y axis (M_{y}) between restraints preventing flexural buckling about the y–y axis (i.e. preventing movement in the z direction). For bridge beams where M_{y} causes bending in a vertical plane, the relevant length between restraints will typically be equal to the span length.

  $$C_{\mathrm{mLT}} = C_{\mathrm{my}}^2 \frac{a_{\mathrm{LT}}}{\sqrt{\left(1 - \dfrac{N_{\mathrm{Ed}}}{N_{\mathrm{cr,z}}}\right)\left(1 - \dfrac{N_{\mathrm{Ed}}}{N_{\mathrm{cr,T}}}\right)}}$$

relates to the magnification of lateral and torsional displacements discussed above and contains the term C_{my} again. In calculating C_{mLT}, C_{my} should this time be based on the M_y moment shape between restraints preventing movement in the y direction.

- The ratio of elastic to plastic section resistance for Class 1 and 2 cross-sections.
- The term

$$\mu_y = \frac{1 - \dfrac{N_{Ed}}{N_{cr,y}}}{1 - \chi_y \dfrac{N_{Ed}}{N_{cr,y}}}$$

is an adjustment to the basic magnifier

$$\frac{1}{1 - (N_{Ed}/N_{cr,y})}$$

in equation (D6.3-29) to account for the problem identified above that the ratio $N_{Ed}/\chi_y N_{pl,Rd}$ usually overestimates the real ratio of extreme fibre stress to yield stress under axial loading alone. This overestimation increases with increasing slenderness and μ_y addresses this by introducing the reduction factor χ_y such that μ_y reduces as the reduction factor reduces.

Where there is biaxial bending, k_{yz} deals with a similar magnification of moment about the z–z axis by the axial load, together with the shape of the moment diagram between restraints, but includes no magnification for any torsional displacements as the beam is not susceptible to lateral torsional buckling when bent about the minor axis. The shape of the moment diagram between points braced in the y direction is used when calculating C_{mz}.

The approach in 3-1-1/Annex B is slightly different and simpler to use, although the intention is similar. In Annex B, k_{yy} depends on the member slenderness for flexural buckling about the y–y axis, the relative axial force according to $N_{Ed}/(\chi_y N_{pl,Rd})$ and the shape of the moment diagram between restraints to flexural buckling about the y–y axis. k_{yz} is similar but depends on the equivalent parameters for flexural buckling about the z–z axis. When calculating the equivalent moment factors the following apply:

- C_{my} relates to the shape of the M_y moment between points braced in the z direction;
- C_{mz} relates to the shape of the M_z moment between points braced in the y direction;
- C_{mLT} relates to the shape of the M_y moment between points braced in the z direction.

Expression 3-1-1/(6.61) considers flexural buckling about the major axis and the magnification of the major axis moment by the axial load. It is also however necessary to consider flexural buckling about the minor axis and magnification of any minor axis moment present by the axial load. To do this, EN 1993-1-1 introduces expression 3-1-1/(6.62):

$$\frac{N_{Ed}}{\dfrac{\chi_z N_{Rk}}{\gamma_{M1}}} + k_{zy}\frac{M_{y,Ed} + \Delta M_{y,Ed}}{\chi_{LT}\dfrac{M_{y,Rk}}{\gamma_{M1}}} + k_{zz}\frac{M_{z,Ed} + \Delta M_{z,Ed}}{\dfrac{M_{z,Rk}}{\gamma_{M1}}} \le 1.0 \qquad \text{3-1-1/(6.62)}$$

Considering again only uniaxial bending about the y–y axis, in 3-1-1/Annex A, k_{zy} deals with the following:

- Amplification of the applied moment about the y–y axis by the factor

$$\frac{1}{1 - (N_{Ed}/N_{cr,y})}$$

- Magnification of the lateral and torsional displacements involved in lateral torsional buckling under axial force by an analogous factor

$$\frac{1}{\sqrt{\left(1 - \dfrac{N_{Ed}}{N_{cr,z}}\right)\left(1 - \dfrac{N_{Ed}}{N_{cr,T}}\right)}}$$

- Shape of the applied first-order moment diagram as discussed above for k_{yy}.
- The ratio of elastic to plastic section resistance for Class 1 and 2 cross-sections.
- The term

$$\mu_z = \frac{1 - \dfrac{N_{Ed}}{N_{cr,z}}}{1 - \chi_z \dfrac{N_{Ed}}{N_{cr,z}}}$$

performs a similar function to μ_y above.

Where there is biaxial bending, k_{zz} deals with a similar magnification of moment about the z–z axis by the axial force, together with the shape of the moment diagram between restraints but includes no magnification for any torsional displacements as the beam is not susceptible to lateral torsional buckling when bent about the minor axis.

The approach in 3-1-1/Annex B is again slightly different and simpler to use. k_{zy} is 80% of k_{yy} where the beam is 'not susceptible to torsional deformations'. 'Susceptibility' to torsional deformations is not defined. It would be reasonable to consider the beam as not susceptible if $\overline{\lambda} \leq 0.2$ in all torsional modes (i.e. lateral torsional buckling under moment and flexural–torsional or torsional buckling under axial load). The simpler alternative is to treat the beam as being susceptible to torsional deformation and to use 3-1-1/Table B.2. In this case, k_{zy} depends on the member slenderness for flexural buckling about the z–z axis, the relative axial force according to $N_{Ed}/(\chi_z N_{pl,Rd})$ and the shape of the moment diagram between points of lateral restraint. k_{zz} similarly depends on the member slenderness for flexural buckling about the z–z axis, the relative axial force and the shape of the moment.

The various k interaction parameters can be greater than 1.0 which differs from previous UK practice where a linear interaction has been used. For small axial force (compared to the elastic buckling force), the parameters are likely to be less than or equal to 1.0. Figure 6.3-15 shows how the shape of the interaction between axial force and moment can change from convex to concave. Worked Example 6.3-3 illustrates numerically how these interaction parameters can exceed 1.0, although the magnitude of these parameters is somewhat exaggerated by the large axial force chosen.

For most beams, the axial force will be relatively small. Following the rule of 3-2/clause 5.2.1(4), second-order effects from axial force may be neglected if $N_{cr}/N_{Ed} \geq 10$. Therefore, providing the lowest elastic critical buckling load under axial force (see section 6.3.1 of this guide) is at least ten times the applied axial force, the magnification by axial force of the moment terms in the interactions of expression 3-1-1/(6.61) and expression 3-1-1/(6.62) could be ignored, i.e. k_{ij} taken as 1.0. If the moments vary considerably between points of restraint, it would be conservative to take k_{ij} as 1.0 in conjunction with

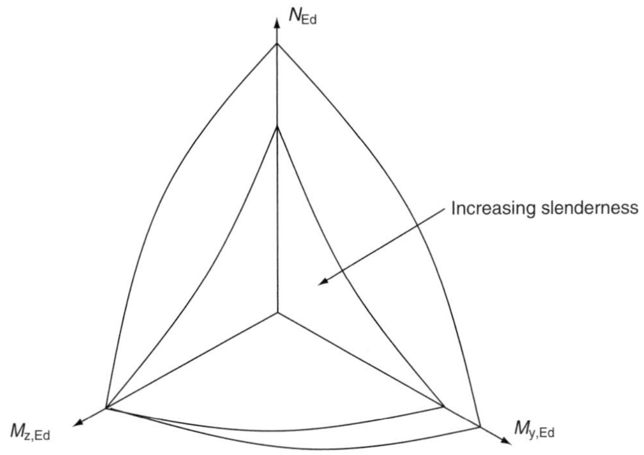

Fig. 6.3-15. Typical shape of interaction diagrams for axial force and moment according to EN 1993-1-1

using the maximum moment values. In this case, k_{ij} could be taken as 1.0 and the moments based on their maximum values within the middle third of the member between restraints, as in previous UK practice. Expression 3-1-1/(6.61) and expression 3-1-1/(6.62) can then be condensed into one equation. The axial force term in this interaction should then be taken as:

$$\frac{N_{Ed}}{\chi N_{Rk}/\gamma_{M1}}$$

where χ is the lowest reduction factor for buckling under axial force from 3-2/clause 6.3.1.

From the limited trial calculations undertaken by the first author, it appears that the interaction parameters in 3-1-1/Annex B generally give the most economic design. Whichever method is chosen, it is likely to require the use of a spreadsheet due to the length of the calculation as illustrated by the length of Worked Example 6.3-3.

6.3.3.2. Simplified interaction in 3-2/clause 6.3.3(1) for uniaxial bending

3-2/clause 6.3.3(1)

A simplified alternative to expression 3-1-1/(6.61) is given in ***3-2/clause 6.3.3(1)*** for the case of uniaxial bending only and flexural buckling in the plane of bending (i.e. no LTB) as follows:

$$\frac{N_{Ed}}{\dfrac{\chi_y N_{Rk}}{\gamma_{M1}}} + C_{mi,0}\frac{M_{y,Ed} + \Delta M_{y,Ed}}{\dfrac{M_{y,Rk}}{\gamma_{M1}}} \leq 0.9 \qquad \text{3-2/(6.9)}$$

where $C_{mi,0}$ accounts for the shape of the moment diagram and is taken as $C_{my,0}$ from 3-1-1/ Table A.2. Expression 3-2/(6.9) does not apply if lateral torsional buckling is possible without some modification, including notably adding χ_{LT} in the denominator. The form of expression 3-2/(6.9) follows simply from the discussions above. If the column slenderness

$$\lambda = \sqrt{\frac{N_{pl,Rk}}{N_{cr,y}}}$$

is introduced, equation (D6.3-26) can be rewritten as:

$$M^{II}_{y,Ed} = M^{I}_{y,Ed}\left(\frac{1}{1 - \dfrac{N_{Ed}}{\chi_y N_{pl,Rd}}\chi_y\bar{\lambda}^2\dfrac{1}{\gamma_{M1}}}\right)$$

If this is substituted into equation (D6.3-28) and rearranged then the following is obtained:

$$\frac{N_{Ed}}{\chi_y N_{pl,Rd}} + \frac{M^{I}_{y,Ed}}{M_{y,Rd}} \leq 1 - \frac{N_{Ed}}{\chi_y N_{pl,Rd}}\left(1 - \frac{N_{Ed}}{\chi_y N_{pl,Rd}}\right)\chi_y\bar{\lambda}^2\frac{1}{\gamma_{M1}}$$

$$\leq 1 - 0.25(\max) \times \chi_y\bar{\lambda}^2\frac{1}{\gamma_{M1}}$$

Noting that from expression 3-1-1/(6.49):

$$\chi \rightarrow \frac{1}{\bar{\lambda}^2} \text{ as } \bar{\lambda} \rightarrow \infty \text{ so } \chi\bar{\lambda}^2 \rightarrow 1.0$$

the minimum value of the above is:

$$1 - 0.25(\max) \times 1.0(\max) \times \frac{1}{\gamma_{M1}} = \frac{0.75}{\gamma_{M1}} = 0.77$$

and hence:

$$\frac{N_{Ed}}{\chi_y N_{pl,Rd}} + \frac{M^{I}_{y,Ed}}{M_{y,Rd}} \leq 0.77$$

which compares with

$$\frac{N_{Ed}}{\chi_y N_{pl,Rd}} + \frac{M^{I}_{y,Ed}}{M_{y,Rd}} \le 0.90$$

in expression 3-2/(6.9) above. The use of '0.9' mitigates the conservatism of the use of the term $N_{Ed}/(\chi_y N_{pl,Rd})$ in the interaction, as discussed above under equation (D6.3-29).

Worked Example 6.3-3: Bending and axial force in a universal beam

A bridge comprises paired simply supported 914 × 305(201) universal beams with span of 30 m. Each beam is subjected to a moment of 1500 kNm at mid-span (varying parabolically to zero at beam ends) and an axial force of 2000 kN. The beams are rigidly braced together transversely at 3 m centres by cross-bracing. Plan bracing is provided to the top flange maintaining the 3 m bay length and the deck is non-composite. The steel is S355 with the yield stress for different thicknesses taken from 3-1-1/Table 3.1 (noting that the UK National Annex requires the values from EN 10025 to be used). The interaction parameters required for use in the interactions of expressions 3-1-1/(6.61) and (6.62) are determined according to 3-1-1/Annexes A and B. The cross-sections are to be designed elastically.

The section properties of the universal beam are taken from section tables as follows:

$A = 2.56 \times 10^4\ \text{mm}^2$

$I_y = 3.26 \times 10^9\ \text{mm}^4$

$W_{el,y} = 7.21 \times 10^6\ \text{mm}^3$

$I_z = 9.43 \times 10^7\ \text{mm}^4$

$I_T = 2.93 \times 10^6\ \text{mm}^4$

$I_w = 18.4 \times 10^{12}\ \text{mm}^6$ (see section 6.3.1.4 of this guide for a calculation method)

$$i_g = \sqrt{\frac{(I_y + I_z)}{A}} = \sqrt{\frac{(3.26 \times 10^9 + 9.43 \times 10^7)}{2.56 \times 10^4}} = 362\ \text{mm}$$

$N_{Rk} = 2.56 \times 10^4 \times 355 = 9088\ \text{kN}$

$M_{y,Rk} = 7.21 \times 10^6 \times 355 = 2559\ \text{kNm}$

Interaction parameters from 3-1-1/Annex A

$N_{cr,T} = (GI_T + \pi^2 EI_w/L_x^2)/i_g^2$ (see section 6.3.1.4 of the guide)

$= (81 \times 10^3 \times 2.93 \times 10^6 + \pi^2 \times 210 \times 10^3 \times 18.4 \times 10^{12}/3000^2)/362^2 = 34\,146\ \text{kN}$

$N_{cr,z} = \pi^2 EI_z/L_z^2 = \pi^2 \times 210 \times 10^3 \times 9.43 \times 10^7/3000^2 = 21\,716\ \text{kN}$

From expression 3-1-1/(6.50):

$$\bar{\lambda}_z = \sqrt{\frac{Af_y}{N_{cr}}} = \sqrt{\frac{2.56 \times 10^4 \times 355}{21\,716 \times 10^3}} = 0.65$$

The reduction factor for minor axis flexural buckling from curve b of 3-1-1/Fig. 6.4 is $\chi_z = 0.81$.

$N_{cr,y} = \pi^2 EI_y/L_y^2 = \pi^2 \times 210 \times 10^3 \times 3.26 \times 10^9/30\,000^2 = 7507\ \text{kN}$

From expression 3-1-1/(6.50):

$$\bar{\lambda}_y = \sqrt{\frac{Af_y}{N_{cr}}} = \sqrt{\frac{2.56 \times 10^4 \times 355}{7507 \times 10^3}} = 1.10$$

The reduction factor for major axis flexural buckling from curve a of 3-1-1/Fig. 6.4 is $\chi_y = 0.59$.

From 3-1-1/Table A.1:

$$k_{yy} = C_{my} C_{mLT} \frac{\mu_y}{1 - \dfrac{N_{Ed}}{N_{cr,y}}}$$

$$\mu_y = \frac{1 - \dfrac{N_{Ed}}{N_{cr,y}}}{1 - \chi_y \dfrac{N_{Ed}}{N_{cr,y}}} = \frac{1 - \dfrac{2000}{7507}}{1 - 0.59 \times \dfrac{2000}{7507}} = 0.87$$

$$\varepsilon_y = \frac{M_{y,Ed}}{N_{Ed}} \frac{A}{W_{el,y}} = \frac{1500 \times 10^6}{2000 \times 10^3} \frac{2.56 \times 10^4}{7.21 \times 10^6} = 2.66$$

$a_{LT} = 1 - I_T/I_y = 1.00$ by inspection

Conservatively and for simplicity the moment is considered here to be uniform throughout the span. Actually the moment diagram is parabolic over the span and close to uniform between transverse restraints. The assumption of uniform moment allows the same value of C_{my} to be used in the calculation of both k_{yy} and k_{zy} as discussed in section 6.3.3.1 above. From 3-1-1/Table A.2:

$$C_{my,0} = 0.79 + 0.21\psi + 0.36(\psi - 0.33)\frac{N_{Ed}}{N_{cr,y}} = 0.79 + 0.21 + 0.36(1 - 0.33)\frac{2000}{7507}$$

$$= 1.06$$

$$C_{my} = C_{my,0} + (1 - C_{my,0})\frac{\sqrt{\varepsilon_y} a_{LT}}{1 + \sqrt{\varepsilon_y} a_{LT}} = 1.06 + (1 - 1.06)\frac{\sqrt{2.66} \times 1.0}{1 + \sqrt{2.66} \times 1.0} = 1.02$$

$$C_{mLT} = C_{my}^2 \frac{a_{LT}}{\sqrt{\left(1 - \dfrac{N_{Ed}}{N_{cr,z}}\right)\left(1 - \dfrac{N_{Ed}}{N_{cr,T}}\right)}} = 1.02^2 \times \frac{1.00}{\sqrt{\left(1 - \dfrac{2000}{21\,716}\right)\left(1 - \dfrac{2000}{34\,146}\right)}}$$

$$= 1.125$$

$$k_{yy} = C_{my} C_{mLT} \frac{\mu_y}{1 - \dfrac{N_{Ed}}{N_{cr,y}}} = 1.02 \times 1.13 \times \frac{0.87}{1 - \dfrac{2000}{7507}} = \mathbf{1.37}$$

$$\mu_z = \frac{1 - \dfrac{N_{Ed}}{N_{cr,z}}}{1 - \chi_z \dfrac{N_{Ed}}{N_{cr,z}}} = \frac{1 - \dfrac{2000}{21\,716}}{1 - 0.81 \times \dfrac{2000}{21\,716}} = 0.98$$

$$k_{zy} = C_{my} C_{mLT} \frac{\mu_z}{1 - \dfrac{N_{Ed}}{N_{cr,y}}} = 1.02 \times 1.13 \times \frac{0.98}{1 - \dfrac{2000}{7507}} = \mathbf{1.54}$$

Interaction parameters from 3-1-1/Annex B

Once again, conservatively consider uniform moment throughout the span. From 3-1-1/Table B.3, $C_{my} = 1.0$ (cf. 0.95 for parabolic distribution).

$$k_{yy} = C_{my}\left(1 + 0.6\bar{\lambda}_y \frac{N_{Ed}}{\chi_y N_{Rk}/\gamma_{M1}}\right) = 1.0\left(1 + 0.6 \times 1.1 \times \frac{2000}{0.59 \times 9088/1.1}\right) = 1.27$$

but not greater than:

$$k_{yy} = C_{my}\left(1 + 0.6\frac{N_{Ed}}{\chi_y N_{Rk}/\gamma_{M1}}\right) = 1.0\left(1 + 0.6 \times \frac{2000}{0.59 \times 9088/1.1}\right) = \mathbf{1.25}$$

$$k_{zy} = 1 - \frac{0.05\bar{\lambda}_z}{(C_{mLT} - 0.25)}\frac{N_{Ed}}{\chi_z N_{Rk}/\gamma_{M1}} = 1 - \frac{0.05 \times 0.65}{(1.0 - 0.25)} \times \frac{2000}{0.81 \times 9088/1.1} = \mathbf{0.99}$$

but not less than:

$$k_{zy} = 1 - \frac{0.05}{(C_{mLT} - 0.25)}\frac{N_{Ed}}{\chi_z N_{Rk}/\gamma_{M1}} = 1 - \frac{0.05}{(1.0 - 0.25)} \times \frac{2000}{0.81 \times 9088/1.1} = 0.98$$

The interaction parameters from Annex B are both smaller than those in Annex A in this instance.

6.3.4. General method for lateral and lateral torsional buckling of structural components

6.3.4.1. General method

The rules presented in 3-1-1/clause 6.3.3 are only intended to be used to check bending and compression in uniform bisymmetric sections so are somewhat limited in their application, although they can be adapted for non-bisymmetric situations. ***3-1-1/clause 6.3.4(1)*** gives a general method of evaluating the combined effect of axial force and bending (applied in the plane of the structure only) without performing an interaction. The method is valid for asymmetric and non-uniform members or for entire plane frames. In principle, this method is more realistic since the structure or member, in reality, buckles in a single mode with a single 'system slenderness'. Interaction formulae assume separate modes under each individual action effect. These each have different slendernesses that have subsequently to be combined to give an overall verification. The disadvantage of the general method is that software capable of elastic critical buckling analysis and second-order analysis is required. Additionally, shell elements will need to be used to determine elastic critical modes resulting from applied bending. *3-1-1/clause 6.3.4(1)*

An alternative simplified method, which will be applicable in many bridge cases, is to consider out-of-plane buckling by treating the compression chord of a beam as a strut. This method, together with its limitations, is discussed in section 6.3.4.2 of this guide. A further alternative is to use second-order analysis with imperfections to cover both in-plane and out-of-plane buckling effects as discussed in sections 5.2 and 5.3 of this guide.

The basic verification in ***3-1-1/clause 6.3.4.1(2)*** is performed by determining a single slenderness for out-of-plane buckling from ***3-1-1/clause 6.3.4.1(3)***, which can include combined lateral and lateral torsional buckling. This slenderness is a slenderness for the whole system and applies to all members included within it. It takes the usual Eurocode form as follows: *3-1-1/clause 6.3.4.1(2)* *3-1-1/clause 6.3.4.1(3)*

$$\lambda_{op} = \sqrt{\frac{\alpha_{ult,k}}{\alpha_{cr,op}}} \qquad \text{3-1-1/(6.64)}$$

where:

$\alpha_{ult,k}$ is the minimum load factor applied to the design loads required to reach the characteristic resistance of the most critical cross-section ignoring out-of-plane buckling but including moments from second-order effects and imperfections in plane;

$\alpha_{cr,op}$ is the minimum load factor applied to the design loads required to give elastic critical buckling in an out-of-plane mode, ignoring in-plane buckling.

The first stage of calculation requires an analysis to be performed to determine $\alpha_{ult,k}$ ignoring any out-of-plane buckling effects but considering in-plane slenderness effects (using second-order analysis if necessary) and imperfections. These can increase the moments which give rise to out-of-plane buckling effects. They must therefore be included in the analysis because second-order effects and imperfections for in-plane behaviour are not otherwise included in the resistance formula used in this method. If the structure is not prone to significant in-plane second-order effects as discussed in section 5.2 of this guide, then first-order analysis may be used.

Each cross-section is then verified using the interactions in section 6.2 of EN 1993-1-1, but using characteristic resistances. The loads are all increased by a factor $\alpha_{\text{ult,k}}$ until the characteristic resistance is reached. The simplest verification is given in expression 3-1-1/(6.2) as:

$$\frac{N_{\text{Ed}}}{N_{\text{Rk}}} + \frac{M_{\text{y,Ed}}}{M_{\text{y,Rk}}} \leq 1.0 \tag{D6.3-30}$$

where N_{Rk} and $M_{\text{y,Rk}}$ include allowance for any reduction necessary due to shear and torsion if separate cross-sectional checks are to be avoided. N_{Ed} and $M_{\text{y,Ed}}$ are the axial forces and moments at a cross-section resulting from the design loads. If first-order analysis is allowable, the critical load factor is then determined from:

$$\alpha_{\text{ult,k}}\left(\frac{N_{\text{Ed}}}{N_{\text{Rk}}} + \frac{M_{\text{y,Ed}}}{M_{\text{y,Rk}}}\right) = 1.0 \tag{D6.3-31}$$

If second-order analysis is necessary, the imposed loads would have had to be increased progressively until one cross-section reaches cross-section failure according to equation (D6.3-30). This is necessary as the system is no longer linear, and results from one analysis cannot simply be factored up when the imposed load is increased. (As an alternative to second-order analysis, $\alpha_{\text{ult,k}}$ could be determined from first-order analysis with a subsequent interaction performed using 3-1-1/clause 6.3.3 rather than equation (D6.3-30) but ignoring out-of-plane buckling).

For a symmetrical I-beam with a Class 1 or 2 cross-section, an alternative cross-sectional check might be that from expression 3-1-1/(6.36) thus:

$$\frac{M_{\text{y,Ed}}}{M_{\text{pl,y,Rk}}\left(1 - \dfrac{N_{\text{Ed}}}{N_{\text{Rk}}}\right)/(1 - 0.5a)} \leq 1.0 \tag{3-1-1/(6.36)}$$

where a depends on the cross-section shape. This leads to the corresponding expression for the critical load factor if first-order analysis is used:

$$\frac{\alpha_{\text{ult,k}} M_{\text{y,Ed}}}{M_{\text{pl,y,Rk}}\left(1 - \dfrac{\alpha_{\text{ult,k}} N_{\text{Ed}}}{N_{\text{Rk}}}\right)/(1 - 0.5a)} = 1.0 \tag{D6.3-32}$$

As an alternative to using cross-section checks to 3-1-1/clause 6.2, global elastic finite-element analysis could be used to determine the load amplifier directly, based on the Von Mises yield criterion. This would be conservative.

The second stage is to determine the lowest load factor $\alpha_{\text{cr,op}}$ to reach elastic critical buckling in an out-of-plane mode but ignoring in-plane buckling modes. This will typically require a finite-element model with shell elements to adequately predict lateral torsional buckling behaviour. If this load factor can only be determined separately for axial forces $\alpha_{\text{cr,N}}$ and bending moments $\alpha_{\text{cr,M}}$, as might be the case if standard text book solutions are used, the overall load factor could be determined from a simple interaction such as:

$$\frac{1}{\alpha_{\text{cr,op}}} = \frac{1}{\alpha_{\text{cr,N}}} + \frac{1}{\alpha_{\text{cr,M}}} \tag{D6.3-33}$$

An overall slenderness is then calculated according to expression 3-1-1/(6.64) for the entire system. This slenderness refers only to out-of-plane effects as discussed above because in-plane effects are separately included in the determination of action effects. A reduction factor χ_{op} for this slenderness must then be determined. This reduction factor depends on whether the mode of buckling is predominantly flexural or lateral torsional as the reduction factor curves can sometimes differ. The simplest solution is to take the lowest reduction factor for either out-of-plane flexural buckling or lateral torsional buckling from 3-1-1/clause 6.3.1 or 6.3.2 respectively. This reduction factor is then applied to the cross-section check performed in stage 1, but this time using design values of the material properties. If the cross-section was verified using the simple interaction in equation (D6.3-30), then the verification taking lateral and lateral torsional buckling into account is given by ***3-1-1/clause 6.3.4(4)a)***: *3-1-1/clause 6.3.4(4)a)*

$$\frac{N_{\text{Ed}}}{N_{\text{Rk}}/\gamma_{\text{M1}}} + \frac{M_{\text{y,Ed}}}{M_{\text{y,Rk}}/\gamma_{\text{M1}}} \leq \chi_{\text{op}} \qquad \text{3-1-1/(6.65)}$$

This follows from the general verification provided in ***3-1-1/clause 6.3.4(2)***, which is written independently of the method of cross-section verification as: *3-1-1/clause 6.3.4(2)*

$$\frac{\chi_{\text{op}}\alpha_{\text{ult,k}}}{\gamma_{\text{M1}}} \geq 1.0 \qquad \text{3-1-1/(6.63)}$$

Alternatively, separate reduction factors can be determined for each effect separately using the same slenderness, so that for axial force the reduction factor is χ and for moment it is χ_{LT}. These are then applied to the section capacities in the cross-section resistance. If the cross-section was verified using the simple interaction in equation (D6.3-30), then the verification taking lateral and lateral torsional buckling into account is given by ***3-1-1/clause 6.3.4(4)b)***: *3-1-1/clause 6.3.4(4)b)*

$$\frac{N_{\text{Ed}}}{\chi N_{\text{Rk}}/\gamma_{\text{M1}}} + \frac{M_{\text{y,Ed}}}{\chi_{\text{LT}} M_{\text{y,Rk}}/\gamma_{\text{M1}}} \leq 1.0 \qquad \text{3-1-1/(6.66)}$$

If the cross-sectional resistance was checked directly using finite-element modelling, expression 3-1-1/(6.63) can be used together with the assumption that χ_{op} takes the minimum value for either flexural or lateral torsional buckling.

It should be noted that this procedure can be conservative where the element governing the cross-section check is not itself significantly affected by the out-of-plane deformations.

The method is illustrated by the following qualitative example.

Worked Example 6.3-4: Plane frame

A plane frame with fabricated I-girder cross-sections is loaded with a uniform load, W, on the horizontal member. The columns are built in at the base but no other transverse restraint is provided. The resistance of the frame for strength and stability is checked using 3-1-1/clause 6.3.4.

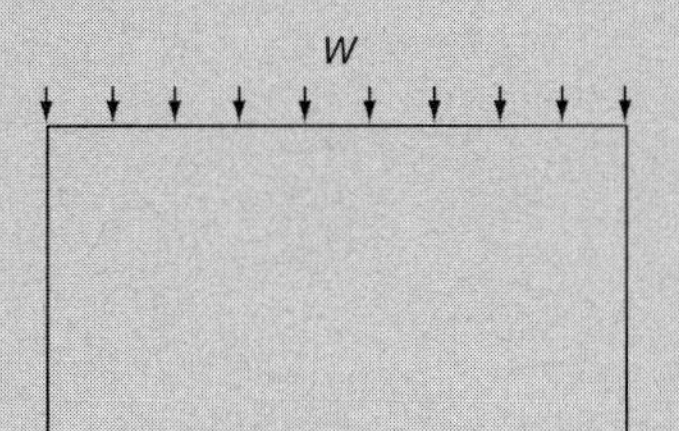

Fig. 6.3-16. Plane frame analysis for determining $\alpha_{\text{ult,k}}$

Step 1: A plane frame model is set up as in Fig. 6.3-16. First-order analysis is used here as this structure is stocky for in-plane effects. Moments $M_{\text{y,Ed},i}$ and axial forces $N_{\text{Ed},i}$ are obtained under the design loads. No out-of-plane imperfections are considered. All cross-sections are checked against their characteristic resistances, for example using equation (D6.3-30), and the most critical section (mid-span here) is determined. The load factor $\alpha_{\text{ult,k}}$ is then determined from equation (D6.3-31) as this system is linear. In this case, $\alpha_{\text{ult,k}} = 1.90$. If second-order analysis had been necessary, the load would have had to be increased progressively to $\alpha_{ult,k} W$ until one cross-section reached its cross-section resistance.

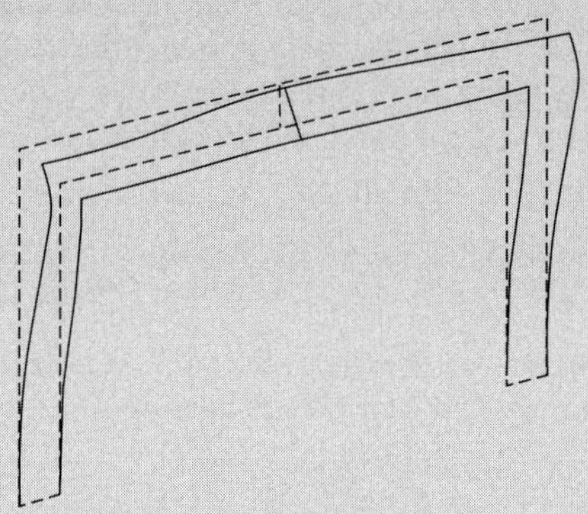

Fig. 6.3-17. Finite-element analysis for determining $\alpha_{cr,LT}$

Step 2: A finite-element model of the frame is set up using shell elements to adequately represent out-of-plane behaviour, including flexural, torsional and distortional deformations. (This model could also have been used for step 1.) Elastic critical buckling analysis gives the combined flexural and lateral torsional buckling mode for out-of-plane buckling as shown in Fig. 6.3-17. The load factor on design loads to give this buckling mode = $\alpha_{cr,op} = 3.50$.

Step 3: The slenderness is computed from expression 3-1-1/(6.64) as:

$$\lambda_{op} = \sqrt{\frac{\alpha_{ult,k}}{\alpha_{cr,op}}} = \sqrt{\frac{1.90}{3.50}} = 0.74$$

Step 4: For flexural buckling, 'curve c' of 3-1-1/Fig. 6.4 applies and for lateral torsional buckling 'curve d' applies. For simplicity, the lowest buckling curve can be used, so from 'curve d' $\chi_{op} = \chi_{LT} = 0.61$. From expression 3-1-1/(6.63):

$$\frac{\chi_{op}\alpha_{ult,k}}{\gamma_{M1}} = \frac{0.61 \times 1.90}{1.1} = 1.05 \geq 1.0$$

so the frame is just adequate. Alternatively, for a slightly less conservative answer, the verification could be done according to expression 3-1-1/(6.66) which is here consistent with the derivation of $\alpha_{ult,k}$.

6.3.4.2. Simplified method

This section covers the simplified method of 3-2/clause 6.3.4.2. It is split into the following additional sub-sections:

- Eigenvalue analysis — *Section 6.3.4.2.1*
- Lengths of beam with U-frames or other intermediate flexible restraints — *Section 6.3.4.2.2*
- Short lengths of beam between rigid bracings — *Section 6.3.4.2.3*
- Rigidity of bracings — *Section 6.3.4.2.4*
- Beams without plan bracing or decking during construction — *Section 6.3.4.2.5*
- Strength of bracings and U-frames — *Section 6.3.4.2.6*

6.3.4.2.1. Eigenvalue analysis

3-2/clause 6.3.4.2(2) The simplified method of ***3-2/clause 6.3.4.2(2)*** is intended for use for beams where one flange is held in position laterally. The method is based on representing lateral torsional buckling (actually lateral distortional buckling since one flange is assumed to be held in position)

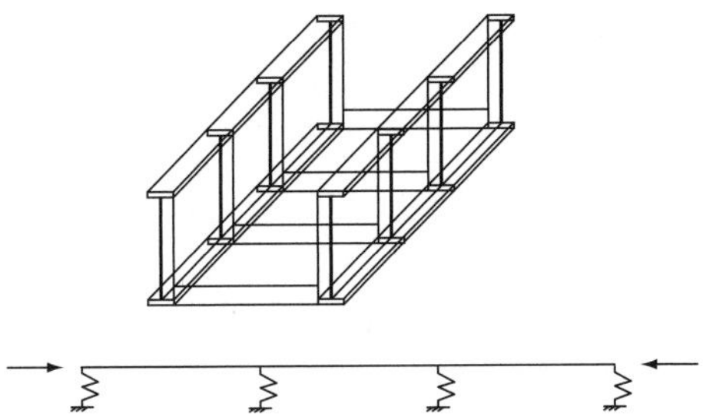

Fig. 6.3-18. Compression chord model for flange stabilised by discrete U-frames

by lateral buckling of the compression flange. All subsequent discussions refer to beam flanges but are equally applicable to truss chords. The method is primarily intended for U-frame-type bridges but can be used for other flexible bracing types as well. It also applies to lengths of girder compression flange between rigid restraints, as found in hogging zones in steel and concrete composite construction – see section 6.3.4.2.3 below. Greater detail is given for its use in composite beams in the *Designers' Guide to EN 1994-2*,[7] including consideration of interaction with axial force. In 3-2/clause 6.3.4.2, the torsional inertia of the beam is ignored. This simplification may become significantly conservative for shallow rolled steel sections but is generally not significant for most fabricated bridge girders.

3-2/clause 6.3.4.2(3)

3-2/clause 6.3.4.2(3) allows the slenderness for lateral buckling to be determined from an elastic critical buckling analysis of the compression chord. The flange (with an attached portion of web in the compression zone) is modelled as a strut with area A_{eff}, supported by springs in the lateral direction representing restraint from bracings (including discrete U-frames) and from any continuous U-frame action. Buckling in the vertical direction is assumed to be prevented by the web in this model, but checks on flange-induced buckling according to 3-1-5/clause 8 should be made to confirm this assumption. Bracings can be flexible, as is the case of bracing by discrete U-frames (in conjunction with plan bracing or a deck slab at the level of the cross-member), or can be rigid, as is likely to be the case for cross-bracing (again in conjunction with plan bracing or a deck slab). Other types of bracing, such as channel bracing mid-height between beams together with plan bracing or deck slab, may be rigid or flexible depending on their stiffness as discussed below.

A typical model for a beam with discrete flexible U-frames is shown in Fig. 6.3-18. Plan bracing provided by the decking is not shown. If smeared springs are used to model the stiffness of discrete restraints such as discrete U-frames, the buckling load should not be taken as larger than that corresponding to the Euler load of a strut between discrete bracings. If computer analysis is used, there would be no particular reason to use smeared springs for discrete restraints. This approximation is generally only made when a hand-calculation approach is used based on beam on elastic foundations theory. This approach is used to derive the equations in this section of the code.

3-2/clause 6.3.4.2(4)

Elastic critical buckling analysis may be performed to calculate the critical buckling load, N_{crit}. The slenderness is then given in ***3-2/clause 6.3.4.2(4)***:

$$\bar{\lambda}_{\text{LT}} = \sqrt{\frac{A_{\text{eff}} f_{\text{y}}}{N_{\text{crit}}}} \qquad \text{3-2/(6.10)}$$

3-2/clause 6.3.4.2(7)

where $A_{\text{eff}} = A_{\text{f}} + A_{\text{wc}}/3$ from ***3-2/clause 6.3.4.2(7)*** as illustrated in Fig. 6.3-19. This approximate definition of A_{eff} (greater than the flange area) is necessary to ensure that the critical stress produced for the strut is the same as that required to produce buckling in the beam under bending moment.

Spring stiffnesses for U-frames may be calculated using 3-2/Table D.3 from 3-2/Annex D, where values of stiffness, C_{d}, can be calculated. A typical case covering trusses with vertical posts and cross-girders or plate girders with stiffeners and cross-girders is shown

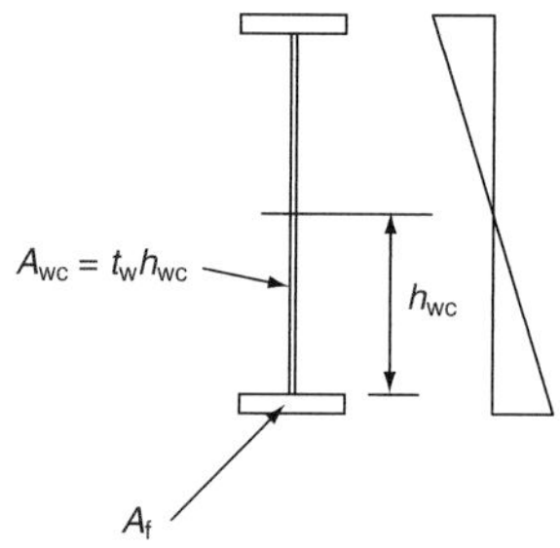

Fig. 6.3-19. Definitions for effective compression zone

in Fig. 6.3-20. The stiffness for this case (under the unit applied forces shown) is:

$$C_d = \frac{EI_v}{\dfrac{h_v^3}{3} + \dfrac{h^2 b_q I_v}{2I_q}} \qquad \text{(D6.3-34)}$$

This case also covers inverted U-frames, such as in steel and concrete composite bridges when the cross-member stiffness is based on the cracked inertia of the deck slab and reinforcement or the cracked composite section of a discrete composite cross-girder. The formula can also be used to derive a stiffness for an unstiffened web acting as the vertical member in a continuous U-frame. Generally, however, inclusion of this small restraint stiffness will have little effect in increasing the buckling resistance, unless the distance between rigid restraints is large, and will necessitate an additional check of the web for the U-frame moments induced – see Worked Example 6.3-5. For multiple girders, the restraint to internal girders may be derived by replacing $2I_q$ by $3I_q$ in the expression for C_d. Section properties for stiffeners should be derived using an attached width of web plate in accordance with 3-1-5/Fig. 9.1 (stiffener width plus $30\varepsilon t_w$).

The above formula makes no allowance for flexibility of joints. Joint flexibility can significantly reduce the effectiveness of U-frames. If the joint was 'semi-continuous' according to 3-1-8/clause 5.2.2, the effect of joint rotational flexibility, S_j, would have to be determined from 3-1-8/clause 6.3 and included in the calculation of C_d. This would typically apply to connections made through unstiffened end plates. BS 5400: Part 3[4] included some generic values of S_j as follows:

(a) 0.5×10^{-10} rad/N mm when the cross-member is bolted or riveted through unstiffened end-plates or cleats;
(b) 0.2×10^{-10} rad/N mm when the cross-member is bolted or riveted through stiffened end plates;
(c) 0.1×10^{-10} rad/N mm when the cross-member is welded right round its cross-section or the connection is by bolting or riveting between stiffened end-plates on the cross-member and a stiffened part of the vertical. This connection flexibility could usually be ignored.

The above values are generally quite conservative as they were derived from studies of shallow members. Rotational stiffness increases with member depth.

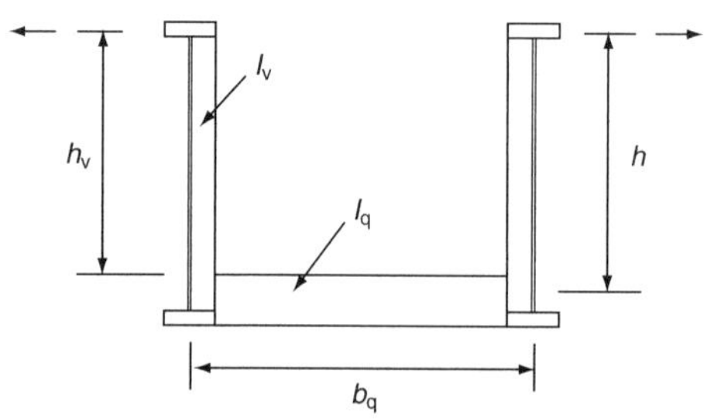

Fig. 6.3-20. Definitions of properties needed to calculate C_d

The stiffness of other restraints, such as a channel section placed between main beams at mid-height, can be derived from a plane frame model of the bracing system. For braced pairs of beams or multiple beams with a common system, it will generally be necessary to consider unit forces applied to the compression flanges such that the displacement of the flange is maximized. For a paired U-frame, the maximum displacement occurs with forces in opposite directions, as in Fig. 6.3-20, but this will not always be the case. For paired beams braced by a horizontal mid-height channel, forces in the same direction will often give greater flange displacement.

6.3.4.2.2. Lengths of beam with U-frames or other intermediate flexible restraints

The above analytical method is useful where, for example, the flange section changes or there is a reversal of the sign of the axial stress in the length of the flange being considered. In other simpler cases (such as in simply supported half through construction or bottom flanges of continuous girders between braces at internal supports), the formulae provided in 3-2/clauses 6.3.4.2(6) and (7) are applicable.

The formula for N_{crit} in ***3-2/clause 6.3.4.2(6)*** is derived from an elastic buckling analysis with continuous springs. From elastic theory (as set out in, for example, Reference 24), the critical load for buckling of such a strut is: *3-2/clause 6.3.4.2(6)*

$$N_{\text{crit}} = n^2\frac{\pi^2 EI}{L^2} + \frac{cL^2}{n^2\pi^2} \qquad \text{(D6.3-35)}$$

where:

- I is the transverse second moment of area of the effective flange and web;
- L is the length between 'rigid' braces;
- c is the stiffness of the restraints smeared per metre; and
- n is the number of half waves in the buckled shape.

By differentiation, this is a minimum when $n^4 = cL^4/(\pi^4 EI)$ which gives:

$$N_{\text{crit}} = 2\sqrt{cEI} \qquad \text{(D6.3-36)}$$

The expression given in 3-2/clause 6.3.4.2(6) is as follows:

$$N_{\text{crit}} = mN_{\text{E}} \qquad \text{3-2/(6.12)}$$

where $N_{\text{E}} = \pi^2 EI/L^2$, $m = 2/\pi^2\sqrt{\gamma} \geq 1.0$, $\gamma = cL^4/EI$ and $c = C_{\text{d}}/l$ with C_{d} equal to the restraint stiffness and l equal to the distance between restraints. When these terms are substituted into expression 3-2/(6.12), the same result as equation (D6.3-36) is produced. As discussed below however, the value of n in equation (D6.3-35) should not be taken less than 1.0; values exceeding 1.0 would imply a buckled length longer than the length between rigid restraints.

Expressions 3-2/(6.10) and 3-2/(6.12) form the basis of the assessment of conventional U-frame bridges, such as half through construction, but they assume that end restraints at supports are 'rigid'. The definition of 'rigid' is discussed in section 6.3.4.2.4 below. Restraints such as cross-bracings will almost certainly be rigid but end U-frames of U-frame decks almost certainly will not be. In this latter case with non-rigid frames, an approach modified from that in BS 5400: Part 3[4] could be used by replacing m in expression 3-2/(6.12) by the following:

$$m = \frac{\sqrt{\gamma}}{\left(\dfrac{\pi}{\sqrt{2}} + \dfrac{0.69}{X + 0.5}\right)^2} \qquad \text{(D6.3-37)}$$

where:

$$X = \frac{C_{\text{e}}}{\sqrt{2}}\left(\frac{l^3}{C_{\text{d}}^3 EI}\right)^{0.25}$$

and C_e is the stiffness of the end support, determined in the same way as the stiffness of intermediate supports, C_d.

Where present, a flexible end U-frame will, however, only reduce the buckling load to the above modified value in the end half wave of buckling. The buckling effective length is given in ***3-2/clause 6.3.4.2(5)*** by:

3-2/clause 6.3.4.2(5)

$$l_k = \pi\sqrt{\frac{EI}{N_{crit}}} \tag{D6.3-38}$$

The buckling effective length will reduce and buckling load will increase with distance from the flexible end U-frame, where the flexibility of the end U-frame has little influence. If the beam is long enough for multiple half-wavelengths to occur, the above buckling load will therefore be overly conservative away from the beam ends. It can be shown that the effective length varies parabolically from the reduced value at the beam end to the value assuming rigid ends over a distance equal to 2.5 times the effective length calculated with rigid ends. Consequently an improved buckling load may be used for the beam away from the end half-wavelengths of buckling, which can be useful in checking mid-span sections of simply supported beams.

6.3.4.2.3. Short lengths of beam between rigid bracings

It should be noted from above that the buckling load according to expression 3-2/(6.12) is independent of the length between rigid restraints. It is possible that for small spring stiffnesses, this value of N_{crit} could correspond to a wavelength greater than L and might therefore be lower than the Euler load over length L. It follows that N_{crit} should not be taken as less than the Euler load over length L, and n in equation (D6.3-35) should not be taken less than 1.0. In this case, the buckling load should be taken as:

$$N_{crit} = \frac{\pi^2 EI}{L^2} + \frac{cL^2}{\pi^2} \tag{D6.3-39}$$

3-2/clause 6.3.4.2(7)

This is the basis of the first of the two equations given under expression 3-2/(6.14) in ***3-2/clause 6.3.4.2(7)*** for short lengths of flange between rigid braces. It implies that the half-wavelength of buckling is restricted to the length between braces, but any flexible restraints included in this length will increase the buckling load from the Euler load for a strut of length L. The expression 3-2/(6.14) formulae also allow the effects of varying end moments and shears to be taken into account, but they are not valid (and are unsafe) for moment reversal cases. m is taken as the minimum value from:

$$m = 1 + 0.44(1+\mu)\Phi^{1.5} + (3+2\Phi)\gamma/(350-50\mu) \text{ or} \tag{3-2/(6.14)}$$

$$m = 1 + 0.44(1+\mu)\Phi^{1.5} + [0.195 + (0.05 + \mu/100)\Phi]\gamma^{0.5}$$

with:

$$\mu = V_2/V_1 \text{ and } \Phi = 2(1 - M_2/M_1)/(1+\mu) \text{ for } M_2 < M_1 \text{ and } V_2 < V_1$$

The first equation corresponds to considerations of buckling in one half-wavelength and the second corresponds to buckling in two half-wavelengths, but is a good approximation for buckling in three and four half-wavelengths also for cases of uniform moment, when compared with the predictions of equation (D6.3-35). For this special case, the second formula for m is accurate for γ up to about 20 000.

If $M_2 = M_1$ and $V_2 = V_1$ in the first equation of expression 3-2/(6.14), the same result as equation (D6.3-39) is obtained for the case of constant flange axial force. The shear ratio, μ, helps to describe the shape of the bending moment diagram between points of restraint. If $\mu = 1.0$ then the moment diagram is linear between points of restraint. If $\mu < 1.0$, the moments fall quicker than assumed from a linear distribution as shown in Fig. 6.3-21 and consequently the flange is less susceptible to buckling.

The lack of validity of expression 3-2/(6.14) for moment reversal is a problem for typical construction with a concrete deck slab and cross-bracing adjacent to the internal supports.

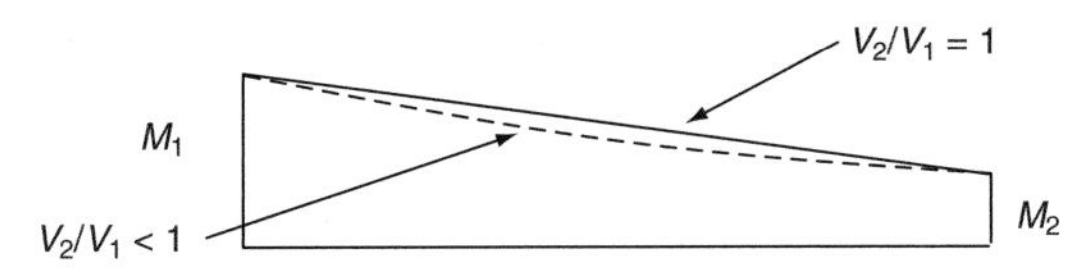

Fig. 6.3-21. Effect of shear ratio on moment diagram shape

Where the most distant brace provided from the pier is still in a hogging zone, the moment in the beam will reverse in the span section between braces as shown in Fig. 6.3-22. In this region, m can conservatively be taken as 1.0 but this is likely to lead to a conservative beam design or the unnecessary specification of additional braces away from the pier to ensure that the section between innermost braces is entirely sagging and the bottom flange is in tension. Alternatively, a higher value can be taken by conservatively taking $M_2 = 0$ as permitted by the note to 3-2/clause 6.3.4.2(7). If benefit from the restraining stiffness of the deck slab is ignored (i.e. $c = 0$), and V_2 is conservatively taken equal to V_1, this leads to $m = 1.88$.

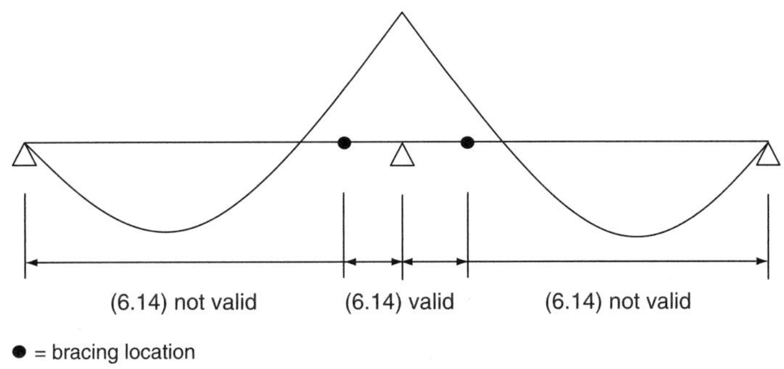

Fig. 6.3-22. Range of validity of equations in expression 3-2/(6.14)

It is important to note that the method of taking $M_2 = 0$ for moment reversal cases is only valid where the tension flange (which becomes the compression flange when the moment reverses) is continuously braced by decking or the top flange may buckle when the moment reverses. This is illustrated in section 6.3.2.3 of this guide. If the top flange is braced at discrete points only, then a separate check of this flange (treated as a strut in the same way) would also become necessary in the sagging zone with appropriate choice of m based on the shape of the moment diagram. $m = 1.0$ would be a conservative value.

Where the top flange is braced continuously by a deck, it is also be possible to 'vary' μ to try to produce a less conservative moment diagram. For the case in Fig. 6.3-23, the use of $V_2/V_1 = 0$, $M_2/M_1 = 0$ achieves the same moment gradient at end 1 as the real set of moments, but the moments lie everywhere else above the real moments so is still conservative. This gives a value of m from expression 3-2/(6.14) of 2.24, again ignoring any U-frame restraint. Providing the top flange is continuously braced, the real m would be greater. Further discussion on this is provided in the *Designers' Guide to EN 1994-2*[7] which shows that a value of 2.24 can also be applicable to cases where the moment reverses twice between rigid restraints.

It is possible to include continuous U-frame action from an unstiffened web between rigid braces in the calculation of c. The benefit is however usually quite small and the web plate,

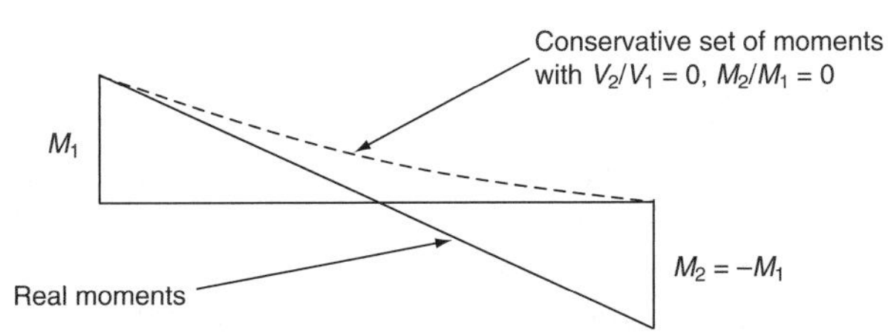

Fig. 6.3-23. Typical calculation of m where moment reverses

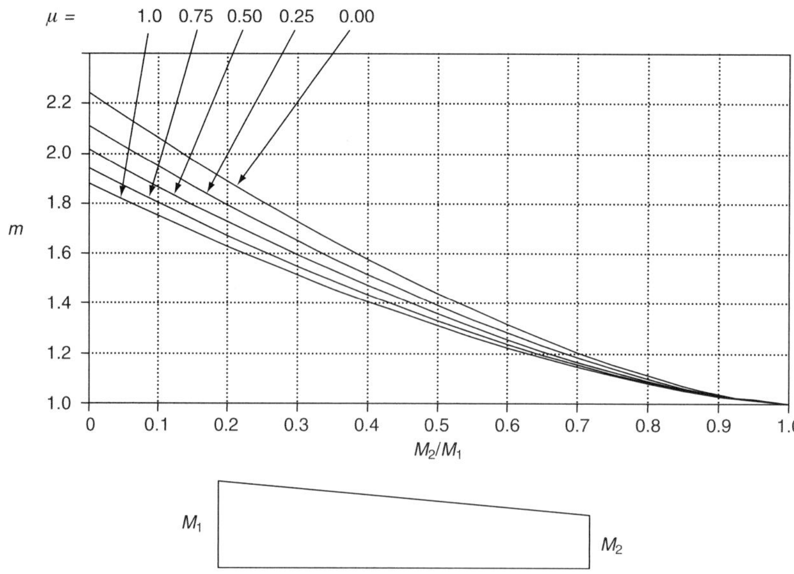

Fig. 6.3-24. Values of 'm' between rigid restraints with $\gamma = 0$

slab and shear studs must be checked for the forces implied by such action if it is considered. Figure 6.3-24 shows a graph of m against M_2/M_1 with $c = 0$ and varying μ.

It is possible to combine expression 3-2/(6.10) and expression 3-2/(6.12) to produce a single formula for slenderness, taking $A_f = bt_f$ for the flange area, as follows:

$$\bar{\lambda}_{LT} = \sqrt{\frac{A_{eff} f_y}{N_{crit}}} = \sqrt{\frac{(A_f + A_{wc}/3) f_y L^2}{m\pi^2 EI}} = L\sqrt{\frac{(1 + A_{wc}/3A_f)(f_y/Em)}{\dfrac{\pi^2}{12}\dfrac{b^3 t_f}{bt_f}}}, \text{ so}$$

$$\bar{\lambda}_{LT} = 1.103\frac{L}{b}\sqrt{\frac{f_y}{Em}}\sqrt{1 + \frac{A_{wc}}{3A_f}} \qquad \text{(D6.3-40)}$$

It will still, however, be necessary to evaluate N_{crit} when checking the strength of bracings as discussed in section 6.3.4.2.6 below.

The formulae in 3-2/clause 6.3.4.2 do not apply directly to haunched girders as they assume that the flange force is distributed in the same way as the bending moment. The general method of using an eigenvalue analysis based on the forces in the compression chord is however still applicable. Alternatively, the formulae provided could be applied using the minimum value of c in the length considered and by using the flange force ratio F_2/F_1 instead of the moment ratio M_2/M_1 with V_2/V_1 taken equal to 1.0 when applying expression 3-2/(6.14).

3-2/clause 6.3.4.2(7) allows the buckling verification to be performed at a distance of $0.25L_k = 0.25L/\sqrt{m}$ (i.e. 25% of the effective length) from the end with the largest moment. (L_k and l_k are both used for effective length in 3-2/clause 6.3.4.2.) At first glance, this appears to be similar to the approximate practice of accounting for the shape of the moment diagram by using the effects within the middle third of the member; it would therefore appear that this double-counted the benefit from moment shape derived in expression 3-2/(6.14). This is, however, not the case. The check at $0.25L_k$ reflects the fact that the peak stress from transverse buckling of the flange occurs some distance away from the rigid flange restraint, whereas the peak stress from overall bending of the beam occurs at the restraint. The beam flange is assumed to be pin-ended at the rigid transverse restraints in this flange model. Since these two peak stresses do not coexist and are not therefore fully additive, the buckling verification can be performed at a 'design' section somewhere between these two locations. The cross-section resistance must still be verified at the point of maximum moment.

There are clearly problems with applying this aspect of 3-2/clause 6.3.4.2(7) where the moment reverses, as the section $0.25L/\sqrt{m}$ from an end may be a point of contraflexure. If the moment reverses, it is recommended here that the design section be taken as 25% of the distance from position of maximum moment to position of zero moment. In addition, if benefit is taken of the verification at the $0.25L_k$ design section, the calculated slenderness above must be modified so that it refers to this design section, as the critical moment value will be less at that section and the slenderness therefore increased. This can be done by defining a new slenderness at the $0.25L_k$ section such that:

$$\bar{\lambda}_{0.25Lk} = \bar{\lambda}_{LT}\sqrt{\frac{M_1}{M_{0.25Lk}}} \qquad \text{(D6.3-41)}$$

where $M_{0.25Lk}$ is the moment at the $0.25L_k$ section. This procedure is illustrated in Worked Example 6.3-5.

6.3.4.2.4. Rigidity of bracings

The formulae in EN 1993-2 discussed above are only valid where the end restraints that define the length L are 'rigid'. It is possible to equate $N_{crit} = 2\sqrt{cEI}$ to $\pi^2 EI/L^2$ to find a limiting stiffness that gives an effective length equal to the distance between rigid restraints, L, but this slightly underestimates the required rigidity. This is because the formulae assume that the restraints are continuously smeared when they are in fact discrete. The former analysis gives a required rigidity for C_d of $\pi^4 EI/(4L^3)$ whereas the 'correct' rigidity is

$$\frac{4\pi^2 EI}{L^3} = \frac{4N_E}{L}$$

as given in 3-2/clause 6.3.4.2(6).

6.3.4.2.5. Beams without plan bracing or decking during construction

During construction it is common to stabilize girders by connecting them in pairs with 'torsional' bracing. Such bracing reduces or prevents torsion of individual beams but does not restrict lateral deflection. 'Torsional' cross-bracing as shown in Fig. 6.3-25 has been considered in the UK for many years to act as a rigid support to the compression flange, thus restricting the effective length to the distance between braces. Computer elastic critical buckling analyses however show that often the effective length is not limited to the spacing of the bracings because a mode of buckling involving rotation of the braced pair over the whole span can occur. BS 5400: Part 3[4] introduced a clause to cover this situation. Its application predicts that such bracing is not fully effective in restricting the effective length to the distance between bracings, although the predictions are somewhat pessimistic. More flexible torsional bracing, such as a horizontal channel between beams acting in bending, will clearly not usually be fully effective.

The method in section 6.3.2.4 of this guide, which refers to BS 5400: Part 3: 2000,[4] can be used to consider buckling during construction, but it may lead to the conclusion in some cases that

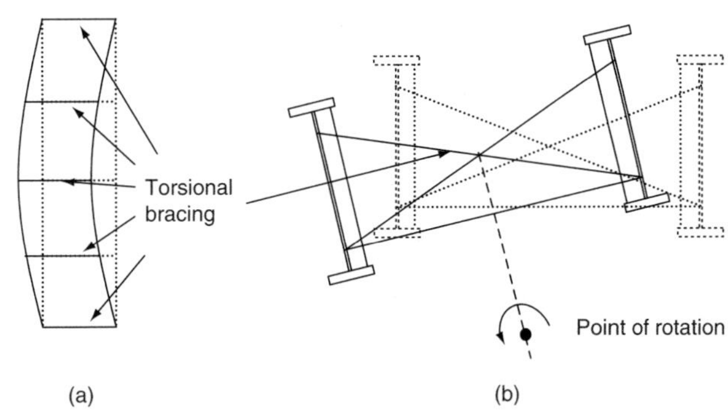

Fig. 6.3-25. Torsional bracing and buckling mode shape for paired beams: (a) plan on braced pair of beams showing buckling mode shape; (b) cross-section through braced pair showing buckling mode shape

either plan bracing or an increase to top flange size is necessary as it is quite conservative. A better estimate of slenderness can be made using a shell finite-element analysis and those familiar with such analysis will probably also complete the check quicker this way.

A finite-element model of a non-composite beam, using shell elements for the paired main beams and beam elements to represent the bracings, can be set up relatively quickly with modern commercially available software. Elastic critical buckling analysis can then be performed and a value of M_{cr} determined directly for use in slenderness calculation to 3-2/clause 6.3.2. Some experience is required however to determine M_{cr} from the output as often the first buckling mode observed does not correspond to the required global buckling mode; there may be many local plate buckling modes for the web and flanges before the first global mode is found. This approach usually demonstrates that the cross-bracings are not fully effective in limiting the effective length of the flange to the distance between bracings, but that it is more effective than is predicted by BS 5400. For simply supported paired girders, a typical lowest global buckling mode under dead load is shown in Fig. 6.3-25.

Worked Example 6.3-5: Steel and concrete composite bridge

A three-span steel and concrete composite bridge in S355 steel is shown schematically in Fig. 6.3-26. It has rigid cross-bracings. The beams have Class 2 cross-section and have the following plate sizes at the internal piers:

Top flange:	400 mm × 25 mm
Web:	1160 mm × 25 mm
Bottom flange:	400 mm × 40 mm

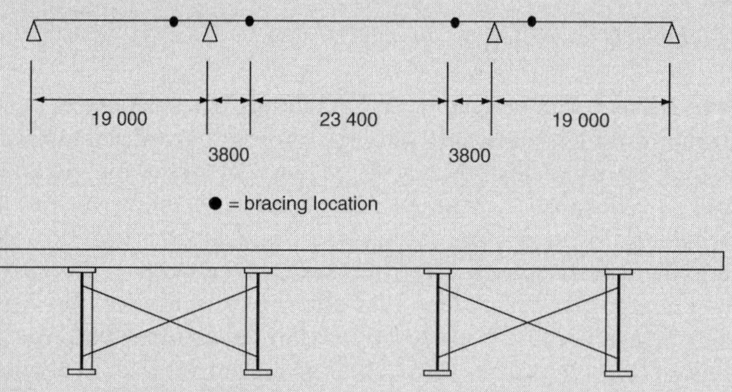

Fig. 6.3-26. Bridge for Worked Example 6.3-5

The beam neutral axis is 735 mm up from the top of the bottom flange and the plastic moment resistance (determined in accordance with EN 1994-2 using γ_{M1}) is $M_{pl,Rd} =$ 10 700 kNm. The moment at the internal support is 8674 kNm and the coexisting moment at the main span bracing is 5212 kNm. The shear at the bracing is 70% of the value at the internal support. Lateral torsional buckling is checked adjacent to the internal support and in the main span beyond the brace, assuming the same cross-section throughout. (A similar example in the *Designers' Guide to EN 1994-2*[7] considers a changing cross-section and also the effects of an axial force.)

Check at internal support

Strictly, the stiffness of the bracing should first be checked (or should later be designed) so that the buckling length is confined to the length between braces. This is done in Worked Example 6.3-7.

The flange area and web compression zone are as follows:

$A_f = 400 \times 40 = 16\,000\,\text{mm}^2$

$A_{wc} = 735 \times 25 = 18\,375\,\text{mm}^2$

The bottom flange transverse second moment of area is:

$$I = \frac{1}{12} \times 40 \times 400^3 = 2.133 \times 10^8\,\text{mm}^4$$

The applied beam moments at each end of the chord are:

$$M_1 = 8674\,\text{kNm}$$

$$M_2 = 5212\,\text{kNm}$$

and so $M_2/M_1 = 0.6$.

$\mu = V_2/V_1 = 0.7$ so from 3-2/clause 6.3.4.2(7):

$$\Phi = 2(1 - M_2/M_1)/(1 + \mu) = 2(1 - 0.60)/(1 + 0.7) = 0.46$$

The deck slab does provide some continuous U-frame stiffness and could have been included using 3-2/Table D.3, case 1a, to calculate a stiffness, c. This contribution has however been ignored to avoid the complexities of designing deck, stiffeners and shear studs for the forces implied, so $\gamma = 0$.

From the first equation in expression 3-2/(6.14):

$$m = 1 + 0.44(1 + \mu)\Phi^{1.5} + (3 + 2\Phi)\gamma/(350 - 50\mu) = 1 + 0.44(1 + 0.7)0.46^{1.5} = 1.23$$

From the second equation:

$$m = 1 + 0.44(1 + \mu)\Phi^{1.5} + [0.195 + (0.05 + \mu/100)\Phi]\gamma^{0.5}$$

$$= 1 + 0.44(1 + 0.7)0.46^{1.5} = 1.23$$

Hence $m = 1.23$.

If the deck slab is considered to provide U-frame restraint, the value of m for this bridge is still only 1.26, so there is no real benefit to stability of the main beams in considering U-frame action over such a short length.

From equation (D6.3-40):

$$\bar{\lambda}_{LT} = 1.1\frac{L}{b}\sqrt{\frac{f_y}{Em}}\sqrt{1 + \frac{A_{wc}}{3A_f}}$$

$$= 1.1 \times \frac{3800}{400}\sqrt{\frac{355}{210 \times 10^3 \times 1.23}} \times \sqrt{1 + \frac{18\,375}{3 \times 16\,000}} = 0.46 > 0.2$$

so the section is therefore prone to lateral torsional buckling.

(The yield stress was taken as 355 MPa from 3-1-1/Table 3.1. However, the UK National Annex requires the value appropriate to thickness to be taken from EN 10025.) The relevant buckling curve from 3-1-1/Table 6.4 is curve d (for $h/b = 1225/400 = 3.1 > 2$) so $\alpha_{LT} = 0.76$ from 3-1-1/Table 6.3.

From expression 3-1-1/(6.56):

$$\Phi_{LT} = 0.5[1 + \alpha_{LT}(\bar{\lambda}_{LT} - 0.2) + \bar{\lambda}_{LT}^2] = 0.5[1 + 0.76(0.46 - 0.2) + 0.46^2] = 0.705$$

$$\chi_{LT} = \frac{1}{\Phi_{LT} + \sqrt{\Phi_{LT}^2 - \bar{\lambda}_{LT}^2}} = \frac{1}{0.705 + \sqrt{0.705^2 - 0.46^2}} = 0.81$$

The reduction factor for LTB according to expression 3-1-1/(6.56) is therefore **0.81**.

The bending resistance is therefore given by:

$$M_{b,Rd} = \chi_{LT}M_{pl,Rd} = 0.81 \times 10\,700 = \mathbf{8667\,kNm}$$

which is about equal to 8674 kNm applied, i.e. a very minor overstress.

According to 3-2/clause 6.3.4.2(7), the check could however be conducted at a design section at $0.25L/\sqrt{m} = 0.25 \times 3800/\sqrt{1.23} = 857\,\text{mm}$ from the support. The moment at this section is:

$$M_{0.25\mathrm{Lk}} \approx 8674 - \frac{857}{3800} \times (8674 - 5212) = 7893\,\mathrm{kNm}$$

The slenderness at this section is, from equation (D6.3-41):

$$\bar{\lambda}_{0.25\mathrm{Lk}} = \bar{\lambda}_{\mathrm{LT}}\sqrt{\frac{M_1}{M_{0.25\mathrm{Lk}}}} = 0.46\sqrt{\frac{8674}{7893}} = 0.48$$

From 3-1-1/Fig. 6.4 curve d, $\chi_{\mathrm{LT}} = 0.79$ so at the design section, $M_{\mathrm{b,Rd}} = \chi_{\mathrm{LT}} M_{\mathrm{pl,Rd}} = 0.79 \times 10\,700 =$ **8453 kNm** $> M_{0.25\mathrm{Lk}} = 7893$ kNm applied. The beam is adequate.

Check the remainder of the main span

Since the moment reverses, the formulae in expression 3-2/(6.13) are not directly applicable.

If, conservatively, m is taken as 1.0 for constant force then:

$$\bar{\lambda}_{\mathrm{LT}} = 1.1\frac{L}{b}\sqrt{\frac{f_y}{Em}}\sqrt{1+\frac{A_{\mathrm{wc}}}{3A_{\mathrm{f}}}}$$

$$= 1.1 \times \frac{23\,400}{400}\sqrt{\frac{355}{210 \times 10^3 \times 1.00}} \times \sqrt{1+\frac{18\,375}{3 \times 16\,000}} = 3.11 > 0.2$$

Using curve d, but this time taking χ_{LT} from 3-1-1/Fig. 6.4 directly, gives $\chi_{\mathrm{LT}} = \mathbf{0.09}$.

If the suggestion of EN 1993-2 is followed and M_2 is taken as 0 (and V_2 is taken as V_1), then $m = 1.88$ and hence:

$$\bar{\lambda}_{\mathrm{LT}} = 1.1\frac{L}{b}\sqrt{\frac{f_y}{Em}}\sqrt{1+\frac{A_{\mathrm{wc}}}{3A_{\mathrm{f}}}}$$

$$= 1.1 \times \frac{23\,400}{400}\sqrt{\frac{355}{210 \times 10^3 \times 1.88}} \times \sqrt{1+\frac{18\,375}{3 \times 16\,000}} = 2.27 > 0.2$$

Using curve d, 3-1-1/Fig. 6.4 gives $\chi_{\mathrm{LT}} = \mathbf{0.14}$.

The hogging moment at the brace is at least 60% of the maximum at the support (the value quoted at the brace is a coexistent value, not a maximum), but the resistance is only approximately 17% of the support hogging resistance. Another bracing would be required. A similar example is presented in the *Designers' Guide to EN 1994-2*.[7] In it, continuous U-frame action from the restraint offered by the web attached to the top slab is considered, as is a value of $m = 2.24$ as discussed in the main text. Such considerations give a significant further improvement here, but are still insufficient to avoid provision of a further bracing. Consideration of continuous U-frame action also has the disadvantage that the web and shear studs would have to be designed for the resulting effects.

Worked Example 6.3-6: Half through bridge

A simply supported half through bridge with 36 m span has the cross-section geometry and section properties shown in Fig. 6.3-27 below. The beams are 2.8 m deep and the webs are 20 mm thick. (The top flange has been idealized but is actually made up from two plates each 60 mm thick.) The elastic neutral axis for the gross cross-section is shown in Fig. 6.3-27 and the section modulus is $2.378 \times 10^8\ \mathrm{mm}^3$ for each flange based on gross cross-section properties. The cross-girders are spaced at 3.0 m centres and are the same throughout. The steel is S355 with yield stress of 335 MPa for the 60 mm thick plate (from EN 10025) which is conservatively used throughout. The resistance moment for LTB is calculated.

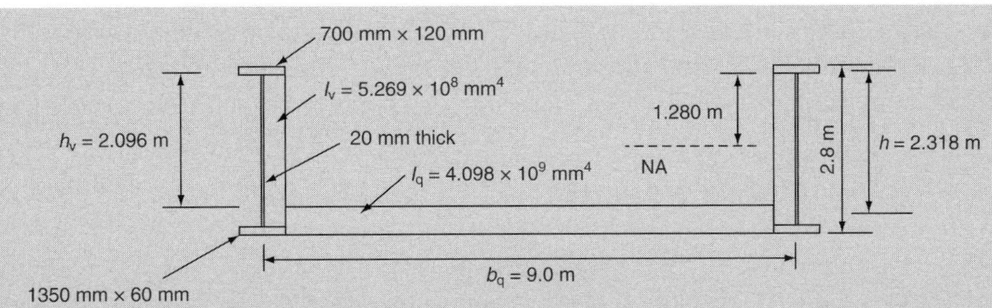

Fig. 6.3-27. Half through bridge for Worked Example 6.3-6

Section classification is first checked. The top flange is Class 1 by inspection. From Fig. 6.3-27, the elastic depth of web in compression = 1280 mm and the depth in tension is 1340 mm so the stress ratio is:

$$\psi = -\frac{1280}{1340} = -0.96$$

From 3-1-1/Table 5.2, the limit for a Class 3 web is:

$$c/t \leq 62\varepsilon(1-\psi)\sqrt{-\psi} = 62 \times \sqrt{\frac{235}{335}} \times (1+0.96)\sqrt{0.96} = 99 < \frac{2620}{20} = 131$$

so the web is actually Class 4. An effective section should therefore be used for the compression zone of the web. From Fig. 6.2-13 in section 6.2.2.5 of this guide, for:

$$b/t\sqrt{\frac{f_y}{235}} = 2620/20\sqrt{\frac{335}{235}} = 156 \text{ and } \psi \approx -1$$

the reduction factor for the compression zone is $\rho = 0.80$. This leads to a small piece of compression zone being ineffective with depth $= (1 - 0.80) \times 1280 = 256$ mm at the location required by 3-1-5/Table 4.1. The section properties now need to be revised to account for this reduction, whereupon the minimum section modulus (at the top flange and conservatively taken at the extreme fibre, rather than the mid-plane of the flange) becomes 2.328×10^8 mm^3. The new centroid is 1298 mm from the top of the web. (The derivation of Class 4 section properties is covered in Worked Example 6.2-3 in section 6.2.2.5.)

The transverse second moment of area of the top flange is $\frac{1}{12} \times 700^3 \times 120 = 3.43 \times 10^9$ mm^4 (ignoring the small contribution from the participating web).

The effective compression area is, from 3-2/clause 6.3.4.2(7):

$$A_{eff} = A_f + A_{wc}/3 = 700 \times 120 + (1298 - 256) \times 20/3 = 90\,947 \text{ mm}^2$$

From 3-2/Annex D, the U-frame stiffness is:

$$C_d = \frac{EI_v}{\frac{h_v^3}{3} + \frac{h^2 b_q I_v}{2I_q}} = \frac{210 \times 10^3 \times 5.269 \times 10^8}{\frac{2096^3}{3} + \frac{2318^2 \times 9000 \times 5.269 \times 10^8}{2 \times 4.098 \times 10^9}} = 17\,910 \text{ Nmm}^{-1}$$

therefore $c = C_d/l = 17\,910/3000 = 5.97$ Nmm^{-2}

This does not make any allowance for joint flexibility in the connection of cross-girder as the type of joint has been assumed to be welded and fully stiffened. If the joint was 'semi-continuous' according to 3-1-8/clause 5.2.2, the joint flexibility, S_j, would have to be determined from 3-1-8/clause 6.3 (or a conservative value assumed as discussed in the main text) and included in the calculation of C_d. This would typically apply to connections made through unstiffened end plates.

By inspection, the end U-frames will not be rigid. The formula for m in expression 3-2/(6.12) is not therefore valid and allowance must be made for the lack of rigidity of the end U-frame using equation (D6.3-37):

$$X = \frac{C_e}{\sqrt{2}}\left(\frac{l^3}{C_d^3 EI}\right)^{0.25} = \frac{17\,910}{\sqrt{2}}\left(\frac{3000^3}{17\,910^3 \times 210 \times 10^3 \times 3.43 \times 10^9}\right)^{0.25} = 0.64\,\text{m}$$

$$= \frac{\sqrt{\gamma}}{\left(\frac{\pi}{\sqrt{2}} + \frac{0.69}{X + 0.5}\right)^2} = \frac{\sqrt{\frac{5.97 \times 36\,000^4}{210 \times 10^3 \times 3.43 \times 10^9}}}{\left(\frac{\pi}{\sqrt{2}} + \frac{0.69}{0.64 + 0.5}\right)^2} = 14.766$$

From expression 3-2/(6.12):

$$N_{crit} = mN_E = 14.766 \times \pi^2 \times 210 \times 10^3 \times 3.43 \times 10^9 / 36\,000^2 = 81\,000\,\text{kN}$$

$$\bar{\lambda}_{LT} = \sqrt{\frac{A_{eff} f_y}{N_{crit}}} = \sqrt{\frac{90\,947 \times 335}{81\,000 \times 10^3}} = 0.61 > 0.2 \text{ from 3-1-1/Fig. 6.4}$$

The section is therefore susceptible to lateral torsional buckling.

The relevant buckling curve for use with expression 3-1-1/(6.56) from Table 6.4 with $h/b = 2800/700 = 4.0 > 2$ is curve d, so from 3-1-1/Fig. 6.4, $\chi_{LT} = 0.70$.

The reduction factor for LTB is therefore **0.70**.

The resistance is next determined using expression 3-1-1/(6.55):

$$M_{b,Rd} = \chi_{LT} W_{el,y} \frac{f_y}{\gamma_{M1}} = 0.70 \times 2.328 \times 10^8 \times \frac{335}{1.1} = \mathbf{49\,416\,kNm}$$

6.3.4.2.6. Strength of bracings and U-frames

3-2/clause 6.3.4.2(5)

Design forces for bracings and U-frame restraints to the compression flange are derived from ***3-2/clause 6.3.4.2(5)***. The formulae given there follow from the beam on elastic foundations model adopted for checking the compression flange itself. Initial bow imperfections in the compression flange give rise to forces in the restraints when the flange is loaded and the bow grows further. This is a second-order effect as discussed in section 5.2.

$$F_{Ed} = \frac{N_{Ed}}{100} \quad \text{if } l_k < 1.2l \qquad \text{3-2/(6.11)}$$

$$F_{Ed} = \frac{l}{l_k}\frac{N_{Ed}}{80}\frac{1}{1 - \frac{N_{Ed}}{N_{crit}}} \quad \text{if } l_k > 1.2l$$

where l is the distance between restraints, whether flexible or rigid and l_k is the effective length. If l is eliminated from the second equation, it can be rewritten as:

$$F_{Ed} = \frac{N_{Ed}}{N_{crit} - N_{Ed}} \times \frac{l_k}{20\pi^2} \times C_d \qquad \text{(D6.3-42)}$$

This effectively represents an increase in initial bow deflection times the stiffness, C_d, of the restraint spring undergoing that deflection. The initial imperfection is $l_k/(20\pi^2)$, which is approximately equal to $l_k/200$, corresponding to the initial bow for type c in Table 3-1-1/clause 5.1. The growth of the bow which is seen by the spring is:

$$\frac{N_{Ed}}{N_{crit} - N_{Ed}} \times \frac{l_k}{20\pi^2}$$

while the total final bow is:

$$\frac{N_{\text{crit}}}{N_{\text{crit}} - N_{\text{Ed}}} \times \frac{l_k}{20\pi^2}$$

from the theory in section 5.2.

The limit of $F_{\text{Ed}} = N_{\text{Ed}}/100$ is given so that the bracing force according to the second equation does not continue to increase beyond that corresponding to a rigid brace. This limit is required when the equations are used with discrete restraints as the mode of buckling changes to a series of half-wavelengths between the rigid restraints and the buckling load no longer increases with increasing restraint stiffness. However, if continuous restraints (such as continuous U-frames) are provided, the first equation is not relevant and the second equation should always be used on the basis of a force per unit length.

No specific guidance is given on the calculation of N_{Ed}, which is assumed to be constant over the whole half-wavelength of buckling. It can always conservatively be based on the greatest value in the span. Alternatively the greatest value in the relevant half-wavelength of buckling could be used.

Where there are two or more interconnected beams, EN 1993-2 does not specify the number of forces, F_{Ed}, to consider. It would be conservative to apply an F_{Ed} force from each beam. It would be reasonable to follow the approach in BS 5400: Part 3[4] which only required forces from any two girders to be considered together, reflecting the fact that it is unlikely that worst-case imperfections would be found in all flanges together. Some care should however be taken if there are very many beams all connected to a single braced pair as consideration of only two forces may not then be safe. In that situation, forces F_{Ed} could be applied from each beam but with the reduction factor $\alpha_m \leq 1.0$ applied to each force in accordance with the formula in 3-1-1/clause 5.3.3(1). The effects should not be taken as less than that from the full force F_{Ed} from any two of the beams. The forces F_{Ed} from each beam should be applied in directions such as to maximize the effect in the element being considered.

In addition to the forces arising from bracing the compression flange, other forces in the bracings should be considered in their relevant combinations. These include the effects of wind and differential deflection between main beams. For the latter actions, the displacements of the main beams obtained ignoring the bracings may be applied to a plane frame model of the braces to determine forces in the bracings. Alternatively, the bracings may be included in the global analysis and the forces determined directly. (If a grillage analysis is used, bracings may be modelled as an equivalent transverse member with a shear area representing the distortional stiffness across the bracing and a bending inertia representing the bending stiffness across the bracing.) Additional forces are also generated in U-frame members (including the chords) by loading on the cross-girders which causes differential deflections between adjacent frames. This effect is referred to in ***3-2/clause 6.3.4.2(2)*** Note 2 but is not directly covered by EN 1993-2. Additional guidance is given in section 6.8 of this guide. *3-2/clause 6.3.4.2(2)*

It is advisable to design the bracing components to elastic limits at the ultimate limit state because plasticity (particularly for restraint members that act in bending) will result in an unmodelled loss of stiffness that could allow buckling of the compression flange. Similarly, bolts should be designed not to slip at the ultimate limit state.

Where the restraint forces are to be transmitted to end supports by a system of plan bracing, the plan bracing system should be designed to resist the more onerous of the forces F_{Ed} from each restraint within a length equal to the half-wavelength of buckling and the forces generated by an overall flange bow in each flange according to clause 5.3.3 of EN 1993-1-1. In the latter case, for a very stiff bracing system with zero first-order transverse deflection, each flange applies a total force of $(N_{\text{Ed}}/62.5)\alpha_m$ uniformly distributed to the plan bracing, where α_m is the reduction factor for the number of interconnected beams in 3-1-1/clause 5.3.3(1).

Worked Example 6.3-7: Stiffness and strength of cross-bracing

The bracing of the continuous bridge in Worked Example 6.3-5 comprises cross-bracing made from $150 \times 150 \times 18$ angle and attached to 100×20 stiffeners on a 25 mm thick web. It is checked that the bracings are 'rigid' and the axial force in them arising from bracing the flanges is determined. It is assumed that the greatest compressive stress in the flange at the internal support is 300 MPa.

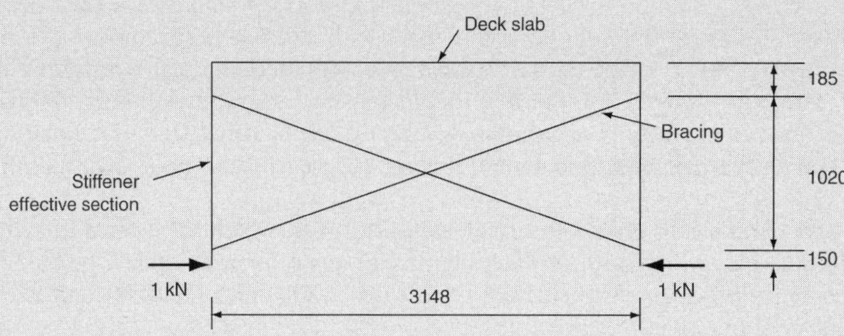

Fig. 6.3-28. Cross-bracing for Worked Example 6.3-7

The stiffness of the bracing was first calculated from a plane frame model as shown in Fig. 6.3-28. (If the cross-bracing had been replaced by a horizontal channel at beam mid-height, acting in bending between the beams, the case of applied forces in the same direction would have given considerably greater deflection than the case with opposing forces.)

Stiffener effective section properties (3-1-5/Fig. 9.1):

$$\text{Attached web width} = 30\varepsilon t_w + t_{\text{stiffener}} = 30 \times 0.81 \times 25 + 20 = 628\,\text{mm}$$

This leads to $A_{st} = 17\,700\,\text{mm}^2$ and $I_{st} = 9.41 \times 10^6\,\text{mm}^4$

Deck slab:

An attached width of deck slab was taken in accordance with the rules for shear lag in EN 1994-2.

The plane frame model gave a deflection of 1.25×10^{-5} m under a 1 kN load.

The brace stiffness is therefore:

$$\frac{1000}{1.25 \times 10^{-2}} = 80\,000\,\text{N}\,\text{mm}^{-1}$$

From expression 3-2/(6.13), the required stiffness for the bracing to be considered as rigid (defining the length $L = 3.8$ m) is:

$$\frac{4\pi^2 EI}{L^3} = \frac{4\pi^2 \times 210 \times 10^3 \times 2.133 \times 10^8}{3800^3} = 32\,227\,\text{Nmm}^{-1} < 80\,000\,\text{Nmm}^{-1}$$

Therefore the bracing is stiff enough to be considered fully rigid and L may be taken as the length between braces. Since the bracings are fully rigid and ℓ_k is restricted to ℓ, the distance between braces, the first equation in expression 3-2/(6.11) is used to determine the force in the bracings. Hence:

$$F_{Ed} = \frac{N_{Ed}}{100} = 300 \times \left(16\,000 + \frac{18\,375}{3}\right)/100 = 66.4\,\text{kN}$$

This force is applied to the bracing by each beam as shown in Fig. 6.3-28.

The axial force in the bracing is then

$$\frac{66.4}{\cos(\tan^{-1} 1020/3148)} = \mathbf{69.8\,kN}$$

6.4. Built-up compression members

Built-up compression members have traditionally been used in large skeletal structures where a fabricated 'solid' member would prove too heavy for the overall structure. Built-up members require considerable fabrication effort, so they generally tend not to be the most economic option. Structurally, the non-continuous lacings or battens create a shear flexible strut. The shear flexibility will cause a reduction in buckling resistance by increasing the second-order moments. EN 1993-2 makes reference directly to EN 1993-1-1 for the design of built-up compression members.

6.4.1. General

3-1-1/clause 6.4 covers only pin-ended uniform columns with length L. For other end connections it would however be possible to use an effective length L_{cr} in place of L. A slightly different approach for checking the buckling resistance of built-up compression members is used compared to the approach in 3-1-1/clause 6.3.1 for solid members. ***3-1-1/clause 6.4.1(1)*** allows the member to be considered as a strut with a shear flexibility, possessing an initial sinusoidal bow imperfection e_0 of $L/500$. The rules only explicitly cover uniaxial bending. Some modifications for biaxial bending are suggested below. *3-1-1/clause 6.4.1(1)*

3-1-1/clause 6.4.1(2) clarifies that the rules presented assume that the lacing and batten centres are constant when deriving shear stiffness. If they are not constant, the design could be based on the greatest spacing unless more detailed calculation is undertaken. A minimum of three bay lengths is also required to allow the transverse flexibility due to the lacings and battens to be idealized as a shear deformation. *3-1-1/clause 6.4.1(2)*

The procedure is to first determine the chord forces, allowing for member global second-order effects, and then check the chords themselves for cross-section resistance and buckling between lacing nodes or batten locations – ***3-1-1/clause 6.4.1(5)*** refers. *3-1-1/clause 6.4.1(5)*

Members with two chords

Where there are only two chords as shown in Fig. 6.4-1(a), and any applied bending moment is about the z–z axis, the force in the chords is calculated from ***3-1-1/clause 6.4.1(6)*** as follows: *3-1-1/clause 6.4.1(6)*

$$N_{ch,Ed} = 0.5N_{Ed} + \frac{M_{Ed}h_0A_{ch}}{2I_{eff}} \qquad \text{3-1-1/(6.69)}$$

with:

$$M_{Ed} = \frac{N_{Ed}e_0 + M_{Ed}^{I}}{1 - \dfrac{N_{Ed}}{N_{cr}} - \dfrac{N_{Ed}}{S_v}}$$

where:

$$N_{cr} = \frac{\pi^2 EI_{eff}}{L^2}$$

which is the effective elastic critical buckling force of the built-up member about the z–z axis;

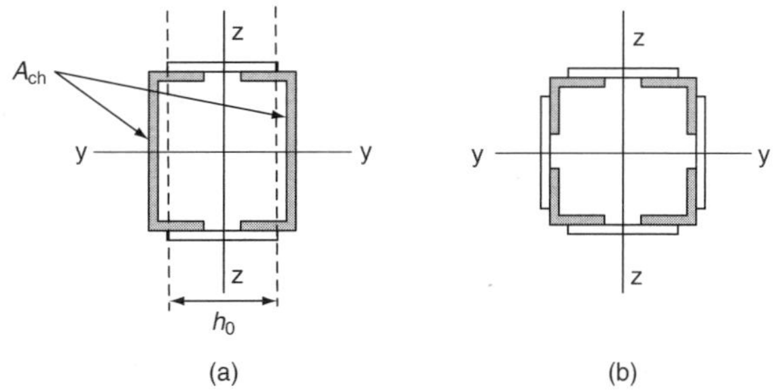

Fig. 6.4-1. Typical built-up members

N_{Ed} is the design value of the compression force on the built-up member;
M^{I}_{Ed} is the design value of the maximum moment about the z–z axis in the middle of the built-up member without considering second-order effects, i.e. the moment from a first-order analysis performed without the bow imperfection;
h_0 is the distance between the centroids of the chords;
A_{ch} is the cross-sectional area of one chord;
I_{eff} and S_v are the effective second moment of area and shear stiffness respectively of the built-up member. These values will be dependent on whether the built-up member is laced or battened. The shear flexibility arises either from the axial shortening of lacing members or from Vierendeel action of battens and chords in the case of battened members.

Expression 3-1-1/(6.69) is effectively the overall buckling check about the z–z axis. The moment M_{Ed} is an amplification of the first-order moment $N_{Ed}e_0 + M^{I}_{Ed}$ by the factor:

$$\frac{1}{1 - \dfrac{N_{Ed}}{N_{cr}} - \dfrac{N_{Ed}}{S_v}}$$

The $1/[1 - (N_{Ed}/N_{cr})]$ factor is discussed in section 5.2 of this guide. The additional term N_{Ed}/S_v contributes a further amplification due to the shear displacement. The chord force in expression 3-1-1/(6.69) assumes that the moment M_{Ed} is carried by opposing forces in the two chords acting at a lever arm of h_0 and the applied axial force, N_{Ed}, is shared by the two chords.

Having determined the chord forces, the chords themselves have to be checked for cross-section resistance and buckling about the z–z axis between lacing nodes or batten locations.

The rules are not written for biaxial bending, so no interaction with any imposed bending moment $M^{I}_{y,Ed}$ about the y–y axis is given in EN 1993-1-1. The effect of such moment is to produce a bending moment $M_{ch,y,Ed}$ in each chord about the y–y axis which would also need to be included in the check of the chords. The global bending moment about the y–y axis needs to allow for global second-order effects where applicable. This can be achieved by multiplying the first-order moment by the factor

$$\frac{1}{1 - \dfrac{N_{Ed}}{N_{cr,y}}}$$

where $N_{cr,y}$ is the elastic critical buckling load for flexural buckling about the y–y axis. It is not necessary to allow for bow imperfections in two directions at once. Consequently when expression 3-1-1/(6.69) is used to determine chord forces, which allows for bow imperfections about the z–z axis, no bow imperfections about the y–y axis need be considered in calculating $M_{ch,y,Ed}$.

Global buckling about the y–y axis should also be checked, but this is again not covered by 3-1-1/clause 6.4. In this case, a bow imperfection is considered about the y–y axis, but not the z–z axis. The chord axial forces can be obtained from:

$$N_{ch,Ed} = 0.5N_{Ed} + \frac{M^{I}_{z,Ed}h_0A_{ch}}{2I_{eff}\left(1 - \dfrac{N_{Ed}}{N_{cr,z}} - \dfrac{N_{Ed}}{S_v}\right)} \qquad \text{(D6.4-1)}$$

where:

$N_{cr,z}$ is the effective elastic critical buckling force of the built-up member about the z–z axis;
$M^{I}_{z,Ed}$ is the design value of the maximum moment about the z–z axis in the middle of the built-up member without considering second-order effects, i.e. the moment from a first-order analysis performed without the bow imperfection.

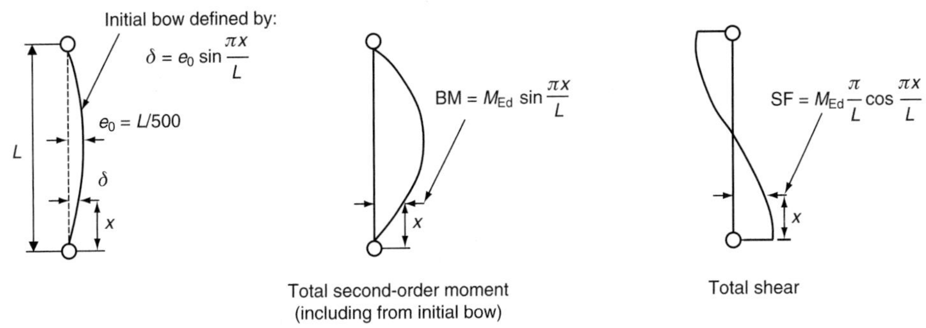

Fig. 6.4-2. Initially curved built-up compression member

The moment in each chord about the y–y axis is:

$$M_{\text{ch,y,Ed}} = 0.5\frac{N_{\text{Ed}}e_0 + M^{\text{I}}_{\text{y,Ed}}}{1 - \dfrac{N_{\text{Ed}}}{N_{\text{cr,y}}}} \qquad \text{(D6.4-2)}$$

Members with four chords

Where there are four chords as shown in Fig. 6.4-1(b), the force in the chords can be calculated in a similar way to that in expression 3-1-1/(6.69) but, for uniaxial bending, $N_{\text{ch,Ed}}$ refers to the force in a pair of chords. Where there is biaxial bending, an additional term for the second moment direction is required. The calculation of chord force then needs to be performed twice with the bow imperfection taken in the two different directions. The chords would then be checked individually for cross-section resistance and buckling between lacing nodes or batten locations. Buckling should be checked about the weakest axis of the chord.

Shear force for checking lacings or battens

In order to design the lacings and their connections, it is necessary to consider the shear force in the built-up column. Shears will arise from both external lateral forces and the bowing of the column under axial load as illustrated in Fig. 6.4-2.

The initial bow is sinusoidal and it is conservatively assumed that any first-order moment from external actions, M^{I}_{Ed}, is also distributed sinusoidally. The magnified bending moment is assumed to be distributed sinusoidally as $M_{\text{Ed}}\sin(\pi x/L)$, although the shear deformation means this assumption is not strictly correct. The shear force is therefore given by:

$$V = \frac{\text{d}}{\text{d}x}\left(M_{\text{Ed}}\sin\frac{\pi x}{L}\right) = M_{\text{Ed}}\frac{\pi}{L}\cos\frac{\pi x}{L}$$

The maximum value of shear will be at the supports where $x = 0$, so the shear force to use in the design of lacings and battens, as given in ***3-1-1/clause 6.4.1(7)***, is: *3-1-1/clause 6.4.1(7)*

$$V_{\text{Ed}} = M_{\text{Ed}}\frac{\pi}{L} \qquad \text{3-1-1/(6.70)}$$

The application of this shear produces different effects in lacings and battens as discussed in the following sections.

6.4.2. Laced compression members

Both chords and lacings must be checked for buckling – ***3-1-1/clause 6.4.2.1(1)*** refers. The chords and overall member can be checked as discussed in section 6.4.1 above. For buckling of chords between lacing nodes, an effective length from 3-1-1/Fig. 6.8 is used in accordance with ***3-1-1/clause 6.4.2.1(2)***. *3-1-1/clause 6.4.2.1(1)* *3-1-1/clause 6.4.2.1(2)*

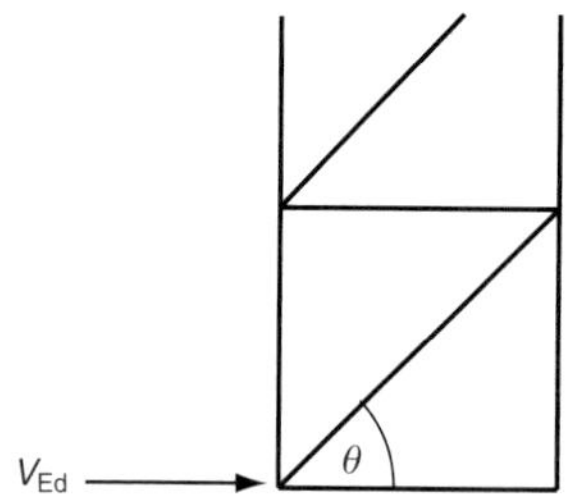

Fig. 6.4-3. Design force for lacings

Laced built-up compression members have a triangulated lattice arrangement joining the individual compression chords as illustrated in 3-1-1/Fig. 6.9. The shear stiffness values S_v given there are derived from the axial shortening of the lacings under axial force. The effective second moment of area of the whole member may be taken as $I_{eff} = 0.5h_0^2A_{ch}$ in accordance with ***3-1-1/clause 6.4.2.1(4)***. This assumes the area of each chord is concentrated at its centroid.

3-1-1/clause 6.4.2.1(4)

The shear force in expression 3-1-1/(6.70) has to be used to design the lacings. The design force on the lacing system is as illustrated in Fig. 6.4-3 and the force in the n planes of lacings is therefore $V_{Ed}/\cos\theta$.

Care needs to be taken if lacings are combined with battens. For a single lacing system, as in the left-hand system of 3-1-1/Fig. 6.9, the chords move apart under axial force and no forces are induced from this effect in the lacings. If battens are introduced, particularly in conjunction with a cross-laced system, the battens prevent the spread of the chords under axial force and forces are generated in the lacings and battens. In situations like this, the lacings and battens should be modelled with the chords in the structural model and the components designed for the resulting actions.

6.4.3. Battened compression members

Battened built-up compression members have horizontal braces ('battens') joining the individual compression chords in a Vierendeel arrangement as illustrated in Fig. 6.4-4 below. Since a Vierendeel truss resists shear loads by combined bending of the braces and chords, the shear stiffness depends on the second moment of area of the battens and chords. The shear stiffness is given in ***3-1-1/clause 6.4.3.1(2)***:

3-1-1/clause 6.4.3.1(2)

$$S_v = \frac{24EI_{ch}}{a^2\left[1 + \dfrac{2I_{ch}h_0}{nI_b a}\right]} \leq \frac{2\pi^2 EI_{ch}}{a^2} \qquad \text{3-1-1/(6.73)}$$

3-1-1/clause 6.4.3.1(3)

The effective second moment of area of a battened built-up column is given in ***3-1-1/clause 6.4.3.1(3)*** by:

$$I_{eff} = 0.5h_0^2A_{ch} + 2\mu I_{ch} \qquad \text{3-1-1/(6.74)}$$

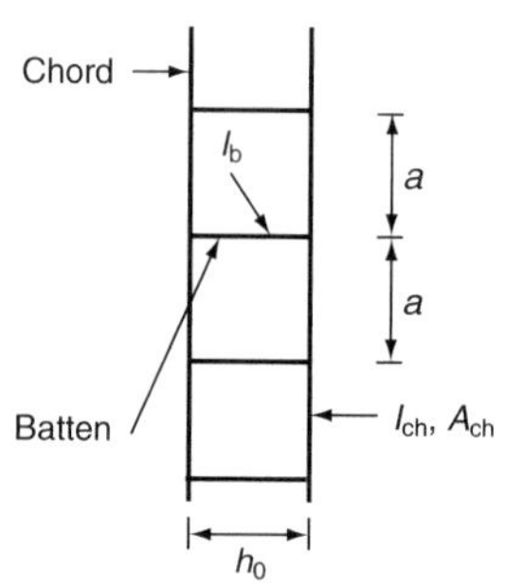

Fig. 6.4-4. Typical batten layout with S_v value

where μ is an efficiency factor from 3-1-1/Table 6.8 which:

$= 0$ for $\lambda \geq 150$

$= 2 - \dfrac{\lambda}{75}$ for $75 < \lambda < 150$

$= 1.0$ for $\lambda \leq 75$

with $\lambda = L/i_0$, $i_0 = \sqrt{I_1/(2A_{ch})}$ and $I_1 = 0.5h_0^2 A_{ch} + 2I_{ch}$.

The chords and overall member must be checked as discussed in section 6.4.1 above with the maximum chord force $N_{ch,Ed}$ derived from expression 3-1-1/(6.69). For battened members, the shear force of expression 3-1-1/(6.70) also produces moments in the chords. The moments in an end bay are shown in 3-1-1/Fig. 6.11 – not reproduced here. The local check of the chords between battens must also therefore include these additional local moments. ***3-1-1/clause 6.4.3.1(1)*** requires the chords and battens to be checked at an end bay (where the shear from expression 3-1-1/(6.70) and hence Vierendeel effects are a maximum but the chord force is a minimum) and at the centre of the column half-wavelength of buckling (where the chord force is a maximum but the Vierendeel effects are a minimum). It is however simplest to combine the greatest chord force with the greatest shear force. It is often desirable for the spacings of battens to be set such that the chord slenderness $\bar{\lambda}$ is less than 0.2; the chords then only need to be checked for cross-section resistance. *3-1-1/clause 6.4.3.1(1)*

6.4.4. Closely spaced built-up members

Built-up compression members which are classed as 'closely spaced' can be designed as ordinary members in accordance with sections 6.2 and 6.3 of EN 1993. This assumes that the compression members are not prone to buckling between the points where they are connected together – defined as 'interconnections' in 3-1-1/Table 6.9. 3-1-1/Table 6.9 defines the maximum interconnection spacing to ensure that local buckling does not occur.

6.5. Buckling of plates

This section of the guide is split into two sub-sections as follows:

- Plates without out-of-plane loading *Section 6.5.1*
- Plates with out-of-plane loading *Section 6.5.2*

6.5.1. Plates without out-of-plane loading

3-2/clause 6.5(1) refers to EN 1993-1-5 for checks involving plate buckling. Local buckling of plates in stiffened girders and unstiffened Class 4 cross-sections under in-plane stresses can be accounted for in one of two ways according to ***3-2/clause 6.5(2)***: *3-2/clause 6.5(1)* *3-2/clause 6.5(2)*

(a) A reduction to the section properties for stress analysis (section 6.2.2.5 of this guide) for checking sections under bending and axial force. Interactions with shear and transverse loads are carried out as for Class 4 cross-sections as discussed in sections 6.2.8 to 6.2.11 of this guide. There are certain geometrical limitations placed on the applicability of this method as discussed in section 6.2.2.5 of this guide.
(b) Separate panel-by-panel buckling checks using stresses obtained on the gross cross-section (reduced stress method – see section 6.2.2.6 of this guide). This method directly incorporates interaction between shear and direct stresses.

6.5.2. Plates with out-of-plane loading

Where there is out-of-plane loading, as will occur for example in a longitudinally stiffened deck plate subjected to traffic loading, the situation is not particularly well covered in EN 1993-2 – ***3-2/clause 6.5(3)*** refers. Similar problems occur with a flange curved in *3-2/clause 6.5(3)*

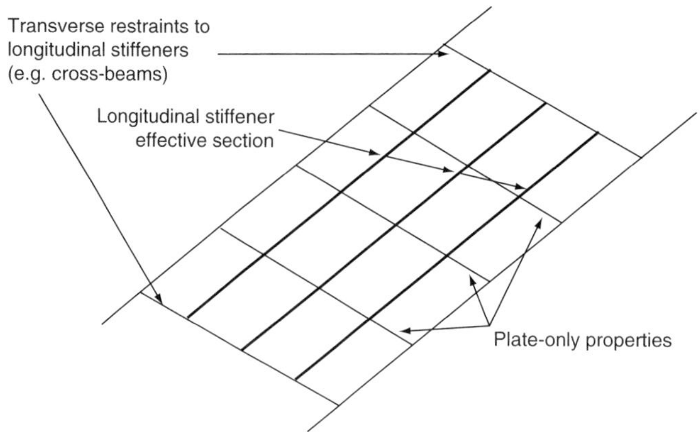

Fig. 6.5-1. Grillage idealization of longitudinally stiffened plate

elevation, which also leads to out-of-plane bending moments as discussed in section 6.10 of this guide. It is intended that EN 1993-1-7 will cover out-of-plane loading but it was not completed at the time of writing this guide. The approach would be different depending on whether method (a) or (b) above were used for designing the stiffened plate. In both methods discussed below (sections 6.5.2.1 and 6.5.2.2), any additional flange moments arising from cambering or curving of the flange would also need to be included – section 6.10 of this guide refers. Only stiffened plate is considered below. A similar method could be developed for unstiffened plate, in which case the magnifier in sections 6.5.2.1(i)b) and 6.5.2.1(ii) below would be based on $\sigma_{cr,p}$ rather than $\sigma_{cr,c}$.

6.5.2.1. Design using effective sections

For longitudinally stiffened panels, the stability of the stiffeners against buckling needs to be checked and the interaction with shear stress and transverse bending moments verified. These are considered in (i), (ii) and (iii) respectively below.

(i) Longitudinal stiffener stability

To determine the out-of-plane longitudinal bending moments in the stiffener, the deck plate can be modelled as a grillage of beam elements as in Fig. 6.5-1 or, more realistically, by shell elements. For grillage modelling, longitudinal members should be placed on the line of each longitudinal stiffener. Transverse members between transverse restraints to the longitudinal stiffeners should represent the deck plate only. Each longitudinal member should represent the stiffener together with an attached width of flange plate. The analysis is unlikely to be strongly influenced by the attached width chosen. A convenient choice of attached width is therefore the effective width derived for sub-panel buckling between the stiffeners in accordance with 3-1-5/clause 4.4, as this is compatible with the proposed strength checks below and gives a reasonable stiffness for the participating deck plate. The effects of shear lag in further reducing attached width for the local moments can also be considered if necessary using EN 1993-1-5 as discussed in section 6.2.2.3 of this guide.

The global longitudinal stresses in the stiffened deck plate should be determined from the effective sections for the bridge for bending and axial force as discussed in section 6.2.2.5. This must include the effects of shear lag as discussed in section 6.2.2.3.

There are then two possibilities for performing the interaction:

(a) Interaction for beam–columns in 3-1-1/clause 6.3.3

3-2/clause 6.5(3) ***3-2/clause 6.5(3)*** suggests that deck plates in compression with bending moments from out-of-plane load can be verified using the interaction between axial force and bending moment given in 3-2/clause 6.3.3 or 3-1-1/clause 6.3.3. To define the beam–column, an effective section for each stiffener acting as a strut can be derived from 3-1-5/clause 4.5.1(4), ignoring the effects of overall flange buckling. The reason for using an effective section ignoring overall stiffened plate buckling is because global buckling is to be considered

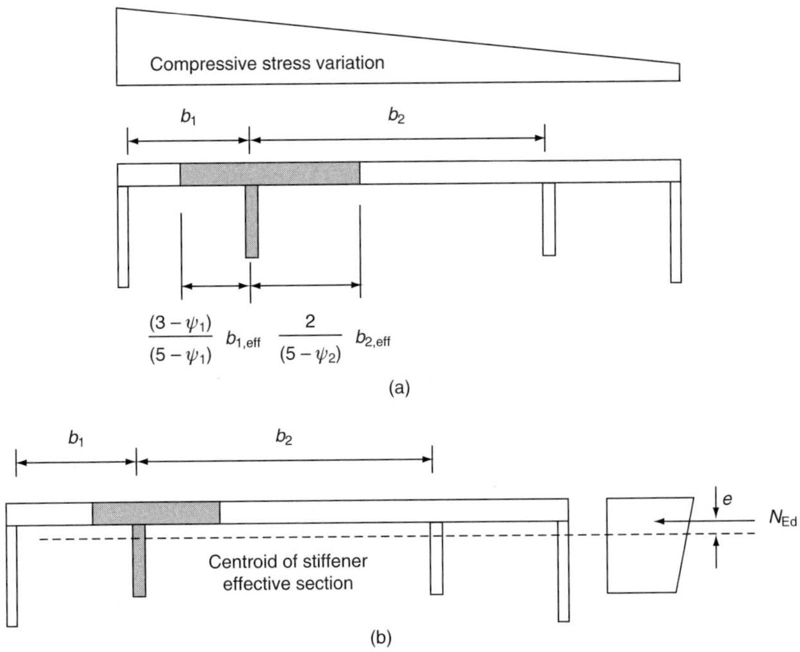

Fig. 6.5-2. Effective section and action derivation for beam–column buckling check: (a) effective section for beam–column buckling check (ignoring overall flange buckling); (b) determination of stiffener force and moment from overall cross-section effective section (including effect of overall flange buckling)

subsequently in the check to 3-2/clause 6.3.3. The effective area is therefore given by:

$$A_{c,eff,loc} = A_{sl,eff} + \rho_{loc} b_{c,loc} t \qquad \text{(D6.5-1)}$$

where:

$A_{sl,eff}$ is the effective cross-sectional area of the stiffener considered in the compression zone reduced for plate buckling if relevant;

$\rho_{loc} b_{c,loc} t$ is the effective cross-sectional area of attached adjacent sub-panels in the compression zone, reduced for local plate buckling as shown in Fig. 6.5-2(a). For closed stiffeners, an effective width of deck plating between the two stiffener attachment points would also be included.

It should be noted that the definition of $A_{c,eff,loc}$ here, as the area of one stiffener effective section, differs from that in 3-1-5/clause 4.5.1(4) where it is the area of the whole compression zone.

It is then necessary to determine the bending moment and axial force acting on the effective cross-section of Fig. 6.5-2(a). The longitudinal local moment from transverse loading is assumed to act on this effective section. The longitudinal force, N_{Ed}, in this effective section can be derived from the flange force in the effective stiffener section **including** the effects of overall flange buckling as in Fig. 6.5-2(b). This effective section is equivalent to the cross-sectional area $A_{c,eff}$ in 3-1-5/clause 4.5.1(3), but relates to the area of one stiffener effective section only and not to the area of the whole compression zone. The flange force is therefore determined from the global flange stress (following the procedure of 3-1-5/clause 4), multiplied by the area of this effective stiffener section **including** the effects of overall buckling in Fig. 6.5-2(b). (If the stress were applied to the cross-section in Fig. 6.5-2(a), the force in the stiffened panel would be overestimated, which is obviously conservative.) If the stress varies significantly through the height of the stiffener effective section, the moment $N_{Ed} \times e$ also needs to be determined from the effective section **including** the effects of overall flange buckling. This moment can then be added to that from the local loading.

The stiffener effective section in Fig. 6.5-2(a) can then be checked for moment and axial force using 3-2/clause 6.3.3 or 3-1-1/clause 6.3.3 for the buckling check. The reduction factor for strut buckling should be calculated using the increased imperfection parameter

$$\alpha_e = \alpha + \frac{0.09}{i/e}$$

appropriate for column buckling in 3-1-5/clause 4.5.3(5). The potential confusion above in having two effective sections can be avoided by using the stiffener effective section in Fig. 6.5-2(b) throughout, but this would be conservative as overall buckling would effectively be considered twice.

(b) Simpler method avoiding the use of the interaction in EN 1993-2

As a simpler method, η_1 could first be calculated as for the case of no local transverse load as discussed in section 6.2.10 of this guide, but an additional term for the local longitudinal bending stress in the stiffener effective section then needs to be added. The maximum longitudinal local bending stress in the stiffened plate, $\sigma_{\text{bend,long}}$, can be calculated from a grillage or finite-element model. This local bending will be amplified by the flange compression. In the absence of a high in-plane direct stress normal to the longitudinal stiffeners, a conservative magnifier is:

$$\frac{\sigma_{\text{cr,c}}}{\sigma_{\text{cr,c}} - \sigma_{\text{Ed}}}$$

so that a conservative criterion is:

$$\eta_{1,\text{mod}} = \frac{N_{\text{Ed}}}{A_{\text{eff}} f_y/\gamma_{M0}} + \frac{M_{y,\text{Ed}} + N_{\text{Ed}} e_{\text{Ny}}}{W_{\text{eff},y} f_y/\gamma_{M0}} + \frac{M_{z,\text{Ed}} + N_{\text{Ed}} e_{\text{Nz}}}{W_{\text{eff},z} f_y/\gamma_{M0}} + \frac{\sigma_{\text{bend,long}}}{f_y/\gamma_{M0}} \frac{\sigma_{\text{cr,c}}}{\sigma_{\text{cr,c}} - \sigma_{\text{Ed}}} \leq 1.0 \qquad \text{(D6.5-2)}$$

where:

- $\sigma_{\text{cr,c}}$ is the critical buckling stress for column-like buckling determined in accordance with 3-1-5/clause 4 as discussed in section 6.2.2.5 of this guide.
- σ_{Ed} is the compressive stress due to global effects at the centroid of the stiffened deck plate at the location of the stiffener being checked (determined using effective section properties as for determination of η_1). This can conservatively be taken as the maximum fibre stress.
- $\sigma_{\text{bend,long}}$ is the maximum longitudinal local bending stress in the stiffened plate, calculated for the fibre of the stiffener effective section which maximizes the value of equation (D6.5-2). This is a conservative alternative to applying a modified version of equation (D6.5-2) separately to each extreme fibre of the effective section.

Other terms have their meanings as in 3-1-5/clause 4.6.

Equation (D6.5-2) assumes that failure occurs when the stress in the most heavily loaded stiffener effective section reaches yield. This can be conservative where there are many longitudinal stiffeners and the local moment affects only a small width of the deck, as global longitudinal stresses can be shed to adjacent less heavily loaded stiffeners. Additionally, the magnifier above can be conservative where $\sigma_{\text{cr,p}}$ is significantly greater than $\sigma_{\text{cr,c}}$ and there is therefore significant restraint to buckling provided by the plating in the transverse direction. The use of axial stress in the flange, σ_{Ed}, based on the effective section is conservative for use in the magnifier above as the column-like buckling stress, $\sigma_{\text{cr,c}}$, is derived on the gross section. σ_{Ed} could therefore be adjusted to be on the gross section (but still including shear lag effects) as used to derive $\sigma_{\text{cr,c}}$ for a less conservative verification.

An in-plane direct stress perpendicular to the stiffeners would give rise to an additional analogous magnification of stiffener moment and a modification to the magnifier in equation (D6.5-2) would be needed.

(ii) Interaction with in-plane shear in stiffened plate

The interaction between in-plane shear and direct stress in the flange must be checked using 3-1-5/clause 7.1 as discussed in section 6.2.9.2.3 of this guide, where a worked example is presented. An additional term for the local longitudinal bending stress in the stiffened plate must however be added into $\bar{\eta}_1$ when there is local transverse load on the flange so that:

$$\bar{\eta}_1 = \eta_{1,\text{mod}} = \frac{N_{\text{Ed}}}{A_{\text{eff}} f_y/\gamma_{M0}} + \frac{M_{y,\text{Ed}} + N_{\text{Ed}} e_{\text{Ny}}}{W_{\text{eff},y} f_y/\gamma_{M0}} + \frac{M_{z,\text{Ed}} + N_{\text{Ed}} e_{\text{Nz}}}{W_{\text{eff},z} f_y/\gamma_{M0}} + \frac{\sigma_{\text{bend,long}}}{f_y/\gamma_{M0}} \frac{\sigma_{\text{cr,c}}}{\sigma_{\text{cr,c}} - \sigma_{\text{Ed}}}$$

and the interaction with shear is then:

$$\eta_{1,\text{mod}} + (2\bar{\eta}_3 - 1)^2 \leq 1.0 \qquad \text{(D6.5-3)}$$

For the check of sub-panel buckling, $\sigma_{\text{bend,long}}$ can be taken as the value at the mid-plane of the flange plate. For overall buckling, the maximum value in the stiffened deck plate should be used. The comments on the use of $\sigma_{\text{cr,c}}$ and σ_{Ed} made in (i)(b) above apply.

It is possible, although very unlikely, that the vertical shear stress from local loading in the stiffener itself might be significant, i.e. greater than 50% of its plastic resistance. In this case, the cross-section resistance of the stiffener effective section should also be verified under bending, axial and shear stress, treating the stiffener as a Class 3 cross-section as discussed in section 6.2.11.1.2 of this guide.

(iii) Deck plate

Where there is direct stress perpendicular to the stiffeners, $\sigma_{z,\text{Ed}}$, a further check is also required of yielding in the parent deck plate. As well as longitudinal local bending stresses in the flange plate, there will also be some transverse bending stress $\sigma_{\text{bend,trans}}$ arising from transverse spanning of the flange between webs and stiffeners, together with in-plane shear. No interaction is given for this combination so the Von Mises check of 3-1-1/clause 6.2.1 (which is similar to the check required in BS 5400: Part 3[4]) could be carried out:

$$\left(\frac{\sigma_{x,\text{Ed}}}{f_y/\gamma_{M0}}\right)^2 + \left(\frac{\sigma_{z,\text{Ed}}}{f_y/\gamma_{M0}}\right)^2 - \left(\frac{\sigma_{x,\text{Ed}}}{f_y/\gamma_{M0}}\right)\left(\frac{\sigma_{z,\text{Ed}}}{f_y/\gamma_{M0}}\right) + 3\left(\frac{\tau_{\text{Ed}}}{f_y/\gamma_{M0}}\right)^2 \leq 1.0 \qquad \text{(D6.5-4)}$$

In this case:

$\sigma_{x,\text{Ed}}$ is the longitudinal direct stress at the mid-plane of the flange plate calculated on the effective section allowing for plate buckling and shear lag and including local bending stress, $\sigma_{\text{bend,long}}$, also calculated at the mid-plane of the deck plate;

$\sigma_{z,\text{Ed}}$ is the transverse direct stress at the extreme fibre of the flange plate including $\sigma_{\text{bend,trans}}$ (extreme fibre stress in the flange is used as there would otherwise be no effect from transverse bending of the flange);

τ_{Ed} is the in-plane shear stress in the flange, based on the elastic shear distribution. In a similar check, BS 5400: Part 3[4] allowed the shear stress to be based on 50% of the maximum shear stress at the web–flange junction due to the beam vertical shear force plus 100% of the torsional shear stress. This is similar to the approach in 3-1-5/clause 7.1 for overall flange buckling, discussed in section 6.2.9.2.3 of this guide, and it would be reasonable to assume this here. Shear stresses from distortional warping and horizontal load should also be included if significant.

The local bending stresses above could conservatively be enhanced by a magnifier similar to that in equation (D6.5-2) but the use of the Von Mises yield criterion is generally sufficiently conservative without doing this. It is possible, although very unlikely, that the shear stress through the thickness of the deck plate might be significant. In this case, the more general Von Mises criterion in section 6.2.1 of this guide could be used to include this additional shear stress component.

6.5.2.2. Design using reduced stress method

If the limiting stress method of 3-1-5/clause 10 is used, the interaction could be modified as follows for overall stiffened panel buckling:

$$\left[\frac{\sigma_{x,Ed}}{\rho_x f_y/\gamma_{M1}} + \frac{\sigma_{bend,long}}{f_y/\gamma_{M1}}\left(\frac{1}{1-1/\alpha_{cr}}\right)\right]^2 + \left[\frac{\sigma_{z,Ed}}{\rho_z f_y/\gamma_{M1}} + \frac{\sigma_{bend,trans}}{f_y/\gamma_{M1}}\left(\frac{1}{1-1/\alpha_{cr}}\right)\right]^2$$
$$- \left[\frac{\sigma_{x,Ed}}{\rho_x f_y/\gamma_{M1}} + \frac{\sigma_{bend,long}}{f_y/\gamma_{M1}}\left(\frac{1}{1-1/\alpha_{cr}}\right)\right]\left[\frac{\sigma_{z,Ed}}{\rho_x f_y/\gamma_{M1}} + \frac{\sigma_{bend,trans}}{f_y/\gamma_{M1}}\left(\frac{1}{1-1/\alpha_{cr}}\right)\right]$$
$$+ 3\left(\frac{\tau_{Ed}}{\chi_v f_y/\gamma_{M1}}\right)^2 \leq 1.0 \qquad \text{(D6.5-5)}$$

where:

α_{cr} is the minimum load factor applied to the design loads required to give elastic critical buckling of the panel considered under all stresses acting together, but excluding the stresses from out-of-plane loading, as discussed in section 6.2.2.6 of this guide. The use of this multiplier will be conservative as not all the in-plane actions will contribute to amplifying the local bending stresses;

$\sigma_{x,Ed}$ is the stress from global analysis (i.e. excluding local moments) in the direction of the deck (parallel to the stiffeners) as defined in 3-1-5/clause 10;

$\sigma_{z,Ed}$ is the stress from global analysis (i.e. excluding local moments) in the direction transverse to the deck (perpendicular to the stiffeners) as defined in 3-1-5/clause 10;

τ_{Ed} is the in-plane shear stress in the flange, taken equal to 50% of the maximum shear stress at the web–flange junction due to the beam vertical shear force plus 100% of the torsional shear stress. Where shear stress from other effects is present, such as from warping or horizontal loading, this shear stress should also be included. This is discussed further in section 6.2.9.2.3. of this guide;

$\sigma_{bend,trans}$ and $\sigma_{bend,long}$ refer to peak bending stress transversely in flange plate and longitudinally in stiffened deck plate respectively. The reduction factors in equation (D6.5-5) should be determined from α_{cr} ignoring the local moments. Gross properties, other than making allowance for shear lag, should be used.

For sub-panel buckling, the method of 3-1-5/clause 10 could be used, modified as follows:

$$\left(\frac{\sigma_{x,Ed}}{\rho_x f_y/\gamma_{M1}}\right)^2 + \left[\frac{\sigma_{z,Ed}}{\rho_z f_y/\gamma_{M1}} + \frac{\sigma_{bend,trans}}{f_y/\gamma_{M1}}\left(\frac{1}{1-1/\alpha_{cr}}\right)\right]^2$$
$$- \left(\frac{\sigma_{x,Ed}}{\rho_x f_y/\gamma_{M1}}\right)\left[\frac{\sigma_{z,Ed}}{\rho_x f_y/\gamma_{M1}} + \frac{\sigma_{bend,trans}}{f_y/\gamma_{M1}}\left(\frac{1}{1-1/\alpha_{cr}}\right)\right]$$
$$+ 3\left(\frac{\tau_{Ed}}{\chi_v f_y/\gamma_{M1}}\right)^2 \leq 1.0 \qquad \text{(D6.5-6)}$$

The definitions are as above except that:

$\sigma_{x,Ed}$ should now include $\sigma_{bend,long}$ calculated at the mid-plane of the flange plate (as it effectively provides an axial force in the flange plate);

τ_{Ed} is the average in-plane shear stress within the sub-panel of the flange, calculated from the elastic shear distribution;

α_{cr} and reduction factors should be calculated for $\sigma_{x,Ed}$, $\sigma_{z,Ed}$ and τ_{Ed}.

6.6. Intermediate transverse stiffeners (additional sub-section)

No rules are given within EN 1993-2 for the design of intermediate transverse stiffeners so reference has to be made to EN 1993-1-5 section 9. This section brings together the relevant rules of EN 1993-1-5. It covers the following:

- Effective section of a stiffener and choice of design method *Section 6.6.1*
- Transverse web stiffeners – general method *Section 6.6.2*
- Transverse web stiffeners not required to contribute to the adequacy of the web under direct stress *Section 6.6.3*
- Additional effects applicable to certain transverse web stiffeners *Section 6.6.4*
- Flange transverse stiffeners *Section 6.6.5*

6.6.1. Effective section of a stiffener and choice of design method

The properties of a stiffener effective section are calculated from ***3-1-5/clause 9.1(2)*** using an attached width of web of $15\varepsilon t$ each side of the stiffener as shown in Fig. 6.6-1, but not greater than the available width, $\varepsilon = \sqrt{235/f_y}$. *3-1-5/clause 9.1(2)*

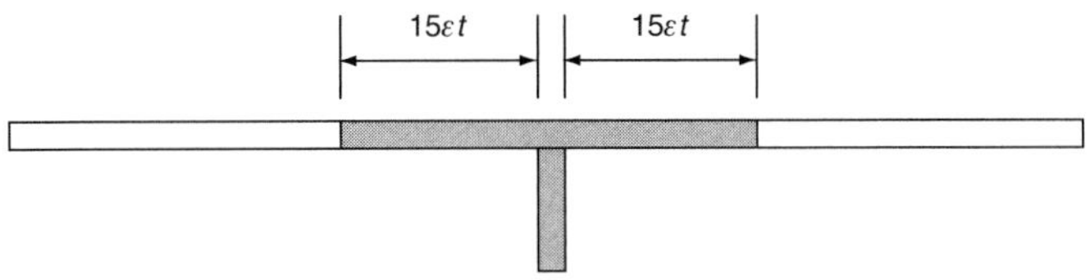

Fig. 6.6-1. Effective section for stiffener

Different methods of design could be used depending on whether the adequacy of the web under direct stress (from axial load and bending moment) is dependent on the presence of the transverse stiffeners. Where transverse stiffeners support longitudinal stiffeners, the method of section 6.6.2 below has to be used. Where there are no longitudinal stiffeners, the choice of method is less clear, although the method in section 6.6.2 is always applicable. The drafters of EN 1993-1-5 did not intend the out-of-plane effects from direct stress in 3-1-5/clause 9.2.1 to be considered, unless the transverse stiffeners are to be considered in deriving the resistance of the web to direct stresses. However, as discussed in section 6.6.2.4(a) below, there are arguments to be made for considering the out-of-plane effects in all cases.

It will usually be found that the out-of-plane forces on a transverse stiffener caused by web direct stresses in 3-1-5/clause 9.2.1 are small for unstiffened webs unless the stiffener spacings are small ($a/b < 1$). Unless the stiffeners contribute to the resistance of the web under direct stress, the stiffness criterion for the transverse stiffeners in 3-1-5/clause 9.2.1 is not relevant; the web is adequate for direct stress without them. In such cases it would be reasonable to use the simplified method of section 6.6.3. This still checks the transverse stiffener for strength under the out-of-plane force from web compression, but omits the stiffness check.

6.6.2. Transverse web stiffeners – general method

The following requirements have to be met:

(i) The outstand should meet the limits in 3-1-5/clause 9.2.1 for preventing torsional buckling.

(ii) The effective section must meet the minimum stiffness requirements for shear in 3-1-5/clause 9.3.3.

(iii) The effective section must resist the force from shear tension field action according to 3-1-5/clause 9.3.3, together with any externally applied forces and moments – 3-1-5/clause 9.1(3) refers. Section 6.6.4 is relevant for the latter.

(iv) The effective section must meet the minimum strength and stiffness requirements in 3-1-5/clause 9.2.1 where the transverse stiffener is assumed to restrain either an unstiffened or stiffened web panel from buckling under direct stress. The forces developed in the stiffener in restraining the web are often said to arise from the 'destabilising influence of the web'. The force from shear tension field action according to 3-1-5/clause 9.3.3, together with any externally applied load or moment, must also be included in these checks of strength and stiffness.

(v) The cross-section resistance at a loaded end should be checked, according to 3-1-5/clause 9.4(2).

These requirements are discussed in turn in sections 6.6.2.1 to 6.6.2.5 respectively.

6.6.2.1. Torsional buckling

Torsional buckling of a stiffener outstand will lead to premature failure of the overall stiffener effective section and must therefore be prevented. Guidance on this is given in section 6.9 of this guide.

6.6.2.2. Minimum stiffness to act as a rigid support to web panels in shear

3-1-5/clause 9.3.3(3)

The following stiffness requirements from ***3-1-5/clause 9.3.3(3)*** for the stiffener effective section have to be met for a stiffener to act as a rigid support to a web panel in shear:

$$I_{st} \geq 1.5h_w^3t^3/a^2 \text{ if } a/h_w < \sqrt{2} \qquad \text{3-1-5/(9.6)}$$

$$I_{st} \geq 0.75h_wt^3 \text{ if } a/h_w \geq \sqrt{2}$$

These equations are the same as were provided in BS 5950: Part 1,[26] although the definition of stiffener second moment of area differed slightly. For shear buckling, the elastic critical shear stress continues to rise as the stiffener stiffness rises. It tends towards a limiting value for rigid stiffeners because nodal lines in the buckling mode along the line of the stiffener are only produced for a fully rigid stiffener. The stiffnesses above are considerably greater than the values necessary to achieve say 95% of the elastic critical buckling for fully rigid boundaries. This is necessary to allow for the presence of imperfections in real plates. The required stiffener inertia is independent of panel length for panels longer than $\sqrt{2}h_w$. For long panels, the mode of buckling changes to multiple buckles, so panel length has little influence.

If the stiffener does not comply with the minimum stiffness requirements for 'rigid' behaviour, it could still be included in the calculation of shear resistance as a 'flexible' stiffener. This is permitted by EN 1993-1-5 but no method for including its contribution to the elastic critical buckling stress, τ_{cr}, is given. This is discussed in section 6.2.6 of this guide.

6.6.2.3. Force from shear tension field plus external vertical loading and moment

3-1-5/clause 9.3.3(3)

A simple provision is proposed for checking the strength of stiffeners which act as rigid restraints to web panels in shear. ***3-1-5/clause 9.3.3(3)*** requires the stiffener to be checked for the difference between the applied shear and the elastic critical shear force of the web panel. This is not strictly compatible with the rotated stress theory used in the shear design, which does not require the stiffeners to carry any load other than the part of the tension field anchored by the flanges, corresponding to the term $V_{bf,Rd}$. In the absence of a stiff flange to contribute to $V_{bf,Rd}$, the stiffeners simply contribute to elevating the elastic critical shear stress of the web.

Despite the EN 1993-1-5 predictions above, stiffeners do in reality develop stresses from compatibility of deflections, because their presence keeps the web flat at the stiffener locations, which changes the state of stress in the web. These stresses vary in a complex manner and a stiffener might not always have adequate post-buckling ductility to shed them in conjunction with the effects of other applied actions and even if it does, a check at serviceability might be necessitated. As a result, 3-1-5/clause 9.3.3(3) effectively requires a stiffener to carry a force equal to the shear force in excess of that required to cause elastic critical buckling. This leads to the stiffener design force being:

$$P_{shear} = V_{Ed} - \frac{h_w t \tau_{cr}}{\gamma_{M1}} \qquad \text{(D6.6-1)}$$

Equation (D6.6-1) follows from a simple truss of the form shown in Fig. 6.6-2. The notation P has been used for the stiffener force to distinguish it from the use of N for web axial force. This equation was not universally agreed at the drafting stage. It was believed by most to be overly conservative. Several European national standards previously provided only a stiffness requirement on the basis that test results indicated that only small forces develop in transverse stiffeners with adequate stiffness.[13] However, the BS 5400: Part 3[4] formula (which is similar to equation (D6.6-1) when beams have equal flanges and no axial force) was compared against tests by Evans and Tang[25] for beams without longitudinal stiffeners and found to be slightly conservative but 'not unreasonably so'. Notably however, no stiffeners actually failed, even in the test designed to produce stiffener failure.

A further criticism that has been made of equation (D6.6-1) by some in the UK, wishing to preserve the BS 5400 rules, was that EN 1993-1-5 does not make allowance for the possibility of elastic critical buckling occurring at a shear stress less than τ_{cr} when direct stresses from bending and axial force are present in the web panels. BS 5400: Part 3[4] considered this effect and reduced τ_{cr} in the presence of direct stress, although it is not clear that this is justified as the buckling modes for shear and axial force are quite different. Such considerations lead to significant discrepancy with EN 1993-1-5 for beams with unequal flanges (and hence significant average web compression). Given the general feeling in mainland Europe that the force produced by equation (D6.6-1) was already too conservative, any further increase in force was rejected by the drafters of EN 1993-1-5.

A non-linear finite-element parametric study of over 40 different cases of varying beam geometries, moment–shear ratios and axial force has been carried out by the first author of this guide and a colleague, Francesco Presta. In all cases, the EN 1993-1-5 rules were shown to be safe. Further, in every case tested, the stiffness requirement of 3-1-5/clause 9.3.3(3) on its own would have sufficed as a design criterion. The behaviour observed was very much as predicted by the rotated stress field theory of Höglund. Up until a shear stress of around the elastic critical value, a linear distribution of bending stress occurred across the depth of the cross-section. Beyond this shear stress, a membrane tension developed which modified the distribution of direct stress in the girder. This gave rise to a net tension in the web which was balanced by opposing compressive forces in the flanges, adding to the flexural compressive stress in one flange and reducing the flexural tensile stress in the other. This behaviour gives an increase in compressive flange force beyond that predicted solely from a cross-section bending analysis, but not from that predicted by the EN 1993-1-5 shear–moment interaction in 3-1-5/clause 7. For cases with strong flanges, some additional tension field was anchored by the flanges and the force transferred to the stiffeners. The conclusion was that the rules of EN 1993-1-5 were somewhat conservative.

Since the slenderness for a panel in shear is:

$$\bar{\lambda}_w = \sqrt{\frac{\tau_y}{\tau_{cr}}} = \sqrt{\frac{f_y}{\sqrt{3}\tau_{cr}}}$$

substitution into equation (D6.6-1) leads to the expression in 3-1-5/clause 9.3.3(3):

$$P_{shear} = V_{Ed} - \frac{1}{\bar{\lambda}_w^2}\frac{f_{yw}h_w t}{\sqrt{3}\gamma_{M1}} \qquad \text{(D6.6-2)}$$

Due to the effect of imperfections, forces may develop in the stiffener slightly before τ_{cr} is reached and γ_{M1} is intended to perform this function. For the truss idealization, this allowance is conservative at high slenderness where imperfections have little effect, but may be slightly unconservative at intermediate slenderness, where their effect is greatest. Given the conservative nature of the whole truss model, this is not of concern.

Notwithstanding the comments on conservatism above, the shear force used to calculate P_{shear} should be based on the value $0.5h_w$ from the most highly stressed end of the panel – 3-1-5/clause 9.3.3(3) refers. The method of calculating λ_w is illustrated in section 6.2.6 of this guide. If the panels are different each side of the stiffener, P_{shear} could be calculated for each adjacent panel and the greater value used in design. This is conservative for buckling of the stiffener as the tension bands in the two panels would produce different forces at the top and bottom of the stiffeners so the value in the middle third would be less than the maximum. BS 5400: Part 3[4] allowed the average of the forces from two panels to be used, but this is not obviously safe for the yield check near stiffener ends. The ENV version of EN 1993-1-1 required the greater force to be used. Where there are longitudinal stiffeners, λ_w could

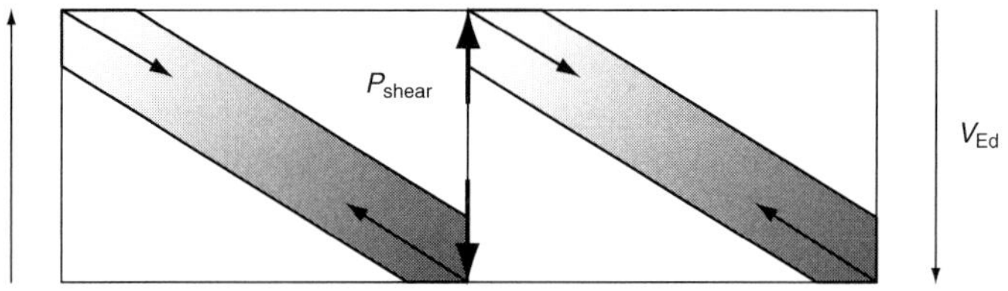

Fig. 6.6-2. Tension band mechanism generating stiffener force

conservatively be taken as the highest value in any sub-panel or from overall web buckling. Where sub-panel buckling governs, the above is clearly conservative when the slenderness is much greater in one panel than the others. BS 5400 allowed τ_{cr} in this situation to be based on the average of the two lowest values of τ_{cr} obtained for sub-panels; this would be reasonable here also.

The stiffener design force P_{shear} acts in the plane of the web (although not explicitly stated in EN 1993-1-5) and is assumed to be constant over the height of the web. Any external axial load, P_{ext}, must be added to the load from shear above so the total axial load to design the stiffener for is:

$$P_{Ed} = P_{shear} + P_{ext} \quad \text{(D6.6-3)}$$

Where the stiffener effective section is asymmetric, the resulting eccentricity should be considered to produce a moment acting on the centroid of the stiffener section in accordance with ***3-1-5/clause 9.4(3)***. Any eccentricity of applied external loads should be similarly considered, together with any applied moments. The resulting stiffener effective section should be checked for combined bending and axial force using the interactions in 3-1-1/clause 6.3.3 or 3-1-1/clause 6.3.4, but assuming that the stiffener is not prone to lateral–torsional buckling. It is not easy to apply 3-1-1/clause 6.3.3 in these circumstances. The use of 3-2/clause 6.3.3 is simpler and is used in Worked Example 6.6-1. A further alternative would be to use the interaction equation (D6.7-2) provided in section 6.7.2 of this guide for bearing stiffeners. In all cases, ***3-1-5/clause 9.4(2)*** requires that the effective length for flexural buckling is not taken less than $0.75h_w$ and that buckling curve c is used.

3-1-5/clause 9.4(3)

3-1-5/clause 9.4(2)

It is recommended here that the check under bending and axial load should be based on elastic section properties, as plastic deformation in a transverse stiffener would be incompatible with the assumptions made for stiffness. If the check in section 6.6.2.4 below is required (necessitated by destabilising direct stress in the web), elastic behaviour is automatically achieved.

Where (iv) above applies, the stiffener axial force from external loads and from shear tension field action must also be considered in the strength and stiffness checks of 3-1-5/clause 9.2.1. This makes the buckling check to 3-2/clause 6.3.3 redundant as 3-1-5/clause 9.2.1 itself includes a buckling check.

6.6.2.4. Strength and stiffness where there is destabilising influence of the web

This section relates to the check in 3-1-5/clause 9.2.1. There are three possible situations, of which (c) is the most general. These are that, acting in conjunction with the web destabilising force, there may be present in the stiffeners:

(a) no vertical load; or
(b) vertical load; or
(c) vertical load and moment.

Cases (a) and (b) lead to the derivation of case (c). They are discussed in turn below.

(a) Design for destabilising influence of web direct stress – no axial load or moment in the stiffeners

When there are longitudinal stiffeners on a web and they are designed to be restrained by transverse stiffeners, the transverse stiffeners must be designed to provide this support using the method of 3-1-5/clause 9.2.1. Transverse stiffeners must also be designed for forces arising from direct stresses in Class 4 webs without longitudinal stiffeners when the presence of transverse stiffeners increases the resistance of the web panel to direct stress, i.e. the resistance is increased from that for an infinitely long panel. This latter case will not be common as transverse stiffeners would have to be spaced with $a < b$ and the above benefit to effective width would have had to be calculated and utilized in the design.

The drafters of EN 1993-1-5 had not intended that this check be applied to transverse stiffeners that do not contribute to the adequacy of the web under direct stress, as the web would still be adequate (for direct stress) if the stiffener were removed. However, a similar check was made in BS 5400: Part 3[4] in all situations, regardless of web adequacy to direct stresses without the stiffener. The reason for this is that, while the stiffener need not be there, its presence is likely to attract loads which it may not be able to shed. These may cause some additional bow in the stiffener which could interact with tension field forces, and which could lead to serviceability problems.

If a check is made to EN 1993-1-5 for a stiffener which does not increase web direct stress resistance, the stresses generated in the stiffener will typically be negligible in any case and the significance of this issue is reduced. This is due to the effect of the ratio $\sigma_{cr,c}/\sigma_{cr,p}$ discussed below. On this basis, it is recommended here that a check according to section 6.6.3, which is simpler, will suffice where the transverse stiffener does not contribute to increasing web direct stress resistance. Both methods are illustrated in Worked Example 6.6-1.

The design criteria specified in ***3-1-5/clause 9.2.1(4)*** are that the stiffener stresses should not exceed yield and that the deflection under load should not exceed $b/300$. The deflection criterion is to ensure adequate stiffness for support of the longitudinal web plating and/or stiffeners. Where there is no vertical stress in the stiffener due to either tension field action under shear force or external load, the simplified check in ***3-1-5/clause 9.2.1(5)*** may be used which covers both strength and stiffness requirements. The requirements for minimum inertia therein are derived below. *3-1-5/clause 9.2.1(4)* *3-1-5/clause 9.2.1(5)*

The stiffener of interest in Fig. 6.6-3 has an initial sinusoidal bow of maximum size w_0. If the adjacent transverse stiffeners are assumed to be straight and rigid (***3-1-5/clause 9.2.1(3)*** makes this assumption) and the longitudinal stiffeners and web plate are assumed to be hinged at the transverse stiffener being checked, then the out-of-plane varying force per metre up the stiffener is given approximately by: *3-1-5/clause 9.2.1(3)*

$$q(x) = w(x)\left[\left(\frac{1}{a_1} + \frac{1}{a_2}\right)\frac{N_{Ed}}{b}\right] \qquad \text{(D6.6-4)}$$

where the various variables are shown in Fig. 6.6-3. N_{Ed} is taken to be the compressive force in the stiffened panel but not less than the maximum compressive stress times half the effective area of the compression zone for webs in bending. The deflection, $w(x)$, is assumed to be sinusoidal and the force, $q(x)$, is also assumed to be sinusoidal despite the

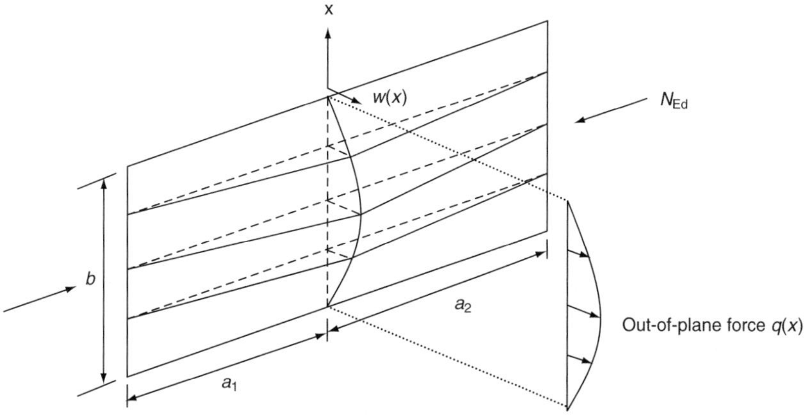

Fig. 6.6-3. Out-of-plane forces acting on a transverse stiffener on a web with direct stress

localized point forces at the levels of any longitudinal stiffeners. For cases with a single large longitudinal stiffener at mid-height, the rules may therefore be slightly unconservative.

The assumption that adjacent stiffeners are straight and rigid differs from the assumption in BS 5400: Part 3,[4] where adjacent stiffeners were assumed to bow in opposite directions, which increases the web kink angle and hence out-of-plane force for a given stiffener bow. The size of initial bow used in 3-1-5/clause 9.2.1(2), together with the low probability that adjacent stiffeners would bow in opposite directions at maximum tolerance, were considered sufficient justification for the EN 1993-1-5 approach by the Project Team.

Since the web plate itself also resists the out-of-plane bowing of the web panel, the force in equation (D6.6-4) may be reduced by introducing the web plate critical buckling stresses as follows:

$$q(x) = w(x)\frac{\sigma_{cr,c}}{\sigma_{cr,p}}\left[\left(\frac{1}{a_1}+\frac{1}{a_2}\right)\frac{N_{Ed}}{b}\right] = w(x)\sigma_m \qquad \text{(D6.6-5)}$$

with

$$\sigma_m = \frac{\sigma_{cr,c}}{\sigma_{cr,p}}\left[\left(\frac{1}{a_1}+\frac{1}{a_2}\right)\frac{N_{Ed}}{b}\right]$$

$\sigma_{cr,c}/\sigma_{cr,p}$ is the ratio of column-like critical buckling stress to plate critical buckling stress. The calculation of these terms is discussed in section 6.2.2.5 of this guide. It will always be conservative to take $\sigma_{cr,c}/\sigma_{cr,p} = 1.0$ but for webs without longitudinal stiffeners, this simplification will usually be excessively conservative. EN 1993-1-5 does not clarify over what panel length to calculate $\sigma_{cr,c}$ and $\sigma_{cr,p}$. A length of $a_1 + a_2$ would be appropriate for the mode in Fig. 6.6-3, but a mode with alternate stiffeners moving in opposite directions is also possible. This latter mode would suggest a length equal to 0.5 $(a_1 + a_2)$ would be appropriate. It will be conservative, and recommended here, to always use the length of the shorter panel in calculating the critical stresses as this will maximize the ratio $\sigma_{cr,c}/\sigma_{cr,p}$. The critical stresses for this length of panel are likely to be available from other calculations; they will not have been calculated for the other lengths.

For webs without longitudinal stiffeners, $\sigma_{cr,c}/\sigma_{cr,p}$ can be very small (as seen in Worked Example 6.6-1) and some have suggested that a lower limit should be placed on its value. The argument against a limit is that if the web does not require the stiffener to be present for its adequacy under direct stress, the stiffener can probably shed the stresses induced. The argument for setting a limit is that the out-of-plane deformation produced acts as an increased initial imperfection when considering the effects of stiffener axial force from the tension field force of 3-1-5/clause 9.3.3(3). No limit has been imposed in EN 1993-1-5. The first author has not found any cases in the course of limited non-linear finite-element studies where one would have been necessary.

If the initial sinusoidal bow is $w_0(x)$ with peak mid-height value w_0 and the additional deflection is $\delta(x)$ with peak mid-height value δ, then the total deflection is:

$$w(x) = w_0(x) + \delta(x) \qquad \text{(D6.6-6)}$$

and the peak distributed load at mid-height is:

$$q_{max} = (w_0 + \delta)\sigma_m \qquad \text{(D6.6-7)}$$

The peak additional deflection in the stiffener with second moment of area, I_{st}, under the sinusoidal load must satisfy:

$$\delta = \frac{(w_0 + \delta)\sigma_m b^4}{\pi^4 E I_{st}} \qquad \text{(D6.6-8)}$$

The stress in the stiffener under the sinusoidal load is:

$$\sigma_s = \frac{(w_0 + \delta)\sigma_m b^2 e_{max}}{\pi^2 I_{st}} \qquad \text{(D6.6-9)}$$

where e_{max} is the greatest distance from stiffener effective section neutral axis to an extreme fibre of the effective section.

From equations (D6.6-8) and (D6.6-9) and setting the stiffener stress to design yield ($f_{yd} = f_y/\gamma_{M1}$), the extra deflection at which yield occurs is:

$$\delta = \frac{b^2 f_{yd}}{\pi^2 E e_{max}} \tag{D6.6-10}$$

If equation (D6.6-10) is substituted into equation (D6.6-8), the following inequality based on limiting stress is produced:

$$I_{st} = \frac{\sigma_m}{E}\left(\frac{b}{\pi}\right)^4 \left(1 + w_0 \frac{\pi^2 E e_{max}}{b^2 f_{yd}}\right) \tag{D6.6-11}$$

Since the additional deflection also has to be limited to $b/300$, using this in equation (D6.6-6) gives:

$$I_{st} = \frac{\sigma_m}{E}\left(\frac{b}{\pi}\right)^4 \left(1 + w_0 \frac{300}{b}\right) \tag{D6.6-12}$$

Both equations (D6.6-11) and (D6.6-12) have to be satisfied, but a single equation can be presented if it is noted by comparing them that:

$$\frac{\pi^2 E e_{max}}{b^2 f_{yd}} \geq 300/b$$

and hence:

$$u = \frac{\pi^2 E e_{max}}{f_{yd} 300 b} \geq 1.0 \tag{D6.6-13}$$

Incorporating equation (D6.6-13) into equation (D6.6-11) leads to the expression in 3-1-5/clause 9.2.1(5):

$$I_{st} \geq \frac{\sigma_m}{E}\left(\frac{b}{\pi}\right)^4 \left(1 + w_0 \frac{300}{b} u\right) \qquad \text{3-1-5/(9.1)}$$

where u is obtained from equation (D6.6-13), but must not be taken as less than 1.0 for deflection control, and σ_m is obtained from equation (D6.6-5).

Since the initial imperfection, w_0, is the lesser of $b/300$ or $a/300$, the problem found in a similar clause in BS 5400: Part 3, where the kink force for closely spaced stiffeners tended to infinity as the stiffener spacing tended to zero, does not occur.

(b) Design for destabilising influence of web direct stress – axial force in stiffeners without eccentricity or other moment

Where there is either external axial force acting on the stiffener or the stiffener carries axial force from shear tension field action according to 3-1-5/clause 9.3.3(3), it is not adequate to verify the above minimum second moment of area and the resistance to axial force separately. In this case, it is necessary to satisfy the basic requirements for deflection and stress given in 3-1-5/clause 9.2.1(4), accounting for the magnifying effect of the axial force in the stiffener. This may be done from a large deflection computer analysis following the assumptions given in 3-1-5/clause 9.2.1. Shell elements could be used or the web could be idealized as a series of discrete struts with actual longitudinal stiffener positions represented. Alternatively, a modified version of the above calculation can be used to account for the magnifying action of the stiffener axial force. It is no longer possible to provide a single expression for the required stiffener second moment of area as the stiffener cross-sectional area also becomes relevant.

A possible hand method, again assuming a sinusoidal force variation from the web, is suggested below. It is the basis of ***3-1-5/clause 9.2.1(6)***. Under the presence of stiffener axial force, P_{Ed}, and with stiffener effective length, $L > 0.75b$ (L will usually equal b for consistency with the end restraints assumed in the analysis above), equation (D6.6-8) *3-1-5/clause 9.2.1(6)* becomes:

$$\delta = \frac{(w_0 + \delta)\sigma_m b^4}{\pi^4 EI_{st}} + \frac{(w_0 + \delta)P_{Ed}L^2}{\pi^2 EI_{st}} = \frac{(w_0 + \delta)}{EI_{st}}\left(\frac{\sigma_m b^4}{\pi^4} + \frac{P_{Ed}L^2}{\pi^2}\right) \quad \text{(D6.6-14)}$$

Rearranging equation (D6.6-14), the extra deflection is:

$$\delta = w_0\left(\frac{EI_{st}}{\frac{\sigma_m b^4}{\pi^4} + \frac{P_{Ed}L^2}{\pi^2}} - 1\right)^{-1} \quad \text{(D6.6-15)}$$

Setting the maximum increase in deflection to $b/300$ gives an expression for the required stiffener inertia based on stiffness as follows:

$$I_{st} \geq \frac{1}{E}\left(1 + w_0\frac{300}{b}\right)\left(\frac{\sigma_m b^4}{\pi^4} + \frac{P_{Ed}L^2}{\pi^2}\right) \quad \text{(D6.6-16)}$$

In order to limit the extreme fibre stress of the stiffener to yield, the expression for stress becomes:

$$\sigma_s = \frac{(w_0 + \delta)\sigma_m b^2 e_{max}}{\pi^2 I_{st}} + \frac{P_{Ed}}{A_{st}} + \frac{P_{Ed}(w_0 + \delta)e_{max}}{I_{st}}$$

$$= \frac{(w_0 + \delta)e_{max}}{I_{st}}\left(\frac{\sigma_m b^2}{\pi^2} + P_{Ed}\right) + \frac{P_{Ed}}{A_{st}} \leq f_{yd} \quad \text{(D6.6-17)}$$

The procedure would thus be to calculate the minimum second moment of area required for deflection according to equation (D6.6-16) and then check stresses using equation (D6.6-17). For the case of zero axial load in the stiffeners, these equations give the same result as presented in expression 3-1-5/(9.1). For the case of zero direct force in the web, they are equivalent to the growth of the initial deflection w_0 and moment $P_{Ed}w_0$ by the magnifier:

$$\frac{1}{1 - P_{Ed}/P_{cr}}$$

as discussed in section 5.2 of this guide.

The above is the basis of 3-1-5/clause 9.2.1(6) which allows the axial force in the stiffener to be taken as:

$$P_{Ed} + \frac{\sigma_m b^2}{\pi^2}$$

to allow for both in-plane web forces and stiffener forces. The term:

$$P_{Ed} + \frac{\sigma_m b^2}{\pi^2}$$

is visible in equation (D6.6-17) where $\sigma_m b^2/\pi^2$ can be seen to contribute only to the bending term and not the axial force term. The *increase* in deflection in a strut from this fictitious axial force would be:

$$\delta = w_0\left(\frac{P_{cr}}{P_{Ed} + \frac{\sigma_m b^2}{\pi^2}} - 1\right)^{-1} = w_0\left[\frac{\pi^2 EI}{\left(P_{Ed} + \frac{\sigma_m b^2}{\pi^2}\right)b^2} - 1\right]^{-1} = w_0\left(\frac{EI_{st}}{\frac{\sigma_m b^4}{\pi^4} + \frac{P_{Ed}b^2}{\pi^2}} - 1\right)^{-1}$$

which is as equation (D6.6-15) with $L = b$.

The axial force in the stiffener is assumed constant throughout the stiffener height in the analysis above. This is conservative for externally applied load on one end of the stiffener only; the force in the stiffener from such a load could be considered to vary from a maximum at the loaded end to zero at the other. In such cases, the value of P_{Ed} in the middle third of the stiffener height could reasonably be used. If this is done, a cross-section check should also be made of stress at the stiffener ends under the maximum effects. Any component of the axial force from tension field action (section 6.6.2.3) should be considered to be constant over the stiffener height.

(c) Design for destabilising influence of web direct stress – axial force in stiffeners with eccentricity and/or other moment

A further limitation of EN 1993-1-5 is that the above does not include the effects of any eccentricity of the axial load (as occurs with typical single-sided stiffeners). The effects of initial moment from eccentricity of axial force, or other applied moments, would have to be added to the above. This is not covered by EN 1993-1-5. Uniform end moments, M_0, could be included by adding the first-order deflection, $M_0b^2/(8EI_{st})$ for pin-ended conditions, onto w_0. Since this first-order deflection is itself also an increase in deflection which occurs under load, this should be added to δ in equation (D6.6-15) and the total compared to the deflection limit of $b/300$. Equation (D6.6-16) should not therefore be used when there are end moments without similar amendment. An additional term, M_0e_{max}/I_{st}, would also have to be introduced into equation (D6.6-17) to allow for the initial moment.

The above discussion assumes that the moment M_0 is either reversible or acts in a direction so as to put the fibre with lowest section modulus (at distance e_{max} from the section centroid) into compression, as the bow direction in the above analysis under web longitudinal stress is chosen to put compression into this fibre. If this is not the case, equation (D6.6-20) developed below is conservative and the method following it could be used to get a less conservative answer.

For stiffeners with end moments M_0, whether from eccentric load or applied moments, assumed constant throughout the height of the stiffener, the procedure is therefore as follows:

$$\delta = w_0'\left(\frac{EI_{st}}{\dfrac{\sigma_m b^4}{\pi^4}+\dfrac{P_{Ed}L^2}{\pi^2}}-1\right)^{-1} \tag{D6.6-18}$$

with

$$w_0' = w_0 + \frac{M_0b^2}{8EI_{st}}$$

Check that total deflection is less than $b/300$ thus:

$$\delta + \frac{M_0b^2}{8EI_{st}} \le \frac{b}{300} \tag{D6.6-19}$$

Check that the stress is less than the design yield stress:

$$\sigma_s = \frac{(w_0'+\delta)e_{max}}{I_{st}}\left(\frac{\sigma_m b^2}{\pi^2}+P_{Ed}\right)+\frac{P_{Ed}}{A_{st}}+\frac{M_0e_{max}}{I_{st}} \le f_{yd} \tag{D6.6-20}$$

This is approximate and slightly underestimates deflections and stresses as the analysis method assumes the initial distribution of M_0 is sinusoidal rather than uniform (except in the calculation of maximum deflection $M_0b^2/(8EI_{st})$ from M_0). The error is however small.

Where the stiffener axial force and moment varies over the height of the stiffener, the values in the middle third could reasonably be used as discussed in (b) above.

Generally, the moment M_0 is not likely to act in a direction which puts the fibre with lowest section modulus (at distance e_{max} from the section centroid) into compression. It is more likely to relieve stresses in most cases, as moment usually arises from load applied at the web position in single-sided stiffeners such that the stiffener outstand is put into tension. Strictly, if a moment **does not** act so as at to put the lowest section modulus fibre into compression, then a number of checks are needed. The stiffener needs to be considered to bow in either direction and equation (D6.6-20) modified accordingly.

For bowing in the direction of a moment producing compression in the higher section modulus fibre, the compression fibre is checked as follows (treating P and M_0 positive throughout):

$$\sigma_s = \frac{(w_0' + \delta)e_{min}}{I_{st}}\left(\frac{\sigma_m b^2}{\pi^2} + P_{Ed}\right) + \frac{P_{Ed}}{A_{st}} + \frac{M_0 e_{min}}{I_{st}} \leq f_{yd} \text{ (compressive stress +ve)}$$

with $w_0' = w_0 + M_0 b^2/(8EI_{st})$.

The tension fibre would be checked with:

$$\sigma_s = \frac{(w_0' + \delta)e_{max}}{I_{st}}\left(\frac{\sigma_m b^2}{\pi^2} + P_{Ed}\right) - \frac{P_{Ed}}{A_{st}} + \frac{M_0 e_{max}}{I_{st}} \leq f_{yd} \text{ (tensile stress +ve)}$$

For bowing in the opposite direction to a moment producing compression in the higher section modulus fibre, the compression fibre (defined as that in compression under the moment $P_{Ed}w_0$) can be checked as follows:

$$\sigma_s = \frac{(w_0' + \delta)e_{max}}{I_{st}}\left(\frac{\sigma_m b^2}{\pi^2} + P_{Ed}\right) + \frac{P_{Ed}}{A_{st}} - \frac{M_0 e_{max}}{I_{st}} \leq f_{yd} \text{ (compressive stress +ve)}$$

with $w_0' = w_0 - M_0 b^2/(8EI_{st})$.

The tension fibre would be checked with:

$$\sigma_s = \frac{(w_0' + \delta)e_{min}}{I_{st}}\left(\frac{\sigma_m b^2}{\pi^2} + P_{Ed}\right) - \frac{P_{Ed}}{A_{st}} - \frac{M_0 e_{min}}{I_{st}} \leq f_{yd} \text{ (tensile stress +ve)}$$

Clearly, using equation (D6.6-20) ignoring the actual sign of the moment is conservative in all cases. This method is illustrated in Worked Example 6.6-1.

6.6.2.5. Bearing stress at loaded end

3-1-5/clause 9.4(2)

The cross-section resistance at a loaded end should be checked where there are cut-outs in the stiffener, according to ***3-1-5/clause 9.4(2)***. It would also seem reasonable to check bearing pressure where the contact area is less than the stiffener effective section area. A check is similarly required at the ends if the axial force in the stiffener used in the checks in section 6.6.2.4 above has been based on the value on the middle third.

6.6.3. Transverse web stiffeners not required to contribute to the adequacy of the web under direct stress

As discussed in section 6.6.1, where there are no longitudinal stiffeners and the transverse stiffeners have not been assumed to contribute to the web effective section for axial force and bending, the adequacy of the web is not dependent on the transverse stiffener for these effects. The requirement for the stiffener to provide a rigid support to the web for direct stresses in 3-1-5/clause 9.2.1 is therefore not relevant as the web is adequate under direct stress without the stiffener. The stiffener deflection requirement therein is therefore also not relevant but the strength requirement still is, as discussed in section 6.6.2.4(a).

In such cases, allowance could be made for the destabilising influence of the web on the stiffener by considering an additional equivalent vertical force of $\sigma_m b^2/\pi^2$, as determined in section 6.6.2.4(b), in the stiffener check of 6.6.2.3. This force is also identified in 3-1-5/clause 9.2.1(6). Its use in a buckling check is conservative as it is intended only to produce a bending moment in the bent strut and not an axial stress as discussed in section 6.6.2.4(b). The force should be considered to act along the centroid of the stiffener effective section. The checks given in sections 6.6.2.1, 6.6.2.2 and 6.6.2.5 should also be performed, but not that in 6.6.2.4. The force $\sigma_m b^2/\pi^2$ should not be considered in cross-section checks at the free ends of a stiffener as it is not a real axial force and the effect induces no stress at the stiffener ends. The methods of sections 6.6.2 and 6.6.3 are illustrated in Worked Example 6.6-1. The former is the more general.

6.6.4. Additional effects applicable to certain transverse web stiffeners

There are some additional effects that could occur within a transverse stiffener which are not explicitly covered by EN 1993-1-5 but which should also be considered:

(i) Where a stiffener participates as part of a U–frame, forces will be developed as a result of its bracing action to the compression flange. For a method of allowing for this see 3-2/clause 6.3.4.2(2) and section 6.3.4.2 of this guide. The resulting moments need to be added into the check of bending and axial force performed for other effects.

(ii) Where there is loading on a cross-member that forms part of a U-frame with a transverse stiffener, the differential deflection between adjacent frames leads to additional forces in the stiffener and the main beam compression flange. This is not covered explicitly in EN 1993 but guidance is given in section 6.8 of this guide. The resulting moments again need to be added into the stiffener check.

(iii) External axial forces applied to stiffeners should also include effects from a change of direction of a flange.

It will also be noticed that there is no check of effective stress presented in EN 1993-1-5 for the attached web plating forming part of the stiffener effective section, which also experiences global stresses from participation in main beam bending and shear. The drafters considered that test evidence suggests that this behaviour is covered by the basic check of shear and moment interaction in the main beam and as the axial force in the stiffener from external load contributes to the shear in the web, it should not be double-counted. A check was however required in BS 5400: Part 3.[4] A check might be necessary where there is a significant eccentricity of an axial load in the stiffener which could give rise to significant direct stresses in the web not implicit in the shear–moment interaction. The Von Mises equivalent stress relationship given in expression 3-1-1/(6.1) could be used to combine shear, longitudinal direct stress and transverse direct stress in the web, but it does not allow for a partial plastic bending stress distribution (as did BS 5400: Part 3) so it would be somewhat conservative.

Worked Example 6.6-1: Girder without longitudinal stiffeners

A continuous girder in S355 steel with Class 3 cross-section has plate sizes as shown in Fig. 6.6-4. Stiffeners are provided every 4000 mm. The maximum shear force in a panel is 1700 kN and the direct stresses vary as shown. There is no significant external load acting on each stiffener. The adequacy of the intermediate transverse stiffeners is checked.

The stiffener effective section of 3-1-5/clause 9.1 is shown in Fig. 6.6-4, for which $I_{st} = 1.343 \times 10^7\,\text{mm}^4$, $A_{st} = 5934\,\text{mm}^2$ and the centroid is 36.7 mm from the back of the web. The section moduli for the web and outstand are $3.658 \times 10^5\,\text{mm}^3$ and $1.072 \times 10^5\,\text{mm}^3$ respectively.

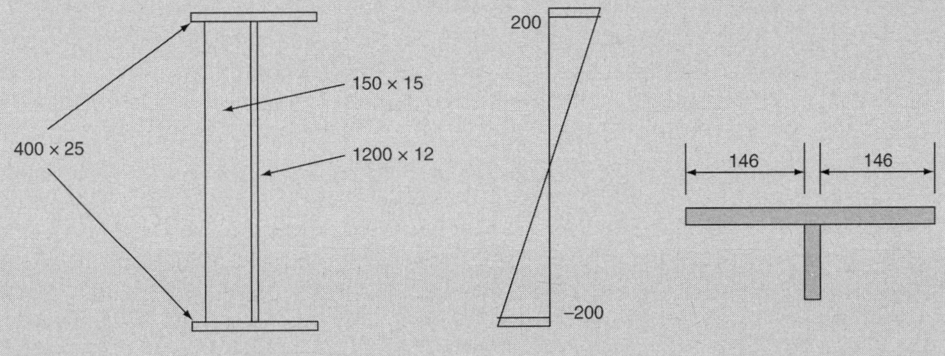

Fig. 6.6-4. Girder for worked example: (a) girder; (b) stresses; (c) stiffener effective section

(i) Torsional buckling (section 6.6.2.1)
The height-to-thickness ratio of the stiffener is 10 which is satisfactory for an S355 stiffener – see section 6.9 of this guide.

(ii) Minimum stiffness for shear according to 3-1-5/clause 9.3.3 (section 6.6.2.2)
For $a/h_w \geq \sqrt{2}$, $I_{st} \geq 0.75 h_w t^3 = 0.75 \times 1200 \times 12^3 = 1.56 \times 10^6\,\text{mm}^4$.
Actual $I_{st} = 1.343 \times 10^7\,\text{mm}^4$, so there is adequate stiffness.

(iii) Axial force in the stiffener due to shear (section 6.6.2.3)
It is first necessary to calculate the shear slenderness.
From 3-1-5/clause 5.2:

$$k_\tau = 5.34 + 4.00\left(\frac{b}{a}\right)^2 + k_{\tau\text{st}} = 5.34 + 4.00\left(\frac{1200}{4000}\right)^2 + 0 = 5.70$$

$$\tau_{cr} = \frac{k_\tau \pi^2 E t^2}{12(1-\nu^2)b^2} = \frac{5.7 \times \pi^2 \times 210\,000 \times 12^2}{12(1-0.3^2) \times 1200^2} = 108.2\,\text{MPa and}$$

$$\bar{\lambda}_w = 0.76\sqrt{\frac{f_{yw}}{\tau_{cr}}} = 0.76\sqrt{\frac{355}{108.2}} = 1.377$$

This leads to a shear resistance, ignoring any contribution from the flanges and using the rigid end-post case, of 1771 kN. The vertical force generated in the stiffener is given by:

$$P_{\text{shear}} = V_{Ed} - \frac{1}{\bar{\lambda}_w^2}\frac{f_{yw} h_w t}{\sqrt{3}\gamma_{M1}} = 1700 \times 10^3 - \frac{1}{1.377^2}\frac{355 \times 1200 \times 12}{\sqrt{3} \times 1.1} = 285\,\text{kN}$$

There is no external load so:

$$P_{Ed} = P_{\text{shear}} + P_{\text{ext}} = 285 + 0 = \mathbf{285\,kN}$$

The stiffener is checked for this axial force in conjunction with the destabilising effect of the web below.

(iv) Destabilising effect of the web (section 6.6.2.4 and section 6.6.3)
As discussed in section 6.6.1 of this guide, it would be reasonable to use the simplified method in section 6.6.3 here as the method of 3-1-5/clause 9.2.1 was not intended to be required where the stiffeners do not contribute to the resistance of the web under direct stress. Since the girder has a Class 3 cross-section, clearly the presence of the stiffeners will not improve the web resistance to direct stress. The stiffeners are however checked here following both methods (section 6.6.2.4 and 6.3) for illustration. In both cases, it is necessary to calculate $\sigma_m b^2/\pi^2$ from 3-1-5/clause 9.2.1(6).
From 3-1-5/clause 4:

$$\sigma_{cr,p} = \frac{k_\sigma \pi^2 E t^2}{12(1-\upsilon^2)b^2} = \frac{23.9 \times \pi^2 \times 210\,000 \times 12^2}{12(1-0.3^2) \times 1200^2} = 453.6\,\text{MPa}$$

$$\sigma_{cr,c} = \frac{\pi^2 E t^2}{12(1-\nu^2)a^2} = \frac{\pi^2 \times 210\,000 \times 12^2}{12(1-0.3^2) \times 4000^2} = 1.71\,\text{MPa}$$

(The critical stresses are based on a single panel as discussed in the main text above.)
From 3-1-5/clause 9.2.1(5):

$$\sigma_m = \frac{\sigma_{cr,c}}{\sigma_{cr,p}}\left[\left(\frac{1}{a_1} + \frac{1}{a_2}\right)\frac{N_{Ed}}{b}\right] = \frac{1.71}{453.6}\left[\left(\frac{1}{4000} + \frac{1}{4000}\right)\frac{200 \times 0.5 \times 12 \times 1200/2}{1200}\right]$$

$$= 1.13 \times 10^{-3}\,\text{MPa}$$

The equivalent axial force in the stiffener (see discussion on equation (D6.6-17)) is therefore:

$$\frac{\sigma_m b^2}{\pi^2} = \frac{1.13 \times 10^{-3} \times 1200^2}{\pi^2} = 0.17\,\text{kN}$$

which is much less than that for shear. This will typically be the case for unstiffened webs with $a/b < 1$ where there is no benefit to the web stability under direct stress from the transverse stiffener. Comment on the ratio $\sigma_{cr,c}/\sigma_{cr,p}$ is made in the main text.

(a) Method of section 6.3.3
The equivalent axial force representing the destabilising effect of the web = 0.17 kN. The axial force due to shear tension field action = 285 kN. The total axial force is therefore:

$$P_{Ed} + \frac{\sigma_m b^2}{\pi^2} = 285 + 0.17 = 285\,\text{kN, i.e. the former force is negligible}$$

Stiffener moment:

$$285 \times 10^3 \times (36.71 - 12/2) = 8.75\,\text{kNm (constant over web depth)}$$

The following interaction from 3-2/clause 6.3.3 has to be satisfied:

$$\frac{N_{Ed}}{\dfrac{\chi_y N_{Rk}}{\gamma_{M1}}} + \beta_m \frac{M_{y,Ed} + \Delta M_{y,Ed}}{\dfrac{M_{y,Rk}}{\gamma_{M1}}} \leq 0.9$$

(noting that N_{Ed} is the stiffener axial force in this expression).

$$\bar{\lambda} = \sqrt{\frac{A f_y}{N_{cr}}} = \sqrt{\frac{5934 \times 355}{19\,330 \times 10^3}} = 0.33$$

with

$$N_{cr} = N_{cr,y} = \frac{\pi^2 EI}{L_{cr,y}^2} = \frac{\pi^2 \times 210 \times 10^3 \times 1.343 \times 10^7}{1200^2} = 19\,330\,\text{kN}$$

From curve c of 3-1-1/Fig. 6.4, $\chi_y = 0.93$
From 3-1-1/Table A.2 for uniform moment with $\psi = 1.0$:

$$\beta_m = C_{my,0} = 0.79 + 0.21\psi + 0.36(\psi - 0.33)N_{Ed}/N_{cr,y}$$

$$= 0.79 + 0.21 + 0.36(1 - 0.33)285/19\,330 = 1.00$$

$$M_{y,Rk} = W_{el,min} f_y = 1.072 \times 10^5 \times 355 = 38.1\,\text{kNm}$$

This is conservative as it is based on the stiffener outstand when actually the applied moment induces compression in the web plate. This problem arises because the formula in 3-1-1/clause 6.3.3 was intended for bisymmetric sections. It would be reasonable here to use the section modulus for the web plate in this application.

$$N_{Rk} = 5934 \times 355 = 2107\,\text{kN}$$

$$\frac{N_{Ed}}{\dfrac{\chi_y N_{Rk}}{\gamma_{M1}}} + \beta_m \frac{M_{y,Ed} + \Delta M_{y,Ed}}{\dfrac{M_{y,Rk}}{\gamma_{M1}}} = \frac{285}{\dfrac{0.93 \times 2107}{1.1}} + 1.00 \times \frac{8.75}{\dfrac{38.1}{1.1}} = 0.16 + 0.25$$

$$= 0.41 < 0.9$$

Therefore stiffener is adequate, with a **usage of 46%**.

(b) Method of section 6.6.2.4
Conservatively assume that the moment acts to put the fibre with lowest section modulus into compression, even though the opposite is true here. For a less conservative method,

see the main text. The axial force and moment are $P_{Ed} = 285\,kN$ and $M_0 = 8.75\,kNm$ respectively from above.

From equation (D6.6-18), the initial bow is:

$$w_0' = w_0 + \frac{M_0 b^2}{8EI_{st}} = 1200/300 + \frac{8.75 \times 10^6 \times 1200^2}{8 \times 210 \times 10^3 \times 1.343 \times 10^7} = 4 + 0.56 = 4.56\,mm$$

and the additional deflection:

$$\delta = w_0' \left(\frac{EI_{st}}{\dfrac{\sigma_m b^4}{\pi^4} + \dfrac{P_{Ed} L^2}{\pi^2}} - 1 \right)^{-1}$$

$$= 4.56 \left(\frac{210 \times 10^3 \times 1.343 \times 10^7}{\dfrac{1.13 \times 10^{-3} \times 1200^4}{\pi^4} + \dfrac{285 \times 10^3 \times 1200^2}{\pi^2}} - 1 \right)^{-1} = 0.068\,mm$$

From equation (D6.6-19), check that total additional deflection is less than $b/300$:

$$\delta + \frac{M_0 b^2}{8EI_{st}} = 0.068 + 0.56 = 0.63\,mm \le \frac{b}{300} = 4\,mm$$

so deflection is acceptable.

From equation (D6.6-20), check that the stress is less than the design yield stress:

$$\sigma_s = \frac{(w_0' + \delta)e_{max}}{I_{st}} \left(\frac{\sigma_m b^2}{\pi^2} + P_{Ed} \right) + \frac{P_{Ed}}{A_{st}} + \frac{M_0 e_{max}}{I_{st}}$$

$$= \frac{(4.56 + 0.068) \times 125.3}{1.343 \times 10^7} \left(\frac{1.13 \times 10^{-3} \times 1200^2}{\pi^2} + 285 \times 10^3 \right)$$

$$+ \frac{285 \times 10^3}{5934} + \frac{8.75 \times 10^6 \times 125.3}{1.343 \times 10^7}$$

$$= 12.3 + 48.0 + 81.6 = 141.9\,MPa < 355/1.1 = 322.7\,MPa$$

Therefore stiffener is adequate, with a **usage of 44%**.

(v) Bearing stress at loaded end (section 6.6.2.5)

The stress on the stiffener effective section at cut-outs at stiffener ends should also be checked according to 3-1-5/clause 9.4(2) but this will clearly be adequate here.

6.6.5. Flange transverse stiffeners

Flange transverse stiffeners on compression flanges can, in principle, be designed in the same way as web transverse stiffeners using an effective section in accordance with 3-1-5/clause 9.1 and the design method of 3-1-5/clause 9.2.1 as expanded upon in section 6.6.2 above. Three additional loadings will however typically be required:

(i) local transverse load from traffic where the transverse stiffener is on a deck plate;
(ii) local transverse load from any flange vertical curvature;
(iii) weight of wet concrete in composite flanges and other construction loads.

The determination of forces from (ii) is discussed in section 6.10 of this guide. When these effects are added, equations (D6.6-18) to (D6.6-20) can be used directly by setting the moment M_0 equal to the peak first-order moment from the above transverse loading. This is likely to be slightly conservative as this moment is unlikely to be uniform across the stiffener and therefore the first-order deflection caused by it will be less than $M_0 b^2/(8EI_{st})$. If the moment distribution is more parabolic or triangular, equations (D6.6-18) to (D6.6-20) could be modified as follows:

$$\delta = w_0' \left(\frac{EI_{st}}{\frac{\sigma_m b^4}{\pi^4} + \frac{P_{Ed} L^2}{\pi^2}} - 1 \right)^{-1} \qquad \text{(D6.6-21)}$$

with $w_0' = w_0 + \Delta$ where Δ is the peak first-order deflection due to the transverse loads. The total deflection must be less than $b/300$ thus:

$$\delta + \Delta \leq \frac{b}{300} \qquad \text{(D6.6-22)}$$

The final stress check remains as in equation (D6.6-20). The comments regarding the sign of the applied moment in relation to the lowest section modulus fibre made in section 6.6.2 also apply here.

6.7. Bearing stiffeners and beam torsional restraint (additional sub-section)

No rules are given within EN 1993-2 for the design of bearing stiffeners so reference has to be made to EN 1993-1-5 section 9. This section brings together the relevant rules of EN 1993-1-5. It covers the following:

- Effective section of a bearing stiffener *Section 6.7.1*
- Design requirements for bearing stiffeners at simply supported ends of beams *Section 6.7.2*
- Design requirements for bearing stiffeners at intermediate supports *Section 6.7.3*
- Bearing fit *Section 6.7.4*
- Additional effects applicable to certain bearing stiffeners *Section 6.7.5*
- Beam torsional restraint at supports *Section 6.7.6*

6.7.1. Effective section of a bearing stiffener

The properties of a stiffener effective section are calculated using an attached width of web of $15\varepsilon t$ each side of the stiffener as shown in Fig. 6.7-1 (with $\varepsilon = \sqrt{235/f_y}$), but not greater than the available width – 3-1-5/clause 9.1(2) refers. If attached widths from a pair of adjacent stiffeners overlap then the adjacent stiffeners could be treated as acting together.

6.7.2. Design requirements for bearing stiffeners at simply supported ends of beams

The following requirements have to be met:

(i) The outstand should meet the limits in 3-1-5/clause 9.2.1 for preventing torsional buckling. This is discussed in section 6.9 of this guide.
(ii) The effective section must resist the bearing reaction, according to 3-1-5/clause 9.4(2).
(iii) The cross-section resistance at a loaded end should be checked where there are cut-outs in the stiffener, according to 3-1-5/clause 9.4(2).

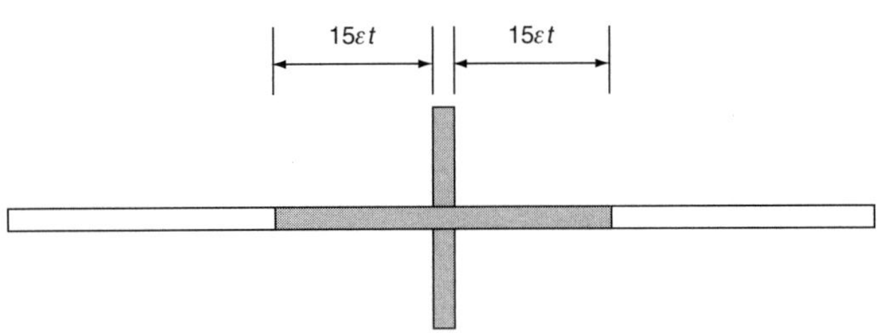

Fig. 6.7-1. Effective section for stiffener

(iv) Where the shear design has been based on 'rigid end-post' conditions, the stiffener must also be designed to resist the membrane forces resulting from tension field action and satisfy a minimum stiffness.

Requirements (ii) to (iv) are discussed below.

6.7.2.1. Buckling check under bearing reaction

3-1-5/clause 9.4(2)

A bearing stiffener has to be designed as a strut to ***3-1-5/clause 9.4(2)***, resisting the bearing reaction together with any eccentricities resulting from temperature movement and any movement in the point of contact as the deck rotates. 3-1-5/clause 9.4 does not mention load eccentricities, other than from stiffener asymmetry. Movements due to temperature can be calculated using the method in Annex A of EN 1993-2 (which is scheduled to be moved to EN 1990 as it is not specific to steel bridges) and guidance on eccentricity from varying point of contact can be found in EN 1337-4 for roller bearings.

Eccentricity from tolerances in positioning bearings and uneven seating on flat surfaces should additionally be included but no guidance is given in EN 1993-2. It would be reasonable to use the values which were contained in BS 5400: Part 3[4] to cover tolerances and uneven seating as follows:

- half the width of the flat bearing surface plus 10 mm for flat-topped rocker bearing in contact with flat bearing surface
- 3 mm for radiused upper bearing on flat or radiused lower part
- 10 mm for flat upper bearing on radiused lower part.

3-1-5/clause 9.4(3)

If a stiffener effective section is asymmetric about the web, the resulting eccentricity should also be considered to produce a moment acting on the centroid of the stiffener section in accordance with ***3-1-5/clause 9.4(3)***. Bearing stiffeners should normally be made symmetric wherever possible.

3-1-5/clause 9.4(2) requires the bearing stiffener effective section to be checked for combined bending moment and axial force using the interactions in 3-1-1/clause 6.3.3 or 3-1-1/clause 6.3.4, allowing for the fact that the stiffener cannot buckle in the plane of the web. It is not easy to apply 3-1-1/clause 6.3.3 in these circumstances. The simplified check in 3-2/clause 6.3.3 is easier to apply, but it needs to be extended to cover moments in the plane of the web as follows:

$$\frac{N_{Ed}}{\dfrac{\chi_y N_{Rk}}{\gamma_{M1}}} + C_{mi,o}\frac{M_{y,Ed}}{\dfrac{M_{y,Rk}}{\gamma_{M1}}} + \frac{M_{z,Ed}}{\dfrac{M_{z,Rk}}{\gamma_{M1}}} \leq 0.9 \qquad \text{(D6.7-1)}$$

$M_{y,Ed}$ is based on the peak moment in the stiffener and the shape of the moment diagram is allowed for by the factor $C_{mi,o}$ as discussed in section 6.3.3 of this guide. A similar factor could be used with the $M_{z,Ed}$ term or $M_{z,Ed}$ could be taken as the maximum value within the middle third as has been previous UK practice.

Alternatively, equation (D6.3-29) from section 6.3.3 could be used (which is in any case the origin of the equation in 3-2/clause 6.3.3 for cases where lateral torsional buckling is prevented) with an additional term for the M_z moment. No magnifier is required on the M_z moment as the web prevents buckling in its plane. This leads to the interaction:

$$\frac{N_{Ed}}{\chi_y N_{pl,Rd}} + \frac{1}{1-(N_{Ed}/N_{cr,y})}\frac{M_{y,Ed}}{M_{y,Rd}} + \frac{M_{z,Ed}}{M_{z,Rd}} \leq 1.0 \qquad \text{(D6.7-2)}$$

with

$$N_{pl,Rd} = \frac{Af_y}{\gamma_{M1}},\ M_{y,Rd} = \frac{W_{el,y}f_y}{\gamma_{M1}}$$

and

$$M_{z,Rd} = \frac{W_{el,z} f_y}{\gamma_{M1}}$$

The section moduli should be appropriate to the point on the stiffener being checked. As no factor is included for the shape of the moment diagram here, the maximum values in the middle third of the stiffener height could be used. It is recommended here that the check be based on elastic properties as it would be undesirable to have plastic deformation in a bearing stiffener; it is likely to be incompatible with the assumptions made for its stiffness.

In both equations (D6.7-1) and (D6.7-2) above, the axial force from the bearing reaction is typically not constant up the stiffener and usually varies from a maximum at the loaded end to zero at the top. Assuming the force to be constant throughout the length is conservative for the buckling check. A reasonable approach for pin-ended bearing stiffeners would be to use two-thirds of the reaction (the maximum value within the middle third) in the buckling checks. In such circumstances, a check of cross-section resistance must always be made at the ends of the stiffener. The design effects from bearing reaction must be combined with any moments resulting from the bearing stiffener acting as a rigid end-post as discussed below.

The effective length for flexural buckling cannot be taken as less than $0.75h_w$ and buckling curve c in 3-1-1/Fig. 6.4 has to be used – 3-1-5/clause 9.4(2) refers. Care must be taken with effective length where a bearing stiffener is providing the sole torsional restraint by cantilevering up from the bearing, as might be the case in a U-frame bridge. In this instance, the effective length will be greater than or equal to $2.0h_w$, depending on the restraint provided by the U-frame cross-member.

6.7.2.2. Bearing stress at loaded end and cross-section resistance generally

The cross-section resistance at a loaded end should be checked where there are cut-outs in the stiffener, according to 3-1-5/clause 9.4(2). It would also be appropriate to check bearing pressure where the contact area is less than the stiffener effective section area.

Although not stated, the cross-section resistance should generally be checked, particularly if benefit has been taken in the buckling check of variations in the axial force and moment over the height of the stiffener. This is illustrated in Worked Example 6.7-1.

6.7.2.3. Membrane forces for stiffeners acting as a rigid end-post

If the shear resistance has been produced assuming a rigid end-post (see section 6.2.6 of the guide), then the bearing stiffener should be designed to resist the resulting longitudinal membrane stress in the web by acting as a beam spanning between the flanges according to ***3-1-5/clause 9.3.1(1)***. It must also satisfy a minimum stiffness. *3-1-5/clause 9.3.1(1)*

3-1-5/clause 9.3.1(2) requires rigid end-posts to be designed as two double-sided stiffeners forming the beam as shown in Fig. 6.7-2. The resulting beam is required by ***3-1-5/clause 9.3.1(3)*** to have a minimum section modulus thus: *3-1-5/clause 9.3.1(2)* *3-1-5/clause 9.3.1(3)*

$$z_{min} = 4h_w t^2 \quad \text{(D6.7-3)}$$

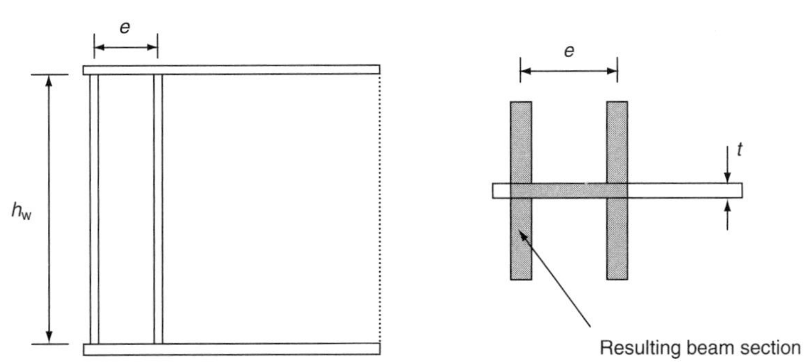

Fig. 6.7-2. Rigid end post

If the stiffeners are flats, this is equivalent to each double-sided stiffener having a minimum cross-sectional area thus:

$$A_{min} = 4h_w t^2/e \tag{D6.7-4}$$

The spacing of the stiffeners, e, must be greater than $0.1h_w$.

The method of calculating the membrane force is not given in EN 1993-1-5 but it can be derived from the shear buckling model as discussed in section 6.2.6.2 where it is shown that, for perfectly flat plates, the membrane force is given by:

$$N_H = h_w t_w \left(\frac{\tau^2}{\tau_{cr}} - \tau_{cr}\right) \geq 0 \tag{D6.7-5}$$

where h_w and t_w are the height and thickness of the web panel respectively. This approach is conservative as the membrane stress is not developed fully over the entire web height; equation (D6.7-5) assumes it is. It can be seen that there is no membrane force to resist until the shear stress reaches the elastic critical value, τ_{cr}. τ_{cr} can be calculated as discussed in section 6.2.6.2 of this guide. Where there are longitudinal stiffeners, τ_{cr} can conservatively be based on the lowest value for either overall stiffened panel buckling or for the weakest sub-panel. Where sub-panel buckling governs, the above is clearly conservative when the slenderness is much greater in one panel than the others. BS 5400 allowed τ_{cr} in this situation to be based on the average of the two lowest values obtained for sub-panels and this might be considered reasonable here.

For real design purposes however, equation (D6.7-5) will lead to a discontinuity with the shear rules at slenderness less than about 1.2 because it is possible for τ_{cr} to exceed the limiting shear stress for a rigid end-post obtained from 3-1-5/Fig. 5.2. This means that although benefit is being taken from the presence of a rigid end-post, equation (D6.7-5) will give no load to apply to its design. The problem arises because the rotated stress field starts to develop in reality at a lower stress than τ_{cr} in this slenderness region due to imperfections in the web plate. To avoid this anomaly, a reduction factor of 1.2 could be applied to τ_{cr} as shown in Fig. 6.7-3. This factor also makes allowance for $\gamma_{M1} = 1.1$ in the shear design and ensures that the membrane force is approximately zero at a slenderness of 1.08 where the shear resistance curves for rigid and non-rigid end-posts separate. For higher slenderness, the web shear resistance is enhanced by the presence of a rigid end-post and the membrane force is greater than zero. The expression for membrane force then becomes:

$$N_H = h_w t_w \left(\frac{\tau^2}{\tau_{cr}/1.2} - \tau_{cr}/1.2\right) \geq 0 \tag{D6.7-6}$$

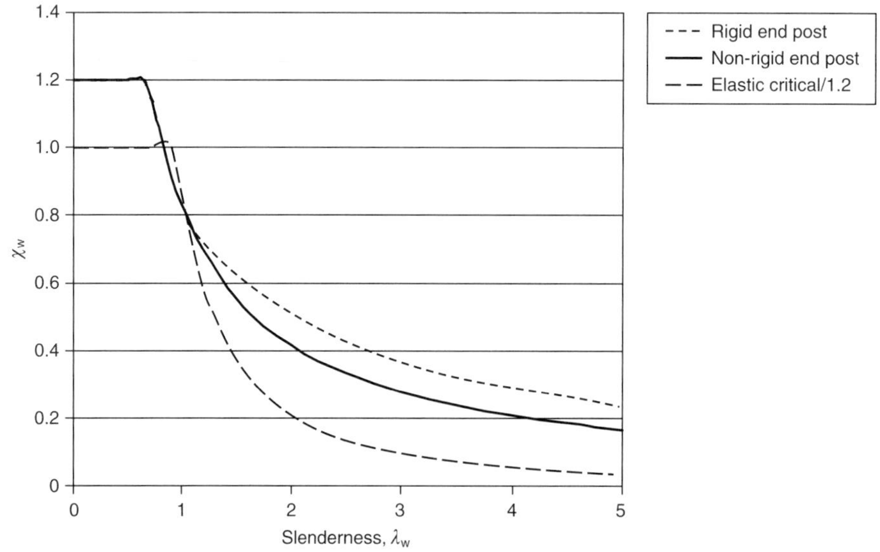

Fig. 6.7-3. Reduction factor on τ_{cr} to avoid discontinuity with rigid end post case

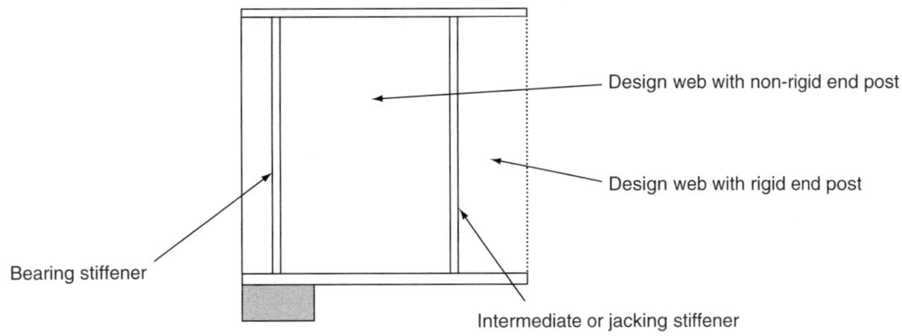

Fig. 6.7-4. Alternative to providing rigid end post while still maintaining rigid end-post conditions in the shear design

At higher slenderness, tension field action will start at approximately τ_{cr}, as imperfections have less effect at high slenderness, so this reduction factor on τ_{cr} will then be very conservative. A reduction factor that reduces with slenderness is really required such that equation (D6.7-5) is used unmodified at greater slenderness. Equation (D6.7-6) is however always conservative.

The membrane force is applied as a uniformly distributed load to the beam section in Fig. 6.7-2 so that the maximum moment to be resisted by the beam bending in the plane of the web is $N_H h_w/8$ at mid-height. For buckling checks, the effect of moment from the membrane force acting on the end post could be added to other effects by simply adding another $M_{z,Ed}/M_{z,Rd}$ term in the buckling interaction and similarly in the cross-section resistance check. It should be noted that the effective section for the rigid end-post (Fig. 6.7-2) is not the same as that for the bearing stiffener (Fig. 6.7-1). This should be taken into account when combining stresses. For simplicity, the stresses in web and stiffener developed on the basis of the two effective sections could simply be added.

The added effort of designing a bearing stiffener as a rigid end-post can be avoided in two ways. First, and obviously, the shear design can be done assuming non-rigid end-posts, as there will be no loss of economy in the web design unless the web slenderness is higher than 1.08 according to 3-1-5/Table 5.1. Second, ***3-1-5/clause 9.3.1(4)*** provides an alternative means of developing rigid end-post conditions by placing an intermediate transverse stiffener sufficiently close to the bearing stiffener so that the panel between transverse stiffener and bearing stiffener is adequate when designed with the non-rigid end-post conditions. Beyond the transverse stiffener, rigid end-post conditions then apply as shown in Fig. 6.7-4. This might be particularly appropriate if a full-height jacking stiffener is going to be provided along the girder in any case. *3-1-5/clause 9.3.1(4)*

6.7.3. Design requirements for bearing stiffeners at intermediate supports

The following requirements have to be met:

(i) The outstand should meet the limits in 3-1-5/clause 9.2.1 for preventing torsional buckling. This is discussed in section 6.9 of this guide.
(ii) The effective section must resist the bearing reaction, according to 3-1-5/clause 9.4(2).
(iii) The cross-section resistance at a loaded end should be checked where there are cut-outs in the stiffener, according to 3-1-5/clause 9.4(2).
(iv) The effective section must meet minimum strength and stiffness requirements in 3-1-5/clause 9.2.1 where the transverse stiffener is assumed to restrain either an unstiffened or stiffened web panel with direct stress. This is in conjunction with resisting the bearing reaction.

The methods of design for items (ii) and (iii) are as discussed above in section 6.7.2. Item (iv) is discussed below.

6.7.3.1. Design for provision of restraint to webs under direct stress

When there are longitudinal stiffeners on a web which are designed to be restrained by bearing stiffeners, the bearing stiffeners must themselves be designed to provide this support in accordance with 3-1-5/clause 9.2.1, in addition to resisting the bearing reaction. The bearing stiffener may also have to be designed for similar forces in unstiffened webs as discussed in section 6.6.1, but these forces will be much smaller. Since the bearing reaction interacts with the out-of-plane forces arising from web longitudinal direct force, the buckling check can be performed using the method discussed in section 6.6.2.4 of this guide. An additional term, $M_{z,Ed}/W_{z,el}$, is necessary in equation (D6.6-20) to allow for moments in the plane of the web thus:

$$\sigma_s = \frac{(w'_0 + \delta)}{W_{y,el}}\left(\frac{\sigma_m b^2}{\pi^2} + P_{Ed}\right) + \frac{P_{Ed}}{A_{st}} + \frac{M_{y,Ed}}{W_{y,el}} + \frac{M_{z,Ed}}{W_{z,el}} \leq f_{yd} \qquad \text{(D6.7-7)}$$

The terms above all have their meanings as defined in section 6.6.2.4 of this guide, with $M_{y,Ed}$ above replacing M_0 in section 6.6. $W_{y,el}$ is used here in place of e_{max}/I_{st}. $W_{y,el}$ and $W_{z,el}$ are the section moduli corresponding to the point being checked. They should be chosen so that they are coexisting values at individual points checked on the stiffener. Since stiffeners are typically cruciform-shaped, the minimum values of $W_{y,el}$ and $W_{z,el}$ do not generally coexist at a single location. If the stiffener is asymmetric about the web, which is undesirable, equation (D6.7-7) can become conservative depending on the direction of the moment, and the comments at the end of section 6.6.2.4 apply. The deflection check of equation (D6.6-19) can be used without modification, other than for M_0 as above.

A check of cross-section resistance would be required in addition to the check of equation (D6.7-7) if the axial force and moments are based on their values in the middle third of the stiffener, as would be reasonable.

Alternatively, and more simply, the buckling check used in section 6.7.2.1 above (either equation (D6.7-1) or (D6.7-2)) could be used and the destabilising influence of the stiffened or unstiffened web allowed for by the use of a fictitious axial force, $\sigma_m b^2/\pi^2$, where the symbols are defined in section 6.6.2.4 of this guide and 3-1-5/clause 9.2.1. The origin of this term is discussed in section 6.6.3 of this guide. Its use is conservative since it is not a real force and is intended to contribute only to generating a moment in the bent strut and not to producing axial stress – 3-1-5/clause 9.2.1(6) and section 6.6.2.4(b) of this guide refer. The deflection check required by 3-1-5/clause 9.2.1 will generally be satisfied by inspection because bearing stiffeners will usually be sufficiently stiff as a result of being strong enough to resist the bearing reaction. The use of the additional axial force $\sigma_m b^2/\pi^2$ is therefore a pragmatic simplification. Both methods are illustrated in Worked Example 6.7-1.

6.7.4. Bearing fit

If a full contact end bearing is specified in accordance with EN 1090, it would be reasonable to take all the direct compression through bearing at ULS, although EN 1993 does not discuss this. If this is done, a fatigue check must still be made of the weld provided, assuming all the compression passes through the weld and none through direct bearing. Weld design at ULS is illustrated in Worked Example 8.2-1 and for fatigue in Worked Example 9-4.

6.7.5. Additional effects applicable to certain bearing stiffeners

There are some additional actions that could be applied to a bearing stiffener, but which are not specifically covered by EN 1993-1-5. Items (i) to (iii) in section 6.6.4 of this guide are relevant. Additionally, where a bearing stiffener assists in providing torsional restraint to main beams, the effects of this participation should be included in the stiffener design.

3-1-5/clause 9.4(3)

This is discussed in section 6.7.6. ***3-1-5/clause 9.4(3)*** also reminds the designer that the stiffness of the bearing stiffener must be consistent with that assumed in the design for lateral torsional buckling.

There is no check of equivalent stress presented in EN 1993-1-5 for the attached web plating forming part of the stiffener effective section as discussed in section 6.6.4 of this guide. Recommendations on when such a check might be conducted are given therein.

Worked Example 6.7-1: Bearing stiffener at beam end

A bearing stiffener above a fixed bearing at the end of a bridge beam has two double-sided stiffeners as shown in Fig. 6.7-5. The beam is held against rotation about its longitudinal axis at its ends by bracing and there are no intermediate stiffeners. It is checked that the bearing stiffener is adequate to carry a reaction commensurate with the full shear resistance of the web, assuming a rigid end-post. S355 steel is used throughout.

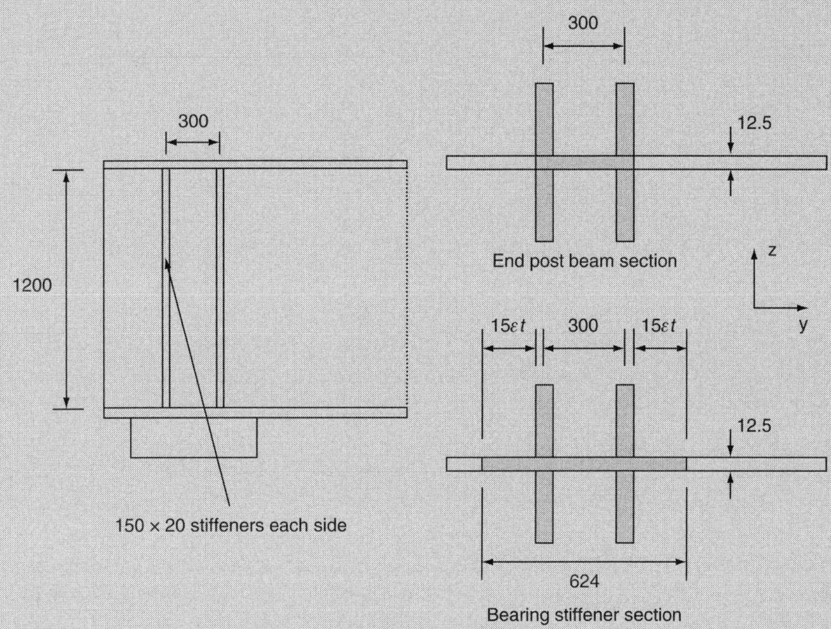

Fig. 6.7-5. Bearing stiffener for Worked Example 6.7-1

For no intermediate stiffeners, the shear slenderness is obtained from expression 3-1-5/(5.5):

$$\bar{\lambda}_w = \frac{h_w}{86.4t\varepsilon} = \frac{1200}{86.4 \times 12.5 \times 0.81} = 1.372$$

The critical shear stress is obtained from expression 3-1-5/(5.4) and for $a/b \gg 1$, $k_\tau = 5.34$:

$$\tau_{cr} = \frac{k_\tau \pi^2 E t^2}{12(1-\nu^2)b^2} = \frac{5.34\pi^2 \times 210 \times 10^3 \times 12.5^2}{12(1-0.3^2)1200^2} = 110\,\text{MPa}$$

The rigid end-post case is used for the shear design, so from 3-1-5/Table 5.1:

$$\chi_w = \frac{1.37}{0.7 + \bar{\lambda}_w} = \frac{1.37}{0.7 + 1.372} = 0.66$$

The contribution from the flanges will be negligible with no intermediate stiffeners, so is ignored. The shear resistance of the web is therefore:

$$V_{bRd} = \frac{\chi_v f_{yw} h_w t}{\sqrt{3}\gamma_{M1}} = \frac{0.66 \times 355 \times 1200 \times 12.5}{\sqrt{3} \times 1.1} = 1845\,\text{kN}$$

The design bearing reaction is therefore 1845 kN.

Action as a bearing stiffener

The bearing stiffener has an attached width of web plate of:

$$30\varepsilon t + 300 + 20 = 30 \times 0.81 \times 12.5 + 300 + 20 = 624\,\text{mm}$$

The properties of the bearing stiffener section as shown in Fig. 6.7-5 are:

$$A = 19\,800\,\text{mm}^2$$

$$I_{yy} = 1.018 \times 10^8\,\text{mm}^4$$

$$I_{zz} = 5.235 \times 10^8\,\text{mm}^4$$

For a fixed spherical bearing, allow say 10 mm eccentricity in each direction. The design actions on the bearing stiffener effective section are then:

$$N_{Ed} = 1845\,\text{kN}$$

$$M_{y,Ed} = M_{z,Ed} = 18.45\,\text{kNm}$$

Worst stress in web plate:

$$= \frac{1845 \times 10^3}{19\,800} + \frac{18.45 \times 10^6}{5.235 \times 10^8/312} = 104\,\text{MPa}$$

Worst stress in stiffener:

$$= \frac{1845 \times 10^3}{19\,800} + \frac{18.45 \times 10^6}{5.235 \times 10^8/160} + \frac{18.45 \times 10^6}{1.018 \times 10^8/156} = 127\,\text{MPa}$$

(The additional force $\sigma_m b^2/\pi^2$ in section 6.7.3.1 is not relevant at a beam end where there is no direct stress in the web.)

Action as a rigid end-post

A slightly different effective section excluding the outer parts of web plate is used as shown in Fig. 6.7-5. Each double-sided stiffener must provide a minimum area according to 3-1-5/clause 9.3.1(3):

$$A_{min} = 4h_w t^2/e = 4 \times 1200 \times 12.5^2/300 = 2500\,\text{mm}^2$$

Actual provided $= 2 \times 150 \times 20 = 6000\,\text{mm}^2 > 2500\,\text{mm}^2$ so area is adequate.

The end post second moment of area $I_{zz} \approx 6000 \times 150^2 \times 2 = 2.700 \times 10^8\,\text{mm}^4$.

The applied shear stress:

$$\tau_{Ed} = \frac{1845 \times 10^3}{1200 \times 12.5} = 123\,\text{MPa}$$

Using equation (D6.7-6), the membrane force is:

$$N_H = h_w t_w \left(\frac{\tau^2}{\tau_{cr}/1.2} - \tau_{cr}/1.2 \right) = 1200 \times 12.5 \left(\frac{123^2}{110/1.2} - 110/1.2 \right) = 1101\,\text{kN}$$

The in-plane moment half way up the beam is then

$$M_{z,Ed} = N_H h_w/8 = 1101 \times 1.200/8 = 165\,\text{kNm}$$

Worst stress in stiffener from membrane action:

$$= \frac{165 \times 10^6}{2.700 \times 10^8/160} = 98\,\text{MPa}$$

According to 3-1-5/clause 9.4, the stiffener should be checked for buckling under combined bending and axial load in accordance with 3-1-1/clause 6.3.3 or 6.3.4. Cross-section resistance should also be checked. The cross-section resistance is checked first, assuming elastic behaviour as discussed in the main text.

Cross-section resistance check

The maximum stresses from membrane action and bearing reaction are conservatively added together here for simplicity, even though they occur at different heights on the stiffener. Consequently, the worst total stress in the stiffener $= 127 + 98 = 225\,\text{MPa} < 345/1.0 = 345\,\text{MPa}$ for S355 steel and 20 mm thick plate according to EN 10025. At present, 3-1-1/Table 3.1 allows the use of 355 MPa for this thickness and this value is used in the buckling check below, although the UK National Annex requires the values in EN 10025 to be used.

Buckling check

The slenderness of the stiffener:

$$\bar{\lambda} = \sqrt{\frac{Af_y}{N_{cr,y}}} = \sqrt{\frac{19\,800 \times 355}{146\,523 \times 10^3}} = 0.22$$

with

$$N_{cr,y} = \frac{\pi^2 EI_{yy}}{L_{cr,y}^2} = \frac{\pi^2 \times 210 \times 10^3 \times 1.018 \times 10^8}{1200^2} = 146\,523\,\text{kN}$$

From curve c of 3-1-1/Fig. 6.4, $\chi_y = 0.99$

$$N_{pl,Rd} = \frac{Af_y}{\gamma_{M1}} = \frac{19\,800 \times 355}{1.1} = 6390\,\text{kN}$$

The section moduli for y and z axis bending will be based on the stiffener outstand as that was found to be critical in the cross-section resistance check. Note that the section modulus for z axis bending is different for the moments arising from membrane action and bearing eccentricity.

$$M_{y,Rd} = \frac{W_{el,y}f_y}{\gamma_{M1}} = \frac{(1.018 \times 10^8/156) \times 355}{1.1} = 210.6\,\text{kNm}$$

$$M_{z,Rd} = \frac{W_{el,z}f_y}{\gamma_{M1}} = \frac{(5.235 \times 10^8/160) \times 355}{1.1} = 1055.9\,\text{kNm for bearing eccentricity}$$

$$M_{z,Rd} = \frac{W_{el,z}f_y}{\gamma_{M1}} = \frac{(2.700 \times 10^8/160) \times 355}{1.1} = 544.6\,\text{kNm for membrane forces}$$

Using the maximum values in the middle third of the stiffener:

$M_{y,Ed} = 0.67 \times 18.45 = 12.4\,\text{kNm}$

$M_{z,Ed} = 0.67 \times 18.45 = 12.4\,\text{kNm}$ for bearing eccentricity

$M_{z,Ed} = 165\,\text{kNm}$ for moment from membrane forces, which is maximum at mid-height.

$N_{Ed} = 0.67 \times 1845 = 1236\,\text{kN}$

Using the simplified interaction of equation (D6.7-2) gives the following verification:

$$\frac{N_{Ed}}{\chi_y N_{pl,Rd}} + \frac{1}{1 - (N_{Ed}/N_{cr,y})}\frac{M_{y,Ed}}{M_{y,Rd}} + \frac{M_{z,Ed}}{M_{z,Rd}}$$

$$= \frac{1236}{0.99 \times 6390} + \frac{1}{1 - (1236/146\,523)}\frac{12.4}{210.6} + \left(\frac{12.4}{1055.9} + \frac{165}{544.6}\right)$$

$$= 0.195 + 0.059 + 0.315 = 0.569 < 1.0$$

The stiffener is therefore adequate.

A check of end bearing stress on the web and stiffeners should also be made if there are cut-outs in the stiffener or if the bearing area is smaller than the effective section area. The check is not included here and would not govern as the cross-section resistance check above was very conservative.

6.7.6. Beam torsional restraint at supports

Beams must be restrained against rotation about their longitudinal axes at supports for overall stability, but no guidance is given on this aspect of design in EN 1993. It is usual to provide vertical bracing or diaphragms at supports for this purpose but it is also possible to utilise the bending stiffness of bearing stiffeners to prevent rotation (as might occur in a U-frame bridge). The design of the restraint should consider forces arising due to initial geometric imperfections in the main beams and due to the effects of skew. All the effects could be determined by second-order analysis with the relevant modelled imperfections if the analysis model is sufficiently detailed. They include:

(i) Force due to the initial bow of the compression flange – the initial bow imperfection of the flange over the span, amplified by second-order effects, gives rise to a reaction at the supports.
(ii) Force due to non-verticality of the web at supports:
   (a) Where each end of a beam has an initial out of verticality in opposite directions, this leads to a further imperfection of the flanges over the length of the span. The growth of this imperfection under compressive load causes a reaction in the restraint that is proportional to the restraint stiffness.
   (b) Where the beam is not vertical at the supports, the restraint must be able to resist the overturning moment generated by the eccentricity of the bearing reaction from the applied load at deck level.
(iii) Forces due to the effects of skew:
   (a) Where end bracing restraints are placed on the skew (not square to the beam), the main beam has to twist about its own longitudinal axis when it rotates under vertical load about its transverse axis. This leads to an additional out of verticality to include in (ii)(b) above.
   (b) The torsional rotation above also twists the beam, generating a torsional reaction whose magnitude depends on the torsional stiffness of the beam.

No further guidance is given here on torsional restraints at beam ends. Reference can be made to BS 5400: Part 3: 2000[4] for more details on applicable design forces for the above and the UK National Annex provides reference to suitable guidance.

6.8. Loading on cross-girders of U-frames (additional sub-section)

The buckling of compression flanges and design forces for stiffeners restraining these flanges were discussed in section 6.3.4.2. However, additional forces are generated in U-frame members (including the flanges) by local loading on the cross-girders which causes differential deflections between adjacent frames. As illustrated in Fig. 6.8-1, loading on a

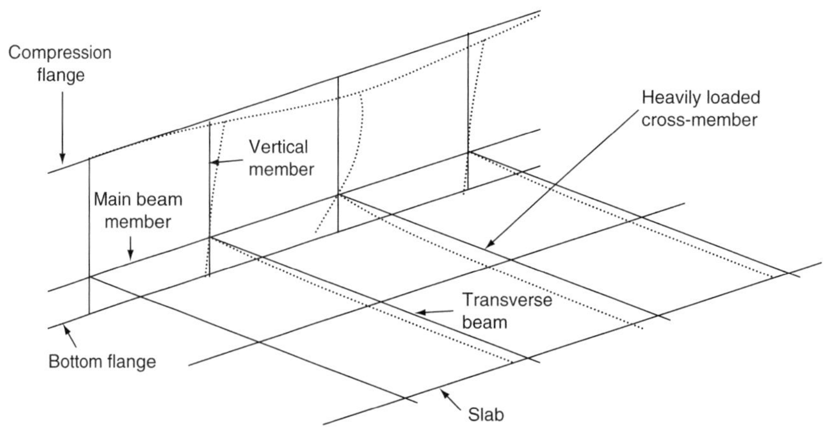

Fig. 6.8-1. Deflected shape of a U-frame bridge under transverse beam loading

Table 6.8-1. Section properties for the U-frame bridge spaceframe in Fig. 6.8-1

Member type	Section property			
	Area (A)	I vertical	I transverse	I_T torsion
Main beam:				
Flanges	*Null*	*Null*	I flange	I_T flange
Vertical stiffener	A stiffener	*Null*	I stiffener	I_T stiffener
Main beam	A beam	I beam	I web	I_T web
Decking:				
Transverse beam	A beam	I beam	I beam	I_T steel beam $+0.5I_T$ slab
Slab	A slab	I slab	I slab	$0.5I_T$ slab

transverse member will cause that transverse member to deflect and rotate at its connection to the vertical stiffener. The stiffener will therefore try to deflect inwards. If all cross-girders are not loaded similarly, the tendency is to produce differential deflections at the tops of the stiffeners but this differential deflection is resisted by the flanges in transverse bending. An outward force is therefore generated at the top of the stiffener that is attached to the cross-member with the local loading. This generates a moment in the stiffener which must be included in its design. The moment produced in the flanges from restraining the stiffener deflections needs to be considered in the stability check of the compression flange.

A simple method of calculation was proposed in BS 5400: Part 3.[4] This essentially assumed that the top flange was fully rigid when considering the force produced in a stiffener forming part of the U-frame with local loading. When the top flange moments were calculated, the assumption was that the top flange spanned between rigid stiffeners either side of the deflecting stiffener, which imparted a displacement to the flange equal to the free deflection of the stiffener. This gave very conservative results, but was easy to do. A less conservative method is to use a spaceframe model as shown in Fig. 6.8-1, using section properties as listed in Table 6.8-1.

Unless a second-order analysis is used, the bending moments obtained from the spaceframe analysis for the top flange and vertical member need to be multiplied by:

$$\frac{1}{1 - N_{Ed}/N_{cr}}$$

to include the destabilising P–δ effect as the top flange bows under compression loading. N_{Ed} is the force in the compression flange and N_{cr} is the elastic critical buckling load of the compression flange determined as in section 6.3.4.2.

6.9. Torsional buckling of stiffeners – outstand limitations (additional sub-section)

Stiffener outstands may buckle locally in a torsional buckling mode transverse to the plane of the parent plate, possibly in combination with an overall global buckling of the stiffener out of the plane of the parent plate. Torsional buckling of a stiffener outstand is illustrated in Fig. 6.9-1. If the stiffener is assumed to be simply supported along its attachment to the parent plate (unlike in Fig. 6.9-1), the elastic critical torsional buckling stress of a general stiffener is as follows:

$$\sigma_{cr} = \frac{1}{I_p}\left(GI_T + \frac{\pi^2 EC_w}{L^2}\right) \qquad \text{(D6.9-1)}$$

where:

I_T is the St Venant torsional constant for the stiffener outstand alone;

I_p is the polar second moment of area of the stiffener outstand alone about the point of attachment to the plate;

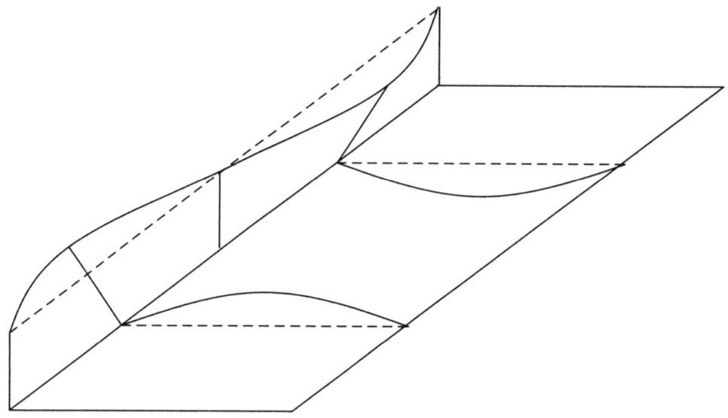

Fig. 6.9-1. Torsional buckling of stiffener

C_w is the warping constant of the stiffenerabout the attachment line;
L is the length between transverse restraints to the stiffener.

With the assumption of a simply supported edge between stiffener and parent plate, the wavelength of buckling would be the length between transverse restraints to the stiffener. The 'true' behaviour is discussed later.

If the warping constant is small, as is the case with flat stiffeners or some bulb flat stiffeners, equation (D6.9-1) may be approximated by:

$$\sigma_{cr} = \frac{GI_T}{I_p} \qquad \text{(D6.9-2)}$$

3-1-5/clause 9.2.1(8)

I_T/I_p is therefore a measure of the elastic critical torsional buckling stress for stiffeners with small warping resistance. ***3-1-5/clause 9.2.1(8)*** gives a limitation for stiffeners with small or zero warping resistance to prevent torsional buckling:

$$\frac{I_T}{I_p} \geq 5.3\frac{f_y}{E} \qquad \text{3-1-5/(9.3)}$$

This is conservative for stiffeners with appreciable warping resistance. The limitations apply equally to transverse stiffeners and longitudinal stiffeners. Since torsional buckling can lead to rapid collapse, it should be prevented when using the effective width method for Class 4 sections in 3-1-5/clause 4.

Substitution of expression 3-1-5/(9.3) into equation (D6.9-2) leads to:

$$\sigma_{cr} = \frac{GI_T}{I_p} \geq 2.04 f_y$$

which is equivalent to a limiting slenderness of:

$$\sqrt{f_y/\sigma_{cr}} = 0.70$$

This is similar to the slenderness limit for outstand plates of approximately 0.75 as implicit in 3-1-5/clause 4.4. This similarity is expected as, for a flat stiffener, the torsional buckling load can easily be shown to be the same as the elastic critical plate buckling load of a plate outstand.

For flat stiffeners:

$$I_p = \tfrac{1}{3}h_s^3 t_s + \tfrac{1}{12}h_s t_s^3 \text{ and } I_T = \tfrac{1}{3}h_s t_s^3$$

where h_s and t_s are the height and thickness of the flat respectively. This leads to the result that:

$$\frac{I_T}{I_p} \approx \left(\frac{t_s}{h_s}\right)^2$$

Substitution of this value into expression 3-1-5/(9.3) gives a limit on h_s/t_s of approximately 10.5 (actually 10.56) for S355 steel. This compares with a limit of 10 to BS 5400: Part 3. The general limit for flat stiffeners therefore becomes:

$$\frac{h_s}{t_s}\sqrt{\frac{f_y}{355}} \leq 10.5 \qquad \text{(D6.9-3)}$$

Stiffeners with warping resistance – Tees and angles

Where stiffeners have significant warping resistance and the length between transverse restraints is small, the use of the simplified criterion provided in EN 1993-1-5 is conservative. It is possible to use equation (D6.9-1) to calculate a critical buckling stress, and hence slenderness, for stiffeners with significant warping resistance, such as angles and tees:

$$\lambda = \sqrt{\frac{f_y}{\sigma_{cr}}} = \sqrt{\frac{f_y}{\frac{1}{I_p}\left[GI_T + \frac{\pi^2 EC_w}{L^2}\right]}} \leq X \qquad \text{(D6.9-4)}$$

The slenderness limit, X, would be 0.2 in accordance with 3-1-1/Fig. 6.4 for column buckling or 0.75 for plate buckling as discussed above. ***3-1-5/clause 9.2.1(9)*** recommends that $\sigma_{cr} \geq \theta f_y$ with $\theta = 6$. This is equivalent to $X = 0.4$ in the above expression, which was considered appropriate as the torsional buckling behaviour of stiffeners with warping resistance is partly plate-like and partly column-like. In the limit where there is no warping stiffness, the use of this lower limiting slenderness of 0.4 would mean that equation (D6.9-4) would produce a lower resistance than would expression 3-1-5/(9.3), which is based on a higher limiting slenderness of 0.7. In this situation, where there is low warping resistance, only the least onerous of the two criteria in 3-1-5/clauses 9.2.1(8) and (9) need be met. *3-1-5/clause 9.2.1(9)*

For an angle section:

$$C_w = \frac{1.3B^3H^2t_f}{3}\left(\frac{\frac{B^3t_f}{3H^2} + Bt_f + \frac{Ht_s}{3}}{Bt_f + Ht_s}\right) \qquad \text{(D6.9-5)}$$

For a Tee section:

$$C_w = \frac{1.1B^3H^2t_f}{12}\left(\frac{\frac{B^3t_f}{12H^2} + Bt_f + \frac{Ht_s}{3}}{Bt_f + Ht_s}\right) \qquad \text{(D6.9-6)}$$

The relevant dimensions are indicated in Fig. 6.9-2.

The above methods ignore any beneficial interaction with the parent plate, as 3-1-5/clause 9.2.1(9) requires the rotational restraint from the plate to be ignored. This is largely because a consensus could not be reached on how to take it into account. In reality, the buckling

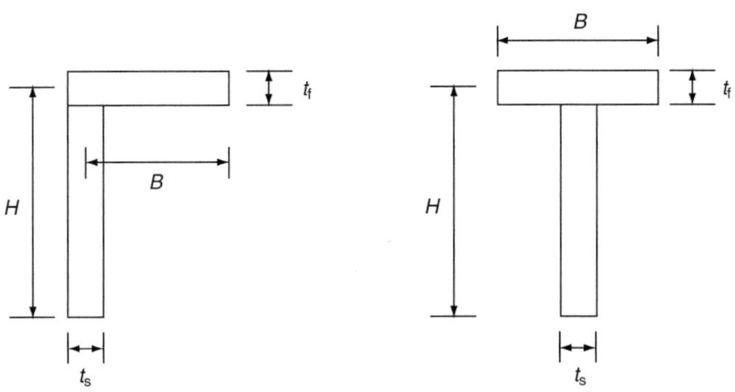

Fig. 6.9-2. Notation for angles and Tees

behaviour is complicated because the buckling wavelengths of simply supported parent plate panels and simply supported stiffeners will generally be different in isolation but must be the same in the actual stiffened plate for compatibility. An early draft of EN 1993-1-5 had a requirement that the elastic critical buckling load of the stiffener should be greater than that of the adjacent plate panels to which it was attached. This, however, led to the stability of stiffeners increasing as the stiffeners moved further apart. This is the opposite behaviour to that generally observed in testing and finite-element analyses, where the rotational restraint afforded to a stiffener by the parent plate can be significant where the span of the plate is small, and the opposite to the relationships which were given in BS 5400: Part 3.

The rules of EN 1993-1-5 can therefore be considered conservative and the neglect of any benefit from rotational restraint afforded to the stiffener by the parent plate means that certain stiffener types, particularly bulb flats, are unlikely to comply. If it is desired to use such stiffeners, it would be necessary to use a more detailed finite-element model, considering the full stiffened plate geometry to check behaviour.

Worked Example 6.9-1: Check of torsional buckling for an angle

An angle stiffener in S355 steel has the cross-section shown in Fig. 6.9-3. The adequacy of the stiffener is checked against torsional buckling for the case of (i) a long length between transverse restraints and (ii) restraints at 1400 mm.

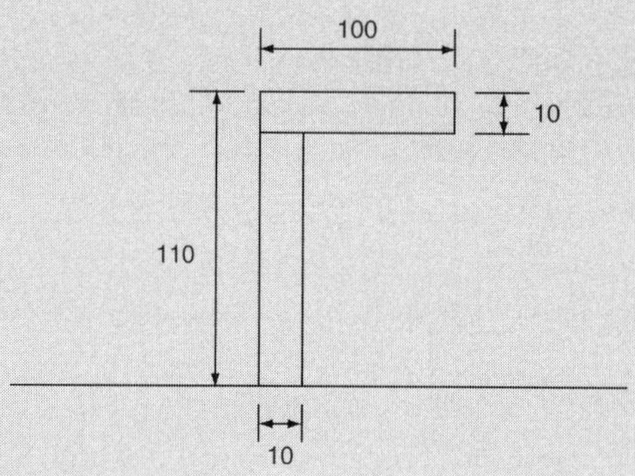

Fig. 6.9-3. Angle stiffener for Worked Example 6.9-1

(i) Restraints a long way apart

For widely spaced restraints, warping resistance will be insignificant and 3-1-5/clause 9.2.1(8) is relevant.

$$I_T = \tfrac{1}{3}(Ht_s^3 + Bt_f^3) = \tfrac{1}{3}(105 \times 10^3 + 95 \times 10^3) = 66.7 \times 10^3\,\text{mm}^4$$

$$I_p = (\tfrac{1}{12} \times 100^3 \times 10 + (100 \times 10 \times 50^2) + \tfrac{1}{12} \times 10^3 \times 100 + 100 \times 10 \times 105^2)$$
$$+ (\tfrac{1}{12} \times 100^3 \times 10 + 100 \times 10 \times 45^2) = 1.723 \times 10^7\,\text{mm}^4$$

$$\frac{I_T}{I_p} = \frac{66.7 \times 10^3}{1.723 \times 10^7} = 3.87 \times 10^{-3} < 5.3\frac{f_y}{E} = 8.96 \times 10^{-3}$$

so the stiffener does not comply.

Try increasing the stiffener thicknesses to 15 mm:

$$I_T = \tfrac{1}{3}(Ht_s^3 + Bt_f^3) = \tfrac{1}{3}(107.5 \times 15^3 + 92.5 \times 15^3) = 2.251 \times 10^5\,\text{mm}^4$$

$$I_p = (\tfrac{1}{12} \times 100^3 \times 15 + 100 \times 15 \times 50^2 + \tfrac{1}{12} \times 15^3 \times 100 + 100 \times 15 \times 107.5^2)$$
$$+ (\tfrac{1}{12} \times 100^3 \times 15 + 100 \times 15 \times 42.5^2) = 2.632 \times 10^7\,\text{mm}^4$$

$$\frac{I_T}{I_p} = \frac{2.251 \times 10^5}{2.632 \times 10^7} = 8.55 \times 10^{-3} < 5.3\frac{f_y}{E} = 8.96 \times 10^{-3}$$

so the stiffener still does not quite comply. A 16 mm thick stiffener would suffice by inspection.

(ii) Restraints at 1400 mm centres

If the original 10 mm thick stiffener is held in place transversely at 1400 mm centres then warping resistance will become significant and 3-1-5/clause 9.2.1(9) is relevant.

From equation (D6.9-5):

$$C_w = \frac{1.3 \times 95^3 \times 105^2 \times 10}{3}\left[\frac{\frac{95^3 \times 10}{3 \times 105^2} + 95 \times 10 + \frac{105 \times 10}{3}}{95 \times 10 + 105 \times 10}\right] = 3.193 \times 10^{10}\ \text{mm}^6$$

and from equation (D6.9-4):

$$\lambda = \sqrt{\frac{355}{\frac{1}{1.723 \times 10^7}\left[80.77 \times 10^3 \times 66.7 \times 10^3 + \frac{\pi^2 \times 210 \times 10^3 \times 3.193 \times 10^{10}}{1400^2}\right]}}$$

$$= 0.395$$

which is less than 0.40 so the stiffener would be prevented from buckling torsionally. It may, however, be impractical to support the stiffeners this closely.

6.10. Flange-induced buckling and effects due to curvature (additional sub-section)

6.10.1. Flange-induced buckling and flange-induced forces on webs and cross-members

6.10.1.1. I-girders without flange longitudinal stiffeners

For I-girders, it is normally assumed that the web provides a rigid linear support to the compression flange against buckling in the plane of the web. If the flange is sufficiently large, however, and the web is very slender, it is possible for the whole flange to buckle into the plane of the web by inducing buckling in the web itself as shown in Fig. 6.10-1.

If the compression flange is continuously curved in elevation, whether because of intended profiling to the soffit or because the whole beam is cambered in elevation, there is a radial force induced in the plane of the web due to the continuous change in direction of the flange force. This force, shown in Fig. 6.10-3, increases the likelihood of flange-induced buckling into the web. The transverse force per unit length of the flange exerted by a curved flange of radius r is given as follows:

$$P_T = F_f/r \quad \text{(D6.10-1)}$$

where F_f is the axial force in the flange.

The flange exerts a similar force due to the curvature from initial imperfections in the beam and from deflections of the beam under loading.

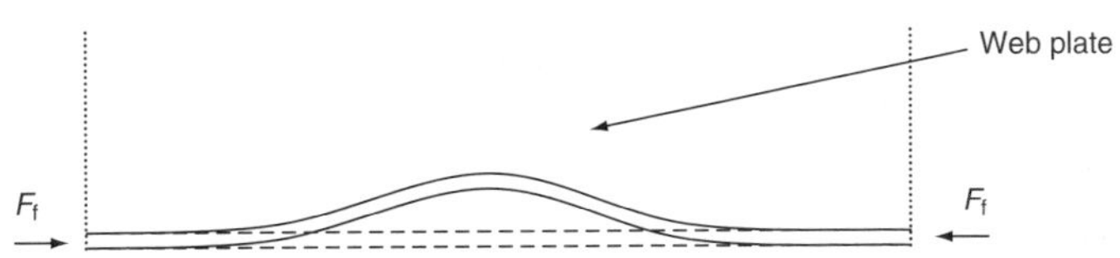

Fig. 6.10-1. Flange-induced buckle in web

3-1-5/clause 8(2)

To prevent flange-induced buckling, a simple limit on web height to thickness is given in ***3-1-5/clause 8(2)*** as follows, which is primarily intended for use with I-girders without stiffeners:

$$\frac{h_w}{t_w} \le \frac{k\dfrac{E}{f_{yf}}\sqrt{\dfrac{A_w}{A_{fc}}}}{\sqrt{1+\dfrac{h_w E}{3r f_{yf}}}} \qquad \text{3-1-5/(8.2)}$$

where:

A_w is the cross-sectional area of the web
A_{fc} is the effective cross-sectional area of the flange
h_w and t_w are the height and thickness of the web respectively
r is the radius of the flange in elevation and
k is a factor which reduces with increasing anticipated strain in the flanges such that:

$k = 0.3$ for plastic global analysis with hinge formation (not generally relevant for bridges as plastic analysis only allowed in certain accidental situations);
$k = 0.4$ for plastic section analysis;
$k = 0.55$ for elastic section analysis.

Expression 3-1-5/(8.2) also assumes that the compression flange is on the concave side. If it is not, the flange induces transverse tension in the web, which cannot cause buckling. Where the flange is not curved in elevation, the simplified expression of expression 3-1-5/(8.1) may be used.

An illustrative derivation of the above can be made considering a vertically curved I-beam with equal flanges, both with assumed radius r_t. This curvature of the two flanges, one in tension and one in compression, leads to the application of equal and opposite compressive transverse forces acting on the web along its top and bottom surfaces. For a long web panel without longitudinal stiffeners and loaded transversely with a uniformly distributed load, the buckling mode is column-like and the critical stress is therefore:

$$\sigma_{cr} = \frac{\pi^2 E}{12(1-\upsilon^2)}\left(\frac{t_w}{h_w}\right)^2 \qquad \text{(D6.10-2)}$$

The applied transverse pressure from a length of curved flange stressed to its yield point is obtained from equation (D6.10-1) as:

$$P_T = \frac{f_{yf}A_{fc}}{r_t}$$

so the transverse pressure on the web at the top and bottom is:

$$\sigma_T = \frac{f_{yf}A_{fc}}{r_t t_w} \qquad \text{(D6.10-3)}$$

To prevent buckling of the web, the critical buckling stress must be greater than the applied transverse stress by some factor α, so that for adequacy:

$$\sigma_{cr} = \frac{\pi^2 E}{12(1-\upsilon^2)}\left(\frac{t_w}{h_w}\right)^2 \ge \alpha\frac{f_{yf}A_{fc}}{r_t t_w} \qquad \text{(D6.10-4)}$$

Rearranging equation (D6.10.4) gives:

$$\frac{h_w}{t_w} \le \sqrt{\frac{\pi^2 E}{12\alpha(1-\upsilon^2)}\frac{r_t A_w}{f_{yf}A_{fc}h_w}} = \sqrt{\frac{\pi^2}{12\alpha(1-\upsilon^2)}}\sqrt{\frac{E r_t A_w}{f_{yf}A_{fc}h_w}}$$

thus:

$$\frac{h_w}{t_w} \le \sqrt{\frac{\pi^2}{12\alpha(1-\upsilon^2)}}\frac{\dfrac{E}{f_{yf}}\sqrt{\dfrac{A_w}{A_{fc}}}}{\sqrt{\dfrac{h_w E}{r_t f_{yf}}}} \qquad \text{(D6.10-5)}$$

The curvature $1/r_t$ comprises an intentional curvature, $1/r$, together with a further curvature from deflection under load and from imperfections. For elastic behaviour, the stress in both flanges is limited to first yield at f_{yf}, so the strain difference across the depth h_w is $2f_{yf}/E$ and the curvature is $2f_{yf}/(h_w E)$. This additional curvature makes no allowance for either flange strains beyond first yield or the effects of flange and member imperfections. These can both be included via an additional factor, β (greater than 1) such that the additional curvature $1/r_i$ can be expressed as:

$$\frac{1}{r_i} = \frac{2\beta f_{yf}}{h_w E} \qquad \text{(D6.10-6)}$$

The total curvature, assuming the curvatures are applied in the same direction, is therefore:

$$1/r_t = 1/r + \frac{2\beta f_{yf}}{h_w E} \qquad \text{(D6.10-7)}$$

Substitution of equation (D6.10-7) into equation (D6.10-5) leads to the following expression:

$$\frac{h_w}{t_w} \le \frac{1}{\sqrt{2\alpha\beta}}\sqrt{\frac{\pi^2}{12(1-\upsilon^2)}}\frac{\dfrac{E}{f_{yf}}\sqrt{\dfrac{A_w}{A_{fc}}}}{\sqrt{1+\dfrac{h_w E}{2\beta r f_{yf}}}}$$

so that:

$$\frac{h_w}{t_w} \le \frac{0.672}{\sqrt{\alpha\beta}}\frac{\dfrac{E}{f_{yf}}\sqrt{\dfrac{A_w}{A_{fc}}}}{\sqrt{1+\dfrac{h_w E}{2\beta r f_{yf}}}} \qquad \text{(D6.10-8)}$$

The factor $0.672/\sqrt{\alpha\beta}$ corresponds to the factor k in expressions 3-1-5/(8.1) and (8.2).

For column-like behaviour and high slenderness, as is usually the case with a web compressed vertically, the real buckling load is very close to the elastic critical buckling load and therefore $\alpha = 1$ can be used in equation (D6.10-4) onwards. For more stocky webs with say $h_w/t_w < 45$ and S355 steel, corresponding to a column slenderness $\overline{\lambda}_c < 2.0$, the real buckling load becomes less than the elastic critical load due to web imperfections and therefore $\alpha > 1$ would be appropriate. (More generally, this approximate limit is $h_w/t_w < 870/\sqrt{f_{yf}}$ for other steel grades.) The criteria in 3-1-5/clause 8 will usually easily be met for straight girders with webs of this stockiness, so the slight lack of conservatism in the choice of α for such cases is not a problem. For curved beams, some caution might be advised where $h_w/t_w < 870/\sqrt{f_{yf}}$, if the criteria in 3-1-5/clause 8 are only just met. It is also noted that for beams with only one flange curved, the whole derivation is conservative, as the compressive force is then applied to one edge of the web only and the corresponding critical buckling load is then much higher.

The factor β for elastic analysis without any geometric imperfections would be 1.0. To allow for imperfections, EN 1993-1-5 appears to take $\beta = 1.5$. Greater flange strains occur for plastic section design and greater still for plastic global analysis and so β, and hence also k in expression 3-1-5/8(1), has a greater value in these cases. For elastic section design, substituting $\beta = 1.5$ and $\alpha = 1$ into equation (D6.10-8) gives:

$$\frac{h_w}{t_w} \leq 0.55 \frac{\dfrac{E}{f_{yf}}\sqrt{\dfrac{A_w}{A_{fc}}}}{\sqrt{1 + \dfrac{h_w E}{3 r f_{yf}}}} \quad \text{(D6.10-9)}$$

which is expression 3-1-5/(8.2) with $k = 0.55$ for elastic section design.

If the flange is not intentionally curved, r is infinite and equation (D6.10-8) becomes:

$$\frac{h_w}{t_w} \leq 0.55 \frac{E}{f_{yf}}\sqrt{\frac{A_w}{A_{fc}}} \quad \text{(D6.10-10)}$$

which is expression 3-1-5/(8.1) with $k = 0.55$ for elastic section design.

As discussed above, the value of α implicit in the derivation of the expressions in 3-1-5/clause 8 appears to be slightly unconservative when h_w/t_w is less than about $870/\sqrt{f_{yf}}$, but for beams with only one flange curved, the rest of the derivation is conservative. Some caution is therefore recommended in applying the expressions in 3-1-5/clause 8 to girders where the whole beam is vertically curved and $h_w/t_w < 870/\sqrt{f_{yf}}$. In such cases, a value of α could be determined as the ratio of the true buckling strength (determined from the column buckling curves in 3-1-1/Fig. 6.4) to the elastic critical buckling load. This value could then be used in conjunction with equation (D6.10-8).

For I-girders, the limit in expression 3-1-5/(8.2) clearly makes no allowance for vertical stiffeners on the webs but these will be of limited benefit on a continuously curved flange unless closely spaced, so can usually be ignored without undue conservatism. There is also no allowance for web longitudinal stiffeners; the formula could however be modified to check buckling of the stiffened panel and weakest sub-panel. For beams with curvature formed from a series of straight panels, it will usually be necessary to place a transverse stiffener at each kink position to carry the concentrated force. In this case, the web should be checked for flange-induced buckling assuming the flange to be straight (with infinite radius) and the stiffeners should be designed for the deviation forces in the flange at each kink.

For beams with straight compression flanges, it is unlikely that flange-induced buckling will govern the web dimensions other than in webs with unusually slender Class 4 section. However, the criterion given for preventing flange-induced buckling is similar to that for torsional buckling of stiffeners in that the section must comply with this limit as there is no method given of taking flange-induced buckling into account in deriving a reduced resistance to other effects.

Interaction of flange-induced web transverse stress with bending, shear and axial force

Despite the applicability of expression 3-1-5/(8.2) to beams with vertically curved flanges, EN 1993 gives no guidance on the interaction of the flange-induced transverse stress on the web with other effects; a check should, however, be made. Strictly, EN 1993-1-5 requires the design of variable-depth curved members to be carried out using 3-1-5/clause 10 by way of the requirements of 3-1-5/clause 2.5 covering non-uniform members. A verification of interaction can therefore be achieved by using the reduced stress method of 3-1-5/clause 10. If it is wanted to ensure that no second-order effects will occur in the web due to flange curvature, one could ensure that $\alpha \geq 10$ in equation (D6.10-4); this is the criterion for neglecting second-order effects in 3-2/clause 5.2.1(4). This could be used as a criterion for when vertically curved beams can be designed as straight to 3-1-5/clause 7.1, but allowing for the effects of flange curling (section 6.10.2.1 of this guide) and bearing stress on the web in deriving a reduced effective yield stress for flange and web respectively to be used in expression 3-1-5/(7.1). The reduced effective yield stress can be derived from the Von Mises equation in section 6.2.1 of this guide.

As an alternative approach to allow for flange-induced web transverse stress, the interaction of 3-1-5/clause 7.2 could be adopted if an equivalent transverse force and resistance can be established in accordance with 3-1-5/clause 6. This approach, while logical, has not been verified by testing. The geometry requirements of EN 1993-1-5 clause 2.3 should be met (other than the requirement for parallel flanges which clearly cannot be met for beams with one flange curved in elevation). In deriving the patch load resistance in 3-1-5/ clause 6, the buckling coefficient for Type (a) in 3-1-5/Fig. 6.1 could be used where only the compression flange is curved in elevation, and the coefficient for Type (b) used where both flanges are curved. There then arises the problem of deciding the applicable length of flange to consider in deriving the patch load and its resistance. Conservative estimates of flange length (e.g. whole length in compression) and flange stress (e.g. greatest stress anywhere in flange) could be used in determining the magnitude of the patch load. Quite large patch loads can be accommodated without reducing the resistance to direct stress when 3-1-5/clause 7.2 is applied, so conservative assumptions may often suffice.

A separate yielding check of the flanges, allowing for the transverse bending induced, should also be made where flanges are curved in elevation. This is discussed in section 6.10.2 below.

6.10.1.2. Box girders

For box girders without longitudinal stiffeners and with widely spaced transverse diaphragms or cross-members, expression 3-1-5/(8.2) can be applied to individual webs and their associated part of attached flange, taken as half the width of effective flange between webs and any additional outstand.

For box girders with longitudinal stiffeners and transverse diaphragms or cross-members at closer centres however, expression 3-1-5/(8.2) may be unduly conservative and does not reflect the real behaviour. In longitudinally stiffened flange panels, the transverse loading induced by a vertically curved flange will tend to be carried longitudinally by the stiffeners spanning between transverse members, as indicated in Fig. 6.10-4.

No check method is provided in this instance but it would be possible to apply the curvature force in equation (D6.10-1) as a distributed load to a computer model of the stiffened panels and determine the bearing pressure on the web and transverse diaphragms as discussed in section 6.10.2 below. Alternatively, a reasonable approximation would be to assume that half the effective width of flange between web and longitudinal stiffener nearest to the web, together with any flange outstand, transmits its force to the web. Expression 3-1-5/(8.2) could then be used to check the web, basing the effective flange area A_f on the above. This is still conservative for longitudinally stiffened webs because the additional buckling resistance they provide is not considered. The comments in section 6.10.1.1 on the lack of an interaction equation for consideration of other effects, and possible methods of considering them, also apply here.

The transverse member also needs to be checked for the force imparted by the curved flange. The force on the transverse member should either be taken from a computer model or, following on from the simplification above, as:

$$\overline{P} = \overline{F}_f a/r \qquad \text{(D6.10-11)}$$

where $\overline{P}$ is the total force distributed across the width of the diaphragm compatible with the area of the constituent parts of the flange, a is the length of the stiffened panel and $\overline{F}_f$ is the force in the flange between webs excluding the force in the half-widths of sub-panels attached to the webs. The transverse member would also have to carry the force from unintentional flange deviation (geometric imperfection). The requirements of 3-1-5/clause 9.2.1 are then applicable. If the cross-member is a transverse stiffener, the force from equation (D6.10-11), together with a further deviation force from 3-1-5/clause 9.2.1, has to be applied. Second-order effects will arise in the transverse stiffener due to its deflection out of plane as discussed in section 6.6.5 of the guide. If the cross-member is a plated diaphragm, the same method can be used but the rigidity of the diaphragm transversely means that the above forces can be applied directly to the diaphragm without consideration of second-order effects.

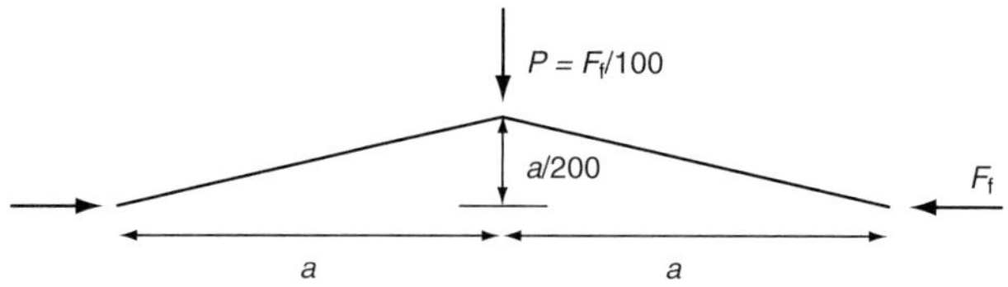

Fig. 6.10-2. Force in diaphragm due to longitudinal stiffener imperfections

An alternative to the use of 3-1-5/clause 9.2.1 for rigid diaphragms is to consider the permissible imperfections in longitudinal stiffeners. From 3-1-5/Table C.2, the imperfection for analysis is $L/400$. If L is taken as the length of two stiffened panels, $2a$, and a kink imperfection of $L/400 = a/200$ is applied at the diaphragm considered, then the transverse force on the diaphragm is given as shown in Fig. 6.10-2. This can be applied in either an upward or downward direction.

6.10.2. Stresses in vertically curved flanges (continuously curved)

No guidance is given in EN 1993 on the design of beams with flanges continuously curved in elevation, mainly because it involves transverse bending in plate panels which is not covered by either EN 1993-2 or EN 1993-1-5. EN 1993-1-7 covers transverse loading (not curved beams specifically) but is not fully applicable to bridge members. Beams with vertical curvature develop out-of-plane bending moments in the flanges. For I-beams, this flange transverse bending is sometimes referred to as 'flange curling' – Fig. 6.10-3(b). It is not covered explicitly by interaction equations, although suggestion for its inclusion in the shear–moment check of expression 3-1-5/(7.1) is made in section 6.10.1.1 above.

The load of equation (D6.10-1) can be applied across the width of the flange as a transverse load to determine the bending effects in the flange (both transverse and longitudinal) and also the bearing stresses on webs, stiffeners and transverse diaphragms.

6.10.2.1. Flanges in girders without longitudinal stiffeners

For an outstand flange of thickness t on a symmetric I-beam with widely spaced transverse stiffeners and without longitudinal stiffeners, equation (D6.10-1) leads to a transverse moment, M_T, at the face of the web as shown in Fig. 6.10-3(b):

$$M_T = F_f/2r \times c/2 = F_f c/(4r) = \sigma_f c^2 t/(2r) \qquad \text{(D6.10-12)}$$

where σ_f is the axial stress in the flange. The transverse bending stress is then as follows:

$$\sigma_T = 3\sigma_f c^2/(rt) \qquad \text{(D6.10-13)}$$

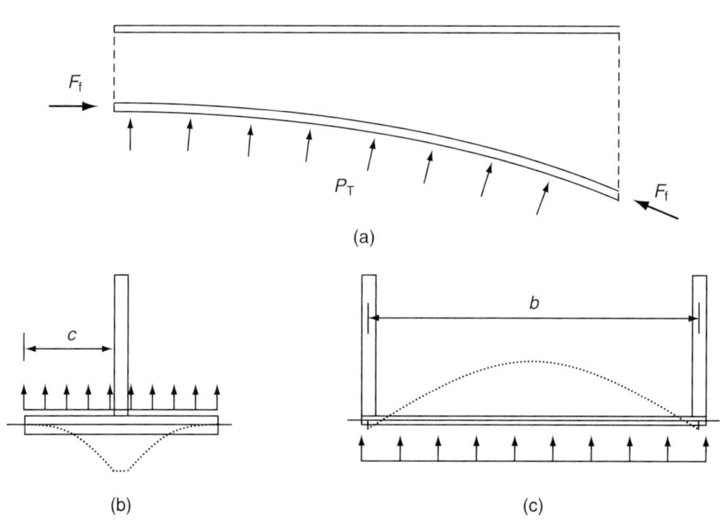

Fig. 6.10-3. Forces and moments from flange curvature: (a) radial force from curved flanges; (b) transverse moments in outstand flange; (c) transverse moments in internal flange

For a flange in a box girder without longitudinal stiffeners and widely spaced cross-girders or diaphragms, the flange spans transversely between webs. The flange moment depends on the flexural stiffness of the webs. However, assuming the web flexural rigidity to be small so that the flange spans simply supported between webs, equation (D6.10-1) leads to a transverse moment, M_T, midway between webs as shown in Fig. 6.10-3(b):

$$M_T = F_f/r \times b/8 = \sigma_f b^2 t/(8r) \qquad \text{(D6.10-14)}$$

The transverse bending stress is then as follows:

$$\sigma_T = 3\sigma_f b^2/(4rt) \qquad \text{(D6.10-15)}$$

If there is a flange outstand, the moments and stresses in the flange can be calculated using the above principles.

The first-order transverse bending stresses and displacements in the flange plate due to vertical curvature are not magnified to any significant extent by the axial force (to give second-order effects) for cases where the transverse restraints are widely spaced. For box-girder cases where the transverse restraints *are* closely spaced, so that the first mode of buckling of the flange plate under axial load is a single half-wavelength in the longitudinal direction between transverse restraints, the flange curvature force will be carried by two-way spanning of the flange. The first-order transverse moment will therefore be less than that predicted by equation (D6.10-14) but some magnification of both longitudinal and transverse bending stresses due to flange compression may then occur. It is unlikely that restraints would be placed this closely in practice but, if they were, it will generally be satisfactory in any case to use the conservative transverse moment from equation (D6.10-14) without magnification.

No interaction is provided to incorporate the effects of transverse bending in checking the flanges, so the Von Mises yield criterion of 3-1-1/clause 6.2.1 could be used – equation (D6.5-4) in section 6.5.2.1 of this guide refers. A reduced effective flange yield stress can also be derived in this way (but ignoring reductions in flange yield stress due to coexisting flange shear stress) for use in shear–moment interaction checks as discussed in section 6.10.1.1.

For overall member buckling checks, it would also be necessary to allow for an effective reduction in flange yield stress in the buckling check. This reduced yield stress could again be derived using the Von Mises criterion, again ignoring coexisting shear stress as is usual in overall member buckling checks.

6.10.2.2. Flanges in girders with longitudinal stiffeners and transverse stiffeners or diaphragms

The determination of out-of-plane bending stresses in longitudinally stiffened plates is more complicated as the stiffened plates will span both longitudinally and transversely. Because of their greater longitudinal flexural rigidity, the stiffened panel will typically mainly span longitudinally as shown in Fig. 6.10-4, with only small transverse global bending moments. However, a local transverse bending action will still develop locally between the longitudinal stiffeners similar to that in Fig. 6.10-3(c). The transverse bending stress in the parent plate from this action can be taken as the same as the stress for unstiffened internal panels above. Use of equation (D6.10-15) would be conservative as the flange sub-panels are not simply supported by the longitudinal stiffeners but rather are continuous over them.

For a better determination of the bending effects in both directions, the transverse load in equation (D6.10-1) should be applied to a grillage or finite-element model as a distributed load over the sub-panels and stiffeners in proportion to their in-plane forces. This will also model continuity of longitudinal stiffeners over the transverse members, thus reducing the moment at mid-span of the longitudinal stiffener.

The stiffener and flange plate can be checked for the combined global and local effects as discussed in section 6.5.2 of this guide, with the first-order stresses from curvature treated as local effects.

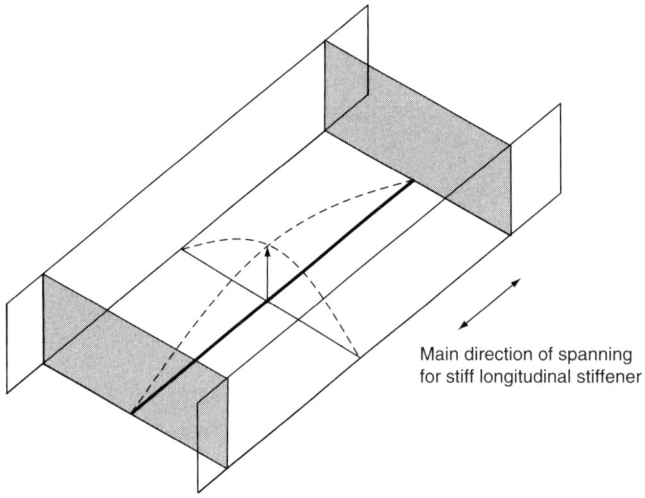

Fig. 6.10-4. Direction of spanning in longitudinally stiffened panel with transverse load

An alternative to the checks of the stiffener under local plus global load presented in section 6.5.2 would be to allow for the initial out-of-straightness in the stiffener caused by the flange curvature directly in the stiffener buckling resistance curve. As discussed in section 6.3.1.2, the strut Perry–Robertson imperfection parameter for geometric imperfections is ye_0/i^2 where y is the maximum distance from stiffener effective section centroidal axis to an extreme fibre of the stiffener effective section, e_0 is the magnitude of imperfection and i is the radius of gyration.

The imperfection parameter in EN 1993 is taken as $\eta = \alpha(\bar{\lambda} - 0.2)$ which also makes allowance for structural imperfections. If an additional imperfection of e_f is considered, representing the largest offset of the stiffener due to curvature from a straight line between transverse restraints, then an additional term in the imperfection parameter of ye_f/i^2 can be added to the imperfection parameter in the strut curves of 3-1-1/clause 6.3.1.2. For longitudinal stiffeners, 3-1-5/clause 4.5.3(5) sets $\alpha = \alpha_e$ for straight stiffeners. Therefore, for curved flange stiffeners, $\eta = \alpha_e(\bar{\lambda} - 0.2)$ needs to be replaced by:

$$\eta = \alpha_e(\bar{\lambda} - 0.2) + ye_f/i^2$$

in expression 3-1-1/(6.49) when deriving the Class 4 section properties in accordance with 3-1-5/clause 4.5.3(5). A yield check of the parent flange plate is still required as in section 6.5.2.1.

The above is conservative, as is the proposal in section 6.5, as the resulting imperfection parameter does not allow consideration of whether the direction of curvature would be adverse or relieving to the critical fibre implicit in the original imperfection parameter. If the longitudinal stiffeners are not in an end bay, such that there is continuity of the stiffeners across transverse restraints, the effect of the curvature bow can be reduced for this continuity. The imperfection parameter could then be taken as:

$$\eta = \alpha_e(\bar{\lambda} - 0.2) + \frac{ye_f}{2i^2}$$

If this additional imperfection approach to modelling curvature is employed, the reduction factor for global plate buckling used in deriving effective cross-section properties should be based on column-type behaviour alone, unless some similar allowance for curvature can be made in considering plate-like buckling – see section 6.2.2.4 of this guide.

6.10.3. Stresses in webs and flanges in beams curved in plan

No guidance is given in EN 1993 on the design of beams which are curved in plan. A beam in bending that is curved in plan will develop similar forces and out-of-plane moments from curvature to those derived above for beams with vertically curved soffits. However, there

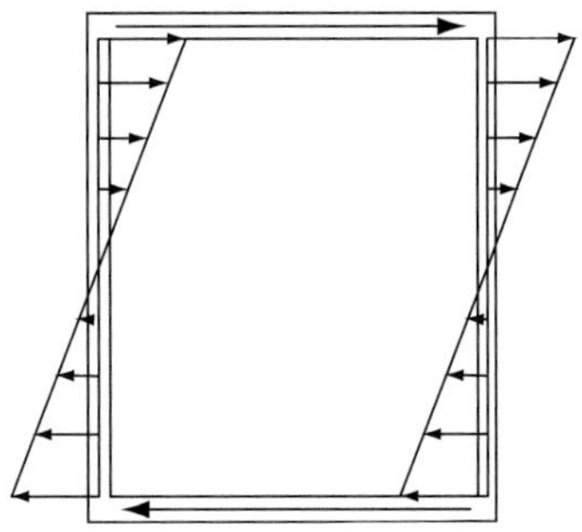

Fig. 6.10-5. Forces acting on a box girder in bending due to plan curvature

will also be in-plane bending of the flanges and a distortion of the cross-section. The curved compression flange and tension flanges give rise to transverse forces in opposite directions giving rise to a torque. A similar transverse force occurs in a web which reverses over its height as shown in Fig. 6.10-5 for a box section. (The effect is the same as that from beam theory whereby a moment about the major axis resolves itself into a torque and a moment on progressing around the curve.)

With very closely spaced rigid diaphragms, the distortion from the forces in Fig. 6.10-5 is controlled by the diaphragms and the torque is carried in pure St Venant torsion. Where there are no diaphragms or more widely spaced diaphragms, there is additional transverse bending of the flange and web plates together with warping of the individual plates in the same way as that due to eccentric loading discussed in section 6.2.7 of this guide. The effects may be modelled in the same way as for eccentric loading, but account has to be taken of the transverse bending that occurs in the webs, even when the box corners are restrained from distorting.

It is simplest to use elastic cross-section analysis when combining effects. The additional warping stresses should be added to other direct stresses. The distortional bending stresses can be combined with other stresses using the Von Mises equivalent stress criterion. This can be done in the same way as the combination of local and global effects discussed in section 6.5.2 of this guide.

CHAPTER 7

Serviceability limit states

This chapter discusses serviceability limit states as covered in section 7 of EN 1993-2 in the following clauses:

- General *Clause 7.1*
- Calculation models *Clause 7.2*
- Limitations for stress *Clause 7.3*
- Limitation of web breathing *Clause 7.4*
- Miscellaneous SLS requirements in clauses *Clauses 7.5–7.12*

7.1. General

The serviceability limit states principally concern the adequate functioning of the bridge, its appearance and the comfort of bridge users. ***3-2/clause 7.1(1)*** refers, by way of EN 1993-1-1, to EN 1990 clause 3.4 which gives the following recommendations for the verification of serviceability limit states: *3-2/clause 7.1(1)*

'*(3) The verification of serviceability limit states should be based on criteria concerning the following aspects:*
a) Deformations that affect
– the appearance (in terms of high deflection and surface cracking)
– the comfort of users
– the functioning of the structure (including the functioning of machines or services)
or that cause damage to finishes or non-structural members;
b) Vibrations
– that cause discomfort to people, or
– that limit the functional effectiveness of the structure;
c) Damage that is likely to adversely effect
– the appearance
– the durability, or
– the functioning of the structure.'

3-2/clause 7.1(4) then relates these general EN 1990 recommendations into specific, although not exhaustive, serviceability limit state recommendations. These are then covered in greater detail in clauses 7.3 to 7.12. Provided the designer follows the recommendations of clauses 7.3 to 7.12, the serviceability limit state recommendations of EN 1990 will be met. *3-2/clause 7.1(4)*

7.2. Calculation models

As with the calculation of fatigue stresses, discussed in Chapter 9, serviceability limit state (SLS) stresses should generally be calculated using an analysis which is as accurate as practically possible, both in terms of structural idealization and the application of loadings.

3-2/clause 7.2(1) This accuracy also applies to deflections, although they are clearly linked closely to stresses.
3-2/clause 7.2(3) ***3-2/clause 7.2(1)*** and ***3-2/clause 7.2(3)***, by reference to EN 1993-1-5, require SLS stresses and deflections to be calculated using a linear elastic analysis and section properties which include the reductions in stiffness due to local plate buckling and shear lag, where relevant. Plate buckling effects will not normally need to be considered in the global analysis as a result of the provisions of 3-1-5/clause 2.2(5). Plate buckling will also generally not need consideration for stress analysis as discussed in section 7.3 below. Section properties for global analysis are discussed in greater detail in section 5.1 of this guide, while the effects of shear lag on cross-section properties for stress analysis are discussed in section 6.2.2.3. The effects of shear lag are usually only significant for members with wide flanges.

If shell finite-element modelling is used for global analysis, the effects of shear lag will automatically be included in part or fully, depending on the detail of the mesh used. Plate buckling effects will only be included if the analysis is second order and initial imperfections have been modelled.

7.3. Limitations for stress

Stresses have to be limited so that yielding does not occur during normal service conditions, principally to avoid excessive permanent deflections and disruption to the corrosion protection system. ***3-2/clause 7.3(1)*** gives limits for serviceability stresses:

3-2/clause 7.3(1)

$$\sigma_{\mathrm{Ed,ser}} \leq \frac{f_y}{\gamma_{\mathrm{M,ser}}} \qquad \text{3-2/(7.1)}$$

$$\tau_{\mathrm{Ed,ser}} \leq \frac{f_y}{\sqrt{3}\gamma_{\mathrm{M,ser}}} \qquad \text{3-2/(7.2)}$$

$$\sqrt{\sigma_{\mathrm{Ed,ser}}^2 + 3\tau_{\mathrm{Ed,ser}}^2} \leq \frac{f_y}{\gamma_{\mathrm{M,ser}}} \qquad \text{3-2/(7.3)}$$

where:

$\sigma_{\mathrm{Ed,ser}}$ is the direct stress obtained from the characteristic load combination;
$\tau_{\mathrm{Ed,ser}}$ is the shear stress obtained from the characteristic load combination;
$\gamma_{\mathrm{M,ser}}$ is a partial safety factor; the recommended value in Note 2 of 3-2/clause 7.3(1) is 1.0.

$\sigma_{\mathrm{Ed,ser}}$ and $\tau_{\mathrm{Ed,ser}}$ must include the effects of shear lag and any secondary effects 'caused by deflections', such as the moments generated from joint stiffness in trusses. This is important to note because the same effects could legitimately be ignored at ULS by idealising the joints as pinned.

3-2/clause (7.3) assumes only uniaxial direct stress and a single plane of shear are present. For more general stress fields, expression 3-2/(7.3) can be extended to the general Von Mises expression provided in section 6.2.1 of this guide. If there is local transverse load applied to the bridge members, such as from concentrated wheel loads applied at deck level, the resulting stress $\sigma_{z,\mathrm{Ed}}$ may be calculated using the dispersion rule in 3-1-5/clause 3.2.3 as discussed in section 6.2.2.3.2 of this guide.

SLS verifications of stress are usually necessary even for Class 3 and 4 cross-sections, even though they are checked elastically at the ultimate limit state. This is because some effects may be ignored at ULS if they are dissipated through a little yielding. If torsional warping or St Venant torsional effects have been neglected at ULS, as allowed by 3-2/clause 6.2.7, SLS stresses should be checked taking these torsional effects into account as they might result in yielding occurring. Shear lag may also cause yielding at SLS; the effective flange widths are greater at ULS because they make allowance for plastic redistribution. Plate buckling effects usually will not need to be considered. If the ULS reduction factor for plate buckling, ρ, exceeds 0.5, 3-1-5/clause 2.3(2) allows stresses at SLS and for fatigue to be calculated on the gross cross-section, but making allowance for shear lag. Note 3 of 3-2/clause 7.3(1) gives a similar recommendation. If this criterion is not satisfied, either the ULS effective cross-section for plate buckling can conservatively be used, or a less onerous effective cross-section can be derived using 3-1-5/Annex E.

The fatigue verifications in 3-2/clause 9.5.1(1) are only valid, according to 3-1-9/clause 8(1), if the direct stress and shear stress ranges due to frequent loads are less than $1.5f_y$ and $1.5f_y/\sqrt{3}$ respectively. ***3-2/clause 7.3(2)*** reinforces this by requiring that the stress range $\Delta\sigma_{fre}$ caused by variable loads within the frequent combination should be limited to $1.5f_y/\gamma_{M,ser}$. The equivalent limit for shear stresses should also be observed. *3-2/clause 7.3(2)*

3-2/clause 7.3(3) requires the SLS force in non-preloaded bolts, derived from the characteristic combination of actions, to be limited as follows to avoid large displacements from occurring due to bolt bearing: *3-2/clause 7.3(3)*

$$F_{b,Rd,ser} \leq 0.7F_{b,Rd} \qquad \text{3-2/(7.4)}$$

where:

$F_{b,Rd,ser}$ is the bolt force derived from the linear elastic SLS analysis;
$F_{b,Rd}$ is the bolt bearing resistance derived from 3-1-8/Table 3.4.

Bolt forces in category B pre-loaded bolted connections, which are designed not to slip at serviceability, should be checked against the resistance determined in accordance with 3-1-8/clause 3.9.1 – ***3-2/clause 7.3(4)*** refers. The bolt force is calculated using the characteristic load combination. *3-2/clause 7.3(4)*

7.4. Limitation of web breathing

Web breathing is a phenomenon which affects slender plates as noted by ***3-2/clause 7.4(1)***. Initial geometrical imperfections in plate panels grow under load and then reduce again when the load is removed as indicated in Fig. 7-1. The term 'breathing' arises because this cyclic movement of the plate panel out of plane resembles the expansion and contraction of the chest during breathing. It can lead to fatigue damage at plate boundaries, i.e. at or adjacent to connections between web and flange and also between web and stiffeners. Breathing will not, however, usually govern the dimensions of typical bridge types. *3-2/clause 7.4(1)*

To avoid detailed considerations of potential damage from web breathing, either the plate slenderness can be limited through appropriate b/t ratios or an interaction can be performed relating applied stresses to their limiting values for elastic buckling. A distinction is made between road and rail bridges in EN 1993-2 because of the greater susceptibility to fatigue of the latter.

Road bridges

An earlier draft of EN 1993-2 recommended that, if sections were checked at the ultimate limit state using the reduced stress method of 3-1-5/clause 10, no further check of breathing

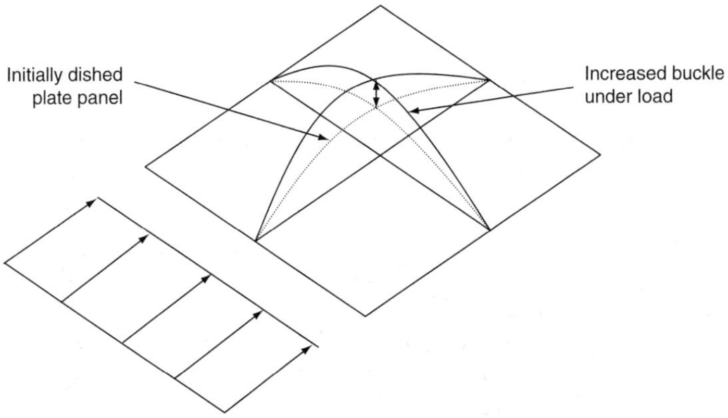

Fig. 7-1. Illustration of web breathing in a plate under axial load

would be required. However, if the effective area method discussed in section 6.2.2.5 was used, then breathing still needed to be checked explicitly. This is because the effective area method can allow considerable load shedding between panels which is not permissible at the serviceability limit state. This guidance was removed in the final draft of EN 1993-2 and the National Annex is permitted to define situations where breathing need not be checked in 3-2/clause 7.4(1). Breathing may be neglected in accordance with ***3-2/clause 7.4(2)*** if the following criterion is satisfied:

3-2/clause 7.4(2)

$$b/t \leq 30 + 4.0L \quad \text{but} \quad b/t \leq 300 \qquad \text{3-2/(7.5)}$$

where:

- b is the depth of the web for a web without longitudinal stiffeners or the depth of the largest sub-panel in a web with longitudinal stiffeners;
- L is the relevant span length of the member, but not taken less than 20 m.

Where there are longitudinal stiffeners, the overall web depth should still be checked for breathing, but no guidance is given in EN 1993-2. Either expression 3-2/(7.5) can conservatively be applied to the entire web depth (which will often still be adequate) or the general check below can be used with the buckling coefficients based on the overall stiffened plate.

Rail bridges

Breathing may be neglected in accordance with 3-2/clause 7.4(2) if the following criterion is satisfied:

$$b/t \leq 55 + 3.3L \quad \text{but} \quad b/t \leq 250 \qquad \text{3-2/(7.6)}$$

where b and L are as defined above.

General interaction

3-2/clause 7.4(3)

If the simple limits on b/t in expression 3-2/(7.5) or expression 3-2/(7.6) cannot be satisfied, the following general interaction given in ***3-2/clause 7.4(3)*** should be checked. This compares applied stresses directly to their elastic critical limiting values (which will often be less than their real ultimate strengths as discussed elsewhere in this guide). For longitudinally stiffened webs, the check should be applied to each sub-panel in turn and also to the overall stiffened plate.

$$\sqrt{\left(\frac{\sigma_{x,Ed,ser}}{k_\sigma \sigma_E}\right)^2 + \left(\frac{1.1\tau_{Ed,ser}}{k_\tau \sigma_E}\right)^2} \leq 1.1 \qquad \text{3-2/(7.7)}$$

where:

$\sigma_{x,Ed,ser}$ and $\tau_{Ed,ser}$ are the stresses from the frequent load combination and

$$k_\sigma \sigma_E = \frac{k_\sigma \pi^2 E t^2}{12(1-\upsilon^2)b^2} \text{ and } k_\tau \sigma_E = \frac{k_\tau \pi^2 E t^2}{12(1-\upsilon^2)b^2}$$

are the linear elastic critical buckling stresses for the panel considered. These critical stresses can be determined from 3-1-5/clause 4 and 3-1-5/clause 5 respectively, as discussed in sections 6.2.2.5 and 6.2.6 of this guide. Where the stress varies along the length of the panel, the Note to 3-2/clause 7.4(3) refers to 3-1-5/clause 4.6(3). This allows the verification to be performed at a distance of $0.4a$ or $0.5b$, whichever is smaller, from the most highly stressed end of the panel.

For panels wholly in tension, it would be reasonable to take $\sigma_{x,Ed,ser}/(k_\sigma \sigma_E)$ as zero since no amount of increase in the tension can lead to buckling. In reality, imperfections still 'breath' under tensile stress by straightening out, but this causes much smaller stresses than breathing under an equivalent magnitude of compressive stress. For similar reasons, if the direct stress in a panel varies with a tensile stress at one edge of greater magnitude than the compressive stress at the other, $\sigma_{x,Ed,ser}/(k_\sigma \sigma_E)$ should still be calculated for the compressive edge. The shear term must be evaluated whether the direct stress is compressive or tensile.

Worked Example 7-1: Web breathing check for unstiffened web panel
A web of a beam forming part of a road bridge with a span of 60 m is 3000 mm deep and 10 mm thick without longitudinal stiffeners. Transverse stiffeners are provided at supports only. The frequent load combination produces a bending stress of 100 MPa at the top of the web and a stress of −100 MPa at the bottom. The shear stress is 50 MPa. The web panel is checked for breathing under these stresses.

Since $b/t = 300 > 30 + 4.0 \times 60 = 270$, the simple criterion of expression 3-2/(7.5) is not satisfied. Consequently the interaction of expression 3-2/(7.7) must be used to check against excessive breathing.

Direct stresses:
From EN 1993-1-5 Table 4.1, for pure bending $\psi = -1$ and $k_\sigma = 23.9$:

$$k_\sigma \sigma_E = \sigma_{cr,x} = \frac{k_\sigma \pi^2 E t^2}{12(1-\upsilon^2)\bar{b}^2} = \frac{23.9 \times \pi^2 \times 210 \times 10^3 \times 10^2}{12(1-0.3^2) \times 3000^2} = 50.4\,\text{MPa}$$

Shear stresses:
From EN 1993-1-5 Annex A.3 for a very long panel:

$$k_\tau \sigma_E = \tau_{cr} = \frac{k_\tau \pi^2 E t^2}{12(1-\nu^2)b^2} = \frac{5.34 \times \pi^2 \times 210 \times 10^3 \times 10^2}{12(1-0.3^2) \times 3000^2} = 11.3\,\text{MPa}$$

where:

$$k_\tau = 5.34 + 4.00\left(\frac{b}{a}\right)^2 = 5.34 + 0 = 5.34$$

From expression 3-2/(7.7):

$$\sqrt{\left(\frac{\sigma_{x,Ed,ser}}{k_\sigma \sigma_E}\right)^2 + \left(\frac{1.1\tau_{Ed,ser}}{k_\tau \sigma_E}\right)^2} = \sqrt{\left(\frac{100}{50.4}\right)^2 + \left(\frac{1.1 \times 50}{11.3}\right)^2} = 5.26 \gg 1.1$$

The web is clearly far too slender. This was of course a rather unrealistic example and ultimate limit state considerations would also have resulted in the beam being unacceptable.

7.5. Miscellaneous SLS requirements in clauses 7.5 to 7.12

3-2/clause 7.5 to 3-2/clause 7.12 give guidance on other serviceability considerations, which are not covered in detail here. The problems covered include:

(i) Inadequate clearance over or under the bridge to allow the safe passage of high-sided vehicles. This can become an ultimate limit state if the structural integrity of the bridge would be undermined in the event of a collision from a high-sided vehicle passing below.
(ii) Excessive sagging deformations that give a visual impression of inadequate strength. This can generally be overcome by precambering.
(iii) Excessive deformations under live load that can damage surfacing, corrosion protection systems, waterproofing, drainage and that can cause dynamic problems.
(iv) Resonance of steel components under either aerodynamic or pedestrian-induced vibrations causing discomfort to users. This can become an ultimate limit state if a fatigue failure, resulting from the excessive vibration of a component, would undermine the structural integrity of the bridge. Divergent wind-induced motion, such as galloping and flutter, can also lead to collapse. Guidance on these is given in EN 1990 and EN 1991-1-4.
(v) Lack of access to details which will require periodic inspection, cleaning and painting.

(vi) Insufficient drainage created by either inadequate drainage systems or by existing drainage systems becoming blocked. This can create corrosion problems.

The guidance and recommendations in EN 1993-2 are reasonably comprehensive, with the exception of problems arising from resonance, which is not a problem specific to steel bridges. The provisions of these clauses are not therefore discussed further here.

CHAPTER 8

Fasteners, welds, connections and joints

This chapter discusses fasteners, welds, connections and joints as covered in section 8 of EN 1993-2 in the following clauses:

- Connections made of bolts, rivets and pins *Clause 8.1*
- Welded connections *Clause 8.2*

Most of the requirements given in the above clauses are by reference to EN 1993-1-8.

8.1. Connections made of bolts, rivets and pins

8.1.1. Categories of bolted connections

3-1-8/clause 3.4 groups connections into five main categories listed below:

8.1.1.1. Shear connections

- **Category A: Bearing type**. *3-1-8/clause 3.4.1(2)* describes Category A as shear connections, without preloading, containing bolts of grades 4.6 up to and including 10.9. This covers the more familiar 'black bolts in shear'. Due to their low fatigue resistance and tendency to work loose under repeated vibration, it is recommended that Category A connections are not used for permanent structural connections in bridges – 3-2/clause 2.1.3.3 refers. At ULS, the bolt shear should not exceed either the bolt shear resistance or the design bearing resistance. *3-1-8/clause 3.4.1(2)*

- **Category B: Slip-resistant at serviceability limit state**. 3-1-8/clause 3.4.1(2) describes Category B as preloaded 8.8 or 10.9 bolts with controlled tightening where slip is to be prevented at the serviceability limit state. This covers the more familiar 'friction grip bolts designed for no slip at SLS'. It is recommended that Category B connections are used for permanent structural connections in bridges where some reduction in the stiffness of the connection at ULS is not important as discussed in section 5.2.1 of this guide under the heading 'slip of bolts'. A suitable example is a main beam splice. At ULS, the bolt shear should not exceed either the bolt shear resistance or the design bearing resistance.

- **Category C: Slip-resistant at ultimate limit state**. 3-1-8/clause 3.4.1(2) describes Category C as preloaded 8.8 or 10.9 bolts with controlled tightening where slip is to be prevented at the ultimate limit state. This covers the more familiar 'friction grip bolts designed for no slip at ULS'. Category C connections are recommended for permanent structural connections on bridges where the stiffness of the connection at ULS is important. A suitable example would be a bracing member connection where the stiffness of the bracing affects the buckling force of a main girder flange at ULS. At ULS, the bolt shear should not exceed either the bolt slip resistance or the design bearing resistance. The check of bearing resistance is required as a fail-safe in case slip does occur in the connection due to, for example, faulty installation of the bolts. (No check is required of bolt shear resistance as it will exceed the slip resistance.)

8.1.1.2. Tension connections

3-1-8/clause 3.4.2(2)

- **Category D: Connections with non-preloaded bolts**. *3-1-8/clause 3.4.2(2)* describes Category D as bolts of grades 4.6 to 10.9 in tension without preload. This covers the more familiar 'black bolts in tension'. Due to the reasons outlined under Category A, they are not recommended for permanent structural connections in bridges.
- **Category E: Connections with preloaded 8.8 or 10.9 bolts**. 3-1-8/clause 3.4.2(2) describes Category E as preloaded 8.8 or 10.9 bolts resisting tension. This covers the more familiar 'friction grip bolts in tension'.

8.1.2. Positioning of holes for bolts and rivets

3-1-8/Table 3.3 gives detailed rules for the maximum and minimum allowable bolt spacings and end and edge distances. The maximum pitch in the transverse direction is a lot smaller than was permitted by BS 5400: Part 3[4] and the maximum pitch in the direction of tensile stress depends on whether or not the steel is exposed to the weather. For bridges, the 'exposed to the weather' case will be the norm. There is also a requirement to check local buckling between bolt holes in compression elements where the pitch in the direction of compression $p_1 \leq 9\varepsilon t$, where t is the thickness of the parent plate. For S355 steel, $\varepsilon = 0.81$. This may therefore become a practical upper limit to the pitch for compression elements, rather than the absolute maximum of the lesser of $14t$ or 200 mm.

The minimum pitch allowed gives room for tightening and effectively limits the amount of reduction to bearing resistance that can occur due to tearing into adjacent holes. The minimum edge and end distances similarly limit the amount of reduction to bearing resistance but this reduction is still considerable if the minimum edge distance is used and the bolt is pulling towards the free edge. In all cases, it is still important to check bearing resistance. The maximum pitch and edge distances ensure that plates in a connection are adequately clamped together so they can be considered to be sealed against corrosion.

8.1.3. Design resistance of individual fasteners

3-1-8/clause 3.6.1 deals with fastener resistances.

8.1.3.1. Bolts (preloaded or non-preloaded) and rivets

(i) Bolt shear resistance

The shear resistance, per shear plane, of a bolt in a normal clearance hole is given in 3-1-8/Table 3.4 by:

$$F_{\text{v,Rd}} = \frac{\alpha_{\text{v}} f_{\text{ub}} A}{\gamma_{\text{M2}}} \tag{D8.1-1}$$

where:

α_{v} is a factor to convert the ultimate tensile stress of the bolt material to the maximum allowable shear stress. The usual bolts specified in the UK are of grades 4.6 and 8.8, for which α_{v} will always be 0.6 regardless of whether the shear failure plane is in the threaded or unthreaded portion of the bolt. If the designer wishes to specify grades of 4.8, 5.8, 6.8 and 10.9 and use $\alpha_{\text{v}} = 0.6$, the maximum allowable thread length of the bolts must be carefully specified to ensure that the shear failure plane does not occur in the threaded area; otherwise $\alpha_{\text{v}} = 0.5$;

f_{ub} is the ultimate tensile strength of the bolt material from 3-1-8/Table 3.1;

A is the tensile stress area of the bolt passing through the shear failure plane. A will be equal to the gross area (A) or threaded area (A_s) depending on whether the shear plane crosses the threaded or unthreaded section of the bolt shank. In the absence of careful thread length specification, it is recommended to always use the threaded area, A_s;

γ_{M2} is the partial safety factor for bolts in shear. 3-2/clause 6.1 recommends a value of 1.25 but this may be amended by the National Annex.

(ii) Bolts in bearing

The bearing resistance of a bolt is obtained from 3-1-8/Table 3.4 as follows:

$$F_{b,Rd} = \frac{k_1 \alpha_b f_u d t}{\gamma_{M2}} \tag{D8.1-2}$$

where f_u is the ultimate tensile strength of the plate material. The reduction factor α_b allows both for the adverse effect on bearing resistance of low end distance and pitch and also for the possibility that the parent plate might actually have an ultimate tensile stress less than the bolt, which would limit the bearing pressure achievable. If an 'end bolt' is pulling away from the free edge, the reduction for end bolts is not applicable and that for internal bolts should be used. The factor k_1 allows for transverse splitting as a function of edge distance and transverse pitch.

For single lap joints with only one row of bolts, the resistance should additionally not exceed $1.5 f_u d t / \gamma_{M2}$ in accordance with 3-1-8/clause 3.6.1(10).

(iii) Bolt tension resistance

The tension resistance of a bolt is obtained from 3-1-8/Table 3.4 as follows:

$$F_{t,Rd} = \frac{k_2 f_{ub} A_s}{\gamma_{M2}} \tag{D8.1-3}$$

The factor k_2 is 0.9 other than for countersunk bolts. This is consistent with the tension resistance of net sections of members in EN 1993-1-1.

(iv) Punching shear resistance for bolted connections

3-1-8/Table 3.4 requires that the parent plate loading bolts in tension is checked for punching shear resistance using a shear resistance of $0.6 f_u$. This is a new check for UK designers and is only likely to govern where the parent plate is unusually thin compared to the bolt diameter. The punching shear resistance of the parent plate is given by:

$$B_{p,Rd} = \frac{0.6 \pi d_m t_p f_u}{\gamma_{M2}} \tag{D8.1-4}$$

where:

$$d_m = \frac{d_{points} + d_{flats}}{2}$$ as shown in Fig. 8.1-1;

t_p is the thickness of the parent plate;

f_u is the ultimate tensile strength of the parent plate.

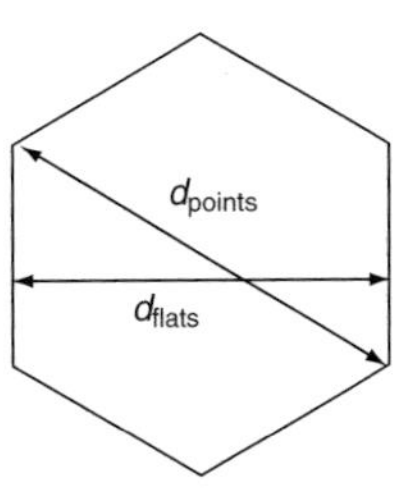

Fig. 8.1-1. Hexagonal bolt head/nut

(v) Combined shear and tension

3-1-8/Table 3.4 gives the following interaction formula for combined shear and tension:

$$\frac{F_{v,Ed}}{F_{v,Rd}} + \frac{F_{t,Ed}}{1.4F_{t,Rd}} \leq 1.0 \qquad \text{(D8.1-5)}$$

where:

$F_{v,Ed}$ is the design shear force per bolt for the ultimate limit state;
$F_{v,Rd}$ is the design shear resistance per bolt;
$F_{t,Ed}$ is the design tensile per bolt for the ultimate limit state;
$F_{t,Rd}$ is the design tension resistance per bolt.

The main point to note is that some tension can be accommodated even when the bolt is stressed in shear to its full shear resistance. The limitations on applicability of this interaction for preloaded bolts are discussed in section 8.1.6 of this guide.

(vi) Countersunk bolts and rivets

3-1-8/clause 3.6 also gives rules for countersunk bolts and rivets. These are not discussed in this guide as they are not commonly used.

8.1.3.2. Injection bolts

3-1-8/clause 3.6.2 gives design guidance for injection bolts. Injection bolts are not discussed further in this guide.

8.1.4. Groups of fasteners

3-1-8/clause 3.7(1)

3-1-8/clause 3.7(1) allows the designer to calculate the resistance of a group of fasteners by summing the resistances $F_{b,Rd}$ of each fastener providing $F_{v,Rd} > F_{b,Rd}$ for each fastener. This is allowed because failure in bearing is ductile and allows redistribution of forces between connectors; failure by bolt shearing is less ductile. If the above requirement is not satisfied, the group resistance has to be taken as the product of the number of fasteners and the resistance of the weakest fastener. In the majority of cases, the bearing resistance of the fasteners will be greater than the shear resistance so the latter will need to be followed.

If a fastener group is required to transmit bending moments then this clause will not apply as the ability of the fastener group to transmit moments will be a function of the fastener arrangement around the centre of rotation and not just the number of fasteners. Further guidance is given in section 8.1.9 of this guide.

8.1.5. Long joints

3-1-8/clause 3.8(1)

3-1-8/clause 3.8(1) requires the total resistance derived for rows of fasteners longer than $15d$ (measured between outermost fasteners in the row) to be multiplied by a reduction factor, β_{Lf}. For very long joints, the reduction factor is 0.75. This reduction applies where the longitudinal strains in the plates being connected do not have the same distribution along their lengths, as this results in unequal forces in the individual fasteners. The reduction therefore applies where the full force of one plate is being transferred to another plate or plates over the length of the connection. It does not apply to bolted connections between the web and flange of a fabricated girder which transmit longitudinal shear, where the various connected parts have the same distribution of longitudinal strain.

8.1.6. Slip resistant connections using grade 8.8 and 10.9 bolts

8.1.6.1. Slip resistance

3-1-8/clause 3.9.1(1)

3-1-8/clause 3.9.1(1) gives the following equation for the design slip resistance $F_{s,Rd}$ of a preloaded 8.8 or 10.9 bolt:

$$F_{s,Rd} = \frac{k_s n \mu}{\gamma_{M3}} F_{p,C} \qquad \text{3-1-8/(3.6)}$$

where:

k_s is a factor for bolt hole size from 3-1-8/Table 3.6. The factor is unity for normal-size holes and less than 1 for larger holes, reflecting the greater consequences of slip;

n is the number of friction interfaces;

μ is the coefficient of friction of the friction (faying) surface taken as 0.5, 0.4, 0.3, 0.2 depending on whether the friction surface is classed as A, B, C, or D respectively. The different classes of friction surface are defined in EN 1090 and are reproduced below for convenience:

Class A: Surfaces blasted with shot or grit, with any loose rust removed, no pitting.
Surfaces blasted with shot or grit, and spray-metallized with aluminium.
Surfaces blasted with shot or grit, and spray-metallized with a zinc-based coating certified to provide a slip factor of not less than 0.5.

Class B: Surfaces blasted with shot or grit, painted with an alkali–zinc silicate paint to produce a coating thickness of 50 to 80 microns.

Class C: Surfaces cleaned by wire brushing or flame cleaning, with any loose rust removed.

Class D: Surfaces not treated.

$F_{p,C}$ is the bolt preload $= 0.7f_{ub}A_s$;

$\gamma_{M3} =$ 1.25 for ultimate limit states as recommended in EN 1993-2 Table 6.1 in line with the recommended value in EN 1993-1-8 where there is fatigue loading, as will usually be the case for bridges. At serviceability, a value of 1.1 is recommended in EN 1993-2 Table 6.1. Slip of bolts can lead to loss of preload (as the shear stress attracted to the bolt causes plastic deformation) and therefore reduction of slip resistance for future service load cases. The potential for more frequent slip is undesirable from fatigue considerations. Poisson's ratio effects can also lead to a reduction in thickness of connected plies in tension which can, in turn, shorten the bolt and reduce preload. The recommended value of $\gamma_{M3,ser} = 1.1$ is lower than the equivalent value of 1.2 in BS 5400: Part 3.[4]

For preloaded bolts which can slip at ULS under shear force alone, it is still not necessary to perform a check of shear and tension (arising from the preload) according to equation (D8.1-5). This is justified from considerations of plasticity. If the bolt yields under the combination of shear and preload, the preload will tend to relax and the full shear can be mobilized. It should be noted that if this was not assumed, high strength friction grip (HSFG) bolts with class A faying surfaces would fail according to equation (D8.1-5) under the combination of shear and preload as soon as sliding occurred.

8.1.6.2. Combined tension and shear

Where external tension is applied, this will mainly go into reducing the clamping force between the connected plates rather than increasing the force in the bolt itself. This is because the stiffness of the plates in the through-thickness direction is much greater than that of the bolt as indicated by the model in Fig. 8.1-2. The bolt tension will however increase slightly. The reduced slip resistance is therefore given in ***3-1-8/clause 3.9.2(1)*** by: *3-1-8/clause 3.9.2(1)*

$$F_{s,Rd} = \frac{k_s n\mu(F_{p,C} - 0.8F_{t,Ed})}{\gamma_{M3}} \qquad \text{3-1-8/(3.1)}$$

where $F_{t,Ed}$ is the applied tensile force on the bolt at the serviceability or ultimate limit state for Category B or C connections respectively. The 0.8 factor makes allowance for some of the external tension going into increasing the force in the bolt, rather than entirely going into reducing the clamping force across the plates. It seems to be a somewhat low factor compared to the conservative value of 1.0 and should be used with care.

In applying the above, it is not generally necessary to make allowance for prying forces as the increase in applied bolt tension tending to unclamp the plates is balanced by an equal and

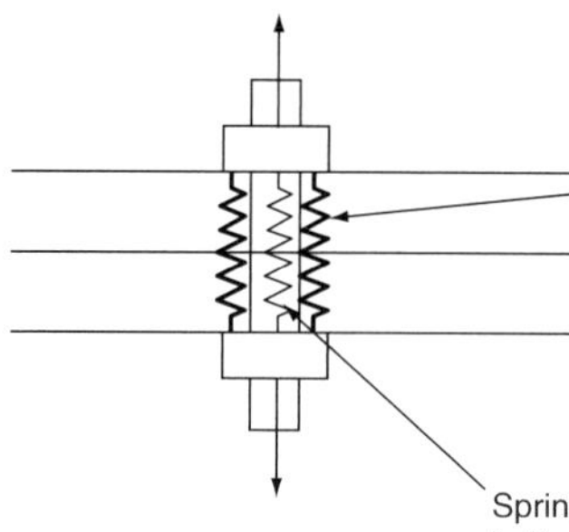

Fig. 8.1-2. Model of stiffness of preloaded joint with external tension

opposite compressive force away from the bolt tending to clamp the plates as discussed in section 8.1.8. There are two occasions however when prying should be considered for slip resistance:

(i) If the faying surface assumed in the calculation does not extend over the entire area of the end plate, the friction developed under the compressive prying reaction may not balance the loss of friction under the bolt. In this situation, it is advisable to include the prying force in the calculation of $F_{t,Ed}$ unless the relevant reduced coefficient of friction can be established.
(ii) If prying action causes the bolt force to exceed the bolt preload (or as an approximation, the external tension plus prying force exceeds the preload), some relaxation of the preload may occur due to plasticity. In this situation, it is advisable to include the prying force in the calculation of $F_{t,Ed}$ to make allowance for this loss of preload.

For preloaded bolts acting in both shear and external tension, it does become necessary to perform a check of shear and tension according to equation (D8.1-5) in case of slip at ULS. The arguments of plasticity made above for preloaded bolts in shear alone can still however be used to justify taking the applicable value of $F_{t,Ed}$ equal to the externally applied tension (not the preload), including any prying force. The external tension should also not exceed the bolt tension resistance. It is generally not recommended to use HSFG grade 10.9 bolts acting in tension and shear. This is because they may not possess adequate ductility to realize the above assumptions.

8.1.6.3. Hybrid connections

3-1-8/clause 2.4(3)
3-1-8/clause 3.9.3(1)

'Hybrid' connections are connections which involve combinations of bolts, welds and other connection components. In such cases, ***3-1-8/clause 2.4(3)*** requires the connectors with greatest stiffness to carry all the load. This means that welded joints, being very stiff, will generally carry the entire load, even if bolts are provided. ***3-1-8/clause 3.9.3(1)*** however allows Category C bolted connections to share loads with welds (provided that the bolts are tightened after completion of welding), as non-slip bolted connections are themselves very stiff. To allow for differences in stiffness and ductility of the different types of connector, the UK National Annex limits the combined resistance achieved in this way to 90% of the resistance obtained by adding the full contributions of bolts and welds together.

8.1.7. Deductions for fastener holes

8.1.7.1. General

The relevant deductions for fastener holes in member design are discussed throughout section 6.2 of this guide.

8.1.7.2. Design for block tearing

3-1-8/clause 3.10.2

In addition to the check of net section of members above, failure can also occur at member end connections by 'block tearing'. This involves a local rip-out of the bolt group in a mechanism involving both shear and tensile failure planes. ***3-1-8/clause 3.10.2*** identifies

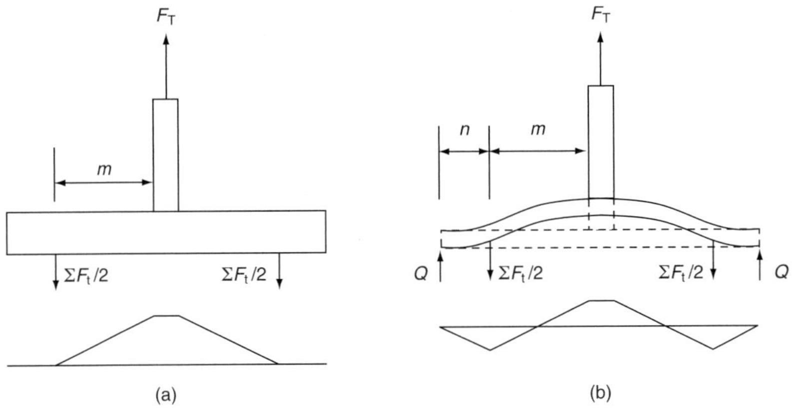

Fig. 8.1-3. Flange connections with and without prying: (a) moments for no prying; (b) moments for maximum prying

typical situations and provides a resistance formula. Failure is most likely to govern where the bolt group occupies a relatively small area of the connected members.

8.1.7.3. Angles connected through one leg

The design of angles bolted through one leg is discussed in section 6.2.3 of this guide.

8.1.8. Prying forces

Prying occurs in end-plate connections (particularly unstiffened ones) that are used for carrying tension. Such connections are particularly prone to fatigue under bridge loading so should be avoided whenever possible. In essence, there are two extremes of behaviour:

(i) No moment in the flange at the bolt line, which occurs with very thick flanges. This gives zero prying force.
(ii) Full plastic moment in the flange at the bolt line, which occurs with thin flanges. This gives the maximum prying force.

These two extremes are shown in Fig. 8.1-3(a) and (b). F_T is the external tension, $\Sigma F_t/2$ is the sum of the bolt forces in a line and Q is the total prying force applicable to the bolts considered.

3-1-8/clause 6.2.4 gives a method for checking the ultimate tensile resistance of such connections, which implicitly allows for prying force. It is referred to as the equivalent T-stub method. When it is applied to real situations, the resistance of the end-plate detail must consider the lowest resistance derived from considerations of failure of both groups of bolts and also single bolt rows. Consequently, the forces Q, F_T and $\Sigma F_t/2$ can apply to either a single bolt row or to a group of bolt rows as appropriate. The method does not however allow the determination of the prying force itself under applied tension for inclusion in other checks, such as HSFG shear resistance – see the comments below. A detailed discussion is beyond the scope of this guide but essentially the procedure is as follows. *3-1-8/clause 6.2.4*

The criteria in 3-1-8/Table 6.2 are first used to determine whether or not prying forces need to develop for adequacy of the end plates. In reality, there will always be some prying force.

No prying force occurs

The resistance is determined as the lowest of either failure by flange failure as in case (a) of Fig. 8.1-3 which gives:

$$F_{T,Rd} = \frac{2M_{pl,Rd}}{m} \quad \text{(Mode 1/2)}$$

or by bolt failure which gives:

$$F_{T,Rd} = \Sigma F_{t,Rd} \quad \text{(Mode 3)}$$

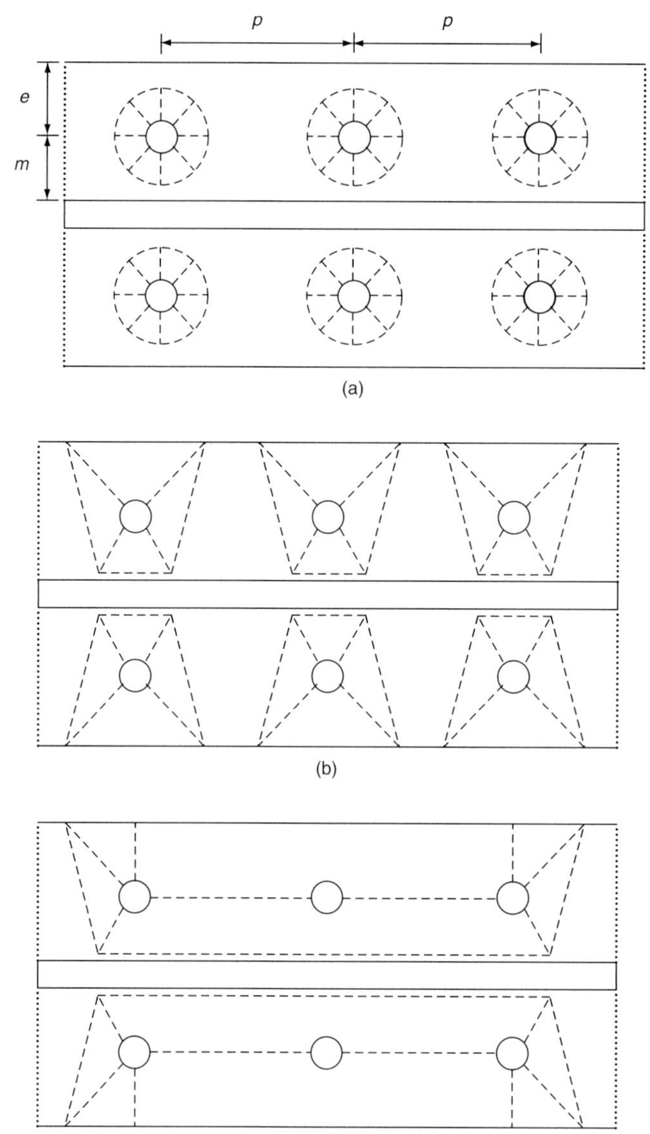

Fig. 8.1-4. Typical yield line patterns used in derivation of T-stub effective lengths: (a) failure of separate bolt lines by circular pattern; (b) failure of separate bolt lines by non-circular pattern; (c) failure of group of bolt lines

$M_{pl,Rd}$ is the plastic resistance of the flange over an effective length which has to be calculated from 3-1-8/Tables 6.4 to 6.6 as appropriate. As discussed above, the resistance of the connection should be taken as the lowest derived from considerations of both failure of groups of bolts and also single bolt rows and this is reflected in the effective lengths derived from these tables. Some typical yield line mechanisms leading to the effective lengths provided are shown in Fig. 8.1-4. $F_{T,Rd}$ is the connection resistance and $\Sigma F_{t,Rd}$ is the tension resistance of all the bolts in the connection effective length.

Prying force occurs

The resistance is determined as the lowest of failure by flange failure at both root and bolt lines as in case (b) of Fig. 8.1-3 which gives:

$$F_{T,Rd} = \frac{4M_{pl,Rd}}{m} \text{ (Mode 1)}$$

by bolt failure which gives:

$$F_{T,Rd} = \Sigma F_{t,Rd} \text{ (Mode 3)}$$

or by simultaneous failure of the flange root and the bolts (Mode 2). In this latter case, the prying force $Q = 0.5(\Sigma F_{t,Rd} - F_T)$ so the root moment is given by:

$$M_{pl} = \frac{F_{T,Rd} m}{2} - 0.5(\Sigma F_{t,Rd} - F_{T,Rd})n$$

and the connection failure load is therefore:

$$F_{T,Rd} = \frac{2M_{pl} + n\Sigma F_{t,Rd}}{m+n}$$

M_{pl} is again the plastic resistance of the flange over a length which has to be calculated and which differs in the different modes.

As for bridges it is not advisable to detail connections with prying force, it is recommended here that connections should be designed to fall into the category of 'no prying force' in 3-1-8/Table 6.2. Since some prying is still inevitable as the flange will not be infinitely rigid, a prying force of 10% of the external tension shared among the bolts could be considered as was previous practice to BS 5400: Part 3.[4] A more detailed study of the calculation of prying forces (which would be needed for intermediate cases between (a) and (b) in Fig. 8.1-3) can be found in Reference 27.

8.1.9. Distribution of forces between fasteners at the ultimate limit state

Plastic analysis is not permitted for bolt or weld groups under moment in bridges. ***3-2/clause 8.1.9(1)*** requires the forces in the individual fasteners of a moment connection to be calculated assuming that the forces are linearly proportional to their distance from the centre of rotation. Web splice bolt groups for beams will therefore need to be designed elastically for the moment carried by the web plate of the girder in addition to the web shear as in previous UK practice. Worked Example 8.1-1 illustrates this. Most connections would however require elastic analysis according to EN 1993-1-8 even without the intervention of EN 1993-2. Category C connections do not possess the required 'ductility' for plastic analysis as, by definition, slip is prevented. Category A or B connections where the shear resistance is less than the bearing resistance cannot use plastic analysis as bolt shear failure similarly does not give sufficient ductility to redistribute forces. EN 1993-1-8 also prohibits plastic analysis where vibration or load reversal occurs, as is the case for most bridges.

3-2/clause 8.1.9(1)

8.1.10. Connections made with pins

Connections made with pins are covered by 3-1-8/clause 3.13 and are not discussed further in this guide.

Worked Example 8.1-1: Design of a plate girder bolted splice

A bolted splice is designed for the plate girder section shown in Fig. 8.1-5. All plates are grade S355 to EN 10025 and the fabricator wishes to use General Grade 8.8 M24 HSFG bolts for all fasteners. The design data at the splice location are as follows for ULS and SLS:

	ULS	SLS
Bending moment (kNm)	1222	870
Shear force (kN)	1000	710
Section modulus, W, for centre of top flange (mm^3)	1.755×10^7	1.755×10^7
Section modulus, W, for top of web (mm^3)	1.792×10^7	1.792×10^7
Section modulus, W, for bottom of web (mm^3)	3.007×10^7	3.007×10^7
Section modulus, W, for centre of bottom flange (mm^3)	2.871×10^7	2.871×10^7

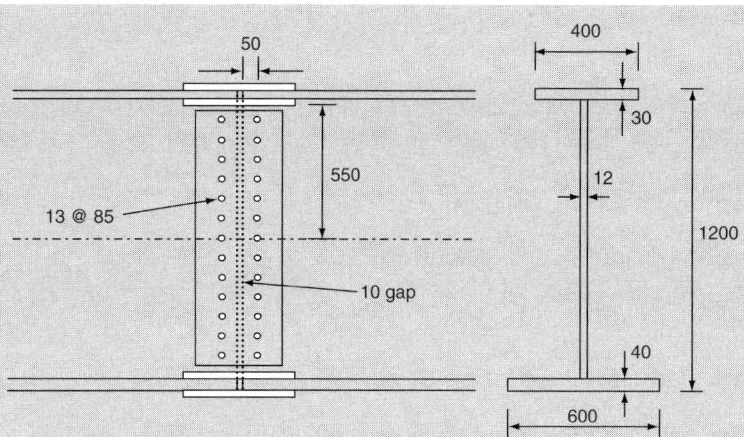

Fig. 8.1-5. Girder for Worked Example 8.1-1

Calculate bolt resistances at both SLS and ULS
As the implications of bolt slippage at ULS are not critical, the bolts can be designed as Category B to 3-1-8/clause 3.4.1.

Friction resistance of bolts at SLS
From expression 3-1-8/(3.6):

$$F_{s,Rd,ser} = \frac{k_s n \mu}{\gamma_{M3}} F_{p,C}$$

where:

$k_s = 1.0$ (3-1-8/Table 3.6 – normal clearance holes);
$n = 2$ (bolts in double shear);
$\mu = 0.50$ (3-1-8/Table 3.7 – Class A surface);
$\gamma_{M3} = 1.1$ (3-2/Table 6.1 which may be amended in the National Annex – see comment in section 8.1.6.1 above);
$F_{p,C}$ = preloading force $= 0.7 f_{ub} A_s$ (expression 3-1-8/(3.7)).

Using preloaded Grade 8.8 bolts $f_{ub} = 800\,\text{MPa}$, $A_s = 358\,\text{mm}^2$ for M24 bolts:

$$F_{p,C} = 0.7 \times 800 \times 358 \times 10^{-3} = 200.5\,\text{kN}$$

Therefore:

$$F_{s,Rd,ser} = \frac{1.0 \times 2 \times 0.5 \times 200.5}{1.1} = \mathbf{182.3\,kN} \text{ in double shear at SLS}$$

Shear resistance per shear plane
Where bolts can slip at ULS, the shear resistance of the bolts must be checked at ULS. From equation (D8.1-1):

$$F_{v,Rd} = \frac{\alpha_v f_{ub} A}{\gamma_{M2}}$$

where:

$\alpha_v = 0.6$ (3-1-8/Table 3.4 – Grade 8.8 bolts);
$f_{ub} = 800\,\text{MPa}$ (3-1-8/Table 3.1);
A = threaded area $= 358\,\text{mm}^2$ for M24 bolts;
$\gamma_{M2} = 1.25$ (3-2/Table 6.1 – may be amended in National Annex);

$$F_{v,Rd} = \frac{0.6 \times 800 \times 358}{1.25} = 137.5\,\text{kN per shear plane.}$$

As the bolts are in double shear, the number of shear planes is 2. Therefore, bolt shear resistance $= 2 \times F_{\mathrm{v,Rd}} = 274.8\,\mathrm{kN}$.

Minimum bolt spacings (3-1-8/Table 3.3)
Edge distance:

$$e_1 \text{ and } e_2 = 1.2d_0 = 1.2 \times 26\,\text{mm dia. hole} = 31.2\,\text{mm}$$

Use minimum edge distance of 40 mm and minimum end distance of 50 mm.
Pitch:

$$p_1 = 2.2d_0 = 2.2 \times 26\,\text{mm dia. hole} = 57.2\,\text{mm}$$

$$p_2 = 2.4d_0 = 2.4 \times 26\,\text{mm dia. hole} = 62.4\,\text{mm}$$

Use minimum bolt pitch of say 75 mm.

Bolt bearing resistance
From expression (D8.1-2):

$$F_{\mathrm{b,Rd}} = \frac{k_1 \alpha_{\mathrm{b}} f_{\mathrm{u}} d t}{\gamma_{\mathrm{M2}}}$$

where:

α_{b} is the smallest of α_{d}, $f_{\mathrm{ub}}/f_{\mathrm{u}}$ or 1.0;
f_{ub} is the ultimate strength of bolt = 800 MPa (3-1-8/Table 3.1);
f_{u} is the ultimate strength of plate = 490 MPa (EN 10025).

$$\frac{f_{\mathrm{ub}}}{f_{\mathrm{u}}} = \frac{800}{490} = 1.63$$

$$\alpha_{\mathrm{d}} = \frac{e_1}{3d_0} \text{ for end bolts} = \frac{50}{3 \times 26} = 0.64$$

$$\alpha_{\mathrm{d}} = \frac{p_1}{3d_0} - \frac{1}{4} \text{ for inner bolts} = \frac{75}{3 \times 26} - \frac{1}{4} = 0.71$$

$$k_1 \text{ is the smallest of 2.5 or } 2.8\frac{e_2}{d_0} - 1.7 \text{ for edge bolts} = 2.8\frac{40}{26} - 1.7 = 2.61 \text{ so } k_1 = 2.5$$

$$k_1 \text{ is the smallest of 2.5 or } 1.4\frac{p_2}{d_0} - 1.7 \text{ for inner bolts} = 1.4\frac{75}{26} - 1.7 = 2.34$$

Conservatively take α_{b} and k_1 as 0.64 and 2.34 respectively for all bolts.

where:

- d is the diameter of bolt (24 mm);
- t is the thickness of plates resisting bearing stress. As bolts are in double shear, t will equal the lesser of the parent plate thickness or the total thickness of the cover plates on either side of the parent plate.

Assuming 12 mm thick cover plates, the bearing resistance for the web splice is governed by the web plate thickness:

$$F_{\mathrm{b,Rd}} = \frac{2.34 \times 0.64 \times 490 \times 24 \times 12}{1.25} = 169\,\mathrm{kN}$$

Therefore, ULS resistance of bolts in web = **169 kN** (bearing critical).

It can be seen here that ULS is critical for the web splice as the ULS bearing resistance is actually less than the SLS slip resistance. It would be possible to make SLS critical here by increasing the bolt end distances but this has not been done in this example. By inspection, bearing would not be critical for the flanges as the parent plate is much thicker. However, for brevity in this example, the number of flange bolts will also be determined at ULS using the same bearing resistance as for the web.

Flange bolts

Force in top flange to be transmitted by bolts:

$$= \frac{1222 \times 10^6 \times 12000}{1.755 \times 10^7} = 836\,\text{kN}$$

Number of bolts required = 836/169 = 4.7 – use minimum of **5 bolts** in top flange.
Force in bottom flange to be transmitted by bolts:

$$= \frac{1222 \times 10^6 \times 24\,000}{2.871 \times 10^7} = 1022\,\text{kN}$$

Number of bolts required = 1022/169 = 6.04 – **6 bolts** in bottom flange just adequate.

Web bolts

Try the bolt arrangement in Fig. 8.1-5. (Note that the vertical pitch is slightly greater than assumed in the bearing resistance calculation but this would not alter the conclusion that ULS is critical.)

$$\text{Z of outer bolt} = \frac{\sum z^2}{z_{\text{max}}} = \frac{2(85^2 + 170^2 + 255^2 + 340^2 + 425^2 + 510^2)}{510} = 2578\,\text{mm}$$

ULS stress at top of web:

$$= -\frac{1222 \times 10^6}{1.792 \times 10^7} = -68.2\,\text{MPa}$$

ULS stress at bottom of web:

$$= \frac{1222 \times 10^6}{3.007 \times 10^7} = 40.6\,\text{MPa}$$

Axial force in web:

$$N_{\text{web}} = 0.5(68.2 - 40.6) \times 1130 \times 12 = 187\,\text{kN}$$

Bending moment in web:

$$M_{\text{web}} = \frac{0.5(68.2 + 40.6) \times 1130^2 \times 12}{6} = 139\,\text{kNm}$$

Maximum horizontal force on outer web bolt:

$$= \frac{M_{\text{web}}}{Z_{\text{bolts}}} + \frac{V_{\text{web}} \times e_{\text{bolts}}}{Z_{\text{bolts}}} + \frac{N_{\text{web}}}{No.\,bolts} = \frac{139 \times 10^3}{2578} + \frac{1000 \times (50 + 5)}{2578} + \frac{187}{13} = 89.6\,\text{kN}$$

Vertical force on web bolts:

$$= \frac{V_{\text{web}}}{No.\,bolts} = \frac{1000}{13} = 76.9\,\text{kN}$$

Resultant maximum bolt force equals vector sum of horizontal and vertical forces:

$$= \sqrt{(89.6)^2 + (76.9)^2} = 118\,\text{kN} < 169\,\text{kN}$$

The web bolts are adequate and there is scope to reduce the number of bolts further.

A check should also be made of the net section at bolt holes in the cover plates and parent plates in both web and tension flange. The tension flange should be checked both for the net section in accordance with 3-1-1/clause 6.2.3 and the block tearing rules in 3-1-8/clause 3.10.2. (This is not performed here but is straightforward.) The check of the web is less clear. The block tearing rules do not apply without modification in this instance, as the bolt group is subject to bending as well as shear. In addition, any tear-out mechanism would have to tear through the flange as well so is unlikely to be critical. Ignoring this latter fact, if the block tearing rules were applied as written, a shear plane passing vertically from web face to face through all the bolts would give a lower resistance than one extending through all the bolts but with a horizontal tension plane extending from an outer bolt to the vertical free edge. It is therefore recommended that a check of bending and shear is performed on the net section of the web and cover plates. If holes are conservatively fully deducted, the section properties of the net web are:

$$A = 9504\,\text{mm}^2$$

$$W = 1.790 \times 10^6\,\text{mm}^3$$

Maximum ULS longitudinal stress in web:

$$= \frac{139 \times 10^6}{1.790 \times 10^6} + \frac{187 \times 10^3}{9504} = 97\,\text{MPa}$$

ULS shear stress:

$$= \frac{1000 \times 10^3}{9504} = 105\,\text{MPa}$$

These stresses are checked using the Von Mises equivalent stress criterion of 3-1-1/clause 6.1:

$$\left(\frac{\sigma_{x,Ed}}{f_y/\gamma_{M0}}\right)^2 + 3\left(\frac{\tau_{Ed}}{f_y/\gamma_{M0}}\right)^2 = \left(\frac{97}{355/1.0}\right)^2 + 3\left(\frac{105}{355/1.0}\right)^2 = 0.34 \le 1.0$$

The web plate is therefore adequate. A similar check should also be performed for the cover plates.

8.2. Welded connections

8.2.1. Geometry and dimensions

Detailed guidance for the allowable geometry and sizes of fillet, butt, plug and flare groove welds are discussed in 3-1-8/clause 4.3. These rules are not discussed further in this guide.

8.2.2. Welds with packings

The rules for the design of weld connecting plates separated by packers are self-explanatory and identical to those in BS 5400: Part 3 and are not therefore discussed here.

8.2.3. Design resistance of a fillet weld

8.2.3.1. Effective length of weld

3-1-8/clause 4.5.1(1)

3-1-8/clause 4.5.1(2)

3-1-8/clause 4.5.1(1) states that the length of a fillet weld is the length over which the weld is full sized. In practice, a weld is often under-size at the start and ends of a run and this can be allowed for in design by deducting twice the throat size, a, from the length of the run. ***3-1-8/clause 4.5.1(2)*** requires structural welds to have an effective length of at least the greater of $6a$ and 30 mm.

8.2.3.2. Effective throat thickness

3-1-8/clause 4.5.2(1)

3-1-8/clause 4.3.2.1(1)

3-1-8/clause 4.5.2(1) defines the effective throat thickness, a, as the height of the largest triangle (with equal or unequal legs) which can be inscribed within the fusion faces and the weld surface, measured perpendicular to the outer side of the triangle as illustrated in Fig. 8.2-1. ***3-1-8/clause 4.3.2.1(1)*** requires fillet welds to generally be used only where the fusion faces form an angle of between 60° and 120°. However, where the fusion faces form an angle of less than 60°, a fillet weld can be designed as a partial penetration butt

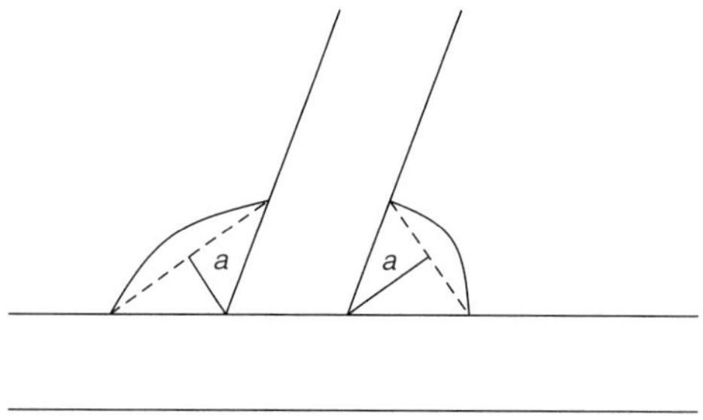

Fig. 8.2-1. Effective throat for fillet welds

weld and the throat thickness obtainable determined by weld procedure trials. This limitation is needed because of the difficulty of guaranteeing full penetration of the weld into the root.

It has generally been UK practice to limit the throat used in calculations to 0.71 times the leg length for angles between fusion faces of less than 90°, again because of the possible lack of penetration to the root. If a greater throat is required, the throat (rather than leg length) can be specified directly on the drawings. In general, benefit of the additional throat from penetration of the weld into the parent plates can only be taken if weld procedure trials show that the required penetration is consistently achieved – ***3-1-8/clause 4.5.2(3)***. Some penetration into the parent plate will always occur but there is also always going to be some 'fit-up' gap between the plates which can reduce the effective throat. It is usually justifiable for the designer to neglect the latter in his specification of weld size as it is offset by the additional penetration. The fabricator may, however, need to increase leg lengths if the root gap exceeds 1 mm.

3-1-8/clause 4.5.2(3)

8.2.3.3. Resistance of fillet welds

3-1-8/clause 4.5.3.2(6)

3-1-8/clause 4.5.3.2(6) requires fillet welds to satisfy the following criteria:

$$[\sigma_{\perp}^2 + 3(\tau_{\perp}^2 + \tau_{||}^2)]^{0.5} \leq \frac{f_u}{\beta_w \gamma_{M2}} \quad \text{and} \quad \sigma_{\perp} \leq \frac{f_u}{\gamma_{M2}} \qquad \text{3-1-8/(4.1)}$$

where:

$\sigma_{\perp}$ is the normal stress perpendicular to the weld throat;
$\tau_{\perp}$ is the shear stress (in the plane of the throat) perpendicular to the axis of the weld;
$\tau_{||}$ is the shear stress (in the plane of the throat) parallel to the axis of the weld;
f_u is the nominal ultimate tensile strength of the weaker part joined;
β_w is a correlation factor obtained from 3-1-8/Table 4.1 which relates the strength of the weld metal to the strength of the parent plate.

The stresses are shown in Fig. 8.2-2. It should be noted that the longitudinal direct stress, $\sigma_{||}$, does not need to be considered.

Expression 3-1-8/(4.1) is not very practical for design purposes as it involves assuming a weld size and then checking it in an iterative procedure. An alternative weld design formula is therefore derived below from expression 3-1-8/(4.1) which permits calculation

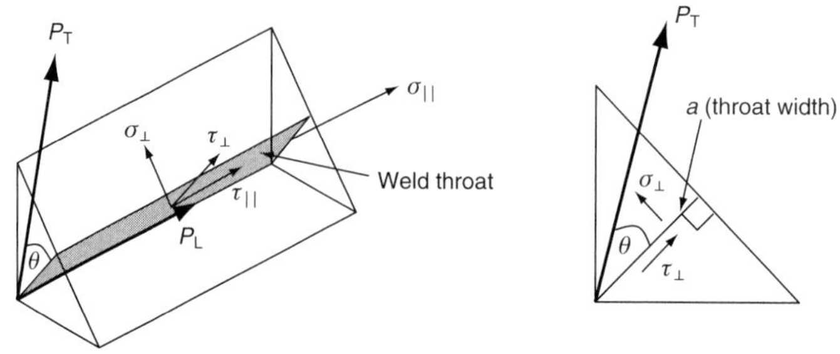

Fig. 8.2-2. Notation for fillet welds

of the weld throat size in one step. Figure 8.2-2 shows resultant forces per unit length on the weld. P_L is the longitudinal force on the weld per unit length, P_T is the resultant transverse force on the weld per unit length and θ is the angle between P_T and the throat of the weld. In terms of these stress resultants, the stresses on the throat are:

$$\sigma_{\perp} = \frac{P_T \sin\theta}{a}$$

$$\tau_{\perp} = \frac{P_T \cos\theta}{a}$$

$$\tau_{||} = \frac{P_L}{a}$$

Substituting the above expressions into expression 3-1-8/(4.1) gives:

$$\left(\frac{P_T^2 \sin^2\theta}{a^2} + \frac{3P_T^2 \cos^2\theta}{a^2} + \frac{3P_L^2}{a^2}\right)^{0.5} \leq \frac{f_u}{\beta_w \gamma_{M2}}$$

which can be rewritten as:

$$\frac{1}{a}\left(\frac{P_T^2}{K^2} + P_L^2\right)^{0.5} = \frac{f_u}{\sqrt{3}\beta_w \gamma_{M2}} \quad \text{where } K = \sqrt{\frac{3}{(1 + 2\cos^2\theta)}} \qquad \text{(D8.2-1)}$$

It is still also necessary to check that:

$$\sigma_{\perp} = \frac{P_T \sin\theta}{a} \leq \frac{f_u}{\gamma_{M2}} \qquad \text{(D8.2-2)}$$

For stiffener design, a bearing fit could be specified to reduce the size of the fillet welds required as discussed in Worked Example 8.2-1.

8.2.3.4. *Simplified method for resistance of fillet welds*

3-1-8/clause 4.5.3.3

3-1-8/clause 4.5.3.3 provides an alternative to using expression 3-1-8/(4.1). The simplified method requires the designer to calculate the resultant force per unit length on the weld and then compare this to the design shear strength such that:

$$F_{w,Ed} \leq F_{w,Rd}$$

where:

$F_{w,Ed}$ is the design value of the weld force per unit length;

$F_{w,Rd}$ is the design weld resistance per unit length $= a\dfrac{f_u/\sqrt{3}}{\beta_w \gamma_{M2}}$.

From Fig. 8.2-2, $F_{w,Ed}$ is the vector sum of the transverse and longitudinal forces on the weld:

$$F_{w,Ed} = (P_T^2 + P_L^2)^{0.5}$$

Combining the above expressions for $F_{w,Ed}$ and $F_{w,Rd}$ gives:

$$\frac{1}{a}(P_T^2 + P_L^2)^{0.5} = \frac{f_u}{\sqrt{3}\beta_w \gamma_{M2}} \qquad \text{(D8.2-2)}$$

This can be compared directly to the weld design expression of (D8.2-1). It can be concluded that the 'simplified method' to 3-1-8/clause 4.5.4 will give identical results to the procedure in 3-1-8/clause 4.5.3 for weld throats loaded in shear only but will give conservative results where weld throats resist direct stresses. The simplified presentation of expression (D8.2-1) means that the simplified rules in 3-1-8/clause 4.5.3.3 should never really need to be used.

8.2.4. Design resistance of fillet welds all round

3-1-8/clause 4.6(1)

3-1-8/clause 4.6(1) allows either of the methods in section 8.2.3 above to be used to check the resistance of a fillet weld all round.

8.2.5. Design resistance of butt welds

8.2.5.1. Full penetration butt welds

3-1-8/clause 4.7.1(1)

3-1-8/clause 4.7.1(1) allows the resistance of full penetration butt welds to be taken as that of the weaker part joined, providing the weld has both an ultimate tensile strength and yield strength at least equal to those of the parent plate.

8.2.5.2. Partial penetration butt welds

3-1-8/clause 4.7.2

3-1-8/clause 4.7.2 requires partial penetration butt welds to be designed as deep penetration fillet weld as discussed in section 8.2.3.2. In the absence of weld procedure trials, BS 5400: Part 3 requires that 3 mm be deducted from the weld preparation depth for V- or bevel-type preparation when calculating throat thickness. This was a precaution for incomplete weld penetration into the root of the preparation. A similar precaution could be adopted when using EN 1993, although it should be the fabricator's responsibility to achieve the specified throat.

8.2.5.3. T butt joints

3-1-8/clause 4.7.3

3-1-8/clause 4.7.3 allows the resistance of two partial penetration butt welds to be determined assuming that they are an effective full penetration butt weld. This condition is only allowed if the combined thicknesses of the weld throats are greater than the thickness off the attached plate, t, and the unwelded gap is less than the smaller of $t/5$ or 3 mm.

8.2.6. Design resistance of plug welds

Plug welds are covered in 3-1-8/clause 4.8 and are not discussed further in this guide.

8.2.7. Distribution of forces

3-1-8/clause 4.9

3-1-8/clause 4.9 gives rules for calculating the distribution of forces in weld groups. Plastic analysis is not prohibited for weld groups, but the deformation capacity of the welds has to be shown to be adequate to develop the assumed distribution of forces.

8.2.8. Connections to unstiffened flanges

3-1-8/clause 4.10

3-1-8/clause 4.10 gives detailed rules for verifying the strength of connections made to unstiffened flanges. These rules are too lengthy for further discussion here.

8.2.9. Long joints

3-1-8/clause 4.11

For similar reasons outlined in section 8.1.5, ***3-1-8/clause 4.11*** requires the designer to reduce the design resistance of long welds by a factor $\beta_{Lw.1}$ to take account of uneven stress distributions along the length of weld. Once again, the rules do not apply to web–flange welds, where the longitudinal state of stress in the weld is the same as in the plates it connects.

8.2.10. Eccentrically loaded single fillet or single-sided partial penetration butt welds

3-1-8/clause 4.12(1)

3-1-8/clause 4.12(3)

3-1-8/clause 4.12 provides guidance for the use of single-sided fillet or single-sided partial penetration butt welds. ***3-1-8/clause 4.12(1)*** recommends that moments about the longitudinal axes of such welds should be avoided wherever possible. Moment may arise from opening of the joint (such as in the web–flange junction of a box girder resisting distortion) or from axial force between the joined parts which has an eccentricity to the weld throat. (In the latter case, moment does not arise if the weld is to the perimeter of a hollow section because the plates are not free to rotate – ***3-1-8/clause 4.12(3)*** refers.) If such welds are subjected to this type of loading, they will undergo bending stresses across the throat which can seriously reduce the fatigue life.

8.2.11. Angles connected by one leg

3-1-8/clause 4.13 gives recommendations for checking the strength of angles connected by one leg. This is discussed in section 6.2.3 of this guide.

8.2.12. Welding in cold-formed zones

3-1-8/clause 4.14(1) gives guidelines for welding in cold-formed areas of steel components. These guidelines are not discussed further in this guide. *3-1-8/clause 4.14(1)*

8.2.13. Analysis of structural joints connecting H- and I-sections

Sections 5 and 6 of EN 1993-1-8 provide lengthy procedures for assessing the strength and stiffness of connections between H- and I-sections. ***3-1-8/clause 5.1*** requires designers to include the rotational behaviour of connections in the global analysis where it is significant. For elastic analysis, connections are classed as: *3-1-8/clause 5.1*

- simple – nominally pinned behaviour
- continuous – fully rigid joint between members
- semi-continuous – not fully rigid so the joint has some rotational flexibility.

Most bridge joints will be continuous and both simple and continuous joints are easy to model. Semi-continuous joints are however less easy to model as they have to be included by using spring elements. As discussed in section 5.1.2 of this guide, semi-continuous joints should generally be avoided for bridges. A possible bridge example of a semi-continuous joint would be in a U-frame deck when the cross-member connected to the main beams through unstiffened end plates.

8.2.14. Hollow section joints

Guidelines for assessing connection strengths between structural hollow sections are given in EN 1993-1-8 section 7. The guidelines are in large part based on the findings of the research of Reference 28 and are too lengthy to cover in this guide.

Worked Example 8.2-1: Design of bearing stiffener welds

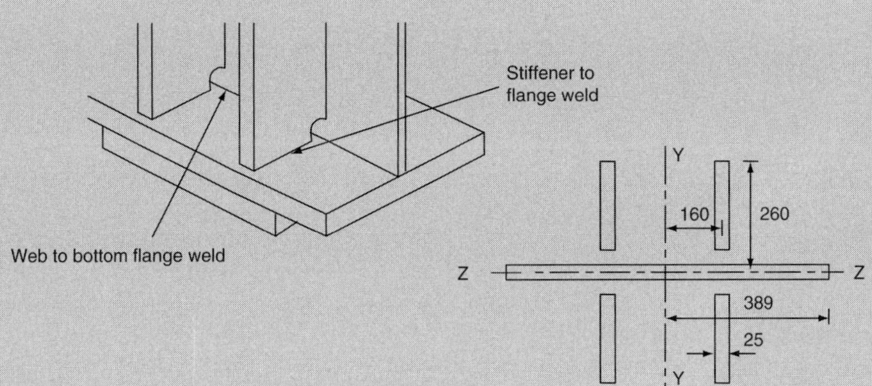

Fig. 8.2-3. Bearing stiffener for Worked Example 8.2-1

The effective section of the bearing area of a bearing stiffener (including cut-outs for the web-flange weld) is shown in Fig. 8.2-3. The stiffeners are fitted to the flange for a full bearing contact in accordance with EN 1090-2, but the flange is not fitted to the web. The bearing stiffener has the following section properties:

Area = 37 040 mm^2
$I_{zz} = 5.83 \times 10^8$ mm^4
$I_{yy} = 1.33 \times 10^9$ mm^4

The forces in the main beam are as follows:

Maximum ULS reaction, $N_{Ed} = 5000\,kN$
Maximum ULS shear force $V_{Ed} = 3000\,kN$
Maximum longitudinal eccentricity = 50 mm
Maximum transverse eccentricity = 20 mm

The elastic shear flow parameter for the bottom flange to web is $Az/I = 0.398 \times 10^{-3}\,mm^{-1}$
The welds for the stiffener to flange and web to flange connections are designed.

Stiffener to flange weld
Maximum stress in stiffener:

$$= \frac{N_{Ed}}{A} + \frac{M_{y,Ed}}{W_y} + \frac{M_{z,Ed}}{W_z}$$

$$= \frac{5000 \times 10^3}{37\,040} + \frac{5000 \times 10^3 \times 160 \times 50}{1.33 \times 10^9} + \frac{5000 \times 10^3 \times 20 \times 260}{5.83 \times 10^8}$$

$$= 135.0 + 30.1 + 44.6$$

$$= 209.7\,MPa$$

If a bearing fit is specified in accordance with EN 1090-2, it would be reasonable to take all the direct compression through bearing, although EN 1993 does not discuss this. If this is done, a fatigue check must still be made of the weld provided, assuming all the compression passes through the weld and none in bearing. In this example, the weld has been designed for the full compression to illustrate the design process.

If fillet welds are placed on either side of the stiffener outstand, the force in each weld per unit length is:

$$P_T = \frac{209.7 \times 25}{2} = 2620\,N/mm$$

From equation (D8.2-1):

$$\frac{1}{a}\left[\frac{P_T^2}{K^2} + P_L^2\right]^{0.5} = \frac{f_u}{\sqrt{3}\beta_w \gamma_{M2}}$$

where:

$$K = \sqrt{\frac{3}{(1 + 2\cos^2\theta)}}$$

and

$f_u = 490\,MPa$ (EN 10025);
$\beta_w = 0.9$ (3-1-8/Table 4.1);
$\gamma_{M2} = 1.25$ (3-2/Table 6.1);
$P_T = 2620\,N/mm$;
$P_L = 0\,N/mm$ (no longitudinal stresses on weld);
$\theta = 45^\circ$ (P_T is applied vertically) therefore $K = 1.225$;
a = required weld throat.

$$\frac{1}{a}\left[\frac{2620^2}{1.225^2} + 0\right]^{0.5} = \frac{490}{\sqrt{3} \times 0.9 \times 1.25}$$

Thus $a = 8.5\,mm$ so specify weld with throat width of 8.5 mm, i.e. leg length = 12 mm. (By inspection equation (D8.2-2) is not critical.)

Web to bottom flange weld

Longitudinal shear force per weld:

$$P_L = \frac{3000 \times 10^3}{2} \times 0.398 \times 10^{-3} = 597\,\text{N/mm}$$

Transverse stress in web:

$$= \frac{N_{Ed}}{A} + \frac{M_{y,Ed}}{W_y} + \frac{M_{z,Ed}}{W_z}$$

$$= \frac{5000 \times 10^3}{37\,040} + \frac{5000 \times 10^3 \times 389 \times 50}{1.33 \times 10^9} + 0$$

$$= 135.0 + 73.1 + 0$$

$$= 208.1\,\text{MPa}$$

Thickness of web = 20 mm, so P_T per weld:

$$= \frac{208.1 \times 20}{2} = 2081\,\text{N/mm}$$

$$\frac{1}{a}\left[\frac{P_T^2}{K^2} + P_L^2\right]^{0.5} = \frac{f_u}{\sqrt{3}\beta_w \gamma_{M2}}$$

$$\frac{1}{a}\left[\frac{2081^2}{1.225^2} + 597^2\right]^{0.5} = \frac{490}{\sqrt{3} \times 0.9 \times 1.25}$$

$a = 7.2$ mm so specify weld with throat width of 7.2 mm, i.e. 11 mm leg.

CHAPTER 9

Fatigue assessment

This chapter discusses fatigue assessment as covered in section 9 of EN 1993-2 in the following clauses:

- General *Clause 9.1*
- Fatigue loading *Clause 9.2*
- Partial factors for fatigue verifications *Clause 9.3*
- Fatigue stress range *Clause 9.4*
- Fatigue assessment procedures *Clause 9.5*
- Fatigue strength *Clause 9.6*
- Post-weld treatment *Clause 9.7*

9.1. General

9.1.1. Requirements for fatigue assessment

Over the lifespan of a bridge, constant road or rail traffic moving over the bridge will produce large numbers of repetitive loading cycles in the steel components. Such components can become susceptible to fatigue damage. As a consequence, ***3-2/clause 9.1.1(1)*** requires fatigue assessment for all steel bridge components, except those given in ***3-2/clause 9.1.1(2)*** as follows: *3-2/clause 9.1.1(1)* *3-2/clause 9.1.1(2)*

(i) pedestrian footbridges not susceptible to pedestrian induced vibration
(ii) bridges carrying canals
(iii) bridges which are predominantly statically loaded
(iv) parts of railway or road bridges that are neither stressed by traffic loads nor likely to be excited by wind loads.

Fatigue assessments are still required in the cases above if bridges are considered to be susceptible to wind-induced excitation. The main cause of wind-induced fatigue, vortex shedding, is covered in EN 1991-1-4 and is not considered further here.

EN 1993-1-9 deals with fatigue in general and EN 1993-2 gives specific rules for bridges. Reference is needed to EN 1993-1-9 for the fatigue strengths of details and for supplementary guidance, as noted in ***3-2/clause 9.1.2(2)***. *3-2/clause 9.1.2(2)*

9.1.2. Design of road bridges for fatigue

In general, ***3-2/clause 9.1.2(1)*** requires all road bridge components to be checked for fatigue unless adequacy can be established by precedent or by testing. The National Annex may give guidance on situations where a fatigue check is not necessary. The UK National Annex does not provide any such exemptions where there is cyclic loading. *3-2/clause 9.1.2(1)*

9.1.3. Design of railway bridges for fatigue

3-2/clause 9.1.3(1)

3-2/clause 9.1.3(1) requires all steel railway bridge components to be checked for fatigue loading. There are no recommendations for exemptions due to the greater susceptibility of railway bridges to fatigue. The National Annex may still however give exceptions.

9.2. Fatigue loading

3-2/clause 9.2.1

3-2/clause 9.2.1 directs the designer to EN 1991-2 for traffic loading models and EN 1991-1-4 for wind excitation.

The basic principle of fatigue assessment is to determine the number of cycles of a particular stress range on a steel component and then ensure that the steel component can withstand this number of stress cycles based on fatigue strengths discussed in section 9.6 below. Worked Example 9-1 demonstrates a simple fatigue check using this principle where there is only one cyclic stress range.

The fatigue life of steel components subjected to varying levels of repetitive stress can be checked with the use of Miner's summation. This is a linear cumulative damage calculation:

$$\sum_{i}^{n} \frac{n_{Ei}}{N_{Ri}} \leq 1.0 \qquad \text{(D9-1)}$$

where n_{Ei} is the number of loading cycles of a particular stress range and N_{Ri} is the number of loading cycles to cause fatigue failure at that particular stress range. Worked Example 9-2 demonstrates the use of equation (D9-1).

For most bridges, the above is a complex calculation because the stress in each steel component usually varies due to the random passage of vehicles from a spectrum. Details on a road or rail bridge could be assessed using the above procedure if the loading regime is known at design. This includes the weight and number of every type of vehicle that will use each lane or track of the bridge throughout its design life, and the correlation between loading in each lane or track. 3-2/clause 9.4.1(6) does allow the designer to carry out fatigue assessment using the above procedure, (with Load Models 4 or 5 specified in EN 1991-1-2), but in the majority of cases this will produce a lengthy calculation because of the large number of different vehicles that will use the bridge during its design lifetime.

3-2/clause 9.2.2
3-2/clause 9.2.3

As an alternative, ***3-2/clause 9.2.2*** and ***3-2/clause 9.2.3*** allow the use of simplified fatigue Load Models 3 and 71 from EN 1991-2, for road and rail bridges respectively, in order to reduce the complexity of the fatigue assessment calculation. It is assumed that the fictitious vehicle/train alone causes the fatigue damage. The calculated stress from the vehicle is then adjusted by factors to give a single stress range which, for 2 million cycles, causes the same damage as the actual traffic during the bridge's lifetime. This is called the 'damage equivalent stress' and is discussed in section 9.4 below.

9.3. Partial factors for fatigue verifications

3-2/clause 9.3(1)P

Fatigue loading must be multiplied by a partial load factor γ_{Ff}. ***3-2/clause 9.3(1)P*** recommends a γ_{Ff} value of 1.0 which may be amended in the National Annex.

Fatigue strength must be divided by a partial factor γ_{Mf} which covers uncertainties in the following:

(i) the size of the detail
(ii) dimensions, shape and proximity of discontinuities
(iii) local stress concentrations due to welding uncertainties
(iv) variable welding processes and metallurgical effects
(v) extent of inspection throughout the design life
(vi) implications of fatigue failure of detail on integrity of whole structure.

3-2/clause 9.3(2)P

Recommended values of γ_{Mf} are obtained from 3-1-9/Table 3.1, reproduced in Table 9-1. These values may be amended by the National Annex as permitted in ***3-2/clause 9.3(2)P***.

Table 9-1. Recommended values of γ_{Mf}

Design concept	Consequence of failure	
	Low consequence	High consequence
Damage tolerance	1.00	1.15
Safe life	1.15	1.35

For the fatigue check of steel components which will have a regular maintenance and inspection programme throughout their design life, the 'damage tolerance' concept can, in theory, be used for the derivation of γ_{Mf} according to 3-1-9/clause 3. The reality is that normal bridge inspections are not carried out in sufficient detail to detect fatigue cracks, unless there has been specific cause for concern in a particular accessible area of the bridge. Such detailed inspections would contribute significantly to the whole-life cost of the bridge. Certain details, such as shear studs in steel–concrete composite bridges, cannot be inspected, making the damage tolerance approach inappropriate. It is likely, therefore, that National Annexes, driven by the major bridge owners, will require the safe life concept to be used, unless agreed otherwise.

UK bridge practice has not previously differentiated between low and high consequences of failure. In Table 9.1, 'High consequence of failure' might be appropriate where fatigue failure of the steel component will result in severe damage or a collapse of the bridge. The possibility of loss of life is also a factor. Low consequence of failure is appropriate where the structure has sufficient redundancy so that a local fatigue failure of a steel component will not be catastrophic due to the presence of alternative load paths. 'High consequence' will often be appropriate for bridges as, although there is usually structural redundancy, it will often not guarantee adequacy in the event of component failure unless the structure is specifically designed to do so.

The worked examples below use the recommended values of γ_{Mf} from Table 9-1. The UK National Annex however always requires a safe life approach and employs a blanket value of $\gamma_{Mf} = 1.1$ to be used as a result of calibration studies.

9.4. Fatigue stress range

9.4.1. General

Where the simplified damage equivalence approach using fatigue Load Models 3 and 71 for road and rail respectively are used, the 'reference stress range' for fatigue assessment is given by ***3-2/clause 9.4.1(3)***: *3-2/clause 9.4.1(3)*

$$\Delta\sigma_p = |\sigma_{p,max} - \sigma_{p,min}| \qquad \text{3-2/(9.1)}$$

It is the maximum change in stress in the detail under the fatigue load model when applied in accordance with EN 1991-2. Previous UK practice in BS 5400: Part 10[29] has been to calculate the stress range by allowing the maximum and minimum effects from the vehicle to come from different lanes. EN 1991-2 clause 4.6.4(2) however implies the maximum stress range should be calculated as the worst stress range produced by the passage of the vehicle along any one lane. At the time of writing, the draft UK National Annex clarifies that the former interpretation should be used (it is the safer of the two) but there is no national provision in EN 1991-2 permitting it to do so and this guidance may have to be removed in the future.

The calculation of stress should be as accurate as possible, based on linear elastic analysis, and should include all effects, even if they have been neglected at ULS (e.g. torsional warping). For Class 4 cross-sections, effective widths may be required for fatigue stress calculation if the ULS reduction to plate area caused by plate buckling exceeds 50% as discussed in section 7.3 of this guide.

Having calculated $\Delta\sigma_p$, the stress needs to be converted into an equivalent stress range for 2×10^6 cycles ($\Delta\sigma_{E,2}$), so that this stress range can be compared directly against the fatigue strengths which all relate to 2×10^6 cycles of a single stress range. ***3-2/clause 9.4.1(4)*** provides an expression for doing this:

3-2/clause 9.4.1(4)

$$\Delta\sigma_{E2} = \lambda\Phi_2\Delta\sigma_p \qquad \text{3-2/(9.2)}$$

where:

λ is the damage equivalence factor discussed in section 9.5.2;
Φ_2 is the damage equivalent impact factor. This may be taken as 1.0 for road bridges (as it is already included in the loading values of fatigue Load Model 3) and derived from EN 1991-2 for rail bridges. For road bridges, however, an additional factor $\Delta\varphi_{fat}$ needs to be included for details at a cross-section within 6 m of an expansion joint. This factor is given in 1-2/clause 4.6.1 and varies linearly from 1.30 at the expansion joint to 1.00 a distance of 6 m away, although the National Annex can vary this.

If the detail incorporates a gross stress concentration that is not included in the basic fatigue Detail Category, this must be included by multiplying the stress range by a stress concentration factor, k_f, according to 3-1-9/clause 6.3. Gross stress concentrations include abrupt changes in cross-section and hard spots at unstiffened connections. (Stress concentration factors for various unreinforced apertures and re-entrant corners can be found in Reference 29 and other standard texts.) If the detail being checked is a welded joint of a hollow section in a truss, 3-1-9/clause 4 allows the joints to be modelled as pinned for the purpose of fatigue stress range calculation, as long as account is taken of moments induced at connections by local load between joints. Secondary moments attracted to joints due to connection stiffness then must be accounted for by multiplying the stress range by an additional factor, k_1, according to 3-1-9/clause 6.4. Obviously, this method is not appropriate for the fatigue analysis of Vierendeel systems.

A similar calculation is performed to determine the shear stress range $\Delta\tau_p = |\tau_{p,max} - \tau_{p,min}|$ so that:

$$\Delta\tau_{E2} = \lambda\Phi_2\Delta\tau_p \qquad \text{(D9-2)}$$

The shear stress should be based on the elastic distribution of shear stress, rather than the average, as indicated in 3-1-9/Table 8.1.

In some cases, principal stresses have to be used in fatigue calculation. These are discussed in section 9.5.1 below.

For the stress range in welds, reference has to be made to 3-1-9/clause 5. Two stress ranges are calculated. The normal stress transverse to the axis of the weld is $\sigma_{wf} = \sqrt{\sigma_{\perp f}^2 + \tau_{\perp f}^2}$ and the shear stress longitudinal to the axis of the weld is $\tau_{wf} = \tau_{||f}$. The various stresses are shown in Fig. 9-1. Calculation of the damage equivalent stress range is performed in the same way as in expression 3-2/(9.2) and equation (D9-2).

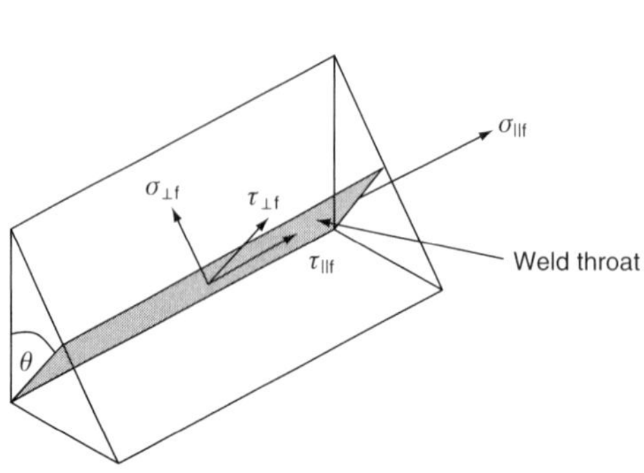

Fig. 9-1. Stress components in welds

9.4.2. Analysis for fatigue

As a general principle, fatigue stresses, like serviceability limit state stresses, should be calculated from as 'realistic' an analysis as practically possible as discussed in sections 7.2 and 7.3. ***3-2/clause 9.4.2.2*** provides detailed analysis guidance for calculating fatigue stresses in steel orthotropic decks, which includes the Vierendeel action in the transverse direction which can lead to fatigue damage in the stiffener-deck connections and in the cross-beams as discussed in Annex C of this guide. *3-2/clause 9.4.2.2*

9.5. Fatigue assessment procedures

9.5.1. Fatigue assessment

The fatigue verifications in ***3-2/clause 9.5.1(1)*** are only valid, according to 3-1-9/clause 8(1), if the direct stress and shear stress ranges due to frequent loads are less than $1.5f_y$ and $1.5f_y/\sqrt{3}$ respectively. The general fatigue assessment equations in EN 1993-2 assume that the damage equivalence method is used. *3-2/clause 9.5.1(1)*

For direct stresses:

$$\gamma_{Ff}\Delta\sigma_{E2} \leq \frac{\Delta\sigma_c}{\gamma_{Mf}} \qquad \text{3-2/(9.7)}$$

For shear stresses:

$$\gamma_{Ff}\Delta\tau_{E2} \leq \frac{\Delta\tau_c}{\gamma_{Mf}} \qquad \text{3-2/(9.8)}$$

where $\Delta\sigma_{E2}$ and $\Delta\tau_{E2}$ are the fatigue equivalent stresses for 2×10^6 cycles for shear and direct stresses respectively as discussed in section 9.4. $\Delta\sigma_c$ and $\Delta\tau_c$ are the direct and shear stress ranges respectively that can be withstood for 2×10^6 cycles without fatigue failure occurring. They are numerically equal to the Detail Category in 3-1-9/Tables 8.1 to 8.10. The determination of these values is discussed in section 9.6.

In the case of combined shear and direct stress, there is a lack of clarity over the required approach in EN 1993-2. It does not provide any combined check, while 3-1-9/8(3) requires a combined check unless otherwise stated in Tables 8.8 and 8.9 of EN 1993-1-9 as follows:

$$\left(\frac{\gamma_{MF}\gamma_{Ff}\Delta\sigma_{E2}}{\Delta\sigma_c}\right)^3 + \left(\frac{\gamma_{MF}\gamma_{Ff}\Delta\tau_{E2}}{\Delta\tau_c}\right)^5 \leq 1.0 \qquad \text{3-1-9/(8.3)}$$

The Note to 3-1-9/clause 8(2) also states that stress ranges should be based on principal stresses where identified in 3-1-9/Tables 8.1 to 8.9. A typical example of a requirement to use principal stresses is for checks of webs where stiffeners curtail in the web, as shown in 3-1-9/Table 8.4. Principal stresses must also be used in the geometric (hot spot) method discussed in section 9.6(vi) below.

Some consideration of the combined effects of shear and direct stress ranges clearly should always be made. To decide whether to use principal stresses or a check via expression 3-1-9/(8.3), the ENV version of EN 1993-1-1[19] is relevant. It recommended the following:

- At locations other than weld throats, if the direct and shear stresses induced by the same loading event vary simultaneously, or if the plane of the maximum principal stress does not change significantly in the course of a loading event, the fatigue check may be performed using the maximum principal stress range in place of the maximum direct stress range.
- If, at a location, the normal and shear stresses vary independently, the fatigue checks for direct and shear stresses according to expressions 3-2/(9.7) and (9.8) should be performed separately and a combined check carried out using expression 3-1-9/(8.3).
- For weld throats, the fatigue checks for direct and shear stresses according to expressions 3-2/(9.7) and (9.8) should be performed separately and a combined check carried out using expression 3-1-9/(8.3).

In the absence of guidance in a National Annex, the use of principal stress ranges is recommended here as a general approach for other cases.

If a particular detail is not contained in 3-1-9/Tables 8.1 to 8.10, the geometric (hot spot) stress method discussed in section 9.6(vi) below should be used.

9.5.2. Damage equivalence factors for road bridges

3-2/clause 9.5.2(1)

3-2/clause 9.5.2(1) defines the damage equivalence factor λ for road bridges with up to 80 m span as follows:

$$\lambda = \lambda_1 \times \lambda_2 \times \lambda_3 \times \lambda_4 \quad \text{but } \lambda \leq \lambda_{max} \qquad \text{3-2/(9.9)}$$

3-2/clause 9.5.2(2)

λ_1 takes into account the damage effect of traffic and depends on the critical length of the influence line or area. Detailed guidance on the appropriate 'span length' to use in 3-2/Fig. 9.5 for reaction, shear, bending moment and other effects is given in ***3-2/clause 9.5.2(2)***. Although the figure is labelled 'for moments', it was intended to be used for shear and reaction also. Reference has to be made to 3-2/Fig. 9.7 to decide whether the 'midspan' or 'support' curve is relevant for the particular location. For example, locations at beam end supports are classified as being 'midspan section' in 3-2/Fig. 9.7.

3-2/clause 9.5.2(3)

λ_2 in ***3-2/clause 9.5.2(3)*** takes the spectrum of traffic frequency and weights into account and is calibrated against the weight of the fatigue Load Model 3 vehicle, which has $Q_0 = 480\,\text{kN}$. It is a fairly crude factor as it makes adjustment to the effects of Load Model 3 on the basis of vehicle weight, rather on the basis of the actual effects of that vehicle. For short spans, the latter is more a function of axle weight and spacing than total weight. Details of the traffic spectrum to be used may be provided in the National Annex in due course, although the clause actually states that data should be provided by the 'competent authority'. A sample calculation for λ_2 on a road bridge is provided in Worked Example 9-3, which is based on traffic data in BS 5400: Part 10: Table 11.[29] Load Model 4 in Table 4.7 of EN 1991-2 could also be used; the UK National Annex for EN 1991-2, however, replaces the recommended Load Model 4 with data very similar to those in BS 5400: Part 10, but also allows the use of BS 5400: Part 10: Table 11. It is not explicitly stated that these data should be used in calculations of λ_2, although it was used in calibration calculations in the UK.

λ_3 is a factor that takes into account the design life of the bridge:

$$\lambda_3 = \left(\frac{t_{Ld}}{100}\right)^{1/5} \qquad \text{3-2/(9.11)}$$

where t_{Ld} is the design life of the bridge (which the UK National Annex makes equal to 120 years).

3-2/clause 9.5.2(6)

λ_4 takes into account traffic on other lanes. Due to the ability of most bridges to transmit load transversely, details will usually attract fatigue stresses from vehicles passing in lanes remote from those directly above them. The equation presented in ***3-2/clause 9.5.2(6)*** therefore includes terms both for the relative magnitude of influence coefficient for adjacent lanes and the number of lorries in these lanes.

λ_{max} is defined as the maximum λ value taking into account the fatigue limit. λ_{max} is calculated from the graphs in 3-2/Fig. 9.6.

The use of these factors is illustrated in Worked Example 9-4.

9.5.3. Damage equivalence factors for railway bridges

The damage equivalence factor λ for railway bridges is determined in a similar way to that for road bridges. For spans up to 100 m:

$$\lambda = \lambda_1 \times \lambda_2 \times \lambda_3 \times \lambda_4 \quad \text{but } \lambda \leq \lambda_{max} \qquad \text{3-2/(9.13)}$$

where:

λ_1 is defined as a factor for different types of girder that takes into account the damage effect of traffic and depends on the length of the influence line or area. 3-2/clause

9.5.3(2) recommends the values in Table 9.3 or Table 9.4 but the National Annex may specify values;

λ_2 takes into account the traffic volume;

λ_3 takes into account the design life of the bridge. λ_3 values are obtained from 3-2/Table 9.6;

λ_4 is a factor that takes into account the extra fatigue damage generated by more than one track loaded at a time. λ_4 values are obtained from 3-2/Table 9.7;

$\lambda_{max} = 1.4$ from 3-2/clause 9.5.3(9).

9.5.4. Combination of damage from local and global stress ranges

Global and local effects on steel deck plates, arising from local wheel loads, should be combined in the calculation of $\Delta\sigma_{E2}$. The effects of local and global loading are particularly significant adjacent to cross-beams and diaphragms where wheel loads cause additional local hogging moments. ***3-2/clause 9.5.4(1)*** provides a conservative interaction where the damage equivalent stress range is determined separately for the global and local actions and then summed to give an overall damage equivalent stress range. *3-2/clause 9.5.4(1)*

9.6. Fatigue strength

3-2/clause 9.6 refers to EN 1993-1-9 for the calculation of fatigue strength. As the relationship between cycle stress range ($\Delta\sigma$) and the number of cycles to failure (N) is exponential, the relationship is normally plotted graphically in the form of a log $\Delta\sigma$–log N curve, commonly abbreviated to 'S–N curve'. 3-1-9/clause 7.1 describes the fatigue performance of steel details in terms of S–N curves. Typical S–N curves for different Detail Categories are provided in 3-1-9/Fig. 7.1, as reproduced with extra explanation in Fig. 9-2 below.

Similar curves are provided in 3-1-9/Fig. 7.2 for shear stress ranges but the exponent on stress range is '5' right up to the cut-off limit.

For the damage equivalent method discussed in section 9.5.1, the fatigue strengths $\Delta\sigma_c$ are required for 2×10^6 cycles in order to be compatible with $\Delta\sigma_{E2}$ and the various limits in Fig. 9-2 need not be considered. All that is required is the actual Detail Category, which is numerically equal to $\Delta\sigma_c$ or $\Delta\tau_c$ as appropriate. EN 1993-1-9 shows typical details in Tables 8.1 to 8.10 together with the Detail Category. For some details, the allowable stress must be reduced by a factor 'k_s' for the 'size effect'. This reflects the fact that

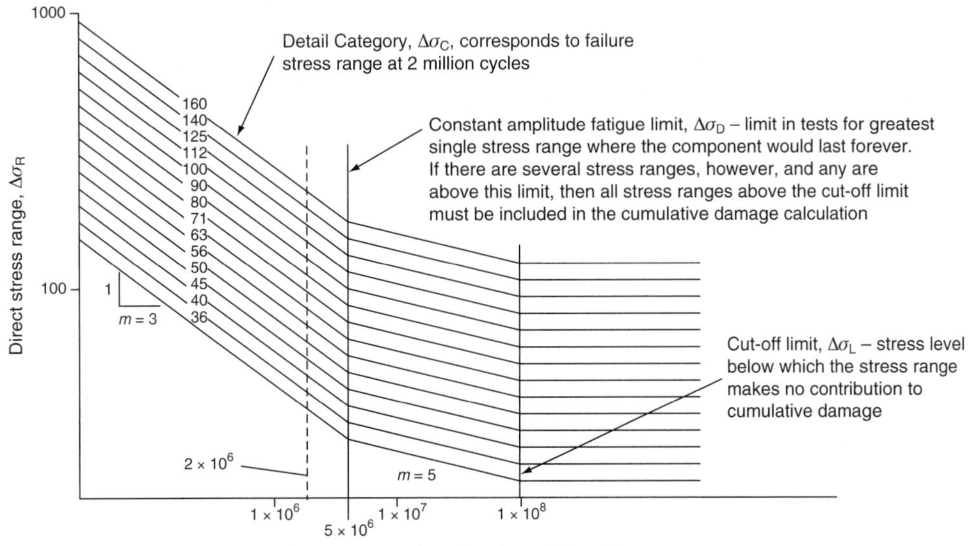

Fig. 9-2. Typical S–N curves

thicker plates may exhibit lower fracture toughness and therefore reduced fatigue performance. Where applicable, 'k_s' is given in the tables. If a particular detail is not covered, the geometric (hot spot) method has to be used as discussed in section 9.6(vi) below.

For calculation methods based on actual traffic spectra, Fig. 9-2 needs to be used. For stress ranges above the constant amplitude fatigue limit, the number of cycles to failure is obtained from $\Delta\sigma_R^3 N_R = \Delta\sigma_C^3 \times 2 \times 10^6$. For stress ranges below the constant amplitude fatigue limit, but above the cut-off limit, the number of cycles to failure for use in a cumulative damage summation is obtained from $\Delta\sigma_R^5 N_R = \Delta\sigma_D^5 \times 5 \times 10^6$. If all stress ranges are below the constant amplitude fatigue limit, there is no need to perform any cumulative damage calculation and the component will not suffer any fatigue damage.

Some typical bridge details and their Detail Categories are discussed below:

(i) Plates and fit bolts in shear

In general, Detail Category 100 applies for all parent metal, full penetration butt welds and bearing-type fitted bolts in shear – 3-1-9/Table 8.1 refers. Where principal stresses are used for the calculation, the Detail Category should be based on that for direct stress.

(ii) Fatigue strength of non-welded details

Fatigue cracks in non-welded details generally tend to occur adjacent to bolt holes or where there are imperfections in the parent plate. 3-1-9/Table 8.1 provides typical Detail Categories. Fatigue damage predominately occurs when steel members are subjected to repetitive cycles of tensile stress. In welded components, welding residual stresses give rise to locked-in tensile stresses so the stress can remain tensile during cycles even if the external imposed stresses are tensile. This is not the case for non-welded components. 3-1-9/clause 7.2.1 takes this fact into account by allowing the designer to reduce any compressive stress portion of a stress cycle by 40% for non-welded details.

(iii) Fatigue strength of welded members with or without welded attachments

Parent plates adjacent to welded details are susceptible to fatigue cracking and 3-1-9/Tables 8.2 to 8.5 provide Detail Categories. Welded details, even if detailed analysis shows that they are only subjected to compressive forces, will invariably have large tensile residual stresses locked into them as a result of weld shrinkage during fabrication. This will also be the case for the parent plate adjacent to the weld. For this reason, EN 1993 does not allow reduction of the compressive stress portion of a stress cycle in a welded detail unless it has been stress-relieved. For some details, the allowable stress must be reduced by the 'size effect' factor, k_s. The classification at transverse butt welds typically requires use of the size effect parameter where the parent plate is greater than 25 mm thick.

Typical details which will limit fatigue strength of members themselves are:

- Stiffeners – provides a Detail Category of 71 or 80 depending on the combined thickness of stiffener and attaching welds (3-1-9/Table 8.4 construction detail 7).
- Longitudinal attachments – can give a Detail Category as low as 56 if the cleat is longer than 100 mm and is not radiused or tapered at its ends (3-1-9/Table 8.4).
- Doubler plates/welded single-sided splice plates – can give a Detail Category as low as 36 for thick doublers where the ends of the plates and welds are not tapered by grinding (3-1-9/Table 8.5 construction detail 6).
- Cruciform joints – can give Detail Categories from 80 down to 40 depending on plate and weld size, but normal geometries will tend to be at the upper end of this range (3-1-9/Table 8.5 construction detail 1).
- Gussets welded to the edge of a flange – can give Detail Categories from 90, if given a sufficiently large radius, down to 40 with no radius (3-1-9/Table 8.4 construction details 4 and 5).

There is no category for stiffeners (or other attachments) with weld toes close to the edge of a flange plate. It is good practice to avoid this and keep weld toes at least 10 mm from the edge of the plate. If it cannot be avoided, it is suggested here that Detail Category 40 be used as for 3-1-9/Table 8.4 construction detail 5.

(iv) Fatigue strength of welds

Load-carrying welds themselves are prone to fatigue cracking. 3-1-9/Tables 8.3 and 8.5 provide Detail Categories for butt and fillet welds respectively. $\Delta\sigma_{E2}$ in butt welds is calculated as for the parent plate adjacent to the weld. $\Delta\sigma_{E2}$ in fillet and partial penetration butt welds is calculated as discussed in section 9.4.1 above.

For longitudinal shear stress on a fillet or partial penetration butt weld, $\Delta\tau_{wf}$, the limiting stress $\Delta\tau_C$ is 80 MPa. For transverse stress on a weld, $\Delta\sigma_{wf}$, the limiting stress $\Delta\sigma_C$ is 36* MPa. The asterisk means that the detail generally has the fatigue performance of the next category up, but the constant amplitude fatigue limit is reached at 10 million cycles rather than 5 million. If a cumulative damage calculation is used, the calculation is either conservatively based on Detail Category 36 or on a modified Detail Category 40 *S–N* curve (with lengthened portion having $m = 3$ slope). For damage equivalent calculations, $\Delta\sigma_C$ can be taken as 40. Where there is transverse stress on a fillet weld, fatigue must also be checked at the weld toe in the parent plate which can give Detail Categories from 80 down to 40 (see 3-1-9/Table 8.4, construction detail 1) depending on plate and weld size, but normal geometries will tend to be at the upper end of this range.

(v) Fatigue strength of hollow section joints

Care should be taken when using the hollow section joint detail classifications in 3-1-9/Table 8.7. The joints shown are two-dimensional, whereas the majority of hollow section connections used in bridge structures are three-dimensional. Designers should only use 3-1-9/Table 8.7 if they are confident that the three-dimensional geometry of the joint does not have a detrimental effect on the fatigue performance. If there is any uncertainty then the geometric (hot spot) stress method should be used.

(vi) Fatigue strength of non-classified details – geometrical (hot spot) stress method

If a particular detail is not contained in 3-1-9/Tables 8.1 to 8.10 (which, for example, includes hollow sections with plate thickness in excess of 12.5 mm), then the principal stress range adjacent to the weld toe must be determined accurately. 'Accurately' includes taking account of the overall geometry including plate misalignments, but excluding the local stress concentration caused by the weld itself. Shell finite-element modelling would be an appropriate modelling technique. This is referred to in EN 1993-1-9 as the geometric (hot spot) stress. This stress range is then compared against the limiting stress ranges for joint types in 3-1-9/Annex B. This applies regardless of whether the damage equivalence method is being used or not.

3-1-9/Table B.1 gives detail categories for fatigue cracks initiating from toes of butt welds, toes of fillet welded attachments and the toes of fillet welds in cruciform joints. The method is not suitable for joints where crack initiation and propagation would occur in the welds themselves.

Worked Example 9-1: Use of the basic fatigue *S–N* curves in EN 1993-1-9

The flange of a welded steel girder is classed as 'Detail Category 125' to EN 1993-1-9. The component is subjected to 500 000 cycles of a stress range of 200 MPa. The fatigue strength is checked for acceptability. For this example, $\gamma_{Mf} = 1.15$ and $\gamma_{Ff} = 1.0$.

$$\gamma_{Ff}\Delta\sigma \leq \frac{\Delta\sigma_R}{\gamma_{Mf}} \quad \text{where } \Delta\sigma = 200\,\text{MPa}$$

$\Delta\sigma_R$ needs to be read from the Detail Category 125 *S–N* curve from EN 1993-1-9 (as shown in Fig. 9-3) or obtained from the expressions in 3-1-9/7.1(2):

$$\Delta\sigma_R^3 N_R = \Delta\sigma_C^3 \times 2 \times 10^6 \text{ so } \Delta\sigma_R^3 N_R = 125^3 \times 2 \times 10^6 \text{ where } N_R = 500\,000$$

$$\Delta\sigma_R = \sqrt[3]{\frac{125^3 \times 2 \times 10^6}{500\,000}} = 198\,\text{MPa}$$

$$\frac{\Delta\sigma_R}{\gamma_{Mf}} = \frac{198\,\text{MPa}}{1.15} = 172\,\text{MPa}$$

$$\gamma_{Ff}\Delta\sigma = 200 \times 1.0 = 200\,\text{MPa}$$

Therefore:

$$\gamma_{Ff}\Delta\sigma > \frac{\Delta\sigma_R}{\gamma_{Mf}}$$

and so the fatigue strength is not acceptable.

The flange would need modifying to reduce the stress per cycle to 172 MPa.

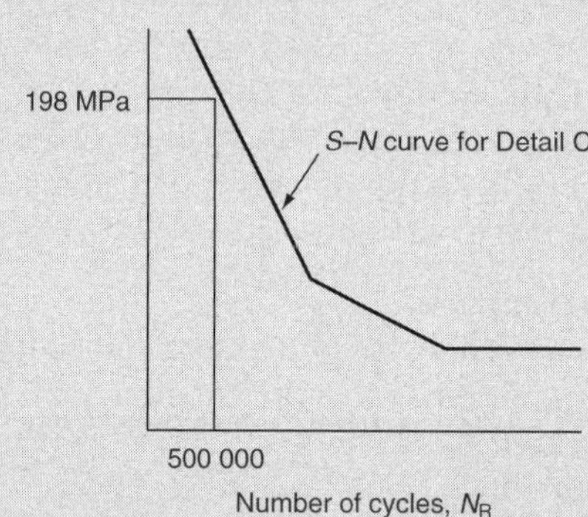

Fig. 9-3. *S–N* curve for Detail Category 125

Worked Example 9-2: Fatigue assessment using Palmgren–Miner summation in 3-1-9/Annex A

The fatigue performance of a welded detail in a steel linkspan structure can be represented by an *S–N* curve corresponding to Detail Category 36 in 3-1-9/Fig. 7.1. The linkspan carries typical vehicles of weight 1, 2 and 5 tonnes off a ferry. The stress ranges in the welded detail under each of the vehicles are as follows:

1-tonne vehicle = 20 MPa
2-tonne vehicle = 40 MPa
5-tonne vehicle = 100 MPa

The ferry carries an average of 50 vehicles – the proportion of 1-, 2- and 5-tonne vehicles being 70%, 28% and 2% respectively – which use the linkspan twice a day. The detail is assessed to determine if it can withstand a service life of 40 years. No more than one vehicle can occupy the linkspan at any one time. Take $\gamma_{Ff} = 1.0$ and $\gamma_{Mf} = 1.15$.

From the Palmgren–Miner Rule in 3-1-9/Annex A:

$$\sum_i^n \frac{n_{Ei}}{N_{Ri}} \leq 1.0$$

The number of cycles to failure for each vehicles load is obtained from the Detail Category 36 *S–N* curve in Fig. 9-4.

The constant amplitude fatigue limit, $\Delta\sigma_D$, occurs at 5 million cycles. Therefore $\Delta\sigma_D^3 N_R = \Delta\sigma_C^3 \times 2 \times 10^6$ so $\Delta\sigma_D^3 \times 5 \times 10^6 = 36^3 \times 2 \times 10^6$ and $\Delta\sigma_D = 26.5\,\text{MPa}$. The 100 MPa and 40 MPa stress cycles therefore cause damage above the constant amplitude fatigue limit according to $\Delta\sigma_R^3 N_R = 36^3 \times 2 \times 10^6$, while the 20 MPa stress cycles cause damage below the constant amplitude fatigue limit according to $\Delta\sigma_R^5 N_R = 26.5^5 \times 5 \times 10^6$. The number of cycles to failure are as follows:

1T vehicles (20 MPa) = 2.05×10^7 cycles to failure
2T vehicles (40 MPa) = 1.458×10^6 cycles to failure
5T vehicles (100 MPa) = 9.331×10^4 cycles to failure

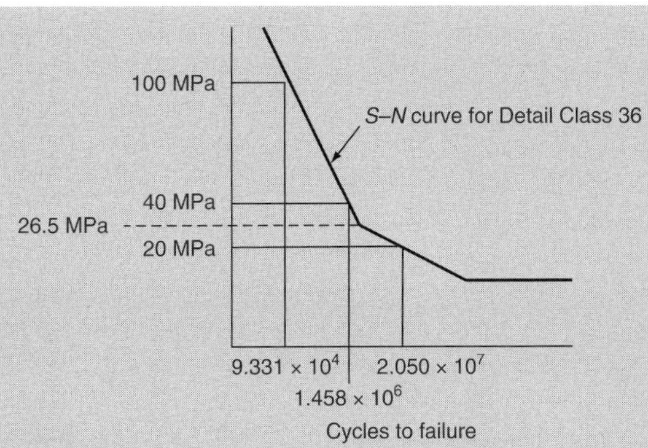

Fig. 9-4. *S–N* curve for Detail Category 36

The next step is to calculate the number of loading cycles for each type of vehicle over the design life.

No. of vehicles to use linkspan over 40-year design life
$= 2$ times a day $\times$ 50 vehicles $\times$ 365 days a year $\times$ 40 years $= 1.460 \times 10^6$ vehicles.

No. of 1-tonne vehicles $= 70\%$ of $1.460 \times 10^6 = 1.022 \times 10^6$ vehicles
No. of 2-tonne vehicles $= 28\%$ of $1.460 \times 10^6 = 4.088 \times 10^5$ vehicles
No. of 5-tonne vehicles $= 2\%$ of $1.460 \times 10^6 = 2.920 \times 10^4$ vehicles

$$\sum_i^n \frac{n_{Ei}}{N_{Ri}} = \frac{n_{1\text{tonne}}\gamma_{Mf}\gamma_{Ff}}{N_{1\text{tonne}}} + \frac{n_{2\text{tonne}}\gamma_{Mf}\gamma_{Ff}}{N_{2\text{tonne}}} + \frac{n_{5\text{tonne}}\gamma_{Mf}\gamma_{Ff}}{N_{5\text{tonne}}}$$

$$= \frac{1.022 \times 10^6 \times 1.15 \times 1.0}{2.05 \times 10^7} + \frac{4.088 \times 10^5 \times 1.15 \times 1.0}{1.458 \times 10^6} + \frac{2.920 \times 10^4 \times 1.15 \times 1.0}{9.331 \times 10^4}$$

$$= 0.057 + 0.322 + 0.360 = 0.74 < 1.0$$

so fatigue life is adequate.

Worked Example 9-3: Calculation of λ_2 for a road bridge

λ_2 is determined for a road bridge using the vehicle spectrum defined in BS 5400: Part 10: Table 11. The slow lane is estimated to carry 500 000 vehicles per year, using Table 4.5 of EN 1991-2.

Vehicle ref.	Weight (kN), Q_i	No./million vehicles, n_i	$n_iQ_i^5$
18GTH	3680	10	6.749×10^{18}
18GTM	1520	30	2.434×10^{17}
9TT-H	1610	20	2.164×10^{17}
9TT-M	750	40	9.492×10^{15}
7GT-H	1310	30	1.157×10^{17}
7GT-M	680	70	1.018×10^{16}
7A-H	790	20	6.154×10^{15}
5A-H	630	280	2.779×10^{16}
5A-M	360	14 500	8.768×10^{16}
5A-L	250	15 000	1.465×10^{16}
4A-H	335	90 000	3.797×10^{17}
4A-M	260	90 000	1.069×10^{17}
4A-L	145	90 000	5.769×10^{15}
4R-H	280	15 000	2.582×10^{16}
4R-M	240	15 000	1.194×10^{16}

Vehicle ref.	Weight (kN), Q_i	No./million vehicles, n_i	$n_i Q_i^5$
4R-L	120	15 000	3.732×10^{14}
3A-H	215	30 000	1.378×10^{16}
3A-M	140	30 000	1.613×10^{15}
3A-L	90	30 000	1.771×10^{14}
3R-H	240	15 000	1.194×10^{16}
3R-M	195	15 000	4.229×10^{15}
3R-L	120	15 000	3.732×10^{14}
2R-H	135	170 000	7.623×10^{15}
2R-M	65	170 000	1.972×10^{14}
2R-L	30	180 000	4.374×10^{12}
Totals		1.000×10^6	8.051×10^{18}

From 3-2/clause 9.5.2(3):

$$Q_{m1} = \left(\frac{\sum n_i Q_i^5}{\sum n_i}\right)^{1/5} = \left(\frac{8.051 \times 10^{18}}{1.000 \times 10^6}\right)^{1/5} = 381.2\,\text{kN}$$

$$N_{\text{Obs}} = 0.5 \times 10^6$$

$$N_{\text{O}} = 0.5 \times 10^6$$

$$Q_{\text{O}} = 480\,\text{kN (weight of Load Model 3)}$$

Therefore:

$$\lambda_2 = \frac{Q_{m1}}{Q_0}\left(\frac{N_{\text{Obs}}}{N_0}\right)^{1/5} = \frac{381.2}{480}\left(\frac{0.5 \times 10^6}{0.5 \times 10^6}\right)^{1/5} = \mathbf{0.794}$$

It will be noted that vehicle 18GTH is the main contributor to λ_2 due to its high weight. It does not mean that this vehicle would necessarily cause the most fatigue damage, as axle weight and spacing may be more important as discussed in the main text, particularly for modest spans.

Worked Example 9-4: Fatigue check of a bearing stiffener and welds to EN 1993-1-9

A two-lane road bridge is shown in Fig. 9-5. Suitable weld sizes are calculated for fatigue for the stiffener to bottom flange weld and the web to bottom flange weld. Fatigue of the web and flange is also checked. The Client requires a design life of 120 years and Table 4.5 of EN 1991-2 classifies the bridge as having 500 000 heavy vehicles per year in each slow lane. The Client has specified the use of the damage tolerance approach as there will be regular inspections, including checks for fatigue cracking. (Note, however, the comments in the main text on the use of the 'damage tolerance' approach regarding reliability of inspections. 'Safe life' will usually be a more appropriate approach.)

Check 1: Calculate suitable weld size for stiffener to bottom flange weld

Bearing reactions from Load Model 3 on bridge:

Max = +401.1 kN
Min = −113.6 kN

Fatigue reaction range = 401.1 − (−113.6) = 514.7 kN

Assume ±10 mm both longitudinally and transversely for average bearing eccentricity under fatigue loading so $M_{\text{Ed,z}} = M_{\text{Ed,y}} = 514.7 \times 0.01 = 5.1\,\text{kNm}$.

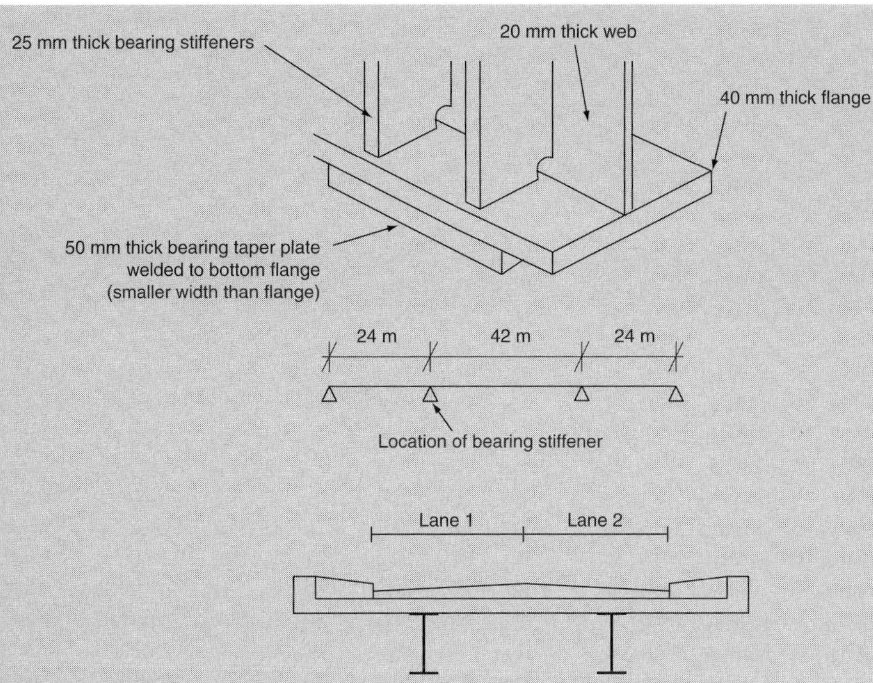

Fig. 9-5. Bridge cross-section, spans and stiffener layout for Worked Example 9-4

Two checks are required according to 3-1-9/Table 8.5. These are toe cracking in the parent plate with Detail Category 80 and root cracking with Detail Category 36*. The latter will clearly be critical.

The vertical stress range in the stiffener outstand = 15.5 MPa (calculation based on stiffener effective section, not shown).

The transverse stress range in the weld is:

$$\sigma_{wf} = \frac{15.5 \times 25}{2a} = 194/a$$

where a is the weld throat on each side of the 25 mm thick stiffener.

Now calculate $\Delta\sigma_{E,2}$ for 2×10^6 cycles.

From 3-2/clause 9.5.2(1):

$$\lambda = \lambda_1 \times \lambda_2 \times \lambda_3 \times \lambda_4 \leq \lambda_{max}$$

From 3-2/Figure 9.5:

$$\lambda_1 = 1.7 + 0.5\left(\frac{L - 30}{50}\right) = 1.7 + 0.5\left(\frac{66 - 30}{50}\right) = 2.03$$

where L is the sum of adjacent spans for reaction from 3-2/clause 9.5.2(2), i.e. $L = 24 + 42 = 66$ m. The 'at support' case applies for intermediate support locations according to 3-2/Fig. 9.7.

From 3-2/clause 9.5.2(3):

$$\lambda_2 = \frac{Q_{m1}}{Q_0}\left(\frac{N_{Obs}}{N_0}\right)^{1/5}$$

For a bridge in the UK with $N_{Obs} = 0.5 \times 10^6$, it would be reasonable to take $\lambda_2 = 0.794$ from Worked Example 9-3.

From 3-2/clause 9.5.2(5):

$$\lambda_3 = \left(\frac{t_{Ld}}{100}\right)^{1/5} = 1.037 \text{ for 120-year design life}$$

From 3-2/clause 9.5.2(6):

$$\lambda_4 = \left[1 + \frac{N_2}{N_1}\left(\frac{\eta_2 Q_{m2}}{\eta_1 Q_{m1}}\right)^5 + \frac{N_3}{N_1}\left(\frac{\eta_3 Q_{m3}}{\eta_1 Q_{m1}}\right)^5 + \ldots + \frac{N_k}{N_1}\left(\frac{\eta_k Q_{mk}}{\eta_1 Q_{m1}}\right)^5\right]^{1/5}$$

The influence coefficient from lane 2 is conservatively taken here as 75% that of lane 1. As both lanes are slow lanes with $N = 0.5 \times 10^6$ vehicles per year:

$$\lambda_4 = \left[1 + \frac{N_2}{N_1}\left(\frac{\eta_2 Q_{m2}}{\eta_1 Q_{m1}}\right)^5\right]^{1/5} = \left[1 + \frac{0.5 \times 10^6}{0.5 \times 10^6}\left(\frac{0.75}{1.0}\right)^5\right]^{1/5} = 1.044$$

From 3-2/Fig. 9.6:

$$\lambda_{max} = 1.80 + 0.90\left(\frac{L - 30}{50}\right) = 1.80 + 0.90\left(\frac{66 - 30}{50}\right) = 2.39$$

$$\lambda = \lambda_1 \times \lambda_2 \times \lambda_3 \times \lambda_4 \leq \lambda_{max} = 2.03 \times 0.794 \times 1.037 \times 1.044 = 1.745 < \lambda_{max}$$

For parent stiffener, $\Delta\sigma_{E2} = \lambda\Phi_2\Delta\sigma_p = 1.745 \times 1.0 \times 15.5 = 27.0\,\text{MPa}$
For weld, $\Delta\sigma_{E2} = \lambda\Phi_2\Delta\sigma_{wf} = 1.745 \times 1.0 \times 194/a = 338/a\,\text{MPa}$
For the weld, Detail Category 36* has $\Delta\sigma_c = 40\,\text{MPa}$ as discussed in the main text. The weld will be regularly inspected throughout its design life. If the weld fails, an alternative load path is provided through direct contact between stiffener and flange so complete failure is unlikely to occur. Therefore, the detail can be classed as 'Damage tolerant, low consequence of failure' and from 3-1-9/Table 3.1, $\gamma_{Mf} = 1.00$.
From 3-2/clause 9.5:

$$\gamma_{Ff}\Delta\sigma_{E2} \leq \frac{\Delta\sigma_c}{\gamma_{Mf}}$$

$$\frac{1.0 \times 338}{a} \leq \frac{40}{1.00} \text{ so } a = \frac{338}{40} = 8.5\,\text{mm}$$

Use fillet welds of leg length $8.5 \times \sqrt{2} \approx 12\,\text{mm}$.

Check 2: Calculate suitable weld size for web to bottom flange weld within the bearing area

The weld must carry both the bearing reaction and longitudinal shear between web and flange. From reactions and eccentricities above, the maximum transverse stress in the web is found to be 15.4 MPa. The transverse stress range in the weld is

$$\sigma_{wf} = \frac{15.4 \times 20}{2a} = 154/a$$

where a is the weld throat on each side of the 20 mm thick web. The maximum shear force range due to Load Model 3 was found to be 329.3 kN.

Bottom flange $A\bar{z}/I = 0.394 \times 10^{-3}/\text{mm}$ (SLS section properties with cracked concrete).

Longitudinal stress in welds:

$$\tau_{wf} = \frac{329.3 \times 10^3 \times 0.394 \times 10^{-3}}{2a} = 64.9/a$$

(i) Transverse stress

For transverse load, the Detail Category is 36* which has $\Delta\sigma_c = 40\,\text{MPa}$ as discussed in the main text. The weld will be regularly inspected throughout its design life but, if the weld fails, an alternative load path is not provided for web–flange longitudinal shear. A crack would however have to propagate over some length before there would be a real problem. The detail here has conservatively been classed as 'Damage tolerant, high consequence of failure' ('low consequence' would probably suffice) and from 3-1-9/Table 3.1, $\gamma_{mf} = 1.15$. The damage equivalent parameters are as before so:

$$\Delta\sigma_{E2} = \lambda\Phi_2\Delta\sigma_{wf} = 1.745 \times 1.0 \times 154/a = 268.7/a$$

From 3-2/clause 9.5.1:

$$\gamma_{Ff}\Delta\sigma_{E2} \leq \frac{\Delta\sigma_c}{\gamma_{Mf}} \text{ so } \frac{1.0 \times 268.7}{a} \leq \frac{40}{1.15} \text{ so } a = 7.7\,\text{mm}$$

A leg length of 12 mm would suffice which provides $a = 8.49$ mm.

(ii) Longitudinal stress

From 3-1-9/Table 8.5, the Detail Category is 80. The length L for calculation of λ_1 is again conservatively taken as the sum of the adjacent spans as a longer length is conservative and load in both spans contributes to maximum shear. Note 3-2/clause 9.5.2(3) suggests using the 'span under consideration', i.e. $L = 42$ m, as an approximation.

$$\Delta\tau_{E2} = \lambda\Phi_2\Delta\sigma_{wf} = 1.745 \times 1.0 \times 64.9/a = 113.3/a$$

From 3-2/clause 9.5.1:

$$\gamma_{Ff}\Delta\tau_{E2} \leq \frac{\Delta\tau_c}{\gamma_{Mf}} \text{ so } \frac{1.0 \times 113.3}{a} \leq \frac{80}{1.15}$$

so $a = 1.6$ mm, i.e. much less critical than for transverse stress.

(iii) Combined check

Assume a weld with 12 mm leg is used.

$$\Delta\sigma_{E2} = 268.7/8.49 = 31.7\,\text{MPa}$$

$$\Delta\tau_{E2} = 113.3/8.49 = 13.4\,\text{MPa}$$

From 3-1-9/clause 8(3):

$$\left(\frac{\gamma_{MF}\gamma_{Ff}\Delta\sigma_{E2}}{\Delta\sigma_c}\right)^3 + \left(\frac{\gamma_{MF}\gamma_{Ff}\Delta\tau_{E2}}{\Delta\tau_c}\right)^5 \leq 1.0$$

$$= \left(\frac{1.15 \times 1.0 \times 31.7}{40}\right)^3 + \left(\frac{1.15 \times 1.0 \times 13.4}{80}\right)^5 = 0.757 + 0.0003 = 0.757 \leq 1.0$$

The longitudinal shear term has very little effect as the Detail Category is quite high and the term $(\Delta\tau_{E2}/\Delta\tau_c)^5$ becomes negligible. This will often be the conclusion for fillet welds with combined longitudinal and transverse stress.

Check 3: Check for cracking in web plate produced by attached stiffener weld

The location of the potential crack is shown in Fig. 9-6. The detail is Detail Category 80 in 3-1-9/Table 8.4 (since the combined width of stiffener and welds is just less than 50 mm) and requires $\Delta\sigma$ to be based on principal stresses as the stiffener terminates in the web.

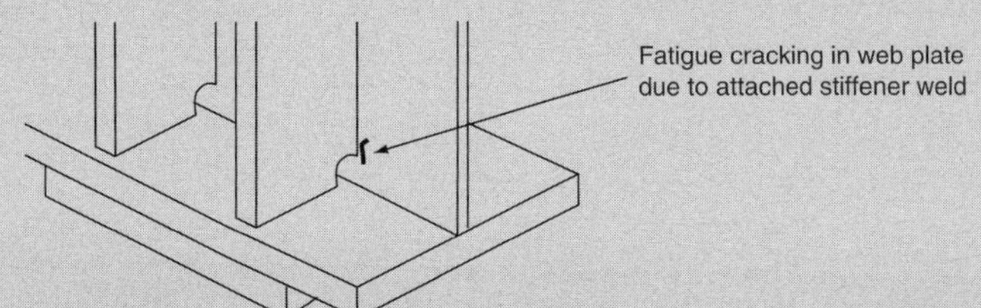

Fig. 9-6. Potential crack in web at stiffener weld

The length L for calculation of λ_1 is different in this case as the stresses will be dominated by bending, for which L is based on the average of the adjacent spans according to 3-2/clause 9.5.2(2)(a), i.e. $L = 0.5 \times (24 + 42) = 33\,\text{m}$

$$\lambda_1 = 1.7 + 0.5\left(\frac{L-30}{50}\right) = 1.7 + 0.5\left(\frac{33-30}{50}\right) = 1.73$$

$$\lambda_{max} = 1.80 + 0.90\left(\frac{L-30}{50}\right) = 1.80 + 0.90\left(\frac{33-30}{50}\right) = 1.85$$

The other parameters can be taken as before. λ_4 remains conservative for bending.

$$\lambda = \lambda_1 \times \lambda_2 \times \lambda_3 \times \lambda_4 \leq \lambda_{max} = 1.73 \times 0.794 \times 1.037 \times 1.044 = 1.487 < \lambda_{max}$$

The bending moment range due to fatigue Load Model 3 in girder at bearing stiffener location was found to be 1567 kNm.

$W_{(\text{bottom web})} = 72.9 \times 10^6\,\text{mm}^3$ (SLS properties – conservatively based on flange location)

Longitudinal stress range, $\Delta\sigma_p$, at bottom of web

$$= \frac{1567 \times 10^6}{72.9 \times 10^6} = 21.5\,\text{MPa}$$

From check 2, bottom flange/web $A\bar{z}/I = 0.394 \times 10^{-3}$/mm and the maximum shear force = 329.3 kN from above.
Therefore shear stress range, $\Delta\tau_p$, in base of web

$$= \frac{329.3 \times 10^3 \times 0.394 \times 10^{-3}}{20} = 6.5\,\text{MPa}$$

Principal stress range at base of web

$$= \frac{\sigma}{2} + \frac{1}{2}\sqrt{\sigma^2 + 4\tau^2} = \frac{21.5}{2} + \frac{1}{2}\sqrt{21.5^2 + 4 \times 6.5^2} = 23.3\,\text{MPa}$$

$$\Delta\sigma_{E2} = \lambda\Phi_2\Delta\sigma_p = 1.487 \times 1.0 \times 23.3 = 34.7\,\text{MPa}$$

From 3-2/clause 9.5.1:

$$\gamma_{Ff}\Delta\sigma_{E2} \leq \frac{\Delta\sigma_c}{\gamma_{Mf}}$$

$$\gamma_{Ff}\Delta\sigma_{E2} = 1.0 \times 34.7 = 34.7 < \frac{\Delta\sigma_c}{\gamma_{Mf}} = \frac{80}{1.15} = 69.6\,\text{MPa}$$

(The web was here classed as 'Damage tolerant, high consequence of failure'.) Therefore the fatigue life of the web plate is adequate.

Check 4: Check for cracking in flange due to welded attachments
The locations of the potential cracks are shown in Fig. 9-7.

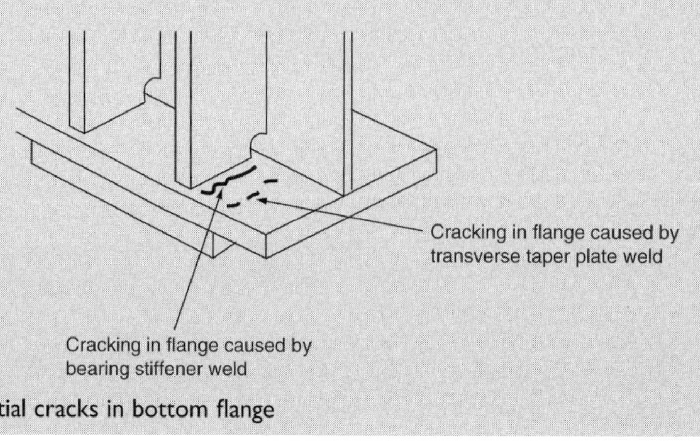

Fig. 9-7. Potential cracks in bottom flange

From Check 3, stress range in bottom flange, $\Delta\sigma_p$, due to Fatigue Load Model 3 = 21.5 MPa. The parameter λ will be the same as for Check 3.

$$\Delta\sigma_{E2} = \lambda\Phi_2\Delta\sigma_p = 1.487 \times 1.0 \times 21.5 = 32.0\,\text{MPa}$$

Cracking in flange caused by transverse taper plate weld is Detail Category 40 for 40 mm flange and 50 mm thick taper plate (3-1-9/Table 8.5, construction detail 6). This is worse than Detail Category 80 obtained at the face of the bearing stiffener. $\gamma_{Mf} = 1.15$ again for 'Damage tolerant, high consequence of failure'.

$$\frac{\Delta\sigma_c}{\gamma_{Mf}} = \frac{40}{1.15} = 34.8\,\text{MPa} > \gamma_{Ff}\Delta\sigma_{web\,E2} = 32.0\,\text{MPa}$$

The fatigue life of the flange is sufficient.

9.7. Post-weld treatment

In order to improve the fatigue life of welded details, ***3-2/clause 9.7(1)*** gives designers scope to specify post-weld treatment techniques. Fatigue strength of fillet welded details can be improved by the following methods: *3-2/clause 9.7(1)*

(i) Reducing stress concentration effects

Techniques such as weld toe grinding, TIG re-melting of the weld toe region and plasma re-melting of the weld toe region are all intended to smooth out the profile of the weld toe. This reduces the stress concentration at the toe of the fillet weld which will improve the fatigue life.

(ii) Introduction of compressive residual stresses

A major factor behind the lower fatigue strength of welded details is the large tensile residual stress present. As a consequence, applied compressive stresses may still lead to cycles of stress which are entirely within the tensile range. If a compressive residual stress can be introduced to the detail then all stresses will cycle in the compressive range which results in a longer fatigue life. Techniques such as hammer peening and shot peening both work on this principle. The toe of the fillet weld is cold worked in compression which introduces a compressive residual stress. This delays the formation of fatigue cracking in the weld which results in a longer fatigue life.

EN 1993-1-9 does not provide any guidance as to how to estimate the increase in fatigue life obtained from the use of post-weld treatment techniques. As a general guide, introducing compressive residual stresses tends to create a greater increase in fatigue life than reducing the stress concentration effects above. However, designers are strongly recommended to seek specialist advice before estimating the increase in fatigue life created by post-weld treatment. In addition, the post-weld treatment needs to be carefully specified and closely supervized as the success of the treatment is highly dependent on the quality of workmanship.

CHAPTER 10

Design assisted by testing

- General
- Types of test
- Verification of aerodynamic effects on bridges by testing

10.1. General

Where tests, for whatever reason, are required to verify design strengths of bridge components, the designer is referred to the requirements of EN 1990 Annex D. As discussion on EN 1990 is beyond the scope of this guide, the testing requirements are not discussed in detail here. Reference should be made to EN 1990 directly and commentary can be found in Reference 1.

10.2. Types of test

3-2/clause 10.2 separates tests into two different types. The first type involves testing to quantify design strengths or parameters for subsequent use in design calculations. The second type involves testing to verify that the values of design strengths or parameters assumed in design calculations are safe. This section is essentially lifted from section D3 of EN 1990 Annex D and is therefore not discussed further here.

10.3. Verification of aerodynamic effects on bridges by testing

Where the aerodynamic performance of a bridge cannot adequately be verified by calculation or, where applicable, by comparison with results from similar structures, testing should be carried out. 3-2/clause 10.3 provides guidance on wind tunnel testing to assess whether the aerodynamic behaviour of steel bridges will be acceptable at the ultimate and serviceability limit states. It is recommended here that expert advice is sought for all wind tunnel testing work to ensure both that the tests are necessary and that the results are interpreted correctly.

ANNEX A

Technical specifications for bearings (informative)

This annex is not specific to steel bridges and it is intended that it be moved to EN 1990. No comments are therefore made on this annex.

ANNEX B

Technical specifications for expansion joints for road bridges (informative)

This annex is not specific to steel bridges and it is intended that it be moved to EN 1990. No comments are therefore made on this annex.

ANNEX C

Recommendations for the structural detailing of steel bridge decks (informative)

C.1. Highway bridges

C.1.1. General

The annex provides recommendations for the detailing of orthotropic (orthogonally anisotropic) steel decks, based mainly on existing German bridges which have been found to perform satisfactorily. It does not cover decks with intermediate transverse stiffeners between cross-frames or diaphragms, although it still provides useful reference in such cases.

The recommendations in Annex C are informative only and there is therefore scope for designers to develop alternative details. They are aimed at providing 'fatigue resistant' details that require no explicit fatigue calculation. However, since steel orthotropic decks are potentially very susceptible to fatigue, it is recommended in this guide that fatigue checks should always be carried out on orthotropic decks, even when compliance with these guidelines is achieved. In any case, Annex C still requires that cross-beams are checked for fatigue. It appears that welds between stiffeners and cross-beams must also be checked, as weld sizes are not directly specified where the stiffener passes through a cut-out – see C.1.3(ii) below.

C.1.2. Deck plate

Guidelines for the thickness of the deck plate, spacing of the stiffeners and minimum stiffness of the longitudinal stiffeners are provided in 3-2/clause C.1.2. Fatigue in the deck plate, and particularly in the welds between stiffener and parent plate, arises principally from local wheel loads applied to the deck. Stresses away from cross-beams arise from a combination of transverse frame bending in the stiffener and deck plate and differential deflection between adjacent stiffeners as shown in Fig. C-1. A particularly severe place for fatigue is, however, at the location of the cross beam where the plate and stiffener cannot deform as shown in Fig. C-1 and additional moment is therefore attracted to the plate and welds.

Fatigue stresses also arise from shear and flexure of the stiffeners spanning between cross-frames and local stresses at the connection between the stiffeners and the webs of the cross-frames supporting the stiffeners.

The susceptibility of the deck to fatigue depends on a number of factors including:

- deck plate thickness
- span of deck plate between webs of stiffeners

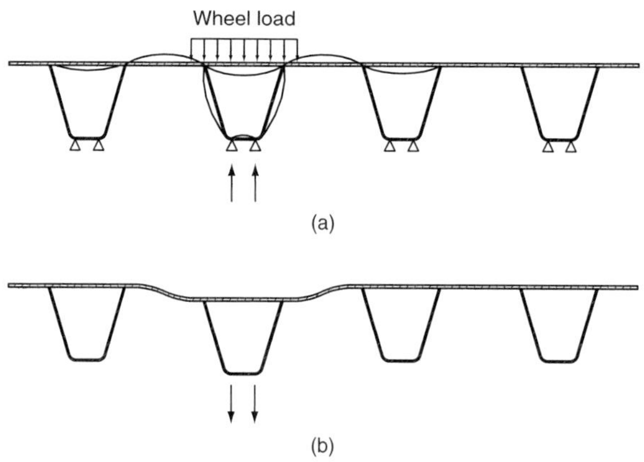

Fig. C-1. Effects of local wheel load away from cross-beams: (a) local frame bending; (b) differential deflections between stiffeners

- thickness of stiffener
- weld detail between deck plate and stiffener.

Guidance is provided on these in Annex C and some of the recommended dimensions reflect greater conservatism than found in existing UK bridges. For example, many existing UK road bridges have deck plates with a thickness of 12 mm and a surfacing thickness of 40 mm. For this surfacing thickness, ***3-2/clause C.1.2.2(1) Note 1*** recommends a deck plate thickness of 16 mm. The reduction in deck plate thickness recommended for greater depths of surfacing in EN 1993-2 reflects the fact that composite action develops between deck plate and surfacing, thus reducing stresses. This composite behaviour is quite complex and may be completely absent during summer periods where the surfacing softens appreciably. For footbridges, fatigue of deck plates is less of a problem. Steel Bridge Group *Guidance Note 2.10*[30] recommends 6 mm thickness for plates spanning up to 550 mm and 8 mm for plates spanning up to 750 mm. Annex C conservatively recommends 10 mm minimum deck plate thickness. This was intended to make allowance for access by maintenance vehicles.

3-2/clause C.1.2.2(1) Note 1

3-2/clause C.1.2.2(1) Note 2 should be treated with some caution. This states that when the various recommendations for deck plate thickness, stiffener spacing and stiffener thickness are satisfied, the bending moments in the deck plate need not be verified. As the majority of steel orthotropic decks are subjected to forces from the global behaviour of the bridge, the deck plate and stiffeners should still be checked for combinations of global and local stresses – section 6.5 of this guide refers. It is, in any case, recommended here that all details are checked for fatigue, despite compliance with the detailing in Annex C. The note was possibly intended to refer only to static transverse moments in the deck plate.

3-2/clause C.1.2.2(1) Note 2

C.1.3. Stiffeners

3-2/clause C.1.3 provides guidelines for designing the welds connecting the longitudinal stiffeners to the deck plate and transverse beams. The main issues are as follows.

(i) Deck plate to longitudinal stiffener weld

3-2/clause C.1.3.3(1)

3-2/clause C.1.3.3(1) states '*For closed section stiffeners under the carriageway the weld between the stiffener and the deck plate should be a butt weld*'. The designer is then referred to Table C.4(3) and (4) where it can be seen that the clause C.1.3.3(1) reference to 'butt weld' actually refers to a partial penetration butt weld with throat thickness at least as large as the stiffener thickness and a maximum of 2 mm lack of penetration to the back of the stiffener. This is shown in Fig. C-2 where a 2 mm gap is also permitted between stiffener and deck plate. This fit-up gap appears to be a little excessive. Current UK best practice would limit the gap between deck plate and stiffener to 0.5 mm instead of the

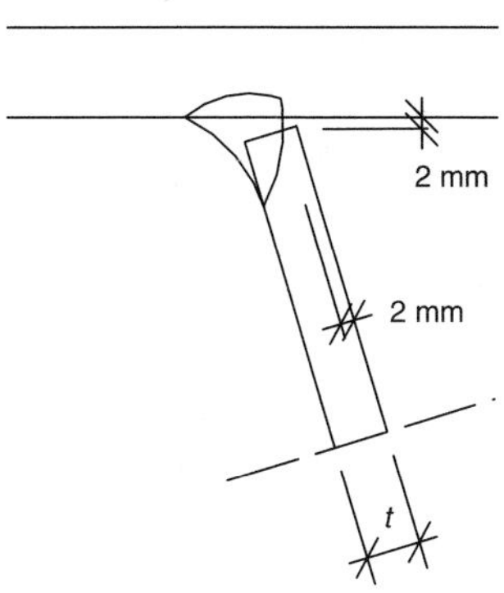

Fig. C-2. Typical weld between closed stiffener and deck plate as recommended by 3-2/Annex C

2.0 mm shown and the penetration would typically be specified as 80% of the stiffener thickness rather than an acceptable gap of 2.0 mm.

3-2/Table C.4(5) allows the designer to specify fillet welds where the longitudinal stiffeners are outside the roadway. In this instance, there is no significant fatigue loading from local wheel loading and the weld size can be determined from considerations of the static loading only.

(ii) Longitudinal stiffener to transverse beam weld

3-2/clause C.1.3.5 provides a large amount of guidance for detailing the longitudinal stiffener connections to transverse beams, but no guidance for the required weld sizes. These welds should therefore be designed for the worst case of static and fatigue actions, using an analysis which takes the effects listed in 3-2/clause C.1.3.5.1 into account. These are:

- Shear forces, torsional moments and restraint to distortional deformations of the stiffeners.
- Rotations of the stiffeners being restrained by the web of the cross-member as shown in Fig. C-3. (Poisson ratio effects also result in transverse deformations of the stiffener cross-section.)
- Flexure and Vierendeel action in a cross-beam, leading to deformations of the stiffener cross-section as shown in Fig. C-4, with stress concentration at the edges of any cope holes provided. The stresses adjacent to the welds can be estimated using the Vierendeel model shown in Fig. C-5.

The stresses from the above items above can be most realistically determined from finite-element analysis, although conservative estimates can be made using simpler models.

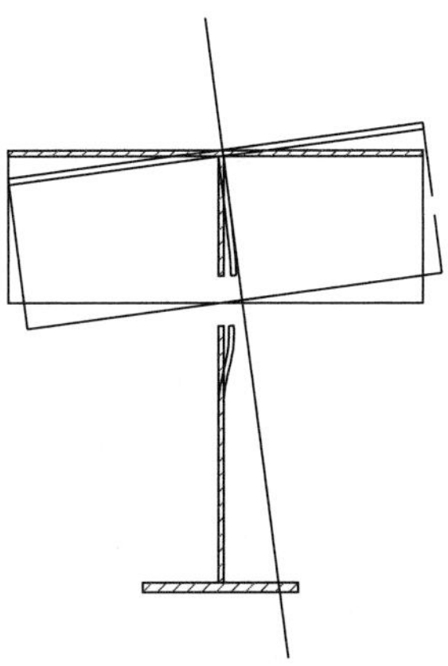

Fig. C-3. Restraint to rotation of longitudinal stiffener by cross-beam or diaphragm

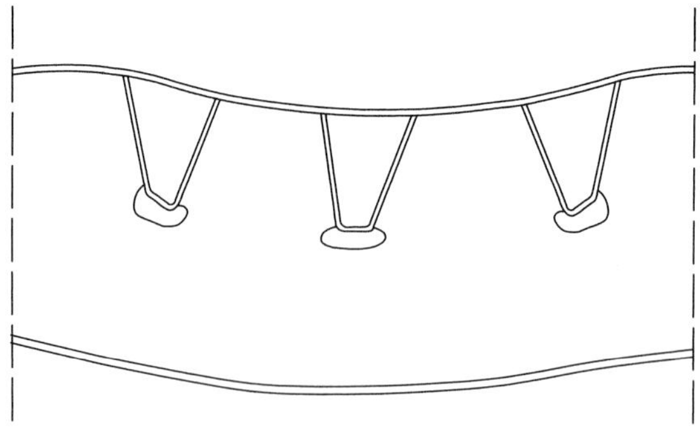

Fig. C-4. Distortion of longitudinal stiffeners and stresses in welded connection due to deformation of a cross-beam

3-2/clause C.1.3.5.1(3)

3-2/clause C.1.3.5.3(1)

3-2/clause C.1.3.5.1(3) recommends that longitudinal stiffeners should pass through cross-beams or diaphragms by way of cut-outs, rather than by abutting the transverse member, as this leads to greater fatigue resistance. An exception may be made, according to ***3-2/clause C.1.3.5.3(1)***, when the stiffeners are not under the traffic lanes, the stiffeners span less than 2.75 m and measures are taken to control/reduce shrinkage. This might be applicable for footbridges. In such cases, a butt weld has to be used between stiffener and cross-beam. Cut-outs can be formed with or without cope holes at the bottom of the longitudinal stiffener, which is often a matter of preference for the fabricator. Generally, cope holes are provided for decks of railway bridges as it keeps the weld away from the lowest point of the stiffener where the stress range from local bending is greatest and thus improves the permissible stress range.

C.1.4. Cross-beams

3-2/clause C.1.4 gives the designer recommendations for suitable detailing of the cross-beams. Detailed fatigue checks are still required on all components. Further guidance on cross-beam fatigue checks is given in 3-2/clause 9.4.2.2. Cross-beams tend to act as Vierendeel frames due to the presence of the cut-outs. The calculation of stresses in cross-beams therefore needs to consider the Vierendeel behaviour and a suitable model is shown in Fig. C-5. The top member comprises an attached width of deck plate acting about its weak axis and the bottom member comprises the cross-beam bottom flange and attached web up to the level of the bottom of the cut-out. Each vertical member comprises a width of cross-beam web plate equal to the distance between cut-outs, acting about its stiff axis.

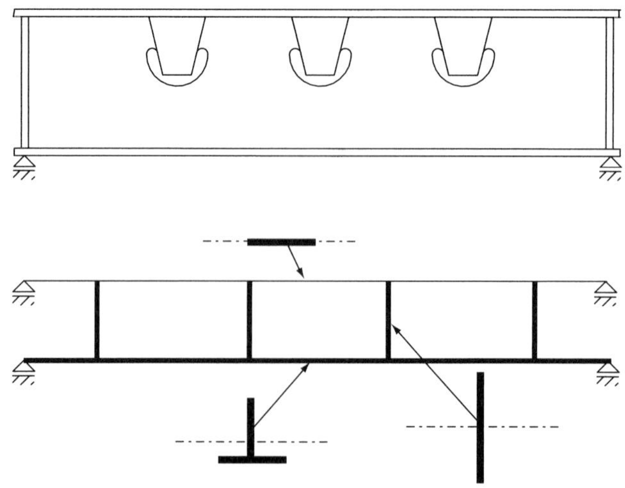

Fig. C-5. Plane frame modelling of cross-beam, allowing for Vierendeel action around cut-outs

C.2. Railway bridges

3-2/clause C.2 gives similar recommendations for designing steel orthotropic decks on railway bridges. As 3-2/clause 9.1.2 requires fatigue checks for all the main orthotropic deck components, the recommendations can be regarded as a guide to good detailing practice only. It is imperative that the fabrication requirements for all details in Table C.4 are followed as the fatigue classifications in EN 1993-1-9 assume they have been met.

ANNEX D

Buckling lengths of members in bridges and assumptions for geometrical imperfections (informative)

D.1. General

As discussed in section 5.2 of this guide, second-order (P–Δ) effects in compression members can be dealt with either by the use of effective lengths in conjunction with the member resistance formulae in section 6.3, or by second-order analysis with initial imperfections included in the structural model. 3-2/Annex D provides useful methods to calculate effective lengths for truss members (including compression chords with U-frame restraint) and arches. The effective lengths for truss members can also be used for analogous situations such as for bracing members between girders. Effective lengths are presented in the form:

$$L_{cr} = l_k = \beta L \qquad \text{3-2/(D.1)}$$

where β is a buckling factor and L is a reference length normally equal to the actual member length between restraints, except in the case of arches. (The term L_{cr} has been introduced in the above as both L_{cr} and l_k are used in EN 1993-2 for effective length.) The annex also gives guidance on imperfections for use in the second-order analysis of arches.

D.2. Trusses

D.2.1. Vertical and diagonal elements with fixed ends

The recommended effective lengths for members between truss chords (without interconnecting members) are given by ***3-2/clause D.2.1(1)*** as $0.9L$ for in-plane buckling and $1.0L$ for out-of-plane buckling. In-plane buckling will involve flexure of the chords to which the members connect at their ends. This provides some rotational resistance, so an effective length less than that for pin-ended conditions is produced. For out-of-plane buckling, the only end rotational stiffness is provided by the twisting stiffness of the chords which, for open sections, is usually small. Gusset plates at connections will further reduce end stiffness. The end restraint therefore is assumed to approximate to pinned conditions. *3-2/clause D.2.1(1)*

These effective lengths can be quite conservative (and are more conservative than in previous UK practice). Reduced effective lengths can be obtained if the rotational stiffness provided by the chord at the member ends is first determined and the formulae in section 5.2.2.3 of this guide used to calculate the effective length. Alternatively, a computer elastic critical buckling analysis could be used. In either case, it is essential to estimate joint flexibility realistically; 3-1-8/clause 6.3 can be used to achieve this.

D.2.2. Vertical elements forming part of a frame

Buckling lengths for the vertical legs of sway frames are given in 3-2/Table D.1. These effective lengths can be useful for frames other than just those in truss bridges. The curves allow for the loading to be applied towards or away from a fixed point as the frame displaces. This effectively allows representation of load applied through a fixed point below the top of the frame, which stabilises the frame, and load applied through a fixed point above the top of the frame, which destabilises the frame.

The fixed point through which loading is applied is defined by a height h_r as shown in Fig. D-1 for the case of a frame with pinned legs. Figure D-1 illustrates some of the different possibilities with the buckling length depending on the ratio h/h_r and the relative stiffness of legs to horizontal member given by $\eta = EIb/(E_0 I_0 h)$.

Two special cases are:

(a) Load remaining vertical but free to translate with the structure. This is the most common case and since there is no fixed point through which the load acts, h_r is infinite and $h/h_r = 0$.
(b) Load always directed towards base of legs so that $h/h_r = 1.0$. This could occur where the load was due to stressed cables attached to the top corners of the frame and anchored at the level of the base of the legs in line with the legs.

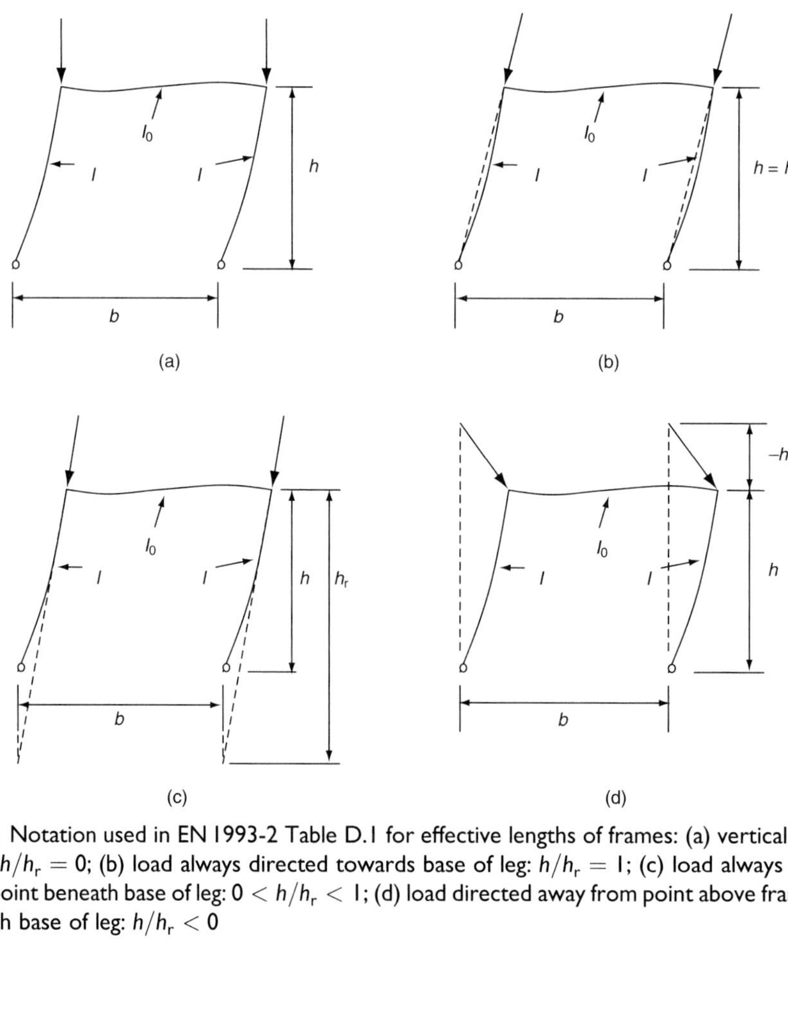

Fig. D-1. Notation used in EN 1993-2 Table D.1 for effective lengths of frames: (a) vertical load free to translate: $h/h_r = 0$; (b) load always directed towards base of leg: $h/h_r = 1$; (c) load always directed towards point beneath base of leg: $0 < h/h_r < 1$; (d) load directed away from point above frame vertically in line with base of leg: $h/h_r < 0$

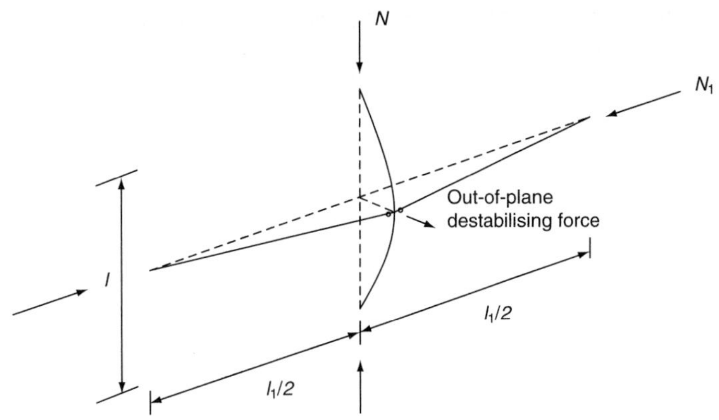

Fig. D-2. Destabilising effect of compression in connecting member

For a stiff horizontal member, such that $\eta = 0$, and pinned feet as in Fig. D-1, case (a) gives an effective length of $2h$ from 3-2/Table D.1 as expected. Similarly, for a stiff horizontal member, case (b) gives an effective length of h from 3-2/Table D.1.

D.2.3. Out-of-plane buckling of diagonals

This section of the annex provides effective lengths for a variety of different configurations of intersecting members presented in 3-2/Table D.2. The key thing to note is that the effective length of a compression member can be reduced by the presence of an interconnecting tension member but it can also be increased by an interconnecting compression member. The latter has not always been recognised by designers or indeed UK codes and occurs because a connecting compression member can destabilise the other compression member as illustrated in Fig. D-2. This is similar to the problem encountered in the design of web stiffeners subject to web compression and transverse load as discussed in section 6.6 of this guide.

A given continuous compression member will not always be destabilised if the interconnecting compression member is also continuous. Its effective length may be reduced to lower than its actual length if the connecting member is stiff and/or relatively lightly loaded. In this case, however, the continuous connecting member will have an effective length greater than its actual length. This reciprocal relationship is illustrated by considering two interconnecting continuous compression members with forces $N = N_1$, lengths $l = l_1$ and second moment of area $I = 2I_1$. The formulae in case 2 of 2-2/Table D.2 give the following:

$$\beta = \sqrt{\frac{1 + \dfrac{N_1 l}{N l_1}}{1 + \dfrac{I_1 l^3}{I l_1^3}}} = \sqrt{\frac{2}{1.5}} = 1.15$$

for the member with greatest second moment of area, and

$$\beta_1 = \sqrt{\frac{1 + \dfrac{N l_1}{N_1 l}}{1 + \dfrac{I l_1^3}{I_1 l^3}}} = \sqrt{\frac{2}{3}} = 0.82$$

for the member with least second moment of area.

The stiffer member is therefore destabilised by the more flexible member.

A given continuous compression member will always be destabilised if the connecting compression member is pin-ended at its connections to the continuous member. This is because a pin-ended connecting member can offer no out-of-plane restraint, and pushes the main member out of plane.

D.2.4. Compression chords of open bridges

Compression chords of trusses can be checked for buckling using the beam on elastic foundation model discussed in section 6.3.4.2 of this guide. 3-2/Table D.3 gives the relevant discrete spring stiffness for the U-frames of trusses and can also be applied to half-through plate girders with stiffeners and cross-girders as shown in Fig. 6.3-20, where the relevant stiffness is:

$$C_d = \frac{EI_v}{\frac{h_v^3}{3} + \frac{h^2 b_q I_v}{2I_q}}$$

This case also covers inverted U-frames such as in steel and concrete composite bridges when the cross-member stiffness is based on the short-term cracked flexural stiffness of the deck slab and reinforcement and any physical cross-girder present. For multiple girders, the restraint to internal girders may be derived by replacing $2I_q$ by $3I_q$ in the expression for C_d. Section properties for stiffeners should be derived using an attached width of web plate in accordance with 3-1-5/Fig. 9.1 (stiffener width plus $30\varepsilon t_w$).

The above formula and others in 3-2/Table D.3 make no allowance for flexibility of joints. Joint flexibility can significantly reduce the effectiveness of U-frames. If the joint was 'semi-continuous' according to 3-1-8/clause 5.2.2, the effect of joint rotational flexibility, S_j, would have to be determined from 3-1-8/clause 6.3 and included in the calculation of C_d. This would typically apply to connections made through unstiffened end plates which are discouraged for bridges. Recommendations for rotational stiffness of certain generic types of connection were made in BS 5400: Part 3[4] but these stiffnesses were not size-dependent and were based on relatively shallow members. They therefore do not correlate well with the predictions of EN 1993-1-8 across a range of connection sizes and are generally conservative (overestimating flexibility) for bridge member connections.

Derivation of the stiffness of other restraint systems is discussed in section 6.3.4.2 of this guide.

D.3. Arch bridges

D.3.1. General

3-2/clause D.3.1 gives effective lengths for in-plane and out-of-plane buckling for **rigid** abutments. The assumption of rigidity is very important as foundation flexibility will allow the arch to spread out and flatten under the arch thrust as shown in Fig. D-3. This flattening of the arch will give rise to an increase in the arch compression and bending moments which will in turn increase the effective length. This effect is more critical in flat arches with a low rise-to-span ratio, f/l.

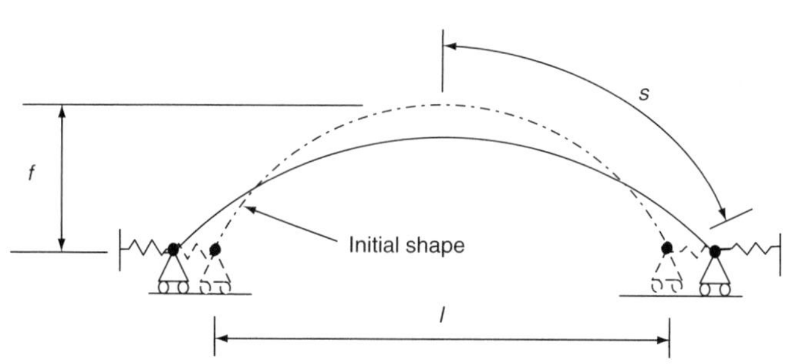

Fig. D-3. Flattening of an arch due to abutment movement

D.3.2. In-plane buckling

Effective length factors for circular, parabolic and catenary arches with different support and crown conditions are given in 3-2/Table D.4 for **rigid** abutments which cannot move apart.

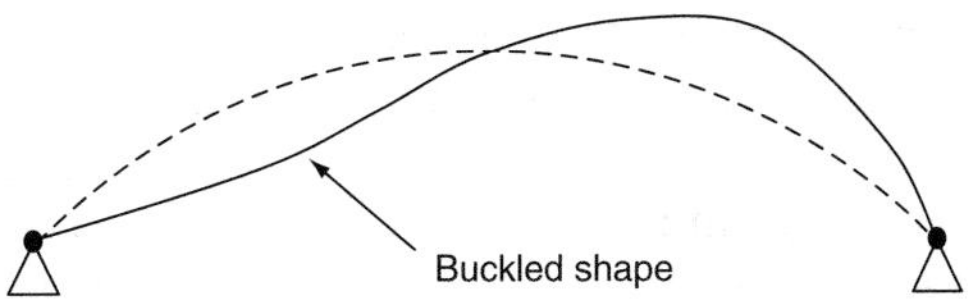

Fig. D-4. Asymmetric in-plane buckling mode of arch

The effective length is given by:

$$L_{cr} = \beta s$$

and, from ***3-2/clause D.3.1(2)***, the critical buckling force measured at the supports under the given load case is: *3-2/clause D.3.1(2)*

$$N_{cr} = \left(\frac{\pi}{\beta s}\right)^2 EI_y \qquad \text{3-2/(D.3)}$$

where s is the half-length of the arch as shown Fig. D-3 and EI_y is the in-plane flexural rigidity of the arch. The effective length or critical force can be used in member buckling checks to 3-2/clause 6.3.1. In addition, the effect of curvature on individual plate components needs to be considered as discussed in section 6.10 of this guide. Where the arch cross-section is not constant, either a conservative value of EI_y should be used (say the lowest value in the middle third of the half-wavelength of buckling) or the elastic critical force should be obtained from a computer elastic critical buckling analysis.

Generally the lowest mode of in-plane buckling is an asymmetric mode shape as shown in Fig. D-4. For arches with a pin at the apex, a symmetric mode of buckling may be the lowest mode.

Where the abutments are not rigid against lateral movement (or at least rigid by comparison with the arch) and the abutments can move apart significantly, 3-2/Table D.4 underestimates the effective lengths. (Elastic critical buckling analysis with rigid supports will similarly underestimate the effective length.) This is most likely to be a problem for arches with low rise-to-span ratio, f/l. Reference 31 recommends that particular caution should be exercised where $f/l < 0.1$. In this situation, it is particularly important to consider the possible increase in arch compression and bending moments from the spreading of the arch and from arch shortening under compression.

To assess the significance of flattening of the arch, Reference 31 recommends that where the vertical deflection of the crown is less than 2.5% of the rise, first-order analysis to determine forces (for use in subsequent buckling checks) is still appropriate. Where the vertical deflection of the crown exceeds 2.5% of the rise, second-order large deflection analysis, capable of considering the change in geometry of the arch (i.e. geometric non-linearity), should be used to determine the increased internal action effects.

Where significant arch flattening can occur, second-order analysis modelling only P–Δ effects (as discussed in section 5.2) will not suffice. This is illustrated for a simple case in Fig. D-5. In Fig. D-5, there are no P–Δ effects but the system support deflections lead to an increase in the axial load due to the flattening of the system geometry. The second-order

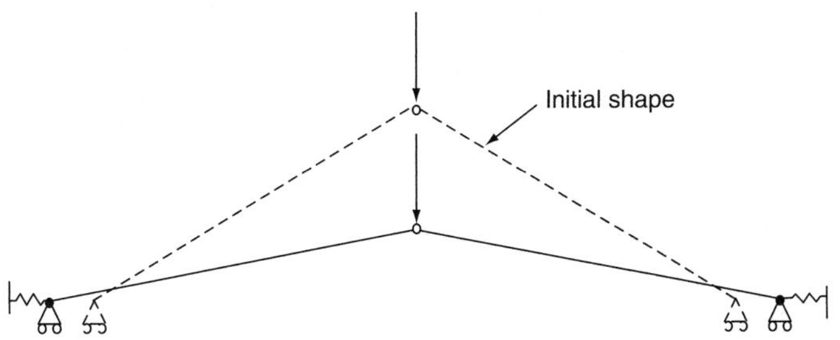

Fig. D-5. Flattening of arch (idealised as two pin-jointed members) due to abutment movement and elastic shortening

large deflection analysis described here must therefore be distinguished from a second-order (P–Δ) analysis with modelled imperfections as described in section 5.2. The latter is appropriate only where changes in overall arch geometry are small, unless the analysis can also consider the effects of change in arch geometry. If effective lengths are to be calculated by computer for arches which can appreciably flatten, the critical buckling analysis must itself consider geometric non-linearity as well as P–Δ effects.

If the rigidity of the abutments is sensitive to the assumed soil stiffness, the analysis should consider a range of soil parameters.

Tied arches

Buckling lengths for tied arches with hangers are given in 3-2/Fig. D.4. The rise-to-span ratio is limited to being greater than 0.1.

Snap-through buckling

Many, if not most, software programs which can perform elastic critical buckling analysis on beam elements can only deal with instability caused by axial force induced bending moments, that is P–Δ effects. Where the arch flattens under load due to elastic shortening or abutment movements, the geometry of the arch changes and the compressive forces increase as illustrated in the simplified system in Fig. D-5. Snap-through buckling then becomes a possibility and this will be undetected by the software mentioned above unless it can include the effects of geometric non-linearity.

3-2/clause D.3.2(3)

To guard against snap-through buckling, ***3-2/clause D.3.2(3)*** provides a limiting criterion relating the cross-sectional properties to the span and rise. Once again, this criterion only applies to rigid supports and the discussions on arch spread above apply.

D.3.3. Out-of-plane buckling for free-standing arches

Effective length factors for circular, parabolic and catenary arches with different support and crown conditions are given in 3-2/Table D.6 and D.7 for **rigid** abutments which cannot move apart. The effective length is this time given by:

$$L_{cr} = \beta l$$

3-2/clause D.3.1(3)

and, from ***3-2/clause D.3.1(3)***, the critical buckling force, measured at the supports under the given load case is:

$$N_{cr} = \left(\frac{\pi}{\beta l}\right)^2 EI_z$$

D.3.4. Out-of-plane buckling of arches with wind bracing and end portals

3-2/clause D.3.4(3)

Buckling of the unbraced length of arch at each end portal can be checked by treating the end portal as a frame according to 3-2/clause D.2.2. ***3-2/clause D.3.4(3)*** gives guidance on the value of h_r to use in the calculation.

D.3.5. Arch imperfections for use in second-order analysis

3-2/clause D.3.5 gives suggested imperfection shapes and magnitudes for in-plane and out-of-plane buckling modes for use in second-order analysis. These are self-explanatory and are based on the approximate buckled shapes of an arch with abutments which resist spread. The magnitudes of imperfections are equivalent to those for simple struts given in 3-1-1/Table 5.1. Imperfections can alternatively be derived from the mode shapes obtained from an elastic critical buckling analysis as discussed in section 5.3.2.1 of this guide.

If abutments are not rigid against moving apart, or the rise-to-span ratio is low, then the discussions on large deflection analysis in section D.3.2 of this guide apply.

ANNEX E

Combination of effects from local wheel and tyre loads and from global loads on road bridges (informative)

Global and local effects on steel deck plates should be combined for SLS and ULS verifications where relevant. The effects of local and global loading are particularly significant adjacent to cross-beams and diaphragms where wheel loads cause additional local hogging moments.

3-2/clause E.1(2) provides a simplified combination rule whereby the maximum global effect and maximum local effect are determined separately and then combined. Two combinations are considered: *3-2/clause E.1(2)*

- full global effect and reduced (by the combination factor ψ) local effect, or
- full local effect and reduced (by the combination factor ψ) global effect.

The combination rules of 3-2/clauses (E.1) and (E.2) are presented in terms of stress. For deck plates with longitudinal stiffeners in global compression, where buckling must be considered, the combination should be performed on the loading before carrying out the resistance checks. Verification of deck plates under local and global loading is discussed in section 6.5 of this guide.

3-2/Annex E is also referenced in EN 1994-2 for checks of deck slabs in composite bridges. For reinforced concrete verifications, care is needed in applying the combination rule as peak global effects causing compression in the slab may increase the resistance of the slab to local effects. The combination therefore generally needs to consider both maximum and minimum global effects in conjunction with local effects. The National Annex has an opportunity to vary the numerical value of the combination factor to be used.

References

1. Gulvanessian, H., Calgaro, J.-A. and Holický, M. (2002) *Designers' Guide to EN 1990, Eurocode: Basis of Structural Design*. Thomas Telford, London.
2. The European Commission (2002) *Guidance Paper L (Concerning the Construction Products Directive – 89/106/EEC). Application and Use of Eurocodes*. EC, Brussels.
3. International Organization for Standardization (1997) *Basis of Design for Structures – Notation – General Symbols*. ISO, Geneva, ISO 3898.
4. British Standards Institution (2000) *Design of Steel Bridges*. BSI, London, BS 5400: Part 3.
5. BD 7/01 (2001) *Weathering Steel for Highway Structures*, Highways Agency, London.
6. Hendy, C. R. and Smith, D. A. (2007) *Designers' Guide to EN 1992-2 – Eurocode 2, Design of Concrete Structures. Part 2, Concrete Bridges, Design and Detailing Rules*. Thomas Telford, London.
7. Hendy, C. R. and Johnson, R. P. (2006) *Designers' Guide to EN 1994-2 – Eurocode 4, Design of Composite Steel and Concrete Structures. Part 2, General Rules and Rules for Bridges*. Thomas Telford, London.
8. Winter, G. (1947) *Strength of Thin Steel Compression Flanges*. Transactions ASCE, Vol. 112, pp. 527–544.
9. Bulson, P. S. (1970) *The Stability of Flat Plates*. Chatto & Windus, London.
10. Inquiry into the Basis of Design and Method of Erection of Steel Box Girder Bridges (1974) *Interim Design and Workmanship Rules*. HMSO, London.
11. Johansson, B. and Veljkovic, M. (2001) *Steel Plated Structures*. Luleå University of Technology, Sweden.
12. Johansson, B., Maquoi, R. and Sedlacek, G. (2001) New design rules for plated structures in Eurocode 3. *Journal of Constructional Steel Research*, **57**, 279–311.
13. Höglund, T. (1981) *Design of Thin Plate I Girders in Shear and Bending, with Special Reference to Web Buckling*. Bulletin No. 94, Division of Building Statics and Structural Engineering, Royal Institute of Technology, Stockholm.
14. Höglund, T. (1995) *Strength of Steel and Aluminium Plate Girders – Shear Buckling and Overall Web Buckling of Plane and Trapezoidal Webs. Comparison with Tests*. Department of Structural Engineering, Royal Institute of Technology, Stockholm, Technical Report 1995:4 Steel Structures.
15. Klöppel, K. and Scheer, J. (1960) *Buckling Coefficients of Stiffened Rectangular Plates*. Verlag, Berlin.
16. Hambly, E. C. (1990) *Bridge Deck Behaviour*. Taylor and Francis, London.
17. Wright, R. N., Abdel-Samad, S. R. and Robinson, A. R. (1968) BEF Analogy for Analysis of Box Girders. Proceedings of the American Society of Civil Engineers, Vol. 94, ST7.
18. Young, W. C. (1989) *Roark's Formulas for Stress and Strain*. McGraw-Hill, Singapore.

19. British Standards Institution (1992) *Design of Steel Structures. Part 1-1, General Rules and Rules for Buildings*. BSI, London, DD ENV 1993-1-1.
20. Lagerqvist, O. (1994) *Patch Loading, Resistance of Steel Girders Subjected to Concentrated Forces*. Doctoral thesis 1994:159 D, Department of Civil and Mining Engineering, Division of Steel Structures Luleå University of Technology, Sweden.
21. Kuhlmann, U. and Seitz, M. (2004) Longitudinally stiffened girder webs subjected to patch loading. *Proceedings of the Steel Bridge 2004 Conference*, Millau.
22. Veljkovic, M. and Johansson, B. (2001) Design for buckling of plates due to direct stress. *Proceedings of the Nordic Steel Construction Conference*, Helsinki.
23. Lääne, A. (2003) *Post-critical Behaviour of Composite Bridges under Negative Moment and Shear*. Thesis No. 2889, EPFL, Lausanne.
24. Trahair, N. S. (1993) *Flexural Torsional Buckling of Structures*. E & FN Spon, London.
25. Evans, H. R. and Tang, K. H. (1984) An experimental study of the collapse behaviour of a plate girder with closely-spaced transverse web stiffeners. *Journal of the Constructional Steel Research*, 4, 253–280.
26. British Standards Institution (2000) *Structural Use of Steelwork in Building*. BSI, London, BS 5950: Part 1.
27. Owens, G. H. and Cheal, B. D. (1989) *Structural Steelwork Connections*. Butterworths, London.
28. CIDECT Monograph No. 6 (1986) *The Strength and Behaviour of Statically Loaded Welded Connections in Structural Hollow Sections*. British Steel Welded Tubes, London.
29. British Standards Institution (1980) *Steel, Concrete and Composite Bridges – Design of Steel Bridges*. BSI, London, BS 5400: Part 10.
30. Steel Bridge Group (1998) *Guidance Notes on Best Practice in Steel Bridge Construction – SCI Publication 185*. Steel Construction Institute, Berkshire.
31. SCI-P281(2001) *Design of Curved Steel*. Steel Construction Institute, Ascot.

Index

Page numbers in italics indicate illustrations. The suffix 'w' refers to a worked example.